Classical and Analytical Mechanics

Classical and Analytical Mechanics

Theory, Applied Examples, and Practice

Alexander S. Poznyak
Automatic Control Department
CINVESTAV-IPN
Cd. de Mexico, Mexico

Elsevier
Radarweg 29, PO Box 211, 1000 AE Amsterdam, Netherlands
The Boulevard, Langford Lane, Kidlington, Oxford OX5 1GB, United Kingdom
50 Hampshire Street, 5th Floor, Cambridge, MA 02139, United States

Copyright © 2021 Elsevier Inc. All rights reserved.

No part of this publication may be reproduced or transmitted in any form or by any means, electronic or mechanical, including photocopying, recording, or any information storage and retrieval system, without permission in writing from the publisher. Details on how to seek permission, further information about the Publisher's permissions policies and our arrangements with organizations such as the Copyright Clearance Center and the Copyright Licensing Agency, can be found at our website: www.elsevier.com/permissions.

This book and the individual contributions contained in it are protected under copyright by the Publisher (other than as may be noted herein).

Notices

Knowledge and best practice in this field are constantly changing. As new research and experience broaden our understanding, changes in research methods, professional practices, or medical treatment may become necessary.

Practitioners and researchers must always rely on their own experience and knowledge in evaluating and using any information, methods, compounds, or experiments described herein. In using such information or methods they should be mindful of their own safety and the safety of others, including parties for whom they have a professional responsibility.

To the fullest extent of the law, neither the Publisher nor the authors, contributors, or editors, assume any liability for any injury and/or damage to persons or property as a matter of products liability, negligence or otherwise, or from any use or operation of any methods, products, instructions, or ideas contained in the material herein.

Library of Congress Cataloging-in-Publication Data
A catalog record for this book is available from the Library of Congress

British Library Cataloguing-in-Publication Data
A catalogue record for this book is available from the British Library

ISBN: 978-0-323-89816-4

For information on all Elsevier publications
visit our website at https://www.elsevier.com/books-and-journals

Publisher: Matthew Deans
Acquisitions Editor: Dennis McGonagle
Editorial Project Manager: Fernanda A. Oliveira
Production Project Manager: Poulouse Joseph
Designer: Alan Studholme

Typeset by VTeX

Dedicated to my alma mater *Moscow Physical Technical Institute (MPhTI)*

Contents

List of figures		xiii
List of tables		xix
About the author		xxi
Preface		xxiii
Notation		xxv
Introduction		xxix

1 Kinematics of a point — 1
  1.1 Products of vectors — 1
    1.1.1 Internal (scalar) product — 2
    1.1.2 Vector product — 3
    1.1.3 Main properties of triple products — 6
  1.2 Generalized coordinates — 8
    1.2.1 Different possible coordinates — 8
    1.2.2 Definition of generalized coordinates — 8
    1.2.3 Relationship of generalized coordinates with Cartesian — 9
    1.2.4 Coefficients of Lamé — 10
  1.3 Kinematics in generalized coordinates — 10
    1.3.1 Velocity in generalized coordinates — 11
    1.3.2 Acceleration in generalized coordinates — 11
  1.4 Movement in the cylindrical and spherical coordinate systems — 14
    1.4.1 Movement in cylindrical coordinates — 14
    1.4.2 Movement in spherical coordinates — 16
  1.5 Normal and tangential accelerations — 18
  1.6 Some examples — 20
  1.7 Exercises — 28

2 Rigid body kinematics — 31
  2.1 Angular velocity — 31
    2.1.1 Definition of a rigid body — 31
    2.1.2 The Euler theorem — 32
    2.1.3 Joint rotation with a common pivot — 35
    2.1.4 Parallel and non-coplanar rotations — 36
  2.2 Complex movements of the rigid body — 39
    2.2.1 General relations — 39
    2.2.2 Plane non-parallel motion and center of velocities — 42
  2.3 Complex movement of a point — 45

		2.3.1	Absolute velocity	46
		2.3.2	Absolute acceleration	46
	2.4	Examples		47
	2.5	Kinematics of a rigid body rotation		59
		2.5.1	Finite rotations	59
		2.5.2	Rotation matrix	61
		2.5.3	Composition of rotations	68
	2.6	Rotations and quaternions		71
		2.6.1	Quaternions	71
		2.6.2	Composition or summation of rotations as a quaternion	80
	2.7	Differential kinematic equations (DKEs)		84
		2.7.1	DKEs in Euler coordinates	84
		2.7.2	DKEs in quaternions: Poisson equation	85
	2.8	Exercises		86
3	**Dynamics**			89
	3.1	Main dynamics characteristics		90
		3.1.1	System of *material points*	90
		3.1.2	Three main dynamics characteristics	91
	3.2	Axioms or Newton's laws		91
		3.2.1	Newton's axioms	91
		3.2.2	Expression for $\dot{\mathbf{Q}}$	92
		3.2.3	Expression for $\dot{\mathbf{K}}_A$	94
	3.3	Force work and potential forces		95
		3.3.1	Elementary and total force work	96
		3.3.2	Potential forces	96
		3.3.3	Force power and expression for $\dot{T}$	98
		3.3.4	Conservative systems	99
	3.4	Virial of a system		100
		3.4.1	Main definition of virial	100
		3.4.2	Virial for homogeneous potential energies	101
	3.5	Properties of the center of mass		103
		3.5.1	Dynamics of the center of inertia (mass)	103
	3.6	"King/König/Rey" theorem		103
		3.6.1	Principle theorem	103
		3.6.2	Moment of inertia and the impulse moment with respect to a pivot	105
		3.6.3	A rigid flat body rotating in the same plane	106
		3.6.4	Calculation of moments of inertia for different rigid bodies	108
		3.6.5	König theorem application	112
		3.6.6	Steiner's theorem on the inertia moment	114
	3.7	Movements with friction		123
	3.8	Exercises		127

4	**Non-inertial and variable-mass systems**		**131**
	4.1	Non-inertial systems	131
		4.1.1 Newton's second law regarding a relative system	132
		4.1.2 Rizal's theorem in a relative system	134
		4.1.3 Kinetic energy and work in a relative system	137
		4.1.4 Some examples dealing with non-inertial systems	139
	4.2	Dynamics of systems with variable mass	142
		4.2.1 Reactive forces and the Meshchersky equation	142
		4.2.2 Tsiolkovsky's rocket formula and other examples	143
	4.3	Exercises	151
5	**Euler's dynamic equations**		**153**
	5.1	Tensor of inertia	153
	5.2	Relative kinetic energy and impulse momentum	156
		5.2.1 Relative kinetic energy	156
		5.2.2 Relative impulse momentum	157
	5.3	Some properties of inertial tensors	158
		5.3.1 Tensor of inertia as a non-negative symmetric matrix	158
		5.3.2 Eigenvalues and eigenvectors of inertial tensors	159
		5.3.3 Examples using tensors of inertia	163
	5.4	Euler's dynamic equations	174
		5.4.1 Special cases of Euler's equations	175
	5.5	Dynamic reactions caused by the gyroscopic moment	182
	5.6	Exercises	185
6	**Dynamic Lagrange equations**		**189**
	6.1	Mechanical connections	189
	6.2	Generalized forces	192
	6.3	Dynamic Lagrange equations	195
	6.4	Normal form of Lagrange equations	204
	6.5	Electrical and electromechanical models	207
		6.5.1 Some physical relations	208
		6.5.2 Table of electromechanical analogies	210
	6.6	Exercises	217
7	**Equilibrium and stability**		**221**
	7.1	Definition of equilibrium	221
	7.2	Equilibrium in conservative systems	223
	7.3	Stability of equilibrium	229
		7.3.1 Definition of local stability	229
		7.3.2 Stability of equilibrium in conservative systems	232
	7.4	Unstable equilibria in conservative systems	236
	7.5	Exercises	242

8	**Oscillations analysis**		**245**
	8.1 Movements in the vicinity of equilibrium points		245
		8.1.1 Small oscillations	245
		8.1.2 Characteristic polynomial	247
		8.1.3 General solution of the characteristic equation	248
	8.2 Oscillations in conservative systems		249
		8.2.1 Some properties of the characteristic equation	249
		8.2.2 Normal coordinates	251
	8.3 Several examples of oscillation analysis		255
		8.3.1 Three masses joined by springs in circular dynamics	255
		8.3.2 Three masses joined by springs with dynamics on a straight line	258
		8.3.3 Four spring-bound masses with restricted linear dynamics	260
		8.3.4 Three identical pendula held by springs	262
		8.3.5 Four-loop LC circuits	264
		8.3.6 Finding one polynomial root using other known roots	266
		8.3.7 Hint: how to resolve analytically cubic equations	268
	8.4 Exercises		270
9	**Linear systems of second order**		**273**
	9.1 Models governed by second order differential equations		273
	9.2 Frequency response		274
	9.3 Examples		277
		9.3.1 Three-variable systems	277
		9.3.2 Electrical circuit	279
		9.3.3 Linear system with input delay	281
		9.3.4 Mechanical system with friction	282
		9.3.5 Electric circuit with variable elements	284
	9.4 Asymptotic stability		286
		9.4.1 Algebraic criteria	286
		9.4.2 Geometric criteria of asymptotic stability	294
	9.5 Polynomial robust stability		304
		9.5.1 Parametric uncertainty and robust stability	304
		9.5.2 The Kharitonov theorem	305
	9.6 Exercises		308
10	**Hamiltonian formalism**		**311**
	10.1 Hamiltonian function		311
	10.2 Hamiltonian canonical form		316
	10.3 First integrals		321
	10.4 Some properties of first integrals		323
		10.4.1 Cyclic coordinates	323
		10.4.2 Some properties of the Poisson brackets	324

		10.4.3	First integrals by inspection	**330**
	10.5	Exercises		**335**

11 The Hamilton–Jacobi equation — **337**
- 11.1 Canonical transformations — **337**
- 11.2 The Hamilton–Jacobi method — **339**
- 11.3 Hamiltonian action and its variation — **339**
- 11.4 Integral invariants — **343**
  - 11.4.1 Integral invariants of Poincaré and Poincaré–Cartan — **343**
  - 11.4.2 The Lee Hwa Chung theorem — **344**
- 11.5 Canonicity criteria — **347**
  - 11.5.1 Poincaré theorem: (c, F)-criterion — **347**
  - 11.5.2 Analytical expression for the Hamiltonian after a coordinate canonical transformation — **350**
  - 11.5.3 Brackets of Lagrange — **353**
  - 11.5.4 Free canonical transformation and the S-canonicity criterion — **356**
- 11.6 The Hamilton–Jacobi equation — **361**
- 11.7 Complete integral of the Hamilton–Jacobi equation — **362**
  - 11.7.1 Complete integral — **362**
  - 11.7.2 Generalized-conservative (stationary) systems with first integrals — **362**
- 11.8 On relations with optimal control — **369**
  - 11.8.1 Problem formulation and value function — **370**
  - 11.8.2 Hamilton–Jacobi–Bellman equation — **370**
  - 11.8.3 Verification rule as a sufficient condition of optimality — **371**
  - 11.8.4 Affine dynamics with a quadratic cost — **372**
  - 11.8.5 The case when the Hamiltonian admits the existence of first integrals — **375**
  - 11.8.6 The deterministic Feynman–Kac formula: the general smooth case — **376**
- 11.9 Exercises — **380**

12 Collection of electromechanical models — **383**
- 12.1 Cylindrical manipulator (2-PJ and 1-R) — **384**
- 12.2 Rectangular (Cartesian) robot manipulator — **387**
- 12.3 Scaffolding type robot manipulator — **389**
- 12.4 Spherical (polar) robot manipulator — **392**
- 12.5 Articulated robot manipulator 1 — **396**
- 12.6 Universal programmable manipulator — **399**
- 12.7 Cincinnati Milacron T^3 manipulator — **404**
- 12.8 CD motor, gear, and load train — **412**
- 12.9 Stanford/JPL robot manipulator — **414**
- 12.10 Unimate 2000 manipulator — **418**

12.11	Robot manipulator with swivel base	**424**
12.12	Cylindrical robot with spring	**427**
12.13	Non-ordinary manipulator with shock absorber	**429**
12.14	Planar manipulator with two joints	**434**
12.15	Double "crank-turn" swivel manipulator	**437**
12.16	Robot manipulator of multicylinder type	**442**
12.17	Arm manipulator with springs	**445**
12.18	Articulated robot manipulator 2	**460**
12.19	Maker 110	**465**
12.20	Manipulator on a horizontal platform	**468**
12.21	Two-arm planar manipulator	**471**
12.22	Manipulator with three degrees of freedom	**476**
12.23	CD motor with load	**478**
12.24	Models of power converters with switching-mode power supply	**479**
	12.24.1 *Buck* type DC-DC converter	**480**
	12.24.2 *Boost* type DC-DC converter	**482**
12.25	Induction motor	**483**

Bibliography   **487**
Index   **489**

List of figures

Fig. 1.1	The projection of the vector **b** to the direction of another vector **a**.	2
Fig. 1.2	Vector product of two vectors **a** and **b**.	3
Fig. 1.3	Right orthogonal system.	5
Fig. 1.4	Cartesian, cylindrical, and spherical vector representations and their relations.	8
Fig. 1.5	Representation of a point's position in cylindrical coordinates.	14
Fig. 1.6	Representation of a point's position in spherical coordinates.	16
Fig. 1.7	Normal and tangential vectors to the trajectory.	18
Fig. 1.8	Speeds, radii, and centers of curvature in two different instants.	19
Fig. 1.9	The detail of the previous figure.	19
Fig. 1.10	Ring moving with a constant magnitude speed.	21
Fig. 1.11	Trajectory of a particle in a Cartesian system.	22
Fig. 1.12	Particle on an elliptical path.	24
Fig. 1.13	Acceleration vector composition.	27
Fig. 2.1	Rigid body with a pivot.	32
Fig. 2.2	Rigid body with pivot O and a coordinate system fixed to the solid body.	33
Fig. 2.3	Rotations with common pivot.	35
Fig. 2.4	Cylinder subject to two parallel rotations.	36
Fig. 2.5	Cylinder rotations.	36
Fig. 2.6	Rotations in oblique planes.	38
Fig. 2.7	A rigid body and its absolute and relative references.	39
Fig. 2.8	Solid cone rotating.	41
Fig. 2.9	Diagram of angular speeds.	41
Fig. 2.10	Plane non-parallel motion and center of velocities.	42
Fig. 2.11	The disk and the shear which are pivotally connected.	44
Fig. 2.12	A point p in the absolute and relative systems.	45
Fig. 2.13	Particle sliding on an oscillating bar.	49
Fig. 2.14	Disk rolling with constant speed v.	51
Fig. 2.15	Accelerations in the points of the disk.	53
Fig. 2.16	Disk rolling with constant magnitude speed on a circular surface.	54
Fig. 2.17	Rotating gear train.	56
Fig. 2.18	The two-gear train with acceleration.	58
Fig. 2.19	Cylinder subject to two rotations with common pivot.	59
Fig. 2.20	A rotation movement with respect to the center of coordinates.	60
Fig. 2.21	Quaternion as a rotation operator.	78
Fig. 2.22	The disk, mounted at right angles to the rod.	87
Fig. 2.23	The "Segner wheel."	87
Fig. 2.24	The rod moving along two mutually perpendicular straight lines.	88
Fig. 3.1	A set of material points referring to a coordinate system.	90
Fig. 3.2	Relationship between the pole A and the origin O.	90

Fig. 3.3	Relationship between a force and elementary displacement.	96
Fig. 3.4	Relationship between the coordinates of an absolute and an auxiliary system for calculating kinetic energy.	104
Fig. 3.5	Flat body rotating in its plane with respect to a pivot O'.	106
Fig. 3.6	Some bodies of simple geometry for the calculation of their moment of inertia.	108
Fig. 3.7	Rigid chain rolling with constant speed.	112
Fig. 3.8	Rolling disc.	113
Fig. 3.9	Solid body rotating around an axis that does not pass through its inertial center.	114
Fig. 3.10	Diagram of distances of a point $i \in S$ to the axes OO' and AA'.	115
Fig. 3.11	Solid cylinder rotating on a transverse axis.	116
Fig. 3.12	Articulated bar about to fall.	117
Fig. 3.13	Bar falling.	117
Fig. 3.14	Vectors $\mathbf{v}_D$ and $[\boldsymbol{\omega}, \overline{DB}]$.	118
Fig. 3.15	Bar detaching from the floor.	119
Fig. 3.16	Set of articulated bars.	121
Fig. 3.17	Left half of the structure of articulated bars.	121
Fig. 3.18	Body subject to friction.	123
Fig. 3.19	How a friction force $\|\mathbf{F}_{fr}\|$ depends on the amplitude of the applied external force $\|\mathbf{Q}\|$.	124
Fig. 3.20	Body on an inclined plane.	124
Fig. 3.21	Sphere on an inclined plane.	125
Fig. 3.22	The physical pendulum, consisting of a homogeneous ball, is suspended on a weightless rod to the fulcrum point O. Another same ball is in a circular groove and rolls along it.	128
Fig. 3.23	A homogeneous cylinder freely rolling from a stationary cylinder.	128
Fig. 4.1	Body on an accelerated inclined plane.	133
Fig. 4.2	Sphere against a step.	136
Fig. 4.3	Ring in rotation with a small ring.	140
Fig. 4.4	Accelerated series of articulated bars.	141
Fig. 4.5	Diagram of forces on a bar.	141
Fig. 4.6	Tank with drain.	143
Fig. 4.7	Rocket shedding mass.	144
Fig. 4.8	Rocket with n fuel tanks.	145
Fig. 4.9	Cylinder driven by the tangential leakage of its contents.	147
Fig. 4.10	Chain with one end falling.	149
Fig. 4.11	A thin flexible inextensible thread in a vertical smooth pipe.	152
Fig. 4.12	A jet vessel driven by a pump.	152
Fig. 5.1	Solid body and a generic axis.	154
Fig. 5.2	A cube with side a and the reference system.	163
Fig. 5.3	A differential element of mass and its position with respect to the x-axis.	164
Fig. 5.4	Cylinder and reference system.	165
Fig. 5.5	Solid cone and hollow cylinder referring to two coordinate systems.	165
Fig. 5.6	Relationship between the radius ρ of the elementary disk and its height w.	166
Fig. 5.7	Elementary disk of radius ρ and height dz.	166
Fig. 5.8	Solid cylinder rotating obliquely.	169
Fig. 5.9	Pendulum.	169

Fig. 5.10	Solid cylinder in eccentric rotation.	171
Fig. 5.11	Disk in eccentric and oblique rotation.	172
Fig. 5.12	Disk subjected to two rotations.	173
Fig. 5.13	Gyroscope.	179
Fig. 5.14	Disk rotating eccentrically.	183
Fig. 5.15	Solid rotating around a given direction.	184
Fig. 5.16	Solid cylinder rotating around a given direction.	185
Fig. 5.17	A homogeneous circular cylinder of mass m, height H, and base radius R.	186
Fig. 5.18	A biaxial gyro platform carries two identical gyroscopes rotating with a constant angular velocity.	186
Fig. 5.19	The servo-gyroscope drive circuit.	187
Fig. 5.20	The frame of the balancing gyroscope mounted on a fixed base using bearings D and E.	188
Fig. 6.1	Simple pendulum with rigid arm.	190
Fig. 6.2	Simple pendulum with extendable arm.	190
Fig. 6.3	Disc rolling without sliding.	190
Fig. 6.4	Point moving with constant magnitude speed.	191
Fig. 6.5	Disk with movement on the xy-plane.	191
Fig. 6.6	A simple pendulum.	195
Fig. 6.7	Disk in movement.	195
Fig. 6.8	Elastic arm pendulum.	198
Fig. 6.9	Pendulum in a block that slides without friction on a horizontal surface.	200
Fig. 6.10	Mechanical system of two masses.	202
Fig. 6.11	Electric circuit in series.	209
Fig. 6.12	Electrical circuit in parallel.	209
Fig. 6.13	Circuit of two loops.	211
Fig. 6.14	Electric transformer.	211
Fig. 6.15	Transformer and variable-capacitance circuit.	213
Fig. 6.16	Electromechanical variable-capacitor system.	215
Fig. 6.17	Electromechanical system of variable inductance.	216
Fig. 6.18	Homogeneous cylinders interconnected by inextensible and weightless threads.	218
Fig. 6.19	Electrical circuits.	218
Fig. 6.20	Electrical circuit modeling the Lagrange system.	219
Fig. 7.1	Capacitor whose plates are held by springs.	224
Fig. 7.2	Series of masses connected by springs.	225
Fig. 7.3	Concept of equilibrium local stability.	230
Fig. 7.4	Linear oscillator.	230
Fig. 7.5	The (ε, δ)-illustration of the local stability of an equilibrium point.	233
Fig. 7.6	A potential electromechanical system.	239
Fig. 7.7	Sliding rings on rotating bars.	241
Fig. 8.1	Three masses joined by springs in circular dynamics.	255
Fig. 8.2	Three masses joined by springs with dynamics on a line.	258
Fig. 8.3	Four spring-bound masses with restricted linear dynamics.	260
Fig. 8.4	Three identical pendula held by springs.	263
Fig. 8.5	Four-loop LC circuit.	265
Fig. 8.6	An inhomogeneous disk rolling without slipping along the horizontal guide.	270
Fig. 8.7	A double mathematical pendulum suspended from a bar.	271

Fig. 8.8	A homogeneous elliptical cylinder rolling without slipping on a horizontal plane.	272
Fig. 8.9	Electric circuit.	272
Fig. 9.1	Input-output system.	276
Fig. 9.2	Hodograph of H_{23}.	278
Fig. 9.3	Amplitude characteristic diagram of H_{23}.	279
Fig. 9.4	Phase characteristic diagram of H_{23}.	279
Fig. 9.5	Electrical circuit.	280
Fig. 9.6	Linear system with input delay.	281
Fig. 9.7	Hodograph of $e^{-j\omega\tau}$.	283
Fig. 9.8	Mechanical system with friction.	283
Fig. 9.9	Electric circuit with variable elements.	284
Fig. 9.10	Illustration of the asymptotic stability concept.	287
Fig. 9.11	Graphics illustrating the zone (9.23).	293
Fig. 9.12	Function graphics of the zone (9.25).	294
Fig. 9.13	$f(j\omega)$ changes when the argument ω varies from $-\infty$ up to ∞.	295
Fig. 9.14	Hodograph of $p(j\omega)$.	297
Fig. 9.15	Zoom of the hodograph of $p(j\omega)$.	297
Fig. 9.16	Hodograph of $p(j\omega)$.	297
Fig. 9.17	Hodograph of $p(j\omega)$.	298
Fig. 9.18	Hodograph of $p(j\omega)$.	299
Fig. 9.19	Hodograph of $p(j\omega)$.	300
Fig. 9.20	Hodograph for the case $k < -3$.	301
Fig. 9.21	Hodograph of $p(j\omega)$ for $k = -3$.	301
Fig. 9.22	Hodograph of $p(j\omega)$ for $-3 < k < 1$.	302
Fig. 9.23	Hodograph of $p(j\omega)$ for $k = 1$.	303
Fig. 9.24	Hodograph of $p(j\omega)$ for $k > 1$.	303
Fig. 9.25	The illustration of the Kharitonov's criterion.	306
Fig. 9.26	Dissipative system of three masses.	308
Fig. 9.27	Electric circuit with 4 capacities, 3 inductions and 1 ohmic resistance.	309
Fig. 10.1	Pendulum with non-negligible-mass arm.	315
Fig. 11.1	A family of trajectories of a Hamiltonian system in the extended state space with two initial and final contours.	340
Fig. 11.2	A family of trajectories of a Hamiltonian system in the extended state space with two initial and final contours: both correspond to constant times.	345
Fig. 12.1	Manipulator with two prismatic joints (PJ) and a rotating joint (R).	384
Fig. 12.2	Rectangular (Cartesian) robot manipulator.	387
Fig. 12.3	Scaffolding robot manipulator.	390
Fig. 12.4	Spherical (polar) manipulator with three rotating joints.	393
Fig. 12.5	Articulated robot manipulator.	396
Fig. 12.6	Universal programmable manipulator (PUMA).	399
Fig. 12.7	Manipulator "Cincinnati Milacron T^3".	404
Fig. 12.8	CD motor, gear train, and load.	412
Fig. 12.9	The Stanford/JPL manipulator.	415
Fig. 12.10	Robot manipulator Unimate 2000.	419
Fig. 12.11	Robot manipulator with swivel base.	424
Fig. 12.12	Cylindrical robot with spring.	427
Fig. 12.13	Non-ordinary manipulator with shock absorber.	429

Fig. 12.14	Two-joint planar manipulator.	**434**
Fig. 12.15	"Crank-turn" robot manipulator.	**438**
Fig. 12.16	Robot manipulator of multicylinder type.	**442**
Fig. 12.17	Arm manipulator with springs.	**445**
Fig. 12.18	Articulated robot.	**460**
Fig. 12.19	"Robot-Maker 110."	**465**
Fig. 12.20	Manipulator on a horizontal platform.	**468**
Fig. 12.21	Two-arms planar manipulator.	**471**
Fig. 12.22	Manipulator with three degrees of freedom.	**476**
Fig. 12.23	CD motor with load.	**478**
Fig. 12.24	*Buck* type DC-DC converter.	**480**
Fig. 12.25	*Boost* type DC-DC converter.	**482**
Fig. 12.26	Model of an induction motor.	**484**

List of tables

Table 6.1	Table of electromechanical analogies.	**210**
Table 9.1	Hodograph crosses for $k < -3$.	**300**
Table 9.2	Hodograph crosses for $k = -3$.	**301**
Table 9.3	Hodograph crosses for $-3 < k < 1$.	**302**
Table 9.4	Hodograph crosses for $k = 1$.	**302**
Table 9.5	Hodograph crosses for $k > 1$.	**303**
Table 9.6	Roots distribution for different values of k.	**304**

About the author

Alexander S. Poznyak graduated from the Moscow Physical Technical Institute (MPhTI) in 1970. He received his PhD and Doctor degrees from the Institute of Control Sciences of the Russian Academy of Sciences in 1978 and 1989, respectively. From 1973 until 1993 he served at this institute as researcher and leading researcher, and in 1993 he accepted a post as full professor (3-F) at CINVESTAV-IPN in Mexico. For 8 years he was the head of the Automatic Control Department. He is the director of **43 PhD** theses (40 in Mexico). He has published more than **260 papers** in different international journals and written **14 books**, which were published by leading publishing houses such as Nauka, Springer, Elsevier, Birkhäuser, and Marcel Decker. **He is Regular Member of the Mexican Academy of Sciences and the System of National Investigators (SNI-Emerito from 2014)**. He is Fellow of the Institute of Mathematics and its Applications (IMA), Essex, UK, and Associate Editor of the *IMA Journal of Mathematical Control and Information* (Oxford), *Kybernetika* (Chech Republic), *Nonlinear Analysis: Hybrid Systems* (IFAC), and the Iberoamerican international journal on "Computations and Systems." He was also Associate Editor of CDC and ACC and Member of the Editorial Board of IEEE CSS. He is a member of the Evaluation Committee of SNI (Ministry of Science and Technology, area 7 and Emeritus Committee), responsible for Engineering Science and Technology Foundation in Mexico, and a member of the Award Committee of the Mexico Prize for Science and Technology. In 2014 he was invited by the USA Government to serve as a member of the NSF committee on "Neuro Sciences and Artificial Intelligence."

Preface

> *One theory is the most impressive,*
> *the simpler are its premises,*
> *the more distinct are the things it connects,*
> *and the broader is its range of applicability.*
> **Albert Einstein**

There are two ways to teach some topic of modern science, namely:
1. the systematic theoretical form and
2. the application-oriented way.

The first means a systematic presentation of the material, governed by the desire for perfection (from a mathematical point of view) and completeness of the presented results. In contrast to the first, the second approach begins with the question *"What is the most important application of the considered topic?"* and then tries to answer this question as quickly as possible without wandering all the good and possibly interesting side roads.

The present book is based on both methods, giving the mathematically precise foundations of mechanics as a natural science, complemented by several practical examples that illustrate the basic enunciations under consideration. The reader feels that the theory is being developed, not only by itself, but by the effective solution of concrete problems. This course deals with different kinds of mechanical, electrical, and electromechanical models providing deep analysis of each one, including the corresponding numerical calculations.

This book is aimed at graduate students (Masters and Doctorate) of the Electrical Engineering faculties, studying mechanics, mechatronics, and control, who wish to learn more about how the elegant theory of classical and analytical mechanics solves different problems that arise in the real world.

The modern word "mechatronics," also called mechatronic engineering, is a multidisciplinary branch of engineering that focuses on the engineering of both electrical and mechanical systems, and also includes a combination of robotics, electronics, computer science, telecommunications, systems control, and product engineering. As technology advances over time, various subfields of engineering have succeeded in both adapting and multiplying. The intention of mechatronics is to produce a design solution that unifies each of these various subfields. Originally, the field of mechatronics was intended to be nothing more than a combination of mechanics and electronics, hence the name being a portmanteau of mechanics and electronics; however, as the complexity of technical systems continued to evolve, the definition was broadened to include more technical areas. Many people treat mechatronics as a modern buzzword synonymous with robotics and electromechanical engineering.[1]

The presented material is based on more than 25 years of teaching experience of the author, initially in Russia (Technical Institute of Physics of Moscow [MFTI] during 1979–1993) and then in Mexico (Center for Research and Advanced Studies of the IPN [CINVESTAV], Automatic Control Department, Mexico City, during 1993–2018). The fundamental concepts of this course were created by world-renowned scientists such as F.R. Gantmacher, M.A. Aizerman, and E.S. Pyatnickii and later developed by I.P. Devyaterikov, G.N. Yakovenko, N.M. Truhan, and Yu.I. Khanukaev.

The author would like to express his wide thanks to his colleagues from MFTI (Russia) and his Mexican ex-PhD students (now doctors) J. Medel, J. Correa-Martinez, Daishi Murano, F. Bejarano, M. Jiménez, and I. Chairez for their kind collaboration and help in the creation of this manuscript.

Alex S. Poznyak
Mexico City and Avándaro, Mexico
2020

---

[1] The word mechatronics originated in Japanese-English and was created by Tetsuro Mori, an engineer of Yaskawa Electric Corporation. The word mechatronics was registered as trademark by the company in Japan with the registration number "46-32714" in 1971. However, afterward the company released the right of using the word to public, the word begun being used across the world. Nowadays, the word is translated into many languages and the word is considered as an essential term for industry.

Notation

Scalars

Scalars are the elements of the real (R) or complex ($\mathbb{C}$) fields and are denoted by lowercase letters in italics, for example, a, b.

Given the scalar a, $|a|$ represents its absolute value if $a \in R$ and its module if $a \in \mathbb{C}$.

Vectors

The vectors are denoted with bold letters, for example, **a**, **K**.

The unit vectors in the Cartesian coordinate directions x, y, z are represented respectively by **i**, **j**, **k**.

Given the vector **a**, its magnitude is denoted by a, or $|\mathbf{a}|$.

(**a**, **b**) denotes the internal or scalar product of vectors **a** and **b**.

[**a**, **b**] denotes the vector product of vectors **a** and **b**.

$\widehat{\mathbf{a}, \mathbf{b}}$ denotes the smallest angle between the vectors **a** and **b**.

a $\parallel$ **b** indicates that the vectors **a** and **b** are parallel.

a $\perp$ **b** indicates that the vectors **a** and **b** are orthogonal.

$\overline{AB}$ denotes the segment with extreme points A and B.

g denotes the gravity acceleration vector.

$\angle AOB$ is the angle in the triangle AOB with the vertex in O.

Matrices

The matrices are tables, represented by uppercase letters in italics, for example,

$$A = \left\| a_{ij} \right\|_{i=1,\ldots,n;\, j=1,\ldots,m}.$$

Given the square matrix A, its determinant is denoted by $\det A$ and its trace by $\mathrm{tr} A$.
Given matrix A, its transpose is represented by $A^T = \left\| a_{ji} \right\|$.
Ker(A) denotes the kernel of matrix A, i.e.,

$$\mathrm{Ker}(A) = \{x : Ax = 0\}.$$

Functions

To represent the functions, the letters are used according to the type to which the value of the function belongs, for example,

$$f_1 : R \to R, \qquad f_2 : R^n \to R,$$
$$\mathbf{g}_1 : R \to R^n, \qquad \mathbf{g}_2 : R^n \to R^m,$$
$$A_1 : R \to R^{n \times m}, \quad A_2 : R^3 \to R^{n \times m},$$

where n, m are positive integers.

Derivatives

$\dot{\mathbf{a}}, \ddot{\mathbf{a}}$ indicate the first and second time derivatives of vector $\mathbf{a}$.

For the real function $f(\mathbf{r}) : R^n \to R$ the following notations are used:

first derivative or gradient,

$$\frac{\partial}{\partial \mathbf{r}} f = \nabla f = \left[\frac{\partial f}{\partial r_1}, \frac{\partial f}{\partial r_2}, \ldots, \frac{\partial f}{\partial r_n} \right]^\mathsf{T};$$

second derivative matrix or Hessian,

$$\frac{\partial^2}{\partial \mathbf{r}^2} f = \nabla^2 f = \begin{bmatrix} \frac{\partial^2 f}{\partial r_1 \partial r_1} & \frac{\partial^2 f}{\partial r_1 \partial r_2} & \cdots & \frac{\partial^2 f}{\partial r_1 \partial r_n} \\ \frac{\partial^2 f}{\partial r_2 \partial r_1} & \frac{\partial^2 f}{\partial r_2 \partial r_2} & \cdots & \frac{\partial^2 f}{\partial r_2 \partial r_n} \\ \vdots & \vdots & \ddots & \vdots \\ \frac{\partial^2 f}{\partial r_n \partial r_1} & \frac{\partial^2 f}{\partial r_n \partial r_2} & \cdots & \frac{\partial^2 f}{\partial r_n \partial r_n} \end{bmatrix}.$$

The derivative of the vector function $\mathbf{g}(\mathbf{r}) : \mathbb{R}^n \to \mathbb{R}^m$ is the functional matrix given by

$$\frac{\partial \mathbf{g}}{\partial \mathbf{r}} = \begin{bmatrix} \frac{\partial g_1}{\partial r_1} & \frac{\partial g_1}{\partial r_2} & \cdots & \frac{\partial g_1}{\partial r_n} \\ \frac{\partial g_2}{\partial r_1} & \frac{\partial g_2}{\partial r_2} & \cdots & \frac{\partial g_2}{\partial r_n} \\ \frac{\partial g_m}{\partial r_1} & \frac{\partial g_m}{\partial r_2} & \cdots & \frac{\partial g_m}{\partial r_n} \end{bmatrix}.$$

Quaternion

A quaternion is a complex number of the form

$$\Lambda := \lambda_0 + \sum_{j=1}^{3} \lambda_j \mathbf{i}_j = \lambda_0 + \boldsymbol{\lambda},$$

where λ_i ($i = 0, ..., 3$) are real numbers and $\mathbf{i}_j$ are imaginary units that satisfy certain conditions.[1]

The product $\Lambda \circ \Delta$ of two quaternions $\Lambda = \lambda_0 + \boldsymbol{\lambda}$ and $\Delta = \delta_0 + \boldsymbol{\delta}$ is defined as

$$\Lambda \circ \Delta := \lambda_0 \delta_0 - (\boldsymbol{\lambda}, \boldsymbol{\delta}) + \lambda_0 \boldsymbol{\delta} + \delta_0 \boldsymbol{\lambda} + [\boldsymbol{\lambda}, \boldsymbol{\delta}],$$

where $(\boldsymbol{\lambda}, \boldsymbol{\delta})$ is the scalar and $[\boldsymbol{\lambda}, \boldsymbol{\delta}]$ is the vector product.

Quadratic forms

We have

$$\|x\|_Q^2 := x^\mathsf{T} Q x = \sum_{i=1}^{n} \sum_{j=1}^{n} q_{ij} x_i x_j,$$

$$Q = \|q_{ij}\|_{i=1,...,n;\ j=1,...,n}.$$

---

[1] Recall that when Hamilton passed from complex numbers to quaternions, multiplication lost one of its normal properties: commutativity.

Introduction

Several good books dedicated to physical, mechanical, electrical, and electromechanical models are well known within the area of technological sciences. Among them we can quote (Rutherford, 1951), (Becker, 1954), (Seely et al., 1958), (Corben and Stehle, 1960), (Kittel et al., 1968), (Symon, 1968), (Landau and Lifshitz, 1969), (Gantmakher, 1970), (Barger and Olsson, 1973), (Kibble, 1973), (Titherington and Rimmer, 1973), (Lawden, 1974), (Bartlett, 1975), (Burghes and Downs, 1975), (Abraham and Marsden, 1978), (Devaney and Nitecki, 1981), (Takwale and Puranik, 1979), (Goldstein, 1980), (Kotkin and Serbo, 1980), (Aizerman, 1980), (Desloge, 1982), (Hestenes, 1986), (Fowles, 1986), (Arnold, 1989), (Matzner and Shepley, 1991), (Marsden, 1992), (Chow, 1995), (Barger et al., 1995), (Bhatia, 1997), (Arya, 1998), (Kwatny and Blankenship, 2000), (Kibble and Berkshire, 2004), (Fowles et al., 2005), (Deriglazov, 2016), (Torres del Castillo, 2018).

This book differs from the aforementioned books in different aspects. Maintaining the precise and rigorous form of mathematical explanation, **this book is basically oriented towards readers in the engineering area**, while the above cited classical books are aimed at specialists in the fields of theoretical and mathematical physics. In this book the most discussed models of practical systems (gyroscopes, robots, and some electrical schemes, in particular, power converters) are considered in details.

The discussion of the contents of the book is presented below.

The *Lagrangian formalism* is presented in Chapters 1–9.

The study of the kinematics of a point as in *Chapter 1* has the purpose of obtaining the expressions that describe the temporal behavior of its position, speed, and acceleration. This chapter introduces the necessary basic concepts and shows how these expressions are deduced. A fundamental aspect of the subject is that concerning the coordinate system used, since the obtained mathematical expressions depend on it. In view of the fact that the most natural system and therefore the employed one is the Cartesian system, the generalized coordinates are defined based on this system. The corresponding relationships that allow the transformation of the kinematic quantities between different types of coordinates are obtained.

In *Chapter 2* the concept of rigid body is introduced and quite general expressions are obtained for the description of the kinematics of this mechanical entity. A fundamental tool for the study of the kinematics of the rigid body is Euler's theorem, with which important concepts such as angular speed and acceleration appear, and which allows the calculation of the speed and acceleration of any point of the rigid body. In particular, this method can be extended to the case of the movement of a point in the presence of a mobile relative coordinate system. This topic is addressed in the final section. The description of rotations, using quaternions as generalized coordinates, is also considered.

The motivation of *Chapter 3* is the study of the relationship between the kinematic quantities of a point system and their causes, that is, the forces. This study leads to the introduction of important concepts, such as kinetic energy, momentum, impulse moment, and force moment, which will allow obtaining very important relations between them.

The relationships obtained in the previous chapters are based on the consideration that the "absolute reference system" is not accelerated. Systems in which this condition is met are called inertial. In *Chapter 4* the dynamics of non-inertial systems are analyzed, that is, systems whose "absolute reference" undergoes an acceleration. Another aspect of this chapter deals with the case when the mass is admitted to be variable, which is another aspect of this chapter.

In *Chapter 5* we continue with the study of the dynamics of solid bodies. The dynamic equations corresponding to the rotation of bodies and referred to as the dynamic Euler equations are obtained. To do that, a fundamental concept of the geometry of the solids is introduced, namely, the inertia tensor, which is key in the description of the equations sought. The inertial tensor will allow calculating fundamental quantities such as kinetic energy and impulse moment with reduced expressions. In the central part of the chapter, the proposed objective is achieved, once some main properties of the inertia tensor have been stated. The chapter concludes with the application (not trivial, but very productive) of Euler's equations to the study of special movements such as the gyroscope and dynamic reactions. Some examples and exercises illustrate the presented theory.

Newton's second law and Euler's dynamic equations are the formalism that allows to obtain the equations of movement in mechanical systems; however, their application is usually complicated if the geometry of the movement is not simple and/or by the presence of restrictions to it. The Lagrange equations, whose study is addressed in *Chapter 6*, are an essential tool for these cases, since they naturally include the constraints, in addition to being based on the concept of generalized coordinates, which allow describing the dynamics in terms of the variables, associated with the degrees of freedom of the system. This particularity also makes it possible to apply the same formalism to electrical and even electromechanical systems. A fundamental part of the Lagrange equations are the generalized forces, which characterize (constitute) the essential part of these equations.

In dynamic systems in general and in mechanical systems in particular, the determination of equilibrium positions and their quality of stability are traditional problems of fundamental importance, which to date have been partially resolved. In *Chapter 7*, based on the concepts and results obtained up to this point (such as coordinates and generalized forces), the study of these topics is addressed and the most important results are reported. As will be seen, the most developed theory is that dealing with conservative systems, which occupy most of the chapter.

The Lagrange equations are an invaluable tool in determining the important properties of mechanical systems. The application of these equations and the study of the consequences derived have been the object of the two preceding chapters. In *Chapter 8* one more application is presented to the study of the important properties of small oscillations of a system around the points of its equilibrium. By the usual technique of linearization around a point of equilibrium, the Lagrange equations can be approximated by a linear expression that describes in sufficient detail the dynamics of the system in a neighborhood sufficiently close to the point of interest. In this approximate expression all known techniques for linear dynamic systems can be applied, leading to very useful conclusions. In addition, if the considered system is restricted to being of the conservative type, then the expression is reduced, which allows to characterize and calculate its solutions in a very simple way.

In *Chapter 9*, the study of the linear systems obtained from the process of linearization of the Lagrange equations is continued. This continuation covers two aspects: first, the consideration of non-potential forces dependent only on time allows the use of the important tool of Fourier transformation, which leads to the consideration of the frequency response of the system; second, dissipative systems are considered, which generalize to those of conservative type and allow the introduction of the concept of asymptotically stable equilibrium, extending the previously discussed idea of equilibrium.

The *Hamiltonian formalism* is presented in Chapters 10 and 11.

In *Chapter 10*, conservative systems are considered and generalized impulses are introduced. Hamilton's variables are also considered and it is demonstrated that they can completely describe the dynamics of a system in the canonical Hamiltonian format. Some properties of these canonical equations are studied as well as their first integrals.

The canonical transformations of the dynamic variables, describing Hamiltonians in new variables, are considered in *Chapter 11*. Several criteria of canonicity (such as the S-criterion) is studying. The Hamilton–Jacobi (HJ) equation (partial differential equation) that corresponds to Hamilton's canonical equations (the system of ordinary differential equations [ODEs]) is also considered.[1] Its complete integrals are found. The considered technique allows to find the solution of the canonical Hamiltonian equation without direct resolution of the corresponding system of ODEs, but resolving only the system of special nonlinear algebraic equations. This chapter also shows the relation between the HJ equation in mechanics of conservative systems and the dynamic programming method in optimal control theory.

---

[1] It is named after William Rowan Hamilton and Carl Gustav Jacob Jacobi.

Some models of electromechanical systems (such as a robot of the PUMA type, the pendubot, DC and induction motors, and also a power converter) are developed in *Chapter 12*.

The formulations of the majority of the presented exercises have been taken from (Pyatnickii et al., 1996).

The basic idea of this book is to build a bridge between theory and practice related to mechanical, electrical, and electromechanical systems.

Kinematics of a point

Contents

1.1 Products of vectors   1
     1.1.1 Internal (scalar) product   2
     1.1.2 Vector product   3
     1.1.3 Main properties of triple products   6
1.2 Generalized coordinates   8
     1.2.1 Different possible coordinates   8
     1.2.2 Definition of generalized coordinates   8
     1.2.3 Relationship of generalized coordinates with Cartesian   9
     1.2.4 Coefficients of Lamé   10
1.3 Kinematics in generalized coordinates   10
     1.3.1 Velocity in generalized coordinates   11
     1.3.2 Acceleration in generalized coordinates   11
1.4 Movement in the cylindrical and spherical coordinate systems   14
     1.4.1 Movement in cylindrical coordinates   14
     1.4.2 Movement in spherical coordinates   16
1.5 Normal and tangential accelerations   18
1.6 Some examples   20
1.7 Exercises   28

The study of point kinematics has the purpose to obtain the expressions that describe the temporal behavior of its position, speed, and acceleration. This chapter introduces the necessary basic concepts and shows how these expressions are deduced. A fundamental aspect of the subject concerns the coordinate system used, since the obtained mathematical expressions depend on it. In view of the fact that the most natural system and therefore the employed one is the Cartesian system, the generalized coordinates are defined based on this concrete system. The relationships that allow the transformation of the kinematic quantities between different types of coordinates are obtained.

1.1 Products of vectors

In the following presentation two operations on vectors are defined. The product, being one of the most interesting both in its scalar and in its vectorial mode, and its most important properties are obtained.

1.1.1 Internal (scalar) product

Fig. 1.1 illustrates the details of the following two definitions related to the so-called internal or scalar product.

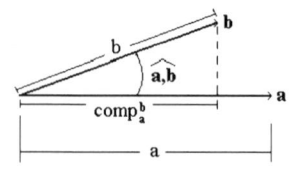

Figure 1.1 The projection of the vector **b** to the direction of another vector **a**.

Definition 1.1. For the two vectors **a** and **b**, the **component** $\text{comp}_{\mathbf{a}}^{\mathbf{b}}$ of **b** over **a** is defined as

$$\text{comp}_{\mathbf{a}}^{\mathbf{b}} := b \cos(\widehat{\mathbf{a}, \mathbf{b}}). \tag{1.1}$$

Definition 1.2. The **internal** or **scalar product** $(\mathbf{a}, \mathbf{b})$ of vectors **a** and **b** is the scalar, defined as

$$(\mathbf{a}, \mathbf{b}) := ab \cos(\widehat{\mathbf{a}, \mathbf{b}}) = a \, \text{comp}_{\mathbf{a}}^{\mathbf{b}}. \tag{1.2}$$

The main properties of the scalar product (1.2) are described in the following lemma.

Lemma 1.1 (The main properties of the **internal** or **scalar product**). *Let* **a**, **b**, **c** *be vectors in* $\mathbb{R}^3$.

1. *Commutativity: We have*

$$(\mathbf{a}, \mathbf{b}) = (\mathbf{b}, \mathbf{a}).$$

2. *Distributivity: We have*

$$(\mathbf{a}, (\mathbf{b} + \mathbf{c})) = (\mathbf{a}, \mathbf{b}) + (\mathbf{a}, \mathbf{c}).$$

3. *Criterion of parallelism: If* $a, b \neq 0$, *then*

$$\frac{(\mathbf{a}, \mathbf{b})}{ab} = \begin{cases} 1, & \text{if and only if the vectors } \mathbf{a} \text{ and } \mathbf{b} \text{ are parallel,} \\ -1, & \text{if and only if the vectors } \mathbf{a} \text{ and } \mathbf{b} \text{ are antiparallel.} \end{cases}$$

4. *Orthogonality criterion: If* $a, b \neq 0$, *then*

$$(\mathbf{a}, \mathbf{b}) = 0$$

if and only if the vectors **a** *and* **b** *are orthogonal, that is,* $\mathbf{a} \perp \mathbf{b}$.

5. *Magnitude (length) of a vector:* This is defined as

$$a = \sqrt{(\mathbf{a}, \mathbf{a})}.$$

Proof. Both vectors **a** and **b** have the following representation:

$$\left.\begin{aligned}\mathbf{a} &= a_{q_1}\mathbf{k}_1 + a_{q_2}\mathbf{k}_2 + a_{q_3}\mathbf{k}_3, \\ \mathbf{b} &= b_{q_1}\mathbf{k}_1 + b_{q_2}\mathbf{k}_2 + b_{q_3}\mathbf{k}_3,\end{aligned}\right\} \quad (1.3)$$

where the unitary vectors $\mathbf{k}_1$, $\mathbf{k}_2$, $\mathbf{k}_3$ satisfy the relation

$$(\mathbf{k}_i, \mathbf{k}_j) = \delta_{i,j} \quad (1.4)$$

with

$$\delta_{i,j} := \begin{cases} 1, & i = j, \\ 0, & i \neq j, \end{cases} \quad i, j = 1, 2, 3,$$

referred to as the Kronecker symbol. So,

$$(\mathbf{a}, \mathbf{b}) = a_{q_1}b_{q_1} + a_{q_2}b_{q_2} + a_{q_3}b_{q_3}. \quad (1.5)$$

Properties 1, 3, 4, and 5 immediately follow from the definition. As for property 2, it results from the projection property:

$$(\mathbf{a}, (\mathbf{b} + \mathbf{c})) = a\mathrm{comp}_\mathbf{a}^{\mathbf{b}+\mathbf{c}} = a\left(\mathrm{comp}_\mathbf{a}^\mathbf{b} + \mathrm{comp}_\mathbf{a}^\mathbf{c}\right) = (\mathbf{a}, \mathbf{b}) + (\mathbf{a}, \mathbf{c}).$$

□

Remark 1.1. By the definition (1.2) and properties 1 and 2 it follows that the scalar product is a **bilinear operation**.

1.1.2 Vector product

Now we present the basic concepts related to the vector product, whose details are illustrated in Fig. 1.2.

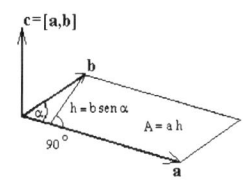

Figure 1.2 Vector product of two vectors **a** and **b**.

Definition 1.3. For $\mathbf{a}, \mathbf{b} \in \mathbb{R}^3$ their vector product, denoted by

$$\mathbf{c} := [\mathbf{a}, \mathbf{b}],$$

is defined as the vector which is orthogonal to the plane, formed by $\mathbf{a}$ and $\mathbf{b}$, with the direction of the advance of a right screw that follows the rotation of $\mathbf{a}$ to $\mathbf{b}$ and with the magnitude

$$c = ab \sin \alpha, \qquad (1.6)$$

where $\alpha := \widehat{\mathbf{a}, \mathbf{b}}$.

As can be seen from Fig. 1.2, the area of the parallelogram, formed by $\mathbf{a}$ and $\mathbf{b}$, is given by

$$A = ah,$$

with

$$h = b \sin \alpha,$$

that is,

$$c = A = \|[\mathbf{a}, \mathbf{b}]\|.$$

From the definition of the vector product it is possible to obtain several consequences, which are given in the following lemma.

Lemma 1.2 (Properties of the vector product). *For any three vectors $\mathbf{a}, \mathbf{b}, \mathbf{c} \in \mathbb{R}^3$ the following properties hold:*

1. *Anticommutativity, i.e.,*

$$[\mathbf{a}, \mathbf{b}] = -[\mathbf{b}, \mathbf{a}].$$

2. *Distributivity on the sum, i.e.,*

$$[\mathbf{a}, (\mathbf{b} + \mathbf{c})] = [\mathbf{a}, \mathbf{b}] + [\mathbf{a}, \mathbf{c}].$$

3. *Criterion of parallelism: If $a, b \neq 0$, then*

$$[\mathbf{a}, \mathbf{b}] = 0$$

if and only if $\mathbf{a}$ and $\mathbf{b}$ are parallel.

4. *Orthogonality criterion: If $a, b \neq 0$, then*

$$\frac{\|[\mathbf{a}, \mathbf{b}]\|}{ab} = 1$$

if and only if $\mathbf{a}$ and $\mathbf{b}$ are orthogonal.

Kinematics of a point

5. *Vector products between the unit vectors:* If the unit vector system $(\mathbf{k}_1, \mathbf{k}_2, \mathbf{k}_3)$, associated with the generalized coordinates $\mathbf{q}$, is orthogonal and has the configuration as in Fig. 1.3, then

$$[\mathbf{k}_1, \mathbf{k}_2] = \mathbf{k}_3, \ [\mathbf{k}_2, \mathbf{k}_3] = \mathbf{k}_1, \ [\mathbf{k}_3, \mathbf{k}_1] = \mathbf{k}_2.$$

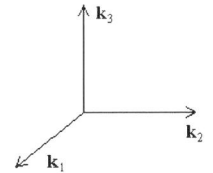

Figure 1.3 Right orthogonal system.

6. *Explicit formula:* If

$$\left.\begin{aligned} \mathbf{a} &= a_{q_1}\mathbf{k}_1 + a_{q_2}\mathbf{k}_2 + a_{q_3}\mathbf{k}_3, \\ \mathbf{b} &= b_{q_1}\mathbf{k}_1 + b_{q_2}\mathbf{k}_2 + b_{q_3}\mathbf{k}_3, \end{aligned}\right\}$$

then

$$\left.\begin{aligned} [\mathbf{a}, \mathbf{b}] = \left(a_{q_2}b_{q_3} - a_{q_3}b_{q_2}\right)\mathbf{k}_1 + \\ \left(a_{q_3}b_{q_1} - a_{q_1}b_{q_3}\right)\mathbf{k}_2 + \left(a_{q_1}b_{q_2} - a_{q_2}b_{q_1}\right)\mathbf{k}_3 \end{aligned}\right\} \quad (1.7)$$

making use of the determinant concept matches as

$$[\mathbf{a}, \mathbf{b}] = \begin{Vmatrix} \mathbf{k}_1 & \mathbf{k}_2 & \mathbf{k}_3 \\ a_{q_1} & a_{q_2} & a_{q_3} \\ b_{q_1} & b_{q_2} & b_{q_3} \end{Vmatrix}.$$

Proof. Properties 1, 3, 4, and 5 are obtained directly from the definition of the vector product. The demonstration of property 2 is left as an exercise to the reader, while property 6 follows immediately from properties 1, 2, and 5. □

Remark 1.2. Again, it is easy to follow the definition (1.6) and properties 1 and 2 and verify that the vector product is **bilinear**.

In the following definitions three vectors are involved, so the considered operations are referred to as triple products.

Definition 1.4. Given vectors $\mathbf{a}, \mathbf{b}, \mathbf{c} \in \mathbb{R}^3$, the scalar

$$(\mathbf{a}, [\mathbf{b}, \mathbf{c}])$$

is called the **triple scalar product**, while the vector

$$[\mathbf{a}, [\mathbf{b}, \mathbf{c}]]$$

is called the **triple vector product**.

1.1.3 Main properties of triple products

The following lemma presents some useful properties of triple products.

Lemma 1.3 (Properties of triple products). *For any vectors*

$$\mathbf{a}, \mathbf{b}, \mathbf{c} \in \mathbb{R}^3$$

the following properties hold:

1. *Alternate formula of the triple scalar product: Under the condition of orthogonality of the unit vectors* $(\mathbf{k}_1, \mathbf{k}_2, \mathbf{k}_3)$ *by the representation*

$$\left.\begin{aligned} \mathbf{a} &= a_{q_1}\mathbf{k}_1 + a_{q_2}\mathbf{k}_2 + a_{q_3}\mathbf{k}_3, \\ \mathbf{b} &= b_{q_1}\mathbf{k}_1 + b_{q_2}\mathbf{k}_2 + b_{q_3}\mathbf{k}_3, \\ \mathbf{c} &= c_{q_1}\mathbf{k}_1 + c_{q_2}\mathbf{k}_2 + c_{q_3}\mathbf{k}_3, \end{aligned}\right\} \tag{1.8}$$

it follows that

$$(\mathbf{a}, [\mathbf{b}, \mathbf{c}]) = \det \begin{Vmatrix} a_{q_1} & a_{q_2} & a_{q_3} \\ b_{q_1} & b_{q_2} & b_{q_3} \\ c_{q_1} & c_{q_2} & c_{q_3} \end{Vmatrix}. \tag{1.9}$$

2. *Cyclic rotation of the triple scalar product: We have*

$$(\mathbf{a}, [\mathbf{b}, \mathbf{c}]) = (\mathbf{b}, [\mathbf{c}, \mathbf{a}]) = (\mathbf{c}, [\mathbf{a}, \mathbf{b}]). \tag{1.10}$$

3. *Alternate formula of the triple vector product: We have*

$$[\mathbf{a}, [\mathbf{b}, \mathbf{c}]] = \mathbf{b}(\mathbf{a}, \mathbf{c}) - \mathbf{c}(\mathbf{a}, \mathbf{b}). \tag{1.11}$$

4. *Jacobi's identity: We have*

$$[\mathbf{a}, [\mathbf{b}, \mathbf{c}]] + [\mathbf{b}, [\mathbf{c}, \mathbf{a}]] + [\mathbf{c}, [\mathbf{a}, \mathbf{b}]] = 0. \tag{1.12}$$

Proof. By (1.5) and (1.7) we have

$$\begin{aligned} (\mathbf{a}, [\mathbf{b}, \mathbf{c}]) &= a_{q_1}\left(b_{q_2}c_{q_3} - b_{q_3}c_{q_2}\right) + \\ &\quad a_{q_2}\left(b_{q_3}c_{q_1} - b_{q_1}c_{q_3}\right) + a_{q_3}\left(b_{q_1}c_{q_2} - b_{q_2}c_{q_1}\right), \end{aligned}$$

which coincides with (1.9). Property 2 follows immediately from the property of the determinants. To show property 3, note that

$$[\mathbf{a},[\mathbf{b},\mathbf{c}]] = \begin{Vmatrix} \mathbf{k}_1 & \mathbf{k}_2 & \mathbf{k}_3 \\ a_{q_1} & a_{q_2} & a_{q_3} \\ b_{q_2}c_{q_3}-b_{q_3}c_{q_2} & b_{q_3}c_{q_1}-b_{q_1}c_{q_3} & b_{q_1}c_{q_2}-b_{q_2}c_{q_1} \end{Vmatrix}$$
$$= \left[b_{q_1}\left(a_{q_2}c_{q_2}+a_{q_3}c_{q_3}\right) - c_{q_1}\left(a_{q_2}b_{q_2}+a_{q_3}b_{q_3}\right) \right]\mathbf{k}_1$$
$$+ \left[b_{q_2}\left(a_{q_1}c_{q_1}+a_{q_3}c_{q_3}\right) - c_{q_2}\left(a_{q_1}b_{q_1}+a_{q_3}b_{q_3}\right) \right]\mathbf{k}_2$$
$$+ \left[b_{q_3}\left(a_{q_1}c_{q_1}+a_{q_2}c_{q_2}\right) - c_{q_3}\left(a_{q_1}b_{q_1}+a_{q_2}b_{q_2}\right) \right]\mathbf{k}_3 \quad (1.13)$$

and

$$\mathbf{b}(\mathbf{a},\mathbf{c}) - \mathbf{c}(\mathbf{a},\mathbf{b}) =$$
$$\left(b_{q_1}\mathbf{k}_1 + b_{q_2}\mathbf{k}_2 + b_{q_3}\mathbf{k}_3\right)\left(a_{q_1}c_{q_1} + a_{q_2}c_{q_2} + a_{q_3}c_{q_3}\right)$$
$$- \left(c_{q_1}\mathbf{k}_1 + c_{q_2}\mathbf{k}_2 + c_{q_3}\mathbf{k}_3\right)\left(a_{q_1}b_{q_1} + a_{q_2}b_{q_2} + a_{q_3}b_{q_3}\right) =$$
$$\left[b_{q_1}\left(a_{q_1}c_{q_1}+a_{q_2}c_{q_2}+a_{q_3}c_{q_3}\right) - c_{q_1}\left(a_{q_1}b_{q_1}+a_{q_2}b_{q_2}+a_{q_3}b_{q_3}\right) \right]\mathbf{k}_1$$
$$+ \left[b_{q_2}\left(a_{q_1}c_{q_1}+a_{q_2}b_{q_2}+a_{q_3}c_{q_3}\right) - c_{q_2}\left(a_{q_1}b_{q_1}+a_{q_2}b_{q_2}+a_{q_3}b_{q_3}\right) \right]\mathbf{k}_2$$
$$+ \left[b_{q_3}\left(a_{q_1}c_{q_1}+a_{q_2}c_{q_2}+a_{q_3}c_{q_3}\right) - c_{q_3}\left(a_{q_1}b_{q_1}+a_{q_2}b_{q_2}+a_{q_3}b_{q_3}\right) \right]\mathbf{k}_3. \quad (1.14)$$

Direct comparison of (1.13) with (1.14) leads to (1.11). The application of property 3 to the left-hand side of (1.12) implies

$$[\mathbf{a},[\mathbf{b},\mathbf{c}]] + [\mathbf{b},[\mathbf{c},\mathbf{a}]] + [\mathbf{c},[\mathbf{a},\mathbf{b}]] =$$
$$\mathbf{b}(\mathbf{a},\mathbf{c}) - \mathbf{c}(\mathbf{a},\mathbf{b}) + \mathbf{c}(\mathbf{a},\mathbf{b}) - (\mathbf{b},\mathbf{c})\mathbf{a} + (\mathbf{b},\mathbf{c})\mathbf{a} - \mathbf{b}(\mathbf{a},\mathbf{c}) = 0.$$

□

Remark 1.3. If in formula (1.11) we put $\mathbf{c} = \mathbf{a}$ with $a \neq 0$, then we get

$$\mathbf{b} = \frac{(\mathbf{a},\mathbf{b})}{a^2}\mathbf{a} + \frac{[\mathbf{a},[\mathbf{b},\mathbf{a}]]}{a^2}, \quad (1.15)$$

that is, any vector $\mathbf{b}$ can be represented as a linear combination of a vector $\mathbf{a}$ and the vector which is perpendicular to $\mathbf{a}$ being contained in the plane formed by $\mathbf{a}$ and $\mathbf{b}$.

The following exercise demonstrates the effectiveness of the direct application of formula (1.15).

Example 1.1. Consider the following system of four algebraic equations with respect to the components of the vector $\mathbf{r} \in \mathbb{R}^3$:

$$\left.\begin{aligned}(\mathbf{r},\mathbf{a}) &= m, \\ [\mathbf{r},\mathbf{a}] &= \mathbf{b},\end{aligned}\right\}$$

where vectors $\mathbf{a}, \mathbf{b} \in \mathbb{R}^3$ and the scalar $m \in \mathbb{R}$ are supposed to be given. We need to find $\mathbf{r}$. Taking in (1.15) $\mathbf{b} = \mathbf{r}$ we get

$$\mathbf{r} = \frac{(\mathbf{a}, \mathbf{r})}{a^2}\mathbf{a} + \frac{[\mathbf{a}, [\mathbf{r}, \mathbf{a}]]}{a^2} = \frac{m}{a^2}\mathbf{a} + \frac{[\mathbf{a}, \mathbf{b}]}{a^2}.$$

1.2 Generalized coordinates

1.2.1 Different possible coordinates

Let p be a moving point in space and let $\mathbf{r}(t)$ be the vector that describes its position at time t with respect to some given reference system. The description of $\mathbf{r}$ can be carried out in as many ways as possible with the reference system. Within these descriptions, the most used reference system is the *Cartesian* system, but in certain problems it may be more natural to use others, for example, representations in cylindrical or spherical coordinates. In Fig. 1.4 these three types of description and their relationships are shown.

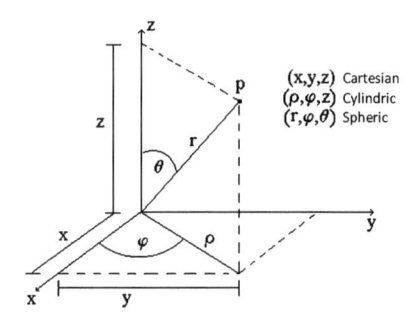

Figure 1.4 Cartesian, cylindrical, and spherical vector representations and their relations.

1.2.2 Definition of generalized coordinates

The relationship between different vectorial descriptions of the position of the point p has special importance in the derivation of the expressions that describe its movement, such as the expressions of position, speed, and acceleration.

Definition 1.5. A triad of numbers $\mathbf{q} = (q_1, q_2, q_3)$ that allow to uniquely specify the position of a point p in space is said to form a set of **generalized coordinates** of p. The set of generalized coordinates $\mathbf{q}$ of all points of a system in space is called the coordinate system corresponding to the generalized coordinates $\mathbf{q}$.

It is clear that the Cartesian coordinates form a system of generalized coordinates too.

1.2.3 Relationship of generalized coordinates with Cartesian

Criterion 1.1. *The set of all possible triads* $\mathbf{q} = (q_1, q_2, q_3)$, *where* $q_i \in S_i \subseteq \mathbb{R}$ ($i = 1, 2, 3$), *forms a **generalized coordinate system** if and only if there is a **one-to-one relationship** between the Cartesian description* $\mathbf{r} = (x, y, z)$ *of each point* $p \in \mathbb{R}^3$ *and a triad* $\mathbf{q} = (q_1, q_2, q_3)$.

In other words, the previous definition states that set

$$S := \{\mathbf{q} = (q_1, q_2, q_3), \quad q_i \in S_i \subseteq \mathbb{R}, \quad i = 1, 2, 3\} \subseteq \mathbb{R}^3$$

forms a generalized coordinate system if and only if the mapping

$$\mathbf{r} : S \to \mathbb{R}^3$$

given by

$$\mathbf{r} = \mathbf{r}(\mathbf{q}) \tag{1.16}$$

with

$$\mathbf{r} = (x, y, z)^\mathsf{T} \tag{1.17}$$

constitutes a **one-to-one transformation**.

The following criterion characterizes this transformation.

Lemma 1.4. *A vector function* $\mathbf{r}(\mathbf{q}) : \mathbb{R}^3 \to \mathbb{R}^3$ *is a one-to-one smooth transformation if and only if the derivative of* $\mathbf{r}$, *called the Jacobian matrix of* $\mathbf{r}$, *given by*

$$\frac{\partial \mathbf{r}}{\partial \mathbf{q}} := \begin{bmatrix} \frac{\partial r_1}{\partial q_1} & \frac{\partial r_1}{\partial q_2} & \frac{\partial r_1}{\partial q_3} \\ \frac{\partial r_2}{\partial q_1} & \frac{\partial r_2}{\partial q_2} & \frac{\partial r_2}{\partial q_3} \\ \frac{\partial r_3}{\partial q_1} & \frac{\partial r_3}{\partial q_2} & \frac{\partial r_3}{\partial q_3} \end{bmatrix},$$

is not singular, which in turn is satisfied if and only if for all x, y, *and* z

$$\det\left(\frac{\partial \mathbf{r}}{\partial \mathbf{q}}\right) \neq 0. \tag{1.18}$$

The existence of the transformation (1.16) ensures that the Cartesian coordinates of a point p can be determined from the triad $\mathbf{q}$ by some functions

$$x = x(q_1, q_2, q_3), \quad y = y(q_1, q_2, q_3), \quad z = z(q_1, q_2, q_3).$$

If a set of coordinates has the property (1.18), you can easily derive the expressions that describe the kinematics of the moving point they represent from those corresponding to the description in Cartesian coordinates and vice versa. This derivation requires some concepts, which are introduced next.

Definition 1.6. Suppose that in the transformation (1.16) two constant generalized coordinates are maintained and the other one is allowed to vary; the generated curve is known as the coordinate curve corresponding to the variable coordinate, and the direction that such a curve follows when the coordinate increases is said to be its **positive direction**. A system of generalized coordinates is called **rectilinear** if its coordinate curves are straight; if they result as curves, then it is called **curvilinear**.

The coordinate curves make it possible to determine the unit vectors of the generalized coordinate system in question. The unitary vector, corresponding to the coordinate q_i, $i = 1, 2, 3$, is given by the positive direction of the tangent to the coordinate curve corresponding to q_i. This concept is formalized in the following two definitions.

1.2.4 Coefficients of Lamé

Definition 1.7. The values

$$H_i := \left| \frac{\partial \mathbf{r}}{\partial q_i} \right| = \sqrt{\left(\frac{\partial x}{\partial q_i} \right)^2 + \left(\frac{\partial y}{\partial q_i} \right)^2 + \left(\frac{\partial z}{\partial q_i} \right)^2}, \quad i = 1, 2, 3, \tag{1.19}$$

are called the **Lamé coefficients** corresponding to the generalized coordinates **q**.

Definition 1.8. The vectors

$$\mathbf{k}_i := \frac{1}{H_i} \frac{\partial \mathbf{r}}{\partial q_i}, \quad i = 1, 2, 3, \tag{1.20}$$

are referred to as **the unit vectors** of the generalized coordinate **q**.

The defined unit vectors constitute the base in the space of the generalized coordinate **q**. Note that in general these bases are not orthogonal; besides, they vary from one point to another. For this last reason they are known as **local bases**. By the previous concepts, given a system of generalized coordinates **q**, the representation of a generic vector **p** with respect to this system is given by

$$\mathbf{p} = p_{q_1} \mathbf{k}_1 + p_{q_2} \mathbf{k}_2 + p_{q_3} \mathbf{k}_3, \tag{1.21}$$

where p_{q_i}, $i = 1, 2, 3$, is the component of the point **p** on the coordinate q_i.

1.3 Kinematics in generalized coordinates

The concepts introduced up to now allow to obtain in a simple way the expressions for the speed and acceleration of a mobile particle p, when the description is made in the generalized coordinates **q**.

Kinematics of a point

1.3.1 Velocity in generalized coordinates

Let $\mathbf{r}(t)$ be the vector that describes the position at time t of the moving point p in the Cartesian coordinate space. The velocity $\mathbf{v}$ of the point p is defined as the temporal derivative of the position vector $\mathbf{r}(t)$:

$$\mathbf{v} := \frac{d\mathbf{r}}{dt}.$$

Now let a system be given in generalized coordinates $\mathbf{q}$. Then there is a transformation

$$\mathbf{r}(t) = \mathbf{r}(\mathbf{q}(t))$$

such that, by the rule of chain differentiation,

$$\mathbf{v} = \frac{d}{dt}\mathbf{r}(\mathbf{q}(t)) = \frac{\partial \mathbf{r}}{\partial q_1}\dot{q}_1 + \frac{\partial \mathbf{r}}{\partial q_2}\dot{q}_2 + \frac{\partial \mathbf{r}}{\partial q_3}\dot{q}_3, \qquad (1.22)$$

where

$$\dot{q}_i := \frac{dq_i}{dt}, \quad i = 1, 2, 3.$$

Now, using the definitions of the coefficients of Lamé (1.19) and of the unit vectors (1.20) for the coordinate system $\mathbf{q}$, expression (1.22) can be rewritten in the form

$$\mathbf{v} = \sum_{i=1}^{3} H_i \dot{q}_i \mathbf{k}_i, \qquad (1.23)$$

where

$$v_{q_i} := H_i \dot{q}_i, \quad i = 1, 2, 3, \qquad (1.24)$$

is the component of vector $\mathbf{v}$ in direction $\mathbf{k}_i$.

Remark 1.4. The magnitude of $\mathbf{v}$ is given by

$$v = \sqrt{\langle \mathbf{v}, \mathbf{v} \rangle}, \qquad (1.25)$$

and by (1.24), if the system $\mathbf{q}$ is orthogonal, formula (1.25) is reduced to

$$v = \sqrt{\sum_{i=1}^{3} H_i^2 \dot{q}_i^2}. \qquad (1.26)$$

1.3.2 Acceleration in generalized coordinates

The acceleration of point p is defined as the temporal derivative of its velocity vector, that is,

$$\mathbf{w} := \frac{d\mathbf{v}}{dt}. \qquad (1.27)$$

Since in the generalized coordinates, **v** is given by (1.22), the expression of **w** in these coordinates is presented as

$$\mathbf{w} = \frac{d}{dt}\left(\frac{\partial \mathbf{r}}{\partial q_1}\dot{q}_1 + \frac{\partial \mathbf{r}}{\partial q_2}\dot{q}_2 + \frac{\partial \mathbf{r}}{\partial q_3}\dot{q}_3\right),$$

whose development leads to an expression of complex structure. A simpler expression can be obtained by an alternate method: note that in the generalized coordinate system **q**, the acceleration vector **w** has an expression of the form

$$\mathbf{w} = w_{q_1}\mathbf{k}_1 + w_{q_2}\mathbf{k}_2 + w_{q_3}\mathbf{k}_3. \tag{1.28}$$

In the case when the generalized coordinate system is orthogonal, satisfying

$$(\mathbf{k}_i, \mathbf{k}_j) = \delta_{ij} := \begin{cases} 1, & \text{if } i = j, \\ 0, & \text{if } i \neq j, \end{cases}$$

where δ_{ij} is the Kronecker symbol, the i-th component w_{q_i} ($i = 1, 2, 3$) may be represented as

$$w_{q_i} = (\mathbf{k}_i, \mathbf{w}), \tag{1.29}$$

or, using the definition (1.20) of $\mathbf{k}_i$,

$$w_{q_i} = \frac{1}{H_i}\left(\frac{\partial \mathbf{r}}{\partial q_i}, \mathbf{w}\right), \tag{1.30}$$

we may conclude that

$$H_i w_{q_i} = \left(\frac{\partial \mathbf{r}}{\partial q_i}, \frac{d\mathbf{v}}{dt}\right). \tag{1.31}$$

The additional steps require some relationships, which are the subject of the following lemma.

Lemma 1.5. *The vector* **v** *(see (1.22)) complies with the following relationships for all $i = 1, 2, 3$:*

$$\frac{\partial \mathbf{v}}{\partial \dot{q}_i} = \frac{\partial \mathbf{r}}{\partial q_i}, \tag{1.32}$$

$$\frac{d}{dt}\frac{\partial \mathbf{r}}{\partial q_i} = \frac{\partial \mathbf{v}}{\partial q_i}. \tag{1.33}$$

Proof. Since **r** is not a function of $\dot{q}_i$ and $\dfrac{\partial \dot{q}_j}{\partial \dot{q}_i} = \delta_{ij}$, equality (1.32) follows directly from (1.22):

$$\frac{\partial \mathbf{v}}{\partial \dot{q}_i} = \sum_{j=1}^{3} \frac{\partial\left(\frac{\partial \mathbf{r}}{\partial q_j}\dot{q}_j\right)}{\partial \dot{q}_i} = \sum_{j=1}^{3}\left(\frac{\partial^2 \mathbf{r}}{\partial \dot{q}_i \partial q_j}\dot{q}_j + \frac{\partial \mathbf{r}}{\partial q_j}\frac{\partial \dot{q}_j}{\partial \dot{q}_i}\right) = \frac{\partial \mathbf{r}}{\partial q_i}, \quad i = 1, 2, 3.$$

Kinematics of a point

Since in mechanics the function $\mathbf{r} = \mathbf{r}(t)$ during any admissible movement is assumed to be smooth we have

$$\frac{\partial^2 \mathbf{r}}{\partial q_j \partial q_i} = \frac{\partial^2 \mathbf{r}}{\partial q_i \partial q_j},$$

which implies

$$\frac{d}{dt}\frac{\partial \mathbf{r}}{\partial q_i} = \sum_{j=1}^{3} \frac{\partial \left(\frac{\partial \mathbf{r}}{\partial q_i}\right)}{\partial q_j} \dot{q}_j = \sum_{j=1}^{3} \frac{\partial^2 \mathbf{r}}{\partial q_i \partial q_j} \dot{q}_j = \frac{\partial}{\partial q_i} \sum_{j=1}^{3} \frac{\partial \mathbf{r}}{\partial q_j} \dot{q}_j = \frac{\partial \mathbf{v}}{\partial q_i}.$$

□

Remark 1.5. The derivative of the scalar product of two vector functions $\mathbf{f}(t)$ and $\boldsymbol{\varphi}(t)$ results in

$$\frac{d}{dt}(\mathbf{f}(t), \boldsymbol{\varphi}(t)) = \left(\frac{d}{dt}\mathbf{f}, \boldsymbol{\varphi}\right) + \left(\mathbf{f}, \frac{d}{dt}\boldsymbol{\varphi}\right),$$

implying

$$\left(\mathbf{f}, \frac{d}{dt}\boldsymbol{\varphi}\right) = \frac{d}{dt}(\mathbf{f}(t), \boldsymbol{\varphi}(t)) - \left(\frac{d}{dt}\mathbf{f}, \boldsymbol{\varphi}\right). \tag{1.34}$$

As the result of (1.34), for the right-hand side of (1.31) we have

$$\left(\frac{\partial \mathbf{r}}{\partial q_i}, \frac{d}{dt}\mathbf{v}\right) = \frac{d}{dt}\left(\frac{\partial \mathbf{r}}{\partial q_i}, \mathbf{v}\right) - \left(\frac{d}{dt}\frac{\partial \mathbf{r}}{\partial q_i}, \mathbf{v}\right), \tag{1.35}$$

and, by the relations (1.32) and (1.33), it follows that

$$H_i w_{q_i} = \frac{d}{dt}\left(\frac{\partial \mathbf{v}}{\partial \dot{q}_i}, \mathbf{v}\right) - \left(\frac{\partial \mathbf{v}}{\partial q_i}, \mathbf{v}\right). \tag{1.36}$$

The following definition is required to obtain the desired final expression for w_{q_i}.

Definition 1.9. Consider a particle of mass m and velocity $\mathbf{v}$. The amount

$$T := \frac{m}{2} v^2 = \frac{m}{2} \langle \mathbf{v}, \mathbf{v} \rangle \tag{1.37}$$

will be referred to as the kinetic energy of this particle.

Using the concept of kinetic energy with $m = 1$, formula (1.36) can be equivalently represented as

$$H_i w_{q_i} = \frac{d}{dt}\left(\frac{\partial T}{\partial \dot{q}_i}\right) - \frac{\partial T}{\partial q_i},$$

which finally leads to

$$w_{q_i} = \frac{1}{H_i}\left[\frac{d}{dt}\left(\frac{\partial T}{\partial \dot{q}_i}\right) - \frac{\partial T}{\partial q_i}\right].$$ (1.38)

In view of (1.38) and using the representation

$$w = \sqrt{(\mathbf{w},\mathbf{w})},$$ (1.39)

we get

$$w = \sqrt{\sum_{i=1}^{3}\frac{1}{H_i^2}\left[\frac{d}{dt}\left(\frac{\partial T}{\partial \dot{q}_i}\right) - \frac{\partial T}{\partial q_i}\right]^2},$$ (1.40)

keeping in mind that coordinate system **q** is orthogonal.

1.4 Movement in the cylindrical and spherical coordinate systems

The importance of the concepts introduced before and the obtained relations are emphasized in the examples presented below.

1.4.1 Movement in cylindrical coordinates

The position of a particle in space is referred to using a cylindrical coordinate system, as illustrated in Fig. 1.5. We need to get the expressions for the velocity and acceleration vectors with respect to these coordinates.

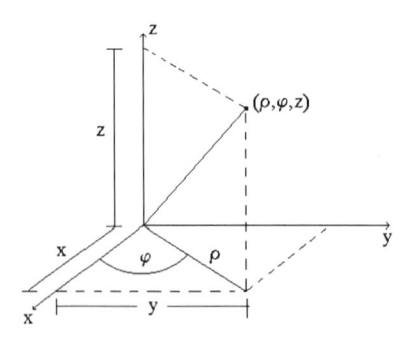

Figure 1.5 Representation of a point's position in cylindrical coordinates.

Following the used notation, we have

$$q_1 = \rho, \quad q_2 = \varphi, \quad q_3 = z,$$

where
$$q_1 \geq 0, \quad 0 \leq q_2 < 2\pi, \quad -\infty < q_3 < \infty,$$
and from Fig. 1.5 it is easy to conclude that
$$\left.\begin{array}{l} x = \rho\cos\varphi = q_1\cos q_2, \\ y\sin\varphi = q_1\sin q_2, \\ z = q_3, \end{array}\right\} \tag{1.41}$$

which defines the component transformation from Euclidian to the generalized coordinates **q**. Using (1.41) and (1.19), it is easy to obtain the Lamé coefficients. Indeed, the partial derivatives of functions (1.41) with respect to q_1 are given by

$$\frac{\partial x}{\partial q_1} = \cos q_2, \quad \frac{\partial y}{\partial q_1} = \sin q_2, \quad \frac{\partial z}{\partial q_1} = 0,$$

which results in

$$H_1 = 1.$$

By the same manner, the partial derivatives of functions (1.41) with respect to q_2 and q_3 are

$$\frac{\partial x}{\partial q_2} = -q_1\sin q_2, \quad \frac{\partial y}{\partial q_2} = q_1\cos q_2, \quad \frac{\partial z}{\partial q_2} = 0,$$
$$\frac{\partial x}{\partial q_3} = 0, \quad \frac{\partial y}{\partial q_3} = 0, \quad \frac{\partial z}{\partial q_3} = 1,$$

implying

$$H_2 = q_1 = \rho$$

and

$$H_3 = 1.$$

Having the expressions for the Lamé coefficients, the velocity components in the coordinates **q** immediately follow from (1.24):

$$\begin{array}{lll} v_{q_1} = \dot{q}_1, & \text{or} & v_\rho = \dot{\rho}, \\ v_{q_2} = q_1\dot{q}_2, & \text{or} & v_\varphi = \rho\dot{\varphi}, \\ v_{q_3} = \dot{q}_3, & \text{or} & v_z = \dot{z}. \end{array}$$

Substitution of these expressions in (1.26) in view of the orthogonality of the system **q** implies that

$$v = \sqrt{\dot{\rho}^2 + \rho^2\dot{\varphi}^2 + \dot{z}^2}.$$

This expression permits to calculate the kinetic energy (1.37), which for $m = 1$ results in

$$T = \frac{1}{2}\left(\dot{\rho}^2 + \rho^2\dot{\varphi}^2 + \dot{z}^2\right).$$

By (1.38) we conclude that

$$w_{q_1} = w_\rho = \frac{d}{dt}(\dot{\rho}) - \rho\dot{\varphi}^2 = \ddot{\rho} - \rho\dot{\varphi}^2,$$
$$w_{q_2} = w_\varphi = \frac{1}{\rho}\left[\frac{d}{dt}\left(\rho^2\dot{\varphi}\right) - 0\right] = \frac{1}{\rho}\left(2\rho\dot{\rho}\dot{\varphi} + \rho^2\ddot{\varphi}\right) = 2\dot{\rho}\dot{\varphi} + \rho\ddot{\varphi},$$
$$w_{q_3} = w_z = \frac{d}{dt}(\dot{z}) - 0 = \ddot{z}.$$

1.4.2 Movement in spherical coordinates

Consider the same particle of the previous example, but now with its position referred to using a spherical coordinate system (the situation is shown in Fig. 1.6). Get again the expressions for velocity and acceleration vectors with respect to these coordinates. Now the coordinates **q** are

$$q_1 = r, \quad q_2 = \varphi, \quad q_3 = \theta,$$

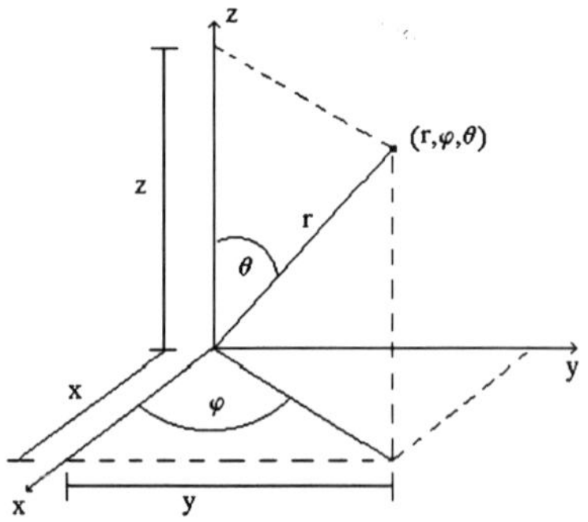

Figure 1.6 Representation of a point's position in spherical coordinates.

where

$$q_1 \geq 0, \quad 0 \leq q_2 < 2\pi, \quad 0 \leq q_3 \leq \pi.$$

Using Fig. 1.6, we have the relations

$$\left.\begin{array}{l} x = r\sin\theta\cos\varphi = q_1\sin q_3\cos q_2, \\ y = r\sin\theta\sin\varphi = q_1\sin q_3\sin q_2, \\ z = r\cos\theta = q_1\cos q_3, \end{array}\right\} \quad (1.42)$$

Kinematics of a point

which, as can be easily checked, constitutes the coordinate transformation, so that **q** composes a system of generalized coordinates. The partial derivatives of the functions in (1.42) with respect to q_1 are as follows:

$$\frac{\partial x}{\partial q_1} = \sin q_3 \cos q_2, \quad \frac{\partial y}{\partial q_1} = \sin q_3 \sin q_2, \quad \frac{\partial z}{\partial q_1} = \cos q_3,$$

which, according to (1.19), gives

$$H_1 = \sqrt{(\sin q_3 \cos q_2)^2 + (\sin q_3 \sin q_2)^2 + (\cos q_3)^2} = 1.$$

Proceeding in a similar way, the partial derivatives of the functions in (1.42) with respect to q_2 and q_3 result in

$$\frac{\partial x}{\partial q_2} = -q_1 \sin q_3 \sin q_2, \quad \frac{\partial y}{\partial q_2} = q_1 \sin q_3 \cos q_2, \quad \frac{\partial z}{\partial q_2} = 0,$$

$$\frac{\partial x}{\partial q_3} = q_1 \cos q_3 \cos q_2, \quad \frac{\partial y}{\partial q_3} = q_1 \cos q_3 \sin q_2, \quad \frac{\partial z}{\partial q_3} = -q_1 \sin q_3,$$

which leads to

$$H_2 = \sqrt{(-q_1 \sin q_3 \sin q_2)^2 + (q_1 \sin q_3 \cos q_2)^2} = q_1 \sin q_3 = r \sin \theta,$$

$$H_3 = \sqrt{(q_1 \cos q_2 \cos q_3)^2 + (q_1 \sin q_2 \cos q_3)^2 + (-q_1 \sin q_3)^2} = q_1 = r.$$

The calculated Lamé coefficients and (1.24) allow to determine the components of velocity with respect to the system **q**:

$$\left. \begin{aligned} v_{q_1} &= \dot{q}_1, \quad \text{or} \quad v_r = \dot{\rho}, \\ v_{q_2} &= q_1 \dot{q}_2 \sin q_3, \quad \text{or} \quad v_\varphi = r\dot{\varphi} \sin \theta, \\ v_{q_3} &= q_1 \dot{q}_3, \quad \text{or} \quad v_\theta = r\dot{\theta}. \end{aligned} \right\} \quad (1.43)$$

The presentation (1.43) together with (1.26) (taking into account that this system of coordinates **q** is orthogonal) permits to conclude that the magnitude v of the velocity **v** is given by

$$v = \sqrt{\dot{\rho}^2 + (r\dot{\varphi} \sin \theta)^2 + (r\dot{\theta})^2}$$

and, by (1.37), we obtain the "kinetic energy" with $m = 1$ as

$$T = \frac{1}{2} \left(\dot{\rho}^2 + r^2 \dot{\varphi}^2 \sin^2 \theta + r^2 \dot{\theta}^2 \right).$$

Based on this representation and applying (1.38), we finally get the formulas for the acceleration components in the spherical coordinates **q**:

$$\left.\begin{aligned}
w_{q_1} &= w_r = \frac{d}{dt}\dot{\rho} - r\dot{\varphi}^2\sin^2\theta - r\dot{\theta}^2 = \ddot{r} - r\left(\dot{\varphi}^2\sin^2\theta + \dot{\theta}^2\right), \\
w_{q_2} &= w_\varphi = \frac{1}{r\sin\theta}\left[\frac{d}{dt}\left(r^2\dot{\varphi}\sin^2\theta\right) - 0\right] = \\
&\quad r\left(\ddot{\varphi}\sin\theta + 2\dot{\varphi}\dot{\theta}\cos\theta\right) + 2\dot{\rho}\dot{\varphi}\sin\theta, \\
w_{q_3} &= w_\theta = \frac{1}{r}\left[\frac{d}{dt}\left(r^2\dot{\theta}\right) - r^2\dot{\varphi}^2\sin\theta\cos\theta\right] = \\
&\quad r\left(\ddot{\theta} - \dot{\varphi}^2\sin\theta\cos\theta\right) + 2\dot{\rho}\dot{\theta}.
\end{aligned}\right\} \quad (1.44)$$

1.5 Normal and tangential accelerations

A particle in motion, in general, is subject to an acceleration that can be seen as consisting of two perpendicular components: one *tangential* to the trajectory, responsible for the change in the magnitude of the moving point speed, and another one *normal*, characterizing its direction change.

Fig. 1.7 shows the directions of these accelerations and their relation to the trajectory, where τ and **n** are unit vectors in the tangential and normal directions, respectively.

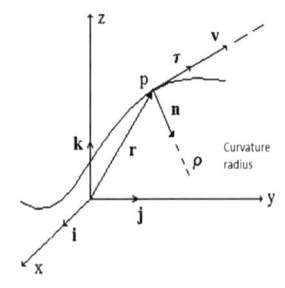

Figure 1.7 Normal and tangential vectors to the trajectory.

In the orthogonal Cartesian system, formed by vectors τ and **n**, the acceleration **w** can be represented as

$$\mathbf{w} = w_\tau\boldsymbol{\tau} + w_\mathbf{n}\mathbf{n}, \quad (1.45)$$

where, in particular,

$$w_\tau = \frac{dv}{dt},$$

Kinematics of a point

with v being the magnitude of the velocity at point p. Regarding the component $w_\mathbf{n}$, its value is given in the following lemma.

Lemma 1.6. *If the radius ρ of the curvature of the trajectory at point p is non-equal to zero, namely,*

$$\rho \neq 0,$$

then the normal component $w_\mathbf{n}$ of the acceleration $\mathbf{w}$ in the point p is expressed by

$$w_\mathbf{n} = \frac{v^2}{\rho}. \tag{1.46}$$

Proof. Since the normal component of the acceleration is only responsible for the change in the direction of the velocity, to determine its normal component we may consider that the magnitude of the velocity is constant.

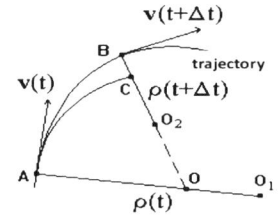

Figure 1.8 Speeds, radii, and centers of curvature in two different instants.

Fig. 1.8 shows a segment of the trajectory of the particle in Fig. 1.7 with the velocity vectors and the radii of curvature in two instants of time: in instant t it is assumed that the particle has velocity $\mathbf{v}(t)$ at point A of the trajectory, where the radius of curvature is denoted by $\rho(t)$ and its center is located at point O_1; at the second instant, $t + \Delta t$, the particle is at point B with velocity $\mathbf{v}(t + \Delta t)$ and the trajectory has radius of curvature $\rho(t + \Delta t)$ and its center at O_2. The point at which the radii of curvature intersect is denoted by O. Points A and C are on a circle with center O, so they are at the corresponding distance from this center: A on $\rho(t)$ and C over $\rho(t + \Delta t)$.

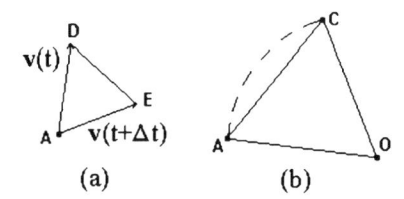

Figure 1.9 The detail of the previous figure.

Part (a) of Fig. 1.9 shows in detail the triangle ADE formed by the velocities $\mathbf{v}(t)$ and $\mathbf{v}(t + \Delta t)$, which is isosceles in view of the fact that the magnitude of speed remains constant. On the other hand, part (b) of the same figure shows in detail the isosceles triangle ACO in Fig. 1.8. Clearly these two triangles are similar, and the following relationship can be established:

$$\frac{DE}{v} = \frac{AC}{AO},$$

where AC, DE, and AO represent the distances between the points in question. From the previous relationship, dividing by the time difference and taking the limit when it tends to zero, we get

$$\lim_{\Delta t \to 0} \frac{DE}{\Delta t} = v \lim_{\Delta t \to 0} \frac{1}{AO} \frac{AC}{\Delta t}.$$

But

$$AC \to \Delta s \text{ and } AO \to \rho(t) \text{ when } \Delta t \to 0,$$

where s is the displacement on the path made by the particle. These considerations and the fact that

$$v = \frac{ds}{dt}$$

lead to the final expression

$$w_{\mathbf{n}} := \lim_{\Delta t \to 0} \frac{DE}{\Delta t} = \frac{v^2}{\rho}.$$

$\square$

Although the test was done considering that the movement is planar, the above is valid in the three-dimensional case, since in the limit the latter tends to the first.

The previous lemma involves the radius of curvature ρ of the trajectory. In the following exercise useful expressions are given for the calculation of this important element.

1.6 Some examples

Example 1.2. The ring in Fig. 1.10 moves with constant velocity of magnitude v on the fixed wire whose configuration is described by the function

$$y = ax^2, \quad a > 0. \tag{1.47}$$

Determine the components of the acceleration $\mathbf{w}$ experienced by the ring, as well as its magnitude, as a function of the abscissa x.

Kinematics of a point

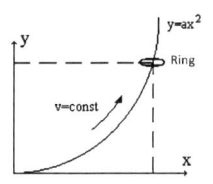

Figure 1.10 Ring moving with a constant magnitude speed.

Since the reference system is Cartesian, it is verified that

$$v_x = \dot{x}, \quad v_y = \dot{y}$$

and

$$w_x = \ddot{x}, \quad w_y = \ddot{y}.$$

Hence

$$v^2 = \dot{x}^2 + \dot{y}^2, \quad w^2 = \ddot{x}^2 + \ddot{y}^2. \tag{1.48}$$

The expression for x is obtained from (1.47), noting that

$$\dot{y} = 2ax\dot{x}. \tag{1.49}$$

This expression, being substituted in the first relation (1.48), transforms it into

$$v^2 = \left(1 + 4a^2x^2\right)\dot{x}^2,$$

implying

$$\dot{x}^2 = \frac{v^2}{1 + 4a^2x^2}. \tag{1.50}$$

The implicit temporal derivation of this expression, considering that v is constant, leads to

$$\ddot{x} = -\frac{4a^2v^2x}{\left(1 + 4a^2x^2\right)^2} \tag{1.51}$$

since $\dot{x} \neq 0$ by (1.50). On the other hand, to obtain the corresponding function $\ddot{y}$ let us derive (1.49):

$$\ddot{y} = 2a\left(\dot{x}^2 + x\ddot{x}\right).$$

This expression, after substitution of (1.50) and (1.51), becomes

$$\ddot{y} = \frac{2av^2}{\left(1 + 4a^2x^2\right)^2}. \tag{1.52}$$

The determination of w is immediate once you have functions (1.51) and (1.52). Substitution of these functions in the second expression in (1.48) results in

$$w = \frac{2av^2}{(1+4a^2x^2)^{3/2}}.$$

Example 1.3. A particle that moves in the three-dimensional Cartesian space in Fig. 1.11 is subject to the acceleration given by

$$\mathbf{w}(t) = [\mathbf{a}, \mathbf{v}(t)], \tag{1.53}$$

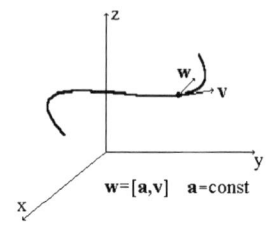

Figure 1.11 Trajectory of a particle in a Cartesian system.

where $\mathbf{a}$ is a constant vector. Find the coordinates of the position as functions of time. In view of the definition $\mathbf{w} = \dot{\mathbf{v}}$ the differentiation of (1.53) leads to

$$\dot{\mathbf{w}} = [\mathbf{a}, \dot{\mathbf{v}}(t)] = [\mathbf{a}, \mathbf{w}],$$

where

$$\dot{\mathbf{w}} = \begin{Vmatrix} \mathbf{i} & \mathbf{j} & \mathbf{k} \\ a_x & a_y & a_z \\ w_x & w_y & w_z \end{Vmatrix} = \begin{Vmatrix} \mathbf{i} & \mathbf{j} & \mathbf{k} \\ a_x & a_y & a_z \\ \ddot{x} & \ddot{y} & \ddot{z} \end{Vmatrix} \tag{1.54}$$

with $\mathbf{i}, \mathbf{j}$, and $\mathbf{k}$ as the unitary vectors in the directions of the coordinate axes x, y, and z, respectively. From the development of expression (1.54) the vector components of the acceleration derivative may be obtained:

$$\left. \begin{array}{l} \dot{w}_x = \dddot{x} = a_y \ddot{z} - a_z \ddot{y}, \\ \dot{w}_y = \dddot{y} = -a_x \ddot{z} + a_z \ddot{x}, \\ \dot{w}_z = \dddot{z} = a_x \ddot{y} - a_y \ddot{x}, \end{array} \right\}$$

or in matrix form,

$$\begin{bmatrix} \dddot{x} \\ \dddot{y} \\ \dddot{z} \end{bmatrix} = \begin{bmatrix} 0 & -a_z & a_y \\ a_z & 0 & -a_x \\ -a_y & a_x & 0 \end{bmatrix} \begin{bmatrix} \ddot{x} \\ \ddot{y} \\ \ddot{z} \end{bmatrix}. \tag{1.55}$$

Using the notation

$$A := \begin{bmatrix} 0 & -a_z & a_y \\ a_z & 0 & -a_x \\ -a_y & a_x & 0 \end{bmatrix}, \quad \Theta := \begin{bmatrix} \ddot{x} \\ \ddot{y} \\ \ddot{z} \end{bmatrix},$$

Eq. (1.55) can be rewritten as

$$\dot{\Theta} = A\Theta, \quad \Theta(0) = \Theta_0,$$

where Θ_0 is the initial condition for the acceleration vector. The solution of this linear differential equation is given by

$$\Theta(t) = \exp(At)\Theta_0,$$

which, after double integration, gives the solution of the initial problem:

$$\begin{bmatrix} x(t) \\ y(t) \\ z(t) \end{bmatrix} = \int_{u=0}^{t} \int_{\tau=0}^{u} \Theta(\tau) d\tau du =$$

$$\int_{u=0}^{t} \int_{\tau=0}^{u} \exp(A\tau) d\tau du \Theta_0 = A^{-1} \int_{u=0}^{t} \int_{\tau=0}^{u} \exp(A\tau) d(A\tau) du \Theta_0 =$$

$$A^{-1} \int_{u=0}^{t} \left[\exp(Au) - I\right] du \Theta_0 = A^{-2} \int_{u=0}^{t} \left[\exp(Au) - I\right] d(Au) \Theta_0 =$$

$$A^{-2} \left[\exp(At) - I - At\right] \Theta_0 = \left[A^{-2} \exp(At) - A^{-2} - A^{-1}t\right] \Theta_0.$$

Note that here matrix A is assumed to be invertible and the function $\exp(At)$ is the matrix exponent defined as

$$\exp(At) = \sum_{k=0}^{\infty} \frac{1}{k!} (At)^k.$$

Example 1.4. Consider the particle from the previous exercise, whose acceleration is given by

$$\mathbf{w}(t) = [\mathbf{a}, \mathbf{v}(t)], \tag{1.56}$$

with $\mathbf{a}$ as a constant. Show that in such a case the magnitudes of $\mathbf{w}(t)$ and $\mathbf{v}(t)$ are constant.

By the condition (1.56) we have

$$\mathbf{w}(t) = \frac{d\mathbf{v}(t)}{dt} = [\mathbf{a}, \mathbf{v}(t)]. \tag{1.57}$$

Multiplying scalarly both sides by $\mathbf{v}(t)$ and taking into account that vectors $\mathbf{v}(t)$ and $[\mathbf{a}, \mathbf{v}(t)]$ are orthogonal, we get

$$\left(\mathbf{v}(t), \frac{d\mathbf{v}(t)}{dt}\right) = (\mathbf{v}(t), [\mathbf{a}, \mathbf{v}(t)]) = 0 \quad \text{for all } t. \tag{1.58}$$

But

$$\left(\mathbf{v}(t), \frac{d\mathbf{v}(t)}{dt}\right) = \frac{1}{2}\frac{d}{dt}\langle\mathbf{v}(t), \mathbf{v}(t)\rangle = \frac{1}{2}\frac{d}{dt}v^2(t) = v(t)\dot{v}(t),$$

and in view of (1.58), it follows that $\dot{v}(t) \equiv 0$, or equivalently, $v(t)$ is a constant. On the other hand, deriving (1.56) leads to

$$\frac{d\mathbf{w}(t)}{dt} = \left[\mathbf{a}, \frac{d\mathbf{v}(t)}{dt}\right] = [\mathbf{a}, \mathbf{w}(t)]$$

and by the same reasoning we may conclude that

$$\left(\mathbf{w}(t), \frac{d\mathbf{w}(t)}{dt}\right) = (\mathbf{w}(t), [\mathbf{a}, \mathbf{w}(t)]) = 0 \quad \forall t.$$

So,

$$\left(\mathbf{w}(t), \frac{d\mathbf{w}(t)}{dt}\right) = \frac{1}{2}\frac{d}{dt}(\mathbf{w}(t), \mathbf{w}(t)) = \frac{1}{2}\frac{d}{dt}w^2(t) = w(t)\dot{w}(t) = 0,$$

from which it follows that necessarily $w(t)$ is constant.

Example 1.5. A point moves following the elliptical path as in Fig. 1.12, in such a way that the following restriction is met:

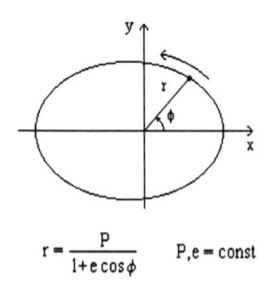

$$r^2(t)\dot{\phi}(t) = k. \tag{1.59}$$

Figure 1.12 Particle on an elliptical path.

If the equation of the ellipse is given by

$$r = \frac{P}{1 + e\cos\phi}, \tag{1.60}$$

Kinematics of a point

with P and e as known constants, we need to find w_r and w_ϕ as functions of r and ϕ, namely,

$$w_r = w_r(r, \phi), \quad w_\phi = w_\phi(r, \phi).$$

Define

$$q_1 = r, \quad q_2 = \phi,$$

where

$$r \geq 0, \quad 0 \leq \phi < 2\pi,$$

and by Fig. 1.12 we have

$$x = r \cos \phi = q_1 \cos q_2,$$
$$y = r \sin \phi = q_1 \sin q_2.$$

As can be verified very easily, these functions make up a one-to-one transformation, and hence, the coordinates (r, ϕ) form a generalized coordinate system. The Lamé coefficients are now determined:

$$H_i := \sqrt{\left(\frac{\partial x}{\partial q_i}\right)^2 + \left(\frac{\partial y}{\partial q_i}\right)^2}, \quad i = 1, 2,$$

and are given by

$$H_1 = 1, \quad H_2 = q_1,$$

taking into account that

$$\frac{\partial x}{\partial q_1} = \cos q_2, \quad \frac{\partial y}{\partial q_1} = \sin q_2,$$
$$\frac{\partial x}{\partial q_2} = -q_1 \sin q_2, \quad \frac{\partial y}{\partial q_2} = q_1 \cos q_2.$$

So, by (1.24), the velocity components are

$$v_{q_i} := H_i \dot{q}_i, \quad i = 1, 2,$$

or equivalently,

$$v_{q_1} = \dot{q}_1, \text{ or } v_r = \dot{r},$$
$$v_{q_2} = q_1 \dot{q}_2, \text{ or } v_\phi = r\dot{\phi}.$$

Since the coordinate system in question is orthogonal, the kinetic energy, with $m = 1$, is given by

$$T = \frac{1}{2} v^2 = \frac{1}{2} \left(\dot{q}_1^2 + q_1^2 \dot{q}_2^2\right),$$

which allows the calculation of the components of the acceleration via expression (1.38):

$$w_{q_i} = \frac{1}{H_i}\left[\frac{d}{dt}\left(\frac{\partial T}{\partial \dot{q}_i}\right) - \frac{\partial T}{\partial q_i}\right], \quad i = 1, 2,$$

implying

$$\begin{aligned} w_{q_1} &= \ddot{q}_1 - q_1\dot{q}_2^2, \text{ or } w_r = \ddot{r} - r\dot{\phi}^2, \\ w_{q_2} &= \frac{1}{q_1}\frac{d}{dt}\left(q_1^2\dot{q}_2\right) = 2\dot{q}_1\dot{q}_2 + q_1\ddot{q}_2, \text{ or } w_\phi = 2\dot{r}\dot{\phi} + r\ddot{\phi}. \end{aligned} \quad (1.61)$$

On the other hand, in view of (1.59) we have

$$\dot{\phi} = \frac{k}{r^2}$$

and by (1.60) it follows that

$$\dot{r} = \frac{Pe\dot{\phi}\sin\phi}{(1+e\cos\phi)^2} = \frac{e}{P}r^2\dot{\phi}\sin\phi = \frac{ke}{P}\sin\phi.$$

Combining these equations we get

$$\ddot{\phi} = -2k\frac{\dot{r}}{r^3} = -\frac{2k^2e}{P}\frac{\sin\phi}{r^3},$$

$$\ddot{r} = \frac{ke}{P}\dot{\phi}\cos\phi = \frac{k^2e}{P}\frac{\cos\phi}{r^2}.$$

Substitution of these relations into (1.61) leads to the final representation which we are interested in:

$$\left.\begin{aligned} w_r &= \frac{k^2}{Pr^2}\left(e\cos\phi - \frac{P}{r}\right) = -\frac{k^2}{Pr^2}, \\ w_\phi &= 0. \end{aligned}\right\} \quad (1.62)$$

Example 1.6. Let the velocity **v** and the acceleration **w** be known. Determine the expression for the calculation of the radius of curvature ρ. The sought expression can be obtained in two different ways.

1) By the relation (1.15), taking

$$\mathbf{b} = \mathbf{w} \text{ and } \mathbf{a} = \mathbf{v},$$

we have

$$\mathbf{w} = \frac{\langle \mathbf{v}, \mathbf{w}\rangle}{v^2}\mathbf{v} + \frac{[\mathbf{v}, [\mathbf{w}, \mathbf{v}]]}{v^2},$$

Kinematics of a point

where the vectors **v** and [**v**, [**w**, **v**]] are orthogonal. Considering the representation

$$\mathbf{v} = v\boldsymbol{\tau}$$

and taking into account that the vector $\dfrac{[\mathbf{v}, [\mathbf{w}, \mathbf{v}]]}{v^2}$ has the direction **n**, by (1.45) it is concluded that the tangential and normal accelerations are given, respectively, by

$$w_\tau = \frac{\langle \mathbf{v}, \mathbf{w} \rangle}{v} \tag{1.63}$$

and

$$w_\mathbf{n} = \frac{|[\mathbf{v}, [\mathbf{w}, \mathbf{v}]]|}{v^2}. \tag{1.64}$$

Here, applying (1.6), we have used the identity

$$|[\mathbf{v}, [\mathbf{w}, \mathbf{v}]]| = v\,|[\mathbf{w}, \mathbf{v}]|,$$

and since vectors **v** and [**w**, **v**] are orthogonal too, we get

$$w_\mathbf{n} = \frac{|[\mathbf{w}, \mathbf{v}]|}{v}. \tag{1.65}$$

Then, using the relation (1.46)

$$w_\mathbf{n} = \frac{v^2}{\rho},$$

we conclude that

$$\rho = \frac{v^3}{|[\mathbf{w}, \mathbf{v}]|}. \tag{1.66}$$

2) From the vector diagram of Fig. 1.13 and considering the Pythagorean theorem, one has

$$w_\mathbf{n} = \sqrt{w^2 - w_\tau^2}. \tag{1.67}$$

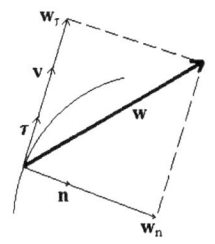

Figure 1.13 Acceleration vector composition.

Now, taking from the first solution the relation (1.63) and replacing it in the previous expression, formula (1.67) can be rewritten as

$$w_\mathbf{n} = \sqrt{w^2 - \frac{\langle \mathbf{v}, \mathbf{w} \rangle^2}{v^2}}.$$

Again using the representation $w_\mathbf{n} = \dfrac{v^2}{\rho}$, it allows to express the radius of curvature as

$$\rho = \frac{v^2}{\sqrt{w^2 - \dfrac{\langle \mathbf{v}, \mathbf{w} \rangle^2}{v^2}}}. \tag{1.68}$$

Although expressions (1.66) and (1.68), obtained in the previous exercise for the calculation of ρ, seem different at first glance, they clearly must be equal. The following steps show this fact. The following equality is verified:

$$\frac{v^3}{|[\mathbf{w}, \mathbf{v}]|} = \frac{v^2}{\sqrt{w^2 - \dfrac{\langle \mathbf{v}, \mathbf{w} \rangle^2}{v^2}}}.$$

Note that

$$\frac{\langle \mathbf{v}, \mathbf{w} \rangle^2}{v^2} = w^2 \cos^2(\widehat{\mathbf{v}, \mathbf{w}}),$$

and hence,

$$\sqrt{w^2 - \frac{\langle \mathbf{v}, \mathbf{w} \rangle^2}{v^2}} = \sqrt{w^2 \left[1 - \cos^2(\widehat{\mathbf{v}, \mathbf{w}})\right]} = w \left|\sin(\widehat{\mathbf{v}, \mathbf{w}})\right| = \frac{|[\mathbf{w}, \mathbf{v}]|}{v},$$

which gives the desired result.

1.7 Exercises

Exercise 1.1. A point describes a circle of radius R. The acceleration of the point forms a constant angle α ($\alpha \neq \pi/2$) with its speed. Show that the speed of the point increases by n times in time,

$$t_n = \frac{n-1}{n} \frac{R}{v_0 \cot \alpha},$$

if at the initial moment it equals v_0.

Kinematics of a point

Exercise 1.2. The motion of a point in a plane is given in the polar coordinate system by the components

$$v_r = \frac{1}{r^2},$$
$$v_\varphi = \frac{1}{ar}, \quad a = \text{const} \neq 0.$$

Show that

$$r = r_0 + a(\varphi - \varphi_0),$$
$$w_r = -r^{-3}\left(\frac{2}{r^2} + \frac{1}{a^2}\right), \quad w_\varphi = 0$$

if at the initial moment $r(0) = r_0$ and $\varphi(0) = \varphi_0$.

Exercise 1.3. The position of the point is determined by the dependence of its radius-vector $\mathbf{r}$ on curvilinear orthogonal coordinates q_1, q_2, q_3, namely,

$$\mathbf{r} = f(q_1, q_2, q_3).$$

Assuming that the coordinates and their derivatives are known (measurable) at any time t, show that the radius ρ of the trajectory curvature is

$$\rho = \frac{\left(\sum_{i=1}^{3} H_i^2 \dot{q}_i^2\right)^{3/2}}{\left[\left(\sum_{i=1}^{3} \frac{L_i^2}{H_i^2}\right)\left(\sum_{i=1}^{3} H_i^2 \dot{q}_i^2\right) - \left(\sum_{i=1}^{3} L_i \dot{q}_i\right)^2\right]^{1/2}},$$

where

$$L_i = \frac{d}{dt}\frac{\partial}{\partial \dot{q}_i}\left(\frac{1}{2}\sum_{s=1}^{3} H_s^2 \dot{q}_s^2\right) - \frac{\partial}{\partial q_i}\left(\frac{1}{2}\sum_{s=1}^{3} H_s^2 \dot{q}_s^2\right).$$

Exercise 1.4. Show that the unit vectors $\boldsymbol{\tau}$, $\mathbf{n}$, and $\mathbf{b} = [\boldsymbol{\tau}, \mathbf{n}]$ of the accompanying trihedron $(\boldsymbol{\tau}, \mathbf{n}, \mathbf{b})$ as functions of the velocity $\mathbf{v}$ and acceleration $\mathbf{w}$ vectors are as follows:

$$\boldsymbol{\tau} = \frac{\mathbf{v}}{v}, \quad \mathbf{n} = \frac{v^2 \mathbf{w} - (\mathbf{w}, \mathbf{v})\mathbf{v}}{v\,\|[\mathbf{v}, \mathbf{w}]\|}, \quad \mathbf{b} = \frac{[\mathbf{v}, \mathbf{w}]}{\|[\mathbf{v}, \mathbf{w}]\|},$$

supposing that

$$[\mathbf{v}, \mathbf{w}] \neq 0, \quad (\boldsymbol{\tau}, \mathbf{v}) > 0.$$

Exercise 1.5. When a point moves, the projection of its speed **v** on the OX-axis has a constant value u. Prove that the relation

$$w = \frac{v^3}{u\rho} \quad \text{(where } \rho \text{ is the curvature radius)}$$

is valid if and only if the trajectory of the point is a plane curve.

Although the speed and acceleration have been obtained by a similar derivation process, they have very different characteristics. In particular, the acceleration is the result of two very important acceleration components. The study of these elements constitutes the motivation of the next chapter.

Rigid body kinematics

Contents

2.1 Angular velocity   31
     2.1.1 Definition of a rigid body   31
     2.1.2 The Euler theorem   32
     2.1.3 Joint rotation with a common pivot   35
     2.1.4 Parallel and non-coplanar rotations   36
2.2 Complex movements of the rigid body   39
     2.2.1 General relations   39
     2.2.2 Plane non-parallel motion and center of velocities   42
2.3 Complex movement of a point   45
     2.3.1 Absolute velocity   46
     2.3.2 Absolute acceleration   46
2.4 Examples   47
2.5 Kinematics of a rigid body rotation   59
     2.5.1 Finite rotations   59
     2.5.2 Rotation matrix   61
     2.5.3 Composition of rotations   68
2.6 Rotations and quaternions   71
     2.6.1 Quaternions   71
     2.6.2 Composition or summation of rotations as a quaternion   80
2.7 Differential kinematic equations (DKEs)   84
     2.7.1 DKEs in Euler coordinates   84
     2.7.2 DKEs in quaternions: Poisson equation   85
2.8 Exercises   86

In this chapter the concept of rigid body is introduced and fairly general expressions are obtained for the description of the kinematics of this mechanical entity. A fundamental tool for the study of the kinematics of the rigid body is Euler's theorem, where important concepts such as angular speed and acceleration appear. This allows the calculation of the speed and acceleration of the points of the body. In particular, this method can be extended to the case of the movement of a point in the presence of a mobile relative coordinate system. Matrix rotations and quaternions are addressed in the final section.

2.1 Angular velocity

2.1.1 Definition of a rigid body

The following concepts are fundamental in this chapter.

Definition 2.1. It is said that a set C of points in space forms **a rigid body** if

$$\|\mathbf{r}_A(t) - \mathbf{r}_B(t)\| = \operatorname*{const}_{t} \quad \text{for any points } A, B \in C, \tag{2.1}$$

where $\mathbf{r}_A(t)$ and $\mathbf{r}_B(t)$ are the position vectors of points A and B, with respect to some reference point O. In other words, C is a rigid body if the distance between any of its points remains invariant in time.

Definition 2.2. Suppose that C is a rigid body. Let the reference point O be such that

$$\|\mathbf{r}_A(t)\| = \operatorname*{const}_{t} \quad \text{for any point } A \in C.$$

In such a case it is said that O is a pivot of the rigid body C. This situation is illustrated in Fig. 2.1.

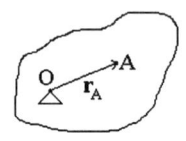

Figure 2.1 Rigid body with a pivot.

In physics, a rigid body is a solid body in which deformation is zero or so small it can be neglected. The distance between any two given points on a rigid body remains constant in time regardless of external forces exerted on it. A rigid body is usually considered as a continuous distribution of mass.

2.1.2 The Euler theorem

The important movement in a rigid body in relation to a pivot is the rotation with respect to this point. The following theorem characterizes this movement.

Theorem 2.1 (Euler). *If O is a pivot of the rigid body C, there exists a vector $\boldsymbol{\omega}(t)$ such that*

$$\frac{d}{dt}\mathbf{r}_A(t) = [\boldsymbol{\omega}(t), \mathbf{r}_A(t)], \quad \forall A \in C, \tag{2.2}$$

where $\mathbf{r}_A(t)$ is the vector from the point O to the point A. It is important that $\boldsymbol{\omega}(t)$ does not depend on the point A.

Proof. Since O is a pivot of C, we have

$$r_A(t) = \operatorname*{const}_{t}, \quad \forall A \in C,$$

and hence,

$$r_A^2(t) = \operatorname*{const}_{t},$$

or equivalently,

$$(\mathbf{r}_A(t), \mathbf{r}_A(t)) = \operatorname*{const}_{t}.$$

The temporal derivative of this expression results in

$$\frac{d}{dt}(\mathbf{r}_A(t), \mathbf{r}_A(t)) = 0, \qquad (2.3)$$

where, by the properties of the internal product,

$$\frac{d}{dt}(\mathbf{r}_A(t), \mathbf{r}_A(t)) = 2(\dot{\mathbf{r}}_A(t), \mathbf{r}_A(t)),$$

implying

$$(\dot{\mathbf{r}}_A(t), \mathbf{r}_A(t)) = 0,$$

which means that the vectors $\dot{\mathbf{r}}_A(t)$ and $\mathbf{r}_A(t)$ are orthogonal, and therefore, there exists a vector $\boldsymbol{\omega}(t)$ such that

$$\dot{\mathbf{r}}_A(t) = [\boldsymbol{\omega}(t), \mathbf{r}_A(t)]. \qquad (2.4)$$

Now, we need to show that $\boldsymbol{\omega}(t)$ does not depend on the point A. By that reason, a Cartesian coordinate system, originating in the pivot O and with respect to which C is fixed, does not experience any movement. These conditions are illustrated in Fig. 2.2. In this system of coordinates we have on one side

$$\boldsymbol{\omega}(t) = \omega_x(t)\mathbf{i}(t) + \omega_y(t)\mathbf{j}(t) + \omega_z(t)\mathbf{k}(t) \qquad (2.5)$$

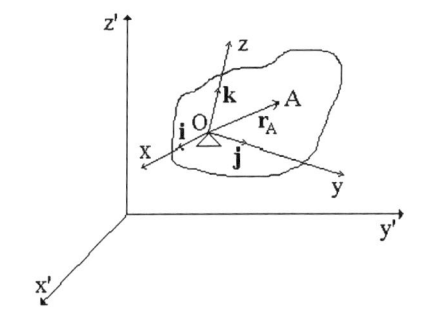

Figure 2.2 Rigid body with pivot O and a coordinate system fixed to the solid body.

and on the other

$$\mathbf{r}_A(t) = x_A \mathbf{i}(t) + y_A \mathbf{j}(t) + z_A \mathbf{k}(t), \qquad (2.6)$$

with x_A, y_A, and z_A fixed as the coordinate system moves with $\mathcal{C}$. So

$$\dot{\mathbf{r}}_A(t) = x_A \frac{d}{dt}\mathbf{i}(t) + y_A \frac{d}{dt}\mathbf{j}(t) + z_A \frac{d}{dt}\mathbf{k}(t). \tag{2.7}$$

In view of the representations (2.6) and (2.5), for the right-hand side of (2.4) we obtain

$$\begin{aligned}{} [\boldsymbol{\omega}(t), \mathbf{r}_A(t)] &= \mathbf{i}\left(\omega_y z_A - \omega_z y_A\right) + \mathbf{j}\left(\omega_z x_A - \omega_x z_A\right) + \mathbf{k}\left(\omega_x y_A - \omega_y x_A\right) \\ &= x_A\left(\omega_z \mathbf{j} - \omega_y \mathbf{k}\right) + y_A\left(\omega_x \mathbf{k} - \omega_z \mathbf{i}\right) + z_A\left(\omega_y \mathbf{i} - \omega_x \mathbf{j}\right), \end{aligned} \tag{2.8}$$

where the dependence of t has not been written in order to simplify the expressions. Substitution of (2.7) and (2.8) in (2.4) allows us to obtain the following relationships:

$$\left. \begin{aligned} \frac{d}{dt}\mathbf{i} &= \omega_z \mathbf{j} - \omega_y \mathbf{k}, \\ \frac{d}{dt}\mathbf{j} &= \omega_x \mathbf{k} - \omega_z \mathbf{i}, \\ \frac{d}{dt}\mathbf{k} &= \omega_y \mathbf{i} - \omega_x \mathbf{j}, \end{aligned} \right\} \tag{2.9}$$

or equivalently,

$$\omega_x = \left(\mathbf{k}, \frac{d}{dt}\mathbf{j}\right), \quad \omega_y = \left(\mathbf{i}, \frac{d}{dt}\mathbf{k}\right), \quad \omega_z = \left(\mathbf{j}, \frac{d}{dt}\mathbf{i}\right),$$

which shows that to characterize $\boldsymbol{\omega}(t)$ it suffices to know the behavior of the unit vectors of the chosen system located in the pivot, without any dependence on point A. □

Definition 2.3. Given a pivot O of a rigid body $\mathcal{C}$, the vector $\boldsymbol{\omega}(t)$ that appears in Euler's theorem, Theorem 2.1, is called **the angular velocity** of the body $\mathcal{C}$ with respect to the point O, and the line, passing through O and coinciding with the direction of $\boldsymbol{\omega}(t)$, is referred to as **the axis of rotation**.

Three interesting conclusions emerge from the previous theorem, which are listed below.

1. As expected by (2.2), from (2.9) it is verified that

$$\frac{d}{dt}\mathbf{i} = [\boldsymbol{\omega}, \mathbf{i}], \quad \frac{d}{dt}\mathbf{j} = [\boldsymbol{\omega}, \mathbf{j}], \quad \frac{d}{dt}\mathbf{k} = [\boldsymbol{\omega}, \mathbf{k}]. \tag{2.10}$$

2. From (2.2) it follows that if $\boldsymbol{\omega} \neq \mathbf{0}$, then $\dot{\mathbf{r}}_A = \mathbf{0}$ if and only if A is on the axis of rotation.
3. A coordinate system can be seen as a rigid body. It is clear that the origin of the system is a pivot. Since a generic vector $\mathbf{r}$ referenced to this coordinate system can be expressed in the form

$$\mathbf{r} = r \mathbf{e}_r,$$

Rigid body kinematics

with $\mathbf{e_r}$ as a unit vector in the direction of $\mathbf{r}$, its derivative is given by

$$\dot{\mathbf{r}} = \dot{r}\mathbf{e_r} + r\dot{\mathbf{e}}_\mathbf{r}.$$

But

$$\dot{\mathbf{e}}_\mathbf{r} = [\boldsymbol{\omega}, \mathbf{e_r}]$$

for some $\boldsymbol{\omega}$, and

$$r[\boldsymbol{\omega}, \mathbf{e_r}] = [\boldsymbol{\omega}, \mathbf{r}]$$

results in the derivative of $\mathbf{r}$:

$$\dot{\mathbf{r}} = \dot{r}\mathbf{e_r} + [\boldsymbol{\omega}, \mathbf{r}]. \tag{2.11}$$

2.1.3 Joint rotation with a common pivot

The result that follows deals with the case presented in Fig. 2.3, in which a body is subjected to rotations around different axes, but it is possible to refer all of them to a common pivot.

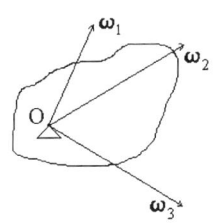

Figure 2.3 Rotations with common pivot.

Lemma 2.1. *If in the movement of a rigid body $\mathcal{C}$ we can distinguish several angular velocities $\boldsymbol{\omega}_i$, $i = 1, ..., n$, all of them referring to a common pivot O, then the vector*

$$\boldsymbol{\Omega} = \boldsymbol{\omega}_1 + \boldsymbol{\omega}_2 + \cdots + \boldsymbol{\omega}_n$$

is such that any point $p \in \mathcal{C}$ has speed

$$\mathbf{v}_p = [\boldsymbol{\Omega}, \mathbf{r}_p], \tag{2.12}$$

where both $\mathbf{v}_p$ and $\mathbf{r}_p$ are referred to O.

Proof. By the superposition principle, the velocity of any point $p \in \mathcal{C}$ may be expressed as

$$\mathbf{v}_p = \sum_{i=1}^{n} \mathbf{v}_{i,p},$$

where, since the pivot is common,

$$\mathbf{v}_{i,p} = [\boldsymbol{\omega}_i, \mathbf{r}_p],$$

and by the distributivity property of the vector product on the sum of vectors

$$\mathbf{v}_p = \left[\sum_{i=1}^n \boldsymbol{\omega}_{i,p}, \mathbf{r}_p\right].$$

So, the result is obtained. □

2.1.4 Parallel and non-coplanar rotations

The following examples illustrate two cases: the first meets the conditions required in the previous lemma, so that the present rotations can be reduced to one; the second one does not.

Parallel rotation

The cylinder in Fig. 2.4 is simultaneously subjected to the coplanar angular velocities $\boldsymbol{\omega}_1$ and $\boldsymbol{\omega}_2$ as illustrated. It goes on to show that these rotations can refer to the same point, the common pivot, so there is an equivalent angular velocity.

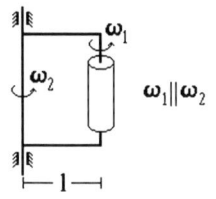

Figure 2.4 Cylinder subject to two parallel rotations.

In Fig. 2.5 the diagram with vectors $\boldsymbol{\omega}_1$ and $\boldsymbol{\omega}_2$ is shown. The effect of this set is not altered if the two vectors designated $\boldsymbol{\omega}_0$ and $-\boldsymbol{\omega}_0$ are added. This gives

$$\boldsymbol{\omega}_1' := \boldsymbol{\omega}_1 + \boldsymbol{\omega}_0, \quad \boldsymbol{\omega}_2' := \boldsymbol{\omega}_2 - \boldsymbol{\omega}_0. \tag{2.13}$$

Figure 2.5 Cylinder rotations.

Immediately it is seen that a common pivot O can be identified for the vectors $\boldsymbol{\omega}_1'$ and $\boldsymbol{\omega}_2'$, so that in view of the previous lemma there is the vector

$$\boldsymbol{\Omega} := \boldsymbol{\omega}_1' + \boldsymbol{\omega}_2',$$

which has the same joint effect of $\boldsymbol{\omega}_1'$ and $\boldsymbol{\omega}_2'$, but by (2.13) it follows that

$$\boldsymbol{\Omega} = \boldsymbol{\omega}_1 + \boldsymbol{\omega}_2.$$

The common pivot found depends on the magnitude of ω_0 chosen, that is, the common pivot for $\boldsymbol{\Omega}$ is not unique, but all possibilities are on the same vertical, whose position is now still determined. With regard to Fig. 2.5 we have

$$\frac{\omega_2}{\omega_0} = \frac{c}{a}, \quad \frac{\omega_1}{\omega_0} = \frac{c}{b},$$

where

$$b = \frac{\omega_2}{\omega_1} a. \tag{2.14}$$

Since

$$l = a + b,$$

in combination with (2.14) this gives

$$l = a\left(1 + \frac{\omega_2}{\omega_1}\right),$$

or equivalently,

$$a = l\left(1 + \frac{\omega_2}{\omega_1}\right)^{-1}.$$

Non-coplanar rotations

Consider a body subject to two non-coplanar angular velocities, referring to pivots O_1 and O_2, respectively, as detailed in Fig. 2.6. Show that there is no rotation that produces the same combined effect as $\boldsymbol{\omega}_1$ and $\boldsymbol{\omega}_2$.

Again, the velocity of a generic point p of the solid is given by the principle of superposition. To simplify suppose that the pivot O_2 is fixed in space, so by Euler's theorem

$$\mathbf{v}_p = [\boldsymbol{\omega}_2, \overline{O_2 p}] + [\boldsymbol{\omega}_1, \overline{O_1 p}].$$

But, note that

$$\overline{O_2 p} = \overline{O_2 O_1} + \overline{O_1 p},$$

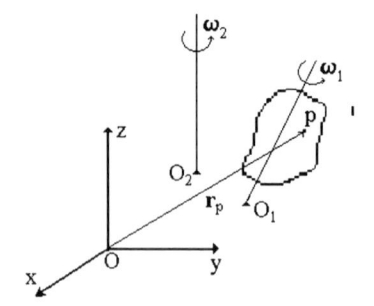

Figure 2.6 Rotations in oblique planes.

which gives
$$\mathbf{v}_p = [\boldsymbol{\omega}_1 + \boldsymbol{\omega}_2, \overline{O_1 p}] + [\boldsymbol{\omega}_2, \overline{O_2 O_1}]. \tag{2.15}$$

This is an expression that **cannot be reduced** to one of the kind
$$\mathbf{v}_p = [\boldsymbol{\omega}_1 + \boldsymbol{\omega}_2, \overline{O' p}]$$
for some pivot O'.

Note, however, that in the case when $\boldsymbol{\omega}_1$ and $\boldsymbol{\omega}_2$ are coplanar, their rotation lines intersect at some point O', so then we have
$$[\boldsymbol{\omega}_2, \overline{O' O_2}] = 0, \quad [\boldsymbol{\omega}_1, \overline{O' O_1}] = 0. \tag{2.16}$$

The sum of expressions (2.16) leads to
$$\mathbf{v}_p = [\boldsymbol{\omega}_1 + \boldsymbol{\omega}_2, \overline{O_1 p}] + [\boldsymbol{\omega}_2, \overline{O' O_2} + \overline{O_2 O_1}] + [\boldsymbol{\omega}_1, \overline{O' O_1}], \tag{2.17}$$

and since
$$\overline{O' O_2} + \overline{O_2 O_1} = \overline{O' O_1},$$

expression (2.17) is reduced to
$$\mathbf{v}_p = [\boldsymbol{\omega}_1 + \boldsymbol{\omega}_2, \overline{O' O_1} + \overline{O_1 p}].$$

In view of the relation
$$\overline{O' O_1} + \overline{O_1 p} = \overline{O' p},$$

we finally have
$$\mathbf{v}_p = [\boldsymbol{\omega}_1 + \boldsymbol{\omega}_2, \overline{O' p}]$$

as expected by the previous lemma. But in the general case, we have only (2.15).

2.2 Complex movements of the rigid body

2.2.1 General relations

Up to this point, we have studied the velocities of the points of a rigid body due to the rotation with respect to one of its pivots, which may be moving too. In this section the velocities and absolute accelerations of the points of the body with respect to an immobile coordinate system external to the body are obtained.

Let $\mathcal{C}$ be a rigid body and O one of its pivots, which has been chosen as the origin of a coordinate system $\mathcal{S}$ that moves being fixed to the body, so that $\mathcal{S}$ receives the name of *relative system*. In turn, this set is referenced to an immobile coordinate system $\mathcal{S}'$, called *absolute system*, originating in a point $O\prime$. Fig. 2.7 illustrates the details.

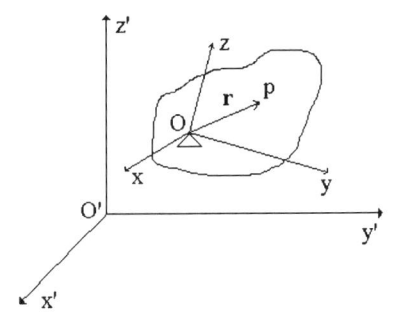

Figure 2.7 A rigid body and its absolute and relative references.

A point $p \in \mathcal{C}$ whose position is a temporal function can be represented by two time-dependent position vectors: $\mathbf{r}_p(t)$ with respect to system $\mathcal{S}$, called *relative position*, and $\overline{O'p}(t)$ with respect to system $\mathcal{S}'$, called *absolute position*. Between these position vectors there is the relationship

$$\overline{O'p} = \overline{O'O} + \mathbf{r}_p,$$

whose temporary derivative is

$$\dot{\overline{O'p}} = \dot{\overline{O'O}} + \dot{\mathbf{r}}_p,$$

or equivalently,

$$\mathbf{v}_p = \mathbf{v}_O + \dot{\mathbf{r}}_p,$$

where

$$\mathbf{v}_p := \dot{\overline{O'p}}, \quad \mathbf{v}_O := \dot{\overline{O'O}}$$

represent, respectively, the velocities of p and O with respect to $\mathcal{S}'$, for which they are called *absolute velocities*. Now, by Euler's theorem there exists a vector $\boldsymbol{\omega}$ such that

$$\dot{\mathbf{r}}_p = [\boldsymbol{\omega}, \mathbf{r}_p].$$

So, the absolute speed (velocity) of p is finally given by

$$\mathbf{v}_p = \mathbf{v}_O + [\boldsymbol{\omega}, \mathbf{r}_p]. \tag{2.18}$$

Remark 2.1. Let $\mathbf{f}, \mathbf{g} : [0, \infty) \to \Re^3$ be temporary functions. It is easily verified, from the definition of the vector product given in Chapter 1, that

$$\frac{d}{dt}[\mathbf{f}(t), \mathbf{g}(t)] = [\dot{\mathbf{f}}(t), \mathbf{g}(t)] + [\mathbf{f}(t), \dot{\mathbf{g}}(t)]. \tag{2.19}$$

Using (2.19), the temporal derivative of expression (2.18) may be obtained:

$$\dot{\mathbf{v}}_p = \dot{\mathbf{v}}_O + [\dot{\boldsymbol{\omega}}, \mathbf{r}_p] + [\boldsymbol{\omega}, \dot{\mathbf{r}}_p],$$

or, in the equivalent form

$$\mathbf{w}_p = \mathbf{w}_O + [\dot{\boldsymbol{\omega}}, \mathbf{r}] + [\boldsymbol{\omega}, \dot{\mathbf{r}}], \tag{2.20}$$

where

$$\mathbf{w}_p := \dot{\mathbf{v}}_p, \quad \mathbf{w}_O := \dot{\mathbf{v}}_O.$$

They receive, respectively, the names of *absolute accelerations* of the points p and O, since they refer to $\mathcal{S}'$.

Definition 2.4. The amount

$$\boldsymbol{\varepsilon} := \dot{\boldsymbol{\omega}}$$

is called **angular acceleration** of the solid body $\mathcal{C}$ with respect to pivot O.

By the above definition and Theorem 2.1, the relation (2.20) may be finally rewritten as

$$\mathbf{w} = \mathbf{w}_O + [\boldsymbol{\varepsilon}, \mathbf{r}] + [\boldsymbol{\omega}, [\boldsymbol{\omega}, \mathbf{r}]]. \tag{2.21}$$

Definition 2.5. In view of the characteristics of the term $[\boldsymbol{\omega}, [\boldsymbol{\omega}, \mathbf{r}]]$, which appears in expression (2.21), it is called **acceleration tending to the axis**.

The following example represents an interesting use of Euler's theorem to calculate $\boldsymbol{\varepsilon}$.

Example 2.1. The solid cone shown in Fig. 2.8 is rolling without sliding at constant speed. Let us calculate $\boldsymbol{\omega}$ and $\boldsymbol{\varepsilon}$. Due to the geometric characteristics of the solid, the

Rigid body kinematics 41

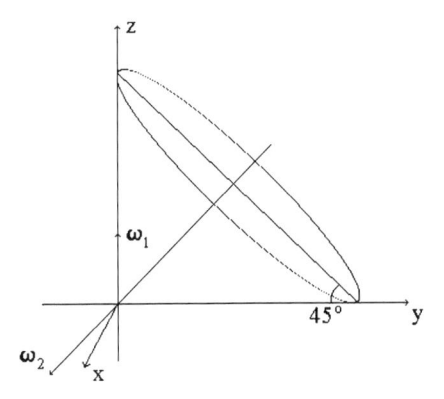

Figure 2.8 Solid cone rotating.

movement can be decomposed in the two rotations shown in Fig. 2.8, where ω_1 is constant in both magnitude and direction. From (2.12) we have

$$\Omega = \omega_1 + \omega_2,$$

where, according to Fig. 2.8,

$$\omega_1 = \omega_1 \mathbf{k}.$$

Since the set of points of the cone on the y-axis has zero velocity (no sliding), it can be considered as the axis of rotation, that is, Ω is on this axis. Therefore (see Fig. 2.9)

$$\Omega = -\omega_1 \mathbf{j}.$$

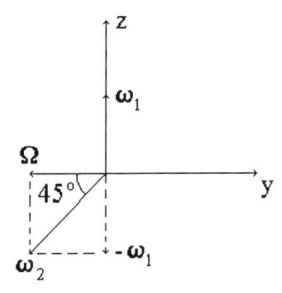

Figure 2.9 Diagram of angular speeds.

Then, since ω_1 is constant, it follows that

$$\boldsymbol{\varepsilon} := \dot{\boldsymbol{\omega}} = -\omega_1 \frac{d}{dt}\mathbf{j}.$$

Then by Euler's theorem

$$\frac{d}{dt}\mathbf{j} = [\boldsymbol{\omega}_1, \mathbf{j}] = \omega_1 \mathbf{i},$$

which finally implies

$$\boldsymbol{\varepsilon} = -\omega_1^2 \mathbf{i}.$$

2.2.2 Plane non-parallel motion and center of velocities

Let us consider a plane solid body, which moves in a plane (two-dimensional) space and realizes a non-parallel motion (see Fig. 2.10).

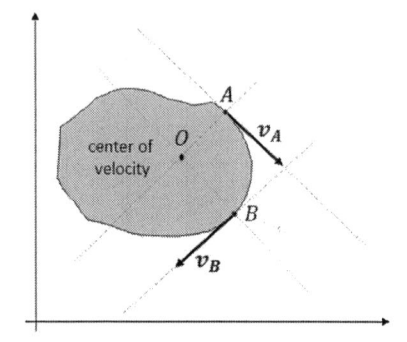

Figure 2.10 Plane non-parallel motion and center of velocities.

Choose two points A and B of a rigid body, which in the case of non-parallel motion will have corresponding non-collinear velocities $\mathbf{v}_A$ and $\mathbf{v}_B$ ($[\mathbf{v}_A, \mathbf{v}_B] \neq 0$). Draw the lines passing through the selected points and perpendicular to the corresponding speed vectors. The intersection point of these lines is denoted by the letter O and will be referred to as the *instantaneous center of velocities* (or simply *the center of velocities*) of a given rigid body.

Lemma 2.2. *The velocity $\mathbf{v}_O$ of the point O is equal to zero, that is,*

$$\mathbf{v}_O = 0. \tag{2.22}$$

Proof. Select the pole in the considered moment in the point O. In view of (2.18) we have

$$\mathbf{v}_P = \mathbf{v}_O + [\boldsymbol{\omega}, \mathbf{r}_P].$$

Applying it for both points $P = A, B$, we get

$$\mathbf{v}_A = \mathbf{v}_O + [\boldsymbol{\omega}, \mathbf{r}_{OA}], \quad \mathbf{v}_B = \mathbf{v}_O + [\boldsymbol{\omega}, \mathbf{r}_{BA}]. \tag{2.23}$$

Rigid body kinematics

Since $\boldsymbol{\omega}$ is orthogonal to both vectors $\mathbf{r}_{OA}$ and $\mathbf{r}_{OB}$, which in turn are orthogonal to $\mathbf{v}_A$ and $\mathbf{v}_B$, it follows that

$$0 = (\mathbf{v}_A, \mathbf{r}_{OA}) = (\mathbf{v}_O, \mathbf{r}_{OA}) + \underbrace{(\mathbf{r}_{OA}, [\boldsymbol{\omega}, \mathbf{r}_{OA}])}_{0} = (\mathbf{v}_O, \mathbf{r}_{OA}),$$

$$0 = (\mathbf{v}_B, \mathbf{r}_{OB}) = (\mathbf{v}_O, \mathbf{r}_{OB}) + \underbrace{(\mathbf{r}_{OB}, [\boldsymbol{\omega}, \mathbf{r}_{OB}])}_{0} = (\mathbf{v}_O, \mathbf{r}_{OB}),$$

which leads to

$$0 = (\mathbf{v}_O, \mathbf{r}_{OA} - \mathbf{r}_{OB}). \tag{2.24}$$

Since the points A and B are chosen arbitrarily, equality (2.24) is possible if and only if $\mathbf{v}_O = 0$. $\square$

This permits to conclude that the vector $\boldsymbol{\omega}$ passes (in the considered moment) through the point O and is perpendicular to the plane. Hence this movement can be considered as a rotation around the point O with the instantaneous rotation rate equal to $\boldsymbol{\omega}$. Therefore from (2.23) we have

$$v_A = \omega r_{OA}, \quad v_B = \omega r_{OB} \tag{2.25}$$

and

$$\omega = \frac{v_A}{r_{OA}} = \frac{v_B}{r_{OB}} = \frac{v_P}{r_{OP}}, \tag{2.26}$$

where P is any point of this solid.

Summary 2.1. a) If $O(t)$ is the center of velocities, by Euler's theorem there exists $\boldsymbol{\omega}(t)$ such that

$$\mathbf{v}_P(t) = [\boldsymbol{\omega}(t), \overline{OP}(t)],$$

where P represents a generic point of the solid.
b) If the body of the previous definition has purely translational movement, then the velocity center is located at infinity.
c) If we know the direction of the velocities of two solid-flat body points, then the center of the velocities C will lie at the intersection of the directions orthogonal to these velocities.

The relations (2.25) and (2.26) turn out to be very useful for the solution of a wide class of problems concerning plane non-parallel motion of rigid bodies.

Example 2.2. The disk of radius ρ and the shear of length l are pivotally connected at point A and, in turn, the end point B of the rod may move along the horizontal line DD' (see Fig. 2.11). Let us try to represent the speed v_B of point B as a function of the angle φ, provided that the disk currently rotates with angular velocity ω_0. Constructing

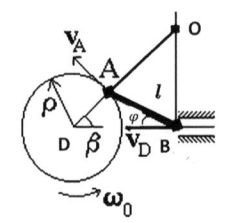

Figure 2.11 The disk and the shear which are pivotally connected.

the center of velocities, from (2.25) we derive

$$v_A = \omega r_{OA} = \omega_0 \rho, \quad v_B = \omega r_{OB},$$

which gives

$$v_B = \omega r_{OB} = \omega_0 \rho \frac{r_{OB}}{r_{OA}}.$$

By the "sinus theorem" we have

$$\left.\begin{aligned}\frac{r_{OA}}{\sin(\angle ABO)} &= \frac{r_{OB}}{\sin(\angle BAO)} = \frac{l}{\sin(\angle AOB)}, \\ \frac{\rho}{\sin\varphi} &= \frac{l}{\sin(\angle ADB)} = \frac{DB}{\sin(\angle DAB)}.\end{aligned}\right\} \quad (2.27)$$

Since

$$\angle ABO = \frac{\pi}{2} - \varphi, \quad \angle BAO = \pi - \angle DAB, \quad \sin(\angle BAO) = \sin(\angle DAB),$$

it follows that

$$\left.\begin{aligned}\frac{r_{OB}}{r_{OA}} &= \frac{\sin(\angle BAO)}{\sin(\angle ABO)} = \frac{\sin(\angle BAO)}{\sin\left(\frac{\pi}{2}-\varphi\right)} = \\ \frac{\sin(\angle BAO)}{\cos\varphi} &= \frac{\sin(\angle DAB)}{\cos\varphi} = \frac{\sin\varphi}{\cos\varphi}\frac{DB}{\rho} = \frac{DB}{\rho}\tan\varphi.\end{aligned}\right\} \quad (2.28)$$

But by the cosine theorem

$$\rho^2 = (DB)^2 + l^2 - 2\rho l \cos(\varphi),$$

implying

$$(DB)^2 = \rho^2 - l^2 + 2\rho l \cos(\varphi). \quad (2.29)$$

Substituting (2.29) into (2.28), finally, gives

$$\frac{r_{OB}}{r_{OA}} = \frac{\sqrt{\rho^2 - l^2 + 2\rho l \cos(\varphi)}}{\rho} \tan \varphi = \sqrt{1 - \left(\frac{l}{\rho}\right)^2 + 2\frac{l}{\rho}\cos(\varphi)} \tan \varphi.$$

2.3 Complex movement of a point

The description of the movement of a point with respect to a fixed system, but using a moving auxiliary system, has very interesting results. These aspects are discussed in this section.

Consider a point p, referred to using a coordinate system S with the origin O, which is going to be called a *relative system*. The system S is a moving system when it is referred to as a fixed coordinate system S' with the origin O'. We will call this system S' absolute. This situation is depicted in Fig. 2.12.

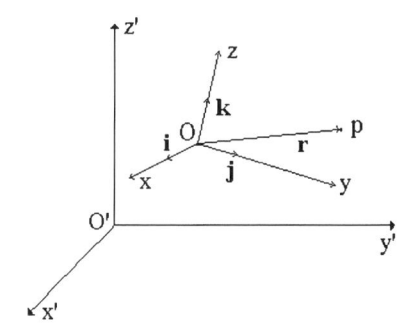

Figure 2.12 A point p in the absolute and relative systems.

Let

$$\mathbf{r} = \begin{bmatrix} x & y & z \end{bmatrix}^T$$

be the position vector of the point p in the system S, called *the relative position vector*. It can be represented as

$$\mathbf{r} = x\mathbf{i} + y\mathbf{j} + z\mathbf{k}, \qquad (2.30)$$

where $\mathbf{i}, \mathbf{j}, \mathbf{k}$ are orthogonal unitary vectors. The absolute position of the point p can be represented as

$$\mathbf{r}_{abs} = \overline{O'O} + \mathbf{r},$$

or by (2.30) as

$$\mathbf{r}_{abs} = \overline{O'O} + x\mathbf{i} + y\mathbf{j} + z\mathbf{k}.$$

2.3.1 Absolute velocity

So, the absolute velocity $\mathbf{v}_{abs}$ of the point p is

$$\mathbf{v}_{abs} := \dot{\mathbf{r}}_{abs} = \mathbf{v}_O + \dot{x}\mathbf{i} + \dot{y}\mathbf{j} + \dot{z}\mathbf{k} + x\frac{d}{dt}\mathbf{i} + y\frac{d}{dt}\mathbf{j} + z\frac{d}{dt}\mathbf{k}, \qquad (2.31)$$

where

$$\mathbf{v}_O := \overline{\dot{O'O}}$$

is the velocity of the origin O in the system $\mathcal{S}'$. Defining the relative velocity of the point p as

$$\mathbf{v}_{rel} := \dot{x}\mathbf{i} + \dot{y}\mathbf{j} + \dot{z}\mathbf{k} \qquad (2.32)$$

and using the relations (2.10),

$$\frac{d}{dt}\mathbf{i} = [\boldsymbol{\omega}, \mathbf{i}], \quad \frac{d}{dt}\mathbf{j} = [\boldsymbol{\omega}, \mathbf{j}], \quad \frac{d}{dt}\mathbf{k} = [\boldsymbol{\omega}, \mathbf{k}], \qquad (2.33)$$

where $\boldsymbol{\omega}$ is a rotation of the system $\mathcal{S}$ around the pivot O, we are able to represent the absolute velocity of the point p (2.31) in the form

$$\mathbf{v}_{abs} = \mathbf{v}_O + \mathbf{v}_{rel} + [\boldsymbol{\omega}, \mathbf{r}]. \qquad (2.34)$$

Moreover, defining the *translation velocity* $\mathbf{v}_{tr}$ of the point p as

$$\mathbf{v}_{tr} := \mathbf{v}_O + [\boldsymbol{\omega}, \mathbf{r}], \qquad (2.35)$$

expression (2.34) results in

$$\mathbf{v}_{abs} = \mathbf{v}_{tr} + \mathbf{v}_{rel}. \qquad (2.36)$$

2.3.2 Absolute acceleration

Through a process similar to the previous one, the temporary derivation (2.34) allows obtaining the expression for the absolute acceleration of the point p with respect to the system $\mathcal{S}'$ in the following form:

$$\mathbf{w}_{abs} := \dot{\mathbf{v}}_{abs} = \mathbf{w}_O + \ddot{x}\mathbf{i} + \ddot{y}\mathbf{j} + \ddot{z}\mathbf{k} + \dot{x}\frac{d}{dt}\mathbf{i} + \dot{y}\frac{d}{dt}\mathbf{j} + \dot{z}\frac{d}{dt}\mathbf{k} + [\dot{\boldsymbol{\omega}}, \mathbf{r}] + [\boldsymbol{\omega}, \dot{\mathbf{r}}], \qquad (2.37)$$

where

$$\mathbf{w}_O := \dot{\mathbf{v}}_O = \overline{\ddot{O'O}}$$

Rigid body kinematics

and the definition (2.32) is used. Note that by (2.33) and the relative velocity definition (2.32) we have

$$\dot{x}\frac{d}{dt}\mathbf{i} + \dot{y}\frac{d}{dt}\mathbf{j} + \dot{z}\frac{d}{dt}\mathbf{k} = [\boldsymbol{\omega}, \mathbf{v}_{rel}]. \tag{2.38}$$

Moreover, in view of (2.11) and using the relation

$$\dot{\mathbf{r}} = \mathbf{v}_{rel} + [\boldsymbol{\omega}, \mathbf{r}],$$

the last term in the right-hand side of (2.37) can be rewritten as

$$[\boldsymbol{\omega}, \dot{\mathbf{r}}] = [\boldsymbol{\omega}, \mathbf{v}_{rel}] + [\boldsymbol{\omega}, [\boldsymbol{\omega}, \mathbf{r}]]. \tag{2.39}$$

Defining the *relative acceleration* of the point p as

$$\mathbf{w}_{rel} = \ddot{x}\mathbf{i} + \ddot{y}\mathbf{j} + \ddot{z}\mathbf{k} \tag{2.40}$$

and substituting the relations (2.38), (2.39), and (2.40) in (2.37) implies

$$\mathbf{w}_{abs} = \mathbf{w}_O + \mathbf{w}_{rel} + [\boldsymbol{\varepsilon}, \mathbf{r}] + 2[\boldsymbol{\omega}, \mathbf{v}_{rel}] + [\boldsymbol{\omega}, [\boldsymbol{\omega}, \mathbf{r}]], \tag{2.41}$$

where we have used the definition of the *angular acceleration* $\boldsymbol{\varepsilon} := \dot{\boldsymbol{\omega}}$, previously introduced.

By the same idea, which the translation speed was defined with in (2.34), the *translation acceleration* of the point p may be defined as

$$\mathbf{w}_{tr} := \mathbf{w}_O + [\boldsymbol{\varepsilon}, \mathbf{r}] + [\boldsymbol{\omega}, [\boldsymbol{\omega}, \mathbf{r}]]. \tag{2.42}$$

So, expression (2.41) can be represented in the form

$$\mathbf{w}_{abs} = \mathbf{w}_{tr} + \mathbf{w}_{rel} + \mathbf{w}_{cor}, \tag{2.43}$$

where

$$\mathbf{w}_{cor} := 2[\boldsymbol{\omega}, \mathbf{v}_{rel}] \tag{2.44}$$

is referred to as the *Coriolis acceleration* of the point p.

2.4 Examples

Although the following example can be easily solved without resorting to the results obtained in the previous section, it is preferred to use them because it is very suitable to illustrate the main concepts introduced before.

Example 2.3. It could be thought that the translational and relative accelerations defined in the previous section are those derived from the respective velocities. The following exercise shows that this is not the case. Let us show that

$$\mathbf{w}_{tr} \neq \dot{\mathbf{v}}_{tr}, \quad \mathbf{w}_{rel} \neq \dot{\mathbf{v}}_{rel}.$$

Recall that

$$\mathbf{v}_{tr} = \mathbf{v}_O + [\boldsymbol{\omega}, \mathbf{r}],$$

and therefore

$$\dot{\mathbf{v}}_{tr} = \mathbf{w}_O + [\dot{\boldsymbol{\omega}}, \mathbf{r}] + [\boldsymbol{\omega}, \dot{\mathbf{r}}].$$

But in view of (2.39) we have

$$\dot{\mathbf{r}} = \mathbf{v}_{rel} + [\boldsymbol{\omega}, \mathbf{r}],$$

which implies

$$\dot{\mathbf{v}}_{tr} = \mathbf{w}_O + [\boldsymbol{\varepsilon}, \mathbf{r}] + [\boldsymbol{\omega}, [\boldsymbol{\omega}, \mathbf{r}]] + [\boldsymbol{\omega}, \mathbf{v}_{rel}],$$

or equivalently, using the definitions (2.42) and (2.44),

$$\dot{\mathbf{v}}_{tr} = \mathbf{w}_{tr} + \frac{1}{2}\mathbf{w}_{cor}.$$

Also we have

$$\mathbf{v}_{rel} = \dot{x}\mathbf{i} + \dot{y}\mathbf{j} + \dot{z}\mathbf{k},$$

so

$$\dot{\mathbf{v}}_{rel} = \ddot{x}\mathbf{i} + \ddot{y}\mathbf{j} + \ddot{z}\mathbf{k} + \dot{x}\frac{d}{dt}\mathbf{i} + \dot{y}\frac{d}{dt}\mathbf{j} + \dot{z}\frac{d}{dt}\mathbf{k},$$

and in view of (2.40) it follows that

$$\dot{\mathbf{v}}_{rel} = \ddot{x}\mathbf{i} + \ddot{y}\mathbf{j} + \ddot{z}\mathbf{k} + [\boldsymbol{\omega}, \mathbf{v}_{rel}].$$

So, finally, combining (2.40) with (2.44) we derive

$$\dot{\mathbf{v}}_{rel} = \mathbf{w}_{rel} + \frac{1}{2}\mathbf{w}_{cor}. \qquad (2.45)$$

Example 2.4. A particle slides on the rod as in Fig. 2.13. The rod oscillates with respect to the vertical forming an angle that evolves according to the law

$$\varphi(t) = \varphi_0 \sin(\omega_0 t),$$

Rigid body kinematics

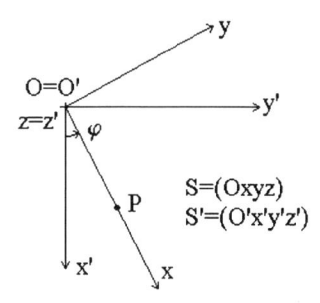

Figure 2.13 Particle sliding on an oscillating bar.

while the particle moves on the rod in such a way that the distance traveled on it follows the rule

$$OP(t) = \frac{1}{2}at^2.$$

Let us try to determine the velocity $\mathbf{v}_{abs}$ and the acceleration $\mathbf{w}_{abs}$ of the particle with respect to the given fixed coordinate system.

Denote by $\mathcal{S}'$ the fixed coordinate system with origin O' as it is illustrated in Fig. 2.13. Let $\mathcal{S}$ be the coordinate system with origin O common with O' that oscillates with the rod and whose axis of abscissas coincides with this one. Denote by $\mathbf{i}, \mathbf{j}, \mathbf{k}$ the unit vectors of $\mathcal{S}$ and by $\mathbf{i}', \mathbf{j}', \mathbf{k}'$ those of $\mathcal{S}'$. In the system $\mathcal{S}$ the relative position vector of the particle has the expression

$$\mathbf{r} = \frac{1}{2}at^2\mathbf{i}.$$

Now, according to (2.36), the absolute velocity of the particle is given by

$$\mathbf{v}_{abs} = \mathbf{v}_{tr} + \mathbf{v}_{rel},$$

where, in view of the conditions of the problem, expression (2.35) results in

$$\mathbf{v}_{tr} = [\boldsymbol{\omega}, \mathbf{r}],$$

with

$$\boldsymbol{\omega}(t) = \omega\mathbf{k}, \quad \omega = \dot{\varphi} = \varphi_0\omega_0\cos(\omega_0 t).$$

So, we have

$$\mathbf{v}_{tr} = \frac{1}{2}at^2\omega\mathbf{j}.$$

On the other hand, for the present case expression (2.32) has the particular form

$$\mathbf{v}_{rel} = at\mathbf{i}, \tag{2.46}$$

so

$$\mathbf{v}_{abs} = at\mathbf{i} + \frac{1}{2}at^2\omega\mathbf{j},$$

or, in view of the relations

$$\mathbf{i} = \cos\varphi\mathbf{i}' + \sin\varphi\mathbf{j}',$$
$$\mathbf{j} = -\sin\varphi\mathbf{i}' + \cos\varphi\mathbf{j}',$$

in the fixed system $\mathbf{i}', \mathbf{j}', \mathbf{k}'$ we have

$$\mathbf{v}_{abs} = at\left(\cos\varphi\mathbf{i}' + \sin\varphi\mathbf{j}'\right) + \frac{1}{2}at^2\omega\left(-\sin\varphi\mathbf{i}' + \cos\varphi\mathbf{j}'\right)$$
$$= at\left(\cos\varphi - \frac{1}{2}t\omega\sin\varphi\right)\mathbf{i}' + at\left(\sin\varphi + \frac{1}{2}t\omega\cos\varphi\right)\mathbf{j}'.$$

The absolute acceleration is determined by (2.43), i.e.,

$$\mathbf{w}_{abs} = \mathbf{w}_{tr} + \mathbf{w}_{rel} + \mathbf{w}_{cor},$$

where in the considered case

$$\mathbf{w}_{tr} = [\boldsymbol{\varepsilon}, \mathbf{r}] + [\boldsymbol{\omega}, [\boldsymbol{\omega}, \mathbf{r}]],$$

with

$$\boldsymbol{\varepsilon} = \varepsilon\mathbf{k}, \quad \varepsilon = \ddot{\varphi} = -\varphi_0\omega_0^2\sin(\omega_0 t).$$

So,

$$[\boldsymbol{\varepsilon}, \mathbf{r}] = \frac{1}{2}at^2\varepsilon\mathbf{j},$$
$$[\boldsymbol{\omega}, [\boldsymbol{\omega}, \mathbf{r}]] = -\frac{1}{2}at^2\omega^2\mathbf{i},$$

implying

$$\mathbf{w}_{tr} = \frac{1}{2}at^2\left(-\omega^2\mathbf{i} + \varepsilon\mathbf{j}\right).$$

Regarding relative acceleration, it is given by (2.40), which in the current case has the expression

$$\mathbf{w}_{rel} = a\mathbf{i},$$

while the Coriolis acceleration is determined by

$$\mathbf{w}_{cor} = 2[\boldsymbol{\omega}, \mathbf{v}_{rel}].$$

Rigid body kinematics

These representations together with (2.46) result in

$$\mathbf{w}_{cor} = 2at\omega\mathbf{j}.$$

Finally, the absolute acceleration may be expressed as

$$\mathbf{w}_{abs} = \left(-\frac{1}{2}at^2\omega^2 + a\right)\mathbf{i} + \left(2at\omega + \frac{1}{2}at^2\varepsilon\right)\mathbf{j},$$

or equivalently,

$$\mathbf{w}_{abs} = a\left[\left(-\frac{1}{2}t^2\omega^2 + 1\right)\cos\varphi - \left(2t\omega + \frac{1}{2}t^2\varepsilon\right)\right]\mathbf{i}' +$$
$$a\left[\left(-\frac{1}{2}t^2\omega^2 + 1\right)\sin\varphi + \left(2t\omega + \frac{1}{2}t^2\varepsilon\right)\cos\varphi\right]\mathbf{j}'.$$

The example that appears next illustrates the application of Euler's theorem, Theorem 2.1.

Example 2.5. Fig. 2.14 shows a disk running at constant speed v and without sliding. Let us calculate the speed and acceleration of the generic point p on the circumference of the disk and, in particular, of the indicated points A, B, C, and D.

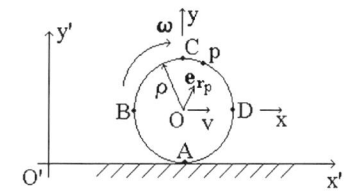

Figure 2.14 Disk rolling with constant speed v.

The most suitable point to serve as a pivot of the disk is its center; that is why it is chosen as the origin of the system with respect to which the body is immobile and whose configuration is chosen as shown in Fig. 2.14. This system is designated as $S = (Oxyz)$ and its unit vector system as $\mathbf{i}$, $\mathbf{j}$, $\mathbf{k}$. The absolute velocity of p with respect to the absolute system $S' = (O'x'y'z')$, with unit vectors $\mathbf{i}'$, $\mathbf{j}'$, $\mathbf{k}'$, is given by (2.18), that is,

$$\mathbf{v}_p = \mathbf{v}_0 + [\boldsymbol{\omega}, \mathbf{r}_p], \tag{2.47}$$

where

$$\mathbf{v}_0 = v\mathbf{i}, \quad \boldsymbol{\omega} = -\omega\mathbf{k}, \quad \mathbf{r}_p = \rho\mathbf{e}_{\mathbf{r}_p}, \tag{2.48}$$

with $\mathbf{e}_{\mathbf{r}_p}$ as the unitary vector in the direction of $\mathbf{r}_p$, which is the position vector of the point p with respect to the system S. To calculate the angular velocity ω, note that the absolute velocity of point A is zero as there is no sliding movement, that is,

$$\mathbf{v}_A = 0,$$

while its position vector $\mathbf{r}_A$ with respect to S is

$$\mathbf{r}_A = -\rho \mathbf{j}. \qquad (2.49)$$

Substitution of these relations into (2.47) leads to

$$v\mathbf{i} + \omega\rho\,[\mathbf{k},\mathbf{j}] = (v - \omega\rho)\mathbf{i} = 0,$$

from which it follows that

$$\omega = \frac{v}{\rho}. \qquad (2.50)$$

In view of (2.48) and (2.50), the relation (2.47) is reduced to

$$\mathbf{v}_p = v\left(\mathbf{i} - \left[\mathbf{k}, \mathbf{e}_{\mathbf{r}_p}\right]\right). \qquad (2.51)$$

Expression (2.51) now allows to calculate the velocities of points B, C, and D, whose unitary position vectors with respect to S are

$$\mathbf{e}_{\mathbf{r}_B} = -\mathbf{i}, \quad \mathbf{e}_{\mathbf{r}_C} = \mathbf{j}, \quad \mathbf{e}_{\mathbf{r}_D} = \mathbf{i}. \qquad (2.52)$$

So, we have

$$\mathbf{v}_B = v\,(\mathbf{i} - [\mathbf{k}, -\mathbf{i}]) = v\,(\mathbf{i} + \mathbf{j}),$$
$$\mathbf{v}_C = v\,(\mathbf{i} - [\mathbf{k}, \mathbf{j}]) = 2v\mathbf{i},$$
$$\mathbf{v}_D = v\,(\mathbf{i} - [\mathbf{k}, \mathbf{i}]) = v\,(\mathbf{i} - \mathbf{j}).$$

Given that the choice of S configuration was made in such a way that

$$\mathbf{i} = \mathbf{i}', \quad \mathbf{j} = \mathbf{j}', \quad \mathbf{k} = \mathbf{k}',$$

in the fixed (absolute) system S' we finally get

$$\mathbf{v}_B = v\left(\mathbf{i}' + \mathbf{j}'\right), \quad \mathbf{v}_C = 2v\mathbf{i}', \quad \mathbf{v}_D = v\left(\mathbf{i}' - \mathbf{j}'\right).$$

By (2.21), the acceleration of the point p is expressed as

$$\mathbf{w}_p = \mathbf{w}_O + \left[\boldsymbol{\varepsilon}, \mathbf{r}_p\right] + \left[\boldsymbol{\omega}, \left[\boldsymbol{\omega}, \mathbf{r}_p\right]\right]. \qquad (2.53)$$

But, taking into account that

$$\mathbf{w}_O = 0, \quad \boldsymbol{\varepsilon} = 0,$$

Rigid body kinematics

in view of (2.48) and (2.50), formula (2.53) is reduced to

$$\mathbf{w}_p = \frac{v^2}{\rho}\left[\mathbf{k},\left[\mathbf{k},\mathbf{e}_{\mathbf{r}_p}\right]\right].$$

By the formula for the triple vector product (see Chapter 1), it follows that

$$\left[\mathbf{k},\left[\mathbf{k},\mathbf{e}_{\mathbf{r}_p}\right]\right] = \left(\mathbf{k},\mathbf{e}_{\mathbf{r}_p}\right)\mathbf{k} - \left(\mathbf{k},\mathbf{k}\right)\mathbf{e}_{\mathbf{r}_p} = -\mathbf{e}_{\mathbf{r}_p}, \tag{2.54}$$

which finally leads to

$$\mathbf{w}_p = -\frac{v^2}{\rho}\mathbf{e}_{\mathbf{r}_p}.$$

Therefore, considering (2.49) and (2.52) for the points A, B, C, and D, we get

$$\mathbf{w}_A = \frac{v^2}{\rho}\mathbf{j}, \quad \mathbf{w}_B = \frac{v^2}{\rho}\mathbf{i}, \quad \mathbf{w}_B = -\frac{v^2}{\rho}\mathbf{j}, \quad \mathbf{w}_D = -\frac{v^2}{\rho}\mathbf{i}.$$

With respect to the system S' we have

$$\mathbf{w}_A = \frac{v^2}{\rho}\mathbf{j}', \quad \mathbf{w}_B = \frac{v^2}{\rho}\mathbf{i}', \quad \mathbf{w}_B = -\frac{v^2}{\rho}\mathbf{j}', \quad \mathbf{w}_D = -\frac{v^2}{\rho}\mathbf{i}'.$$

Fig. 2.15 graphically shows this result.

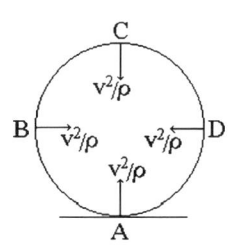

Figure 2.15 Accelerations in the points of the disk.

Example 2.6. Consider the same disk as in the previous example, but now rolling with constant velocity of magnitude and without sliding on a circular surface as seen in Fig. 2.16. Determine the speed and acceleration of the same generic point p on the circumference of the disk and specify for points A, B, C, and D.

1) Again, the center of the disk is chosen as the origin of the coordinate system $S = (Oxyz)$ fixed to the disk, while the absolute reference system $S' = (O'x'y'z')$ is located in the center of the disk, the circumference on which the disk rolls. For purposes of the problem posed, it is convenient to choose for S the configuration shown in Fig. 2.16. The absolute velocity $\mathbf{v}_p$ of the point p is given by the general expression (2.18), namely,

$$\mathbf{v}_p = \mathbf{v}_0 + \left[\boldsymbol{\omega},\mathbf{r}_p\right], \tag{2.55}$$

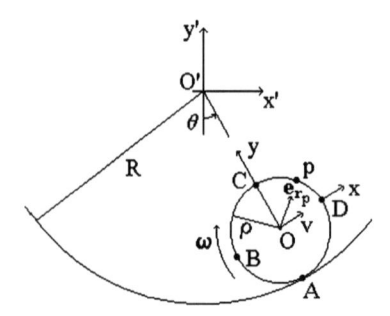

Figure 2.16 Disk rolling with constant magnitude speed on a circular surface.

where

$$\left.\begin{aligned}\mathbf{v}_0 &= v\mathbf{i}, \\ \boldsymbol{\omega} &= -\omega\mathbf{k}, \\ \mathbf{r}_p &= \rho\mathbf{e}_{r_p},\end{aligned}\right\} \tag{2.56}$$

with $\mathbf{e}_{r_p}$ as a unit vector in the direction of $\mathbf{r}_p$, which describes the position of p with respect to the system $\mathcal{S}$, whose unit vectors are denoted as $\mathbf{i}, \mathbf{j}, \mathbf{k}$. The magnitude of the angular velocity ω is determined from the fact that the absolute velocity of point A and its position vector with respect to $\mathcal{S}$ are

$$\left.\begin{aligned}\mathbf{v}_A &= 0, \\ \mathbf{r}_A &= -\rho\mathbf{j},\end{aligned}\right\} \tag{2.57}$$

whereby expression (2.55) gives for this point

$$v\mathbf{i} + \omega\rho\,[\mathbf{k},\mathbf{j}] = (v - \omega\rho)\,\mathbf{i} = 0,$$

implying

$$\omega = \frac{v}{\rho}. \tag{2.58}$$

With (2.56) and (2.58), the relation (2.55) is reduced to

$$\mathbf{v}_p = v\left(\mathbf{i} - \left[\mathbf{k}, \mathbf{e}_{r_p}\right]\right). \tag{2.59}$$

For the points B, C, and D we have

$$\mathbf{e}_{r_B} = -\mathbf{i}, \quad \mathbf{e}_{r_C} = \mathbf{j}, \quad \mathbf{e}_{r_D} = \mathbf{i},$$

Rigid body kinematics

and in view of (2.59), we get

$$\left.\begin{aligned} \mathbf{v}_B &= v\left(\mathbf{i} - [\mathbf{k}, -\mathbf{i}]\right) = v\left(\mathbf{i} + \mathbf{j}\right), \\ \mathbf{v}_C &= v\left(\mathbf{i} - [\mathbf{k}, \mathbf{j}]\right) = 2v\mathbf{i}, \\ \mathbf{v}_D &= v\left(\mathbf{i} - [\mathbf{k}, \mathbf{i}]\right) = v\left(\mathbf{i} - \mathbf{j}\right). \end{aligned}\right\} \qquad (2.60)$$

Now, if we denote by $\mathbf{i}'$, $\mathbf{j}'$, $\mathbf{k}'$ the unit vectors of the fixed system $\mathcal{S}'$, the following relations hold:

$$\left.\begin{aligned} \mathbf{i} &= \cos\varphi\mathbf{i}' + \sin\varphi\mathbf{j}', \\ \mathbf{j} &= -\sin\varphi\mathbf{i}' + \cos\varphi\mathbf{j}', \\ \mathbf{k} &= \mathbf{k}'. \end{aligned}\right\}$$

So, in the fixed system, the relations (2.60) are presented as

$$\left.\begin{aligned} \mathbf{v}_B &= v\left(\cos\varphi\mathbf{i}' + \sin\varphi\mathbf{j}' - \sin\varphi\mathbf{i}' + \cos\varphi\mathbf{j}'\right) \\ &= v\left(\cos\varphi - \sin\varphi\right)\mathbf{i}' + v\left(\sin\varphi + \cos\varphi\right)\mathbf{j}', \\ \mathbf{v}_C &= 2v\left(\cos\varphi\mathbf{i}' + \sin\varphi\mathbf{j}'\right), \\ \mathbf{v}_D &= v\left(\cos\varphi\mathbf{i}' + \sin\varphi\mathbf{j}' + \sin\varphi\mathbf{i}' - \cos\varphi\mathbf{j}'\right) \\ &= v\left(\cos\varphi + \sin\varphi\right)\mathbf{i}' + v\left(\sin\varphi - \cos\varphi\right)\mathbf{j}'. \end{aligned}\right\} \qquad (2.61)$$

2) Recalling (2.21), the acceleration of point p is given by

$$\mathbf{w}_p = \mathbf{w}_O + [\boldsymbol{\varepsilon}, \mathbf{r}_p] + [\boldsymbol{\omega}, [\boldsymbol{\omega}, \mathbf{r}_p]]. \qquad (2.62)$$

Since the angular velocity $\boldsymbol{\omega}$ is constant and the trajectory of O is circular with radius of curvature $R - \rho$, we have

$$\left.\begin{aligned} \boldsymbol{\varepsilon} &= 0, \\ \mathbf{w}_O &= \frac{v^2}{R - \rho}\mathbf{j}. \end{aligned}\right\} \qquad (2.63)$$

The relationships (2.63) together with (2.58) allow to represent (2.62) as

$$\mathbf{w}_p = \frac{v^2}{R - \rho}\mathbf{j} + \frac{v^2}{\rho}\left[\mathbf{k}, [\mathbf{k}, \mathbf{e}_{\mathbf{r}_p}]\right],$$

or, in view of (2.54),

$$\mathbf{w}_p = \frac{v^2}{R - \rho}\mathbf{j} - \frac{v^2}{\rho}\mathbf{e}_{\mathbf{r}_p}. \qquad (2.64)$$

Finally, expression (2.64) together with (2.57) gives for the accelerations of the ordered points A, B, C, and D

$$\begin{aligned}
\mathbf{w}_A &= v^2 \left(\frac{1}{R-\rho} + \frac{1}{\rho} \right) \mathbf{j}, \\
\mathbf{w}_B &= v^2 \left(\frac{1}{\rho} \mathbf{i} + \frac{1}{R-\rho} \mathbf{j} \right), \\
\mathbf{w}_C &= v^2 \left(\frac{1}{R-\rho} - \frac{1}{\rho} \right) \mathbf{j}, \\
\mathbf{w}_D &= v^2 \left(-\frac{1}{\rho} \mathbf{i} + \frac{1}{R-\rho} \mathbf{j} \right),
\end{aligned}$$

or, with respect to the absolute system S',

$$\begin{aligned}
\mathbf{w}_A &= v^2 \left(\frac{1}{R-\rho} + \frac{1}{\rho} \right) \left(-\sin\varphi\, \mathbf{i}' + \cos\varphi\, \mathbf{j}' \right), \\
\mathbf{w}_B &= \frac{v^2}{\rho} \left(\cos\varphi\, \mathbf{i}' + \sin\varphi\, \mathbf{j}' \right) + \frac{v^2}{R-\rho} \left(-\sin\varphi\, \mathbf{i}' + \cos\varphi\, \mathbf{j}' \right) = \\
& v^2 \left(\frac{1}{\rho} \cos\varphi - \frac{1}{R-\rho} \sin\varphi \right) \mathbf{i}' + v^2 \left(\frac{1}{\rho} \sin\varphi + \frac{1}{R-\rho} \cos\varphi \right) \mathbf{j}', \\
\mathbf{w}_C &= v^2 \left(\frac{1}{R-\rho} - \frac{1}{\rho} \right) \left(-\sin\varphi\, \mathbf{i}' + \cos\varphi\, \mathbf{j}' \right), \\
\mathbf{w}_D &= -\frac{v^2}{\rho} \left(\cos\varphi\, \mathbf{i}' + \sin\varphi\, \mathbf{j}' \right) + \frac{v^2}{R-\rho} \left(-\sin\varphi\, \mathbf{i}' + \cos\varphi\, \mathbf{j}' \right) = \\
& v^2 \left(-\frac{1}{\rho} \cos\varphi - \frac{1}{R-\rho} \sin\varphi \right) \mathbf{i}' + v^2 \left(-\frac{1}{\rho} \sin\varphi + \frac{1}{R-\rho} \cos\varphi \right) \mathbf{j}'.
\end{aligned}$$

(2.65)

Example 2.7. In the gear train of Fig. 2.17 the first gear rotates with an angular velocity ω_0 around its axis, while the train as a whole does so with an angular velocity Ω around an axis that coincides with that of the first gear. Determine the angular velocity of each gear ω_i ($i = 1, 2, \cdots, n$). To make the formula

$$\mathbf{v}_P = \mathbf{v}_O + [\boldsymbol{\omega}, \mathbf{r}_P]$$

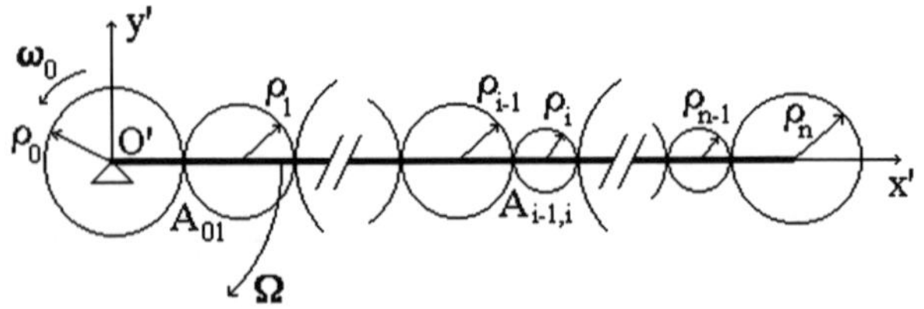

Figure 2.17 Rotating gear train.

to be appropriate to solve this problem, let us choose the coordinate system $\mathcal{S}' = (O'x'y'z')$, shown in Fig. 2.17, as the absolute system. In this system we have

$$\mathbf{\Omega} = -\Omega \mathbf{k}'.$$

For the determination of ω_1 note that the contact point A_{01} between gears 0 and 1 has the same speed on the two gears. To calculate the speed of A_{01} on gear 0, note also that the system $\mathcal{S}_0 = (O_0 x_0 y_0 z_0)$ is fixed to this gear with unit vectors denoted $\mathbf{i}_0, \mathbf{j}_0, \mathbf{k}_0$, coinciding with the unit vectors $\mathbf{i}, \mathbf{j}, \mathbf{k}$ of $\mathcal{S}'$, and with origin in the center of said gear. For these reasons, we have

$$\boldsymbol{\omega}_0 = \omega_0 \mathbf{k}_0,$$

while the position vector of A_{01} with respect to $\mathcal{S}_0$ is

$$\mathbf{r}_1 = -\rho_1 \mathbf{i}_1,$$

so that

$$\mathbf{v}_{A_{01}} = \mathbf{v}_{O_0} + [\boldsymbol{\omega}_0, \mathbf{r}_0] = \omega_0 \rho_0 \mathbf{j}_0, \tag{2.66}$$

with $\mathbf{v}_{O_0} = 0$. On the other hand, to determine the speed of the same point A_{01}, but now on gear 1, note that the system $\mathcal{S}_1 = (O_1 x_1 y_1 z_1)$ is fixed to the center of this gear in a similar way as it was done with gear 0, that is, with the unit vectors $\mathbf{i}_1, \mathbf{j}_1$, and $\mathbf{k}_1$, coinciding with the unit vectors $\mathbf{i}, \mathbf{j}, \mathbf{k}$ of $\mathcal{S}'$. In this situation

$$\boldsymbol{\omega}_1 = -\omega_1 \mathbf{k}_1,$$

while the position vector of A_{01} with respect to $\mathcal{S}_1$ results in

$$\mathbf{r}_1 = -\rho_1 \mathbf{i}_1.$$

Thus

$$\mathbf{v}_{A_{01}} = \mathbf{v}_{O_1} + [\boldsymbol{\omega}_1, \mathbf{r}_1] = [-\Omega(\rho_0 + \rho_1) + \omega_1 \rho_1] \mathbf{j}_1. \tag{2.67}$$

From (2.66) and (2.67) we may conclude that

$$\omega_1 = \omega_0 \frac{\rho_0}{\rho_1} + \Omega \frac{\rho_0 + \rho_1}{\rho_1}.$$

The knowledge of ω_1 allows to determine ω_2 following a similar procedure, and so on. To obtain the general formula, consider the gears $(i-1)$, and i and their contact point $A_{i-1,i}$. Fixing gear $(i-1)$, the system $\mathcal{S}_{i-1} = (O_{i-1} x_{i-1} y_{i-1} z_{i-1})$ with the unit vectors $\mathbf{i}_{i-1}, \mathbf{j}_{i-1}, \mathbf{k}_{i-1}$ coincides with the system $\mathcal{S}'$ of the unit vectors $\mathbf{i}, \mathbf{j}, \mathbf{k}$, where

$$\boldsymbol{\omega}_{i-1} = (-1)^{i-1} \omega_{i-1} \mathbf{k}_{i-1},$$

while the position vector of the point $A_{i-1,i}$ on the gear $(i-1)$ is expressed as $\mathbf{r}_{i-1} = \rho_{i-1}\mathbf{i}_{i-1}$. Therefore

$$\mathbf{v}_{A_{i-1,i}} = \mathbf{v}_{O_{i-1}} + [\boldsymbol{\omega}_{i-1}, \mathbf{r}_{i-1}] = \\ \left[-\Omega \left(\rho_0 + 2\sum_{j=1}^{i-2} \rho_j + \rho_{i-1} \right) + (-1)^{i-1} \omega_{i-1}\rho_{i-1} \right] \mathbf{j}_{i-1}. \quad (2.68)$$

Now, fixing the system $S_i = (O_i x_i y_i z_i)$, again with its matching unit vectors $\mathbf{i}_i, \mathbf{j}_i, \mathbf{k}_i$ with the unit vectors $\mathbf{i}, \mathbf{j}, \mathbf{k}$ of the system S', for gear i we have

$$\boldsymbol{\omega}_i = (-1)^i \omega_i \mathbf{k}_i,$$
$$\mathbf{r}_i = -\rho_i \mathbf{i}_i,$$

and hence,

$$\mathbf{v}_{A_{i-1,i}} = \mathbf{v}_{O_i} + [\boldsymbol{\omega}_i, \mathbf{r}_i] = \\ \left[-\Omega \left(\rho_0 + 2\sum_{j=1}^{i-1} \rho_j + \rho_i \right) + (-1)^{i+1} \omega_i \rho_i \right] \mathbf{j}_i. \quad (2.69)$$

From the equalization of the right sides of (2.68) and (2.69) we get

$$\omega_i = \omega_{i-1}\frac{\rho_{i-1}}{\rho_i} + (-1)^{i+1}\Omega\frac{\rho_{i-1}+\rho_i}{\rho_i}. \quad (2.70)$$

Example 2.8. Consider the two-gear train shown in Fig. 2.18. Assuming that at a given time t the angular velocities and accelerations ω_0, ε_0 of gear 0 and Ω, ε of the train are known, determine the corresponding ω_1, ε_1 of gear 1.

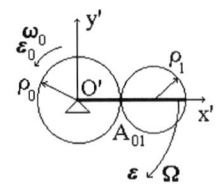

Figure 2.18 The two-gear train with acceleration.

The gear in Fig. 2.18 is a particular case of the one considered in the previous example. So, if the absolute coordinate systems $S' = (O'x'y'z')$ as well as $S_0 = (O_0 x_0 y_0 z_0)$ and $S_1 = (O_1 x_1 y_1 z_1)$ are related to gears 0 and 1 as in the previous example, a procedure similar to that followed allows to conclude that

$$\omega_1 = \omega_0 \frac{\rho_0}{\rho_1} + \Omega \frac{\rho_0 + \rho_1}{\rho_1},$$

Rigid body kinematics

which after differentiation leads to

$$\varepsilon_1 = \varepsilon_0 \frac{\rho_0}{\rho_1} + \varepsilon \frac{\rho_0 + \rho_1}{\rho_1}.$$

Example 2.9. The cylinder (see Fig. 2.19) is subjected to two rotations, one of them around its main axis. Let us calculate the angular acceleration ε that the body experiences. Since O is the common pivot of $\boldsymbol{\omega}_1$ and $\boldsymbol{\omega}_2$, the angular velocity

$$\boldsymbol{\Omega} := \boldsymbol{\omega}_1 + \boldsymbol{\omega}_2$$

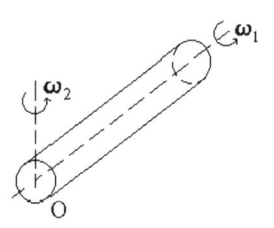

Figure 2.19 Cylinder subject to two rotations with common pivot.

produces the same effects on the cylinder as the separate application of said rotations. So, the angular acceleration ε is given by (2.11)

$$\boldsymbol{\varepsilon} = \dot{\boldsymbol{\Omega}} = \dot{\omega}_1 \mathbf{e}_{\omega_1} + \dot{\omega}_2 \mathbf{e}_{\omega_2} + \omega_1 \dot{\mathbf{e}}_{\omega_1} + \omega_2 \dot{\mathbf{e}}_{\omega_2}.$$

Note now that the configuration shown in Fig. 2.19 rotates with angular velocity given by $\boldsymbol{\omega}_2$. So, by Euler's theorem, we have

$$\omega_1 \dot{\mathbf{e}}_{\omega_1} = \omega_1 \left[\boldsymbol{\omega}_2, \mathbf{e}_{\omega_1} \right] = \left[\boldsymbol{\omega}_2, \boldsymbol{\omega}_1 \right], \quad \dot{\mathbf{e}}_{\omega_2} = \left[\boldsymbol{\omega}_2, \mathbf{e}_{\omega_2} \right] = 0,$$

and hence,

$$\boldsymbol{\varepsilon} = \dot{\omega}_1 \mathbf{e}_{\omega_1} + \dot{\omega}_2 \mathbf{e}_{\omega_2} + \left[\boldsymbol{\omega}_2, \boldsymbol{\omega}_1 \right].$$

2.5 Kinematics of a rigid body rotation

Definition 2.6. If a rigid body moves in such a way that one of its points remains fixed with respect to some coordinate system, it is said that the body realizes a **rotation movement** with respect to the immobile point.

2.5.1 Finite rotations

Suppose that the rigid body shown in Fig. 2.20 is subject to a rotation movement with respect to the center of coordinates. Suppose that a coordinate system is attached to

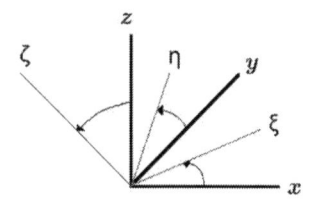

Figure 2.20 A rotation movement with respect to the center of coordinates.

the body, that is, the system is subject to the same rotation movement as the body. Consider that the coordinate systems shown are the positions of the coordinate system fixed to the body at two different times: for example, at time t_1 the position of the body corresponds to that of the coordinate system x, y, and z, while at instant t_2 the position is described by the system coordinate $\xi\eta\zeta$. A movement like the one described in Fig. 2.20 receives the name of a **finite rotation** of the body and it is denoted, for the situation described in the aforementioned figure, by

$$(xyz) \to (\xi\eta\zeta).$$

A finite rotation can be obtained as the sequential application of rotations around the coordinate axes of the system, corresponding to the initial instant (in this case xyz), called **elementary rotations**. The sequence, chosen for the elementary rotations, is referred to as the **description of the rotation**.

Euler's description

This description corresponds to the following sequence:

1. Elemental rotation around the z-axis at an angle ψ, that is,

$$(xyz) \to (x'y'z'), \quad z = z' : \psi\text{-action}, \tag{2.71}$$

where $x'y'z'$ is the configuration presented by the axis coordinate of the system at the end of the turn. The rotated angle ψ is called **the angle of precision**.

2. Elementary rotation around the x'-axis at an angle θ, i.e.,

$$(x'y'z') \to (x''y''z''), \quad x' = x'' : \theta - x\text{-action}, \tag{2.72}$$

where $x''y''z''$ is the configuration presented by the coordinate system at the end of the movement. The rotation angle θ is called **the angle of nutation**.

3. Elementary rotation around the y''-axis at an angle φ, i.e.,

$$(x''y''z'') \to (x'''y'''z'''), \quad y'' = y''' : \varphi\text{-action}, \tag{2.73}$$

where $x'''y'''z'''$ is the coordinate system configuration at the end of the rotation. The angle of rotation is called the **proper rotation angle**.

Definition 2.7. The angles (ψ, θ, φ), corresponding to Euler's description, are called **Euler angles**.

Natural description

The sequence chosen to carry out the movement is as follows:
1. Elemental rotation around the x-axis at an angle α, that is,

$$(xyz) \to (x'y'z'), \quad x = x' : \alpha\text{-action},$$

where $x'y'z'$ is the configuration presented by the coordinate system at the end of the rotation given by the angle α.

2. Elementary rotation around the y'-axis at an angle β, i.e.,

$$(x'y'z') \to (x''y''z''), \quad y' = y'' : \beta\text{-action},$$

where $x''y''z''$ is the configuration that presents the coordinate system at the end of the rotation given by the angle β.

3. Elementary rotation around the z''-axis at an angle γ, i.e.,

$$(x''y''z'') \to (x'''y'''z'''), \quad z'' = z''' : \gamma\text{-action},$$

where $x'''y'''z'''$ is the configuration of the coordinate system at the end of the rotation given by the angle γ.

Definition 2.8. The angles (α, β, γ) of the natural description are called **natural angles**.

2.5.2 Rotation matrix

Definition 2.9. A matrix $A \in \mathbb{R}^{3\times 3}$ such that

$$\|A\mathbf{r}\| = \|\mathbf{r}\| \quad \forall \mathbf{r} \in \mathbb{R}^3 \tag{2.74}$$

is called **rotation matrix**, since the longitude of the vector $A\mathbf{r}$, characterizing the new position, remains the same as the longitude of the initial vector $\mathbf{r}$.

In what follows, the matrix

$$A = [a_{ij}], \quad i, j = 1, 2, 3, \tag{2.75}$$

denotes a **generic rotation matrix**.

Properties of the rotation matrix

The rotation matrix A has a series of properties, which are enunciated and tested in the lines that follow.

P1. We have

$$A^T A = \mathbf{I}, \tag{2.76}$$

where I is the identity matrix of order 3.

Proof. By contradiction, suppose that A is a rotation matrix, but
$$A^T A \neq I.$$
Then for every $\mathbf{r} \in \mathbb{R}^3$ we have
$$\mathbf{r}^T A^T A \mathbf{r} \neq \mathbf{r}^T \mathbf{r},$$
or equivalently,
$$\|A\mathbf{r}\|^2 \neq \|\mathbf{r}\|^2.$$
So by (2.74), A is not a rotation matrix. □

P2. We have
$$\det A = \pm 1.$$
Proof. In view of **P1** it follows that
$$\det\left(A^T A\right) = 1. \tag{2.77}$$
But
$$\det\left(A^T A\right) = \det A^T \det A = (\det A)^2,$$
which proves (2.77). □

Remark 2.2. Based on **P2** matrix A is qualified as:
- **pure rotation** if
$$\det A = 1,$$
- **rotation plus specular reflection** if
$$\det A = -1.$$

P3. We have
$$A A^T = I. \tag{2.78}$$

Proof. By **P2** there exists $\left(A^T\right)^{-1}$; therefore premultiplying by $\left(A^T\right)^{-1}$ and postmultiplying by A^T the relation (2.76) becomes
$$\left(A^T\right)^{-1} A^T A A^T = \left(A^T\right)^{-1} A^T,$$
and since $\left(A^T\right)^{-1} A^T = I$ we obtain (2.78). □

Rigid body kinematics

Corollary 2.1. *From P2 and P3 it follows that*

$$A^T = A^{-1}. \tag{2.79}$$

Moreover, any matrix $A \in \mathbb{R}^{3\times 3}$ satisfying (2.79) is a rotation matrix.

Proof. Postmultiplying (2.79) by A we get

$$A^T A = \mathbf{I}.$$

So, for any $\mathbf{r} \in \mathbb{R}^3$

$$\mathbf{r}^T A^T A \mathbf{r} = \mathbf{r}^T \mathbf{r},$$

or

$$\|A\mathbf{r}\| = \|\mathbf{r}\|,$$

and hence, A is a rotation matrix. $\square$

P4. Representing a matrix A as

$$A = \begin{bmatrix} \bar{a}_1 & \bar{a}_2 & \bar{a}_3 \end{bmatrix}, \quad A = \begin{bmatrix} \underline{a}_1 \\ \underline{a}_2 \\ \underline{a}_3 \end{bmatrix},$$

where $\bar{a}_i$ and $\underline{a}_i$, $i = 1, 2, 3$, denote the i-th column and the i-th row (line) of A, respectively, we have

$$(\bar{a}_i, \bar{a}_j) = \delta_{ij}, \quad (\underline{a}_i, \underline{a}_j) = \delta_{ij},$$

which means that the column vectors of A are orthonormal between them and the same happens with their line vectors.

Proof. The orthonormality of the column vectors follows from **P2** and that of the row vectors from **P3**. $\square$

Remark 2.3. By **P4** the rotation matrix A is said to be **orthonormal**. Now, from the previous corollary it follows that the orthonormality of A is equivalent to the condition (2.79).

To state the following property of A requires remembering a concept presented below.

Definition 2.10. Given the matrix

$$B = \begin{bmatrix} b_{ij} \end{bmatrix}, \quad i, j = 1, 2, ..., n,$$

the product

$$B_{ij} := (-1)^{i+j} M_{ij}$$

is called **algebraic complement** of the element (i, j) of matrix B. Here M_{ij} denotes the (i, j)-**minor** of matrix B which is the determinant obtained from $\det B$ in which the i-th row and the j-th column have been deleted.

Remark 2.4. The previously defined concept has immediate application in the calculation of the inverse matrices, namely, if $\det B \neq 0$, it is verified that

$$B^{-1} = \frac{1}{\det B} \left[B_{ij}^T \right], \tag{2.80}$$

where B_{ij}^T denotes the *algebraic complement* of the element (i, j) of B^T.

P5. If the rotation matrix A defines a pure rotation, then

$$a_{ij} = A_{ij}. \tag{2.81}$$

Proof. From (2.80) and since we deal with a pure rotation ($\det A = 1$), we have

$$A^{-1} = \left[A_{ij}^T \right]. \tag{2.82}$$

But

$$A_{ij}^T = A_{ji}.$$

So by (2.79), the relation (2.82) can be rewritten as

$$A^T = \left[A_{ji} \right],$$

or equivalently,

$$A = \left[A_{ji} \right]^T = \left[A_{ij} \right],$$

which proves (2.81). □

Corollary 2.2. *Property P5 allows to obtain the following relations between the elements of the rotation matrix A:*

$$\left. \begin{array}{l} a_{11} = a_{22}a_{33} - a_{23}a_{32}, \\ a_{12} = -(a_{21}a_{33} - a_{31}a_{23}), \\ \vdots \end{array} \right\} \tag{2.83}$$

As is known, the scalars λ_k and the vectors $\mathbf{r}_k \neq \mathbf{0}$ ($k = 1, 2, 3$) that are solutions of the equation

$$B\mathbf{r}_k = \lambda \mathbf{r}_k \tag{2.84}$$

are called the **eigenvalues** and **eigenvectors** of the square matrix B. The properties that follow deal with the values and eigenvectors of the rotation matrix A given by (2.81).

P6. The eigenvalues λ_k ($k = 1, 2, 3$) of a rotation matrix A are given by

$$\left.\begin{aligned}\lambda_1 &= 1, \\ \lambda_2 &= e^{j\phi}, \\ \lambda_3 &= e^{-j\phi},\end{aligned}\right\} \tag{2.85}$$

where

$$e^{\pm j\phi} = \cos\phi \pm j\sin\phi, \quad j^2 = -1,$$

with

$$\phi := \arccos\left(\frac{\operatorname{tr} A - 1}{2}\right).$$

Proof. By (2.84) the eigenvalues λ_k ($k = 1, 2, 3$) are the solutions of the characteristic equation

$$\det(A - \lambda I) = 0, \tag{2.86}$$

and it is easily verified from the orthonormality of A and the relationships (2.83) that

$$\det(A - \lambda I) = -\lambda^3 + \lambda^2 \operatorname{tr} A - \lambda \operatorname{tr} A + 1.$$

So, it follows immediately that

$$\lambda_1 = 1$$

and that the remaining solutions are given by

$$\lambda^2 - \lambda(\operatorname{tr} A - 1) + 1 = 0,$$

implying

$$\lambda_{2,3} = \frac{\operatorname{tr} A - 1}{2} \pm \sqrt{\left(\frac{\operatorname{tr} A - 1}{2}\right)^2 - 1}. \tag{2.87}$$

On the other hand, since λ_k ($k = 2, 3$) must satisfy (2.84), that is,

$$A\mathbf{r}_k = \lambda_k \mathbf{r}_k, \tag{2.88}$$

for some vectors $\mathbf{r}_k \neq \mathbf{0}$. Therefore,

$$\|A\mathbf{r}_k\|^2 = \|\lambda_k \mathbf{r}_k\|^2 = |\lambda_k|^2 \|\mathbf{r}_k\|^2. \tag{2.89}$$

But by **P1**

$$\|A\mathbf{r}_k\|^2 = (A\mathbf{r}_k, A\mathbf{r}_k) = \left(\mathbf{r}_k, A^T A\mathbf{r}_k\right) = (\mathbf{r}_k, \mathbf{r}_k) = \|\mathbf{r}_k\|^2, \qquad (2.90)$$

which leads to the relation

$$|\lambda_k| = 1, \ k = 2, 3. \qquad (2.91)$$

The condition (2.91) is not met if the discriminant in (2.87) is positive, because in such case both λ_k are real and

$$|\lambda_2| \neq |\lambda_3|.$$

In the case where said discriminant is not positive, (2.87) can be rewritten as

$$\lambda_{2,3} = \frac{\operatorname{tr} A - 1}{2} \pm j\sqrt{1 - \left(\frac{\operatorname{tr} A - 1}{2}\right)^2}, \ j^2 = -1, \qquad (2.92)$$

satisfying the property (2.91). It can be represented as

$$\lambda_{2,3} = e^{j\phi}, \qquad (2.93)$$

such that by the Euler formula we are able to make the following association:

$$\cos\phi = \frac{\operatorname{tr} A - 1}{2}, \quad \sin\phi = \sqrt{1 - \left(\frac{\operatorname{tr} A - 1}{2}\right)^2}. \qquad (2.94)$$

Note that the second relationship (2.94) is another way of writing the first one (2.93) if we take into account that

$$(\sin\phi)^2 + (\cos\phi)^2 = 1.$$

Finally, the two expressions (2.93) and (2.94) lead to the desired result. □

Remark 2.5. If the discriminant in (2.92) is equal to zero, which occurs if

$$\left|\frac{\operatorname{tr} A - 1}{2}\right| = 1,$$

we have

$$\lambda_2 = \lambda_3 = 1 \ \text{or} \ \lambda_2 = \lambda_3 = -1.$$

The case $\lambda_i = 1$, $i = 1, 2, 3$, appears when $A = I$, and is not interesting, because no rotation is properly given.

P7. The eigenvector corresponding to $\lambda_1 = 1$ and denoted $\mathbf{r}_1$ is given by

$$\mathbf{r}_1 = \frac{1}{2\sin\phi} \begin{bmatrix} a_{32} - a_{23} \\ a_{31} - a_{13} \\ a_{12} - a_{21} \end{bmatrix}. \tag{2.95}$$

Proof. Representing $\mathbf{r}_1$ as

$$\mathbf{r}_1 = \begin{bmatrix} x_1 & y_1 & z_1 \end{bmatrix}^T$$

and substituting λ_1 in (2.84) leads to the following system of equations for the determination of the components of $\mathbf{r}_1$:

$$\left. \begin{array}{l} (a_{11} - 1)x_1 + a_{12}y_1 + a_{13}z_1 = 0, \\ a_{21}x_1 + (a_{22} - 1)y_1 + a_{23}z_1 = 0, \\ a_{31}x_1 + a_{32}y_1 + (a_{33} - 1)z_1 = 0. \end{array} \right\} \tag{2.96}$$

Because $\lambda_1 = 1$ is the eigenvalue of the rotation matrix A, assuming that $A \neq I$, it is verified that one of the equations of (2.96) is not independent. However, after normalizing (2.88) by $\|\mathbf{r}_1\|$, without loss of generality, we can consider the case $\|\mathbf{r}_1\| = 1$ only, that is,

$$x_1^2 + y_1^2 + z_1^2 = 1,$$

which is independent of (2.96). So, we may conclude that there is only one solution, given by (2.95). □

P8. It is easy to check that

$$r_2 = \begin{bmatrix} 0 \\ 1 \\ 0 \end{bmatrix}, \quad r_3 = \begin{bmatrix} 0 \\ 0 \\ 1 \end{bmatrix}.$$

Below we need the following definition.

Definition 2.11. Define the **parameters of Rodríguez–Hamilton** as follows:

- $$\lambda_0 := \cos\frac{\phi}{2},$$

  which defines the angle of rotation;

- $$\lambda_1 := x_1 \sin\frac{\phi}{2}, \quad \lambda_2 := y_1 \sin\frac{\phi}{2}, \quad \lambda_3 := z_1 \sin\frac{\phi}{2},$$

  which define the axes of rotation.

It is easy to check that the following property for the parameters of Rodríguez–Hamilton is satisfied:

$$\lambda_0^2 + \lambda_1^2 + \lambda_2^2 + \lambda_3^2 = 1.$$

2.5.3 Composition of rotations

Suppose that A and B are two rotation matrices and that $\mathbf{r} \in R^3$. The notation

$$\mathbf{r} \xrightarrow{A} \mathbf{r}' \xrightarrow{B} \mathbf{r}''$$

indicates that the rotation of $\mathbf{r}$ given by A is applied first, and then the rotation given by B is applied to the obtained vector $\mathbf{r}'$. In other words,

$$\mathbf{r}' = A\mathbf{r}, \quad \mathbf{r}'' = B\mathbf{r}'.$$

So, by immediate substitution we have

$$\mathbf{r}'' = C\mathbf{r},$$

where

$$C := BA. \tag{2.97}$$

The following result characterizes the properties of the matrix C.

Lemma 2.3. *The matrix C is a rotation matrix, that is,*

$$C^T = C^{-1}. \tag{2.98}$$

Proof. By the definition (2.97) and taking into account that both matrices A and B are rotation matrices too, on the one hand,

$$C^T C = (BA)^T BA = A^T B^T BA = A^T A = I,$$

and on the other hand,

$$CC^T = BA(BA)^T = BAA^T B^T = BB^T = I,$$

which leads to (2.98). $\square$

Up to this point we have studied the effect of a rotation matrix A applied to a vector $\mathbf{r} \in R^3$, defined in the original coordinate system. However, the rotation can also be studied in another coordinate system, which has interesting results. Suppose that A is a rotation matrix, acting as

$$\mathbf{i}' = A\mathbf{i}, \quad \mathbf{j}' = A\mathbf{j}, \quad \mathbf{k}' = A\mathbf{k}. \tag{2.99}$$

Rigid body kinematics

The vectors $\mathbf{i}'$, $\mathbf{j}'$, $\mathbf{k}'$ are the unitary vectors obtained by the rotation of the original unitary vectors $\mathbf{i}, \mathbf{j}, \mathbf{k}$. Now, consider a vector $\mathbf{r} \in R^3$ with its representation

$$\mathbf{r} = x\mathbf{i} + y\mathbf{j} + z\mathbf{k}. \tag{2.100}$$

Since there exists A^{-1} (which coincides with A^T), from (2.99) the following relationships can be obtained:

$$\mathbf{i} = A^T \mathbf{i}', \quad \mathbf{j} = A^T \mathbf{j}', \quad \mathbf{k} = A^T \mathbf{k}'.$$

Substitution in (2.100) gives the representation of $\mathbf{r}$ in the unit vector system $\mathbf{i}', \mathbf{j}', \mathbf{k}'$:

$$\mathbf{r} = A^T \left(x\mathbf{i}' + y\mathbf{j}' + z\mathbf{k}' \right). \tag{2.101}$$

The previous expression has important details. If we denote by $\tilde{\mathbf{r}}$ the representation of the vector $\mathbf{r}$ with respect to the rotated system, then the relation (2.101) can be represented as

$$\mathbf{r}' = A^T \mathbf{r},$$

where $\mathbf{r}$ is represented with respect to the original system. So A and A^T can be seen as opposed base changes.

In what follows we investigate the change that a rotation matrix A undergoes when it is represented in the base of the rotated system, given by the same matrix A. Consider matrices A and B, where A defines a first rotation. If S and S' denote, respectively, the coordinate system before and after the application of the rotation A, that is,

$$S \xrightarrow{A} S', \tag{2.102}$$

where the base of S' is given by (2.99), the lemma presented below allows obtaining the representation of B in the base S', denoted $B_{S'}$.

Lemma 2.4. *Suppose that A and B are matrices, where the first is a rotation. Then*

$$B_{S'} = A^T B A, \tag{2.103}$$

with S' given by (2.102).

Proof. Let

$$\mathbf{y} = B\mathbf{x} \tag{2.104}$$

for $\mathbf{x} \in R^3$. Since A^T is non-singular, it can be seen as a base change: in the new base equation (2.104) takes the form

$$A^T \mathbf{y} = A^T B A^{-T} A^T \mathbf{x}, \tag{2.105}$$

where A^{-T} denotes the inverse of A^T. Let $\mathbf{x}'$ and $\mathbf{y}'$ be the representations of the vectors $\mathbf{x}$ and $\mathbf{y}$ in the new base, that is,

$$\mathbf{x}' := A^T \mathbf{x}, \quad \mathbf{y}' := A^T \mathbf{y},$$

whereupon (2.105) may be written as

$$\mathbf{y}' = A^T B A^{-T} \mathbf{x}',$$

and, since $A^{-T} = A$, we get

$$\mathbf{y}' = A^T B A \mathbf{x}'.$$

This expression shows that the matrix $A^T B A$ is the representation of the matrix B in the new base, which proves (2.105). □

Definition 2.12. The representation of the rotation matrix A with respect to the rotated system, obtained from itself, that is, the one given by the base (2.99), is called the **proper (own) matrix** of A, and it is denoted by A^*.

Lemma 2.5. *It is easy to check that*

$$A^* = A.$$

Proof. Note that if $B = A$ in (2.103), then $A_{S'} = A^*$, which immediately gives

$$A^* = A^T A A$$

and the desired result follows if we take into account that $A^T A = I$. □

The previous results, applied to the composition of rotations, lead to the following fact.

Lemma 2.6. *Let A and B be rotational matrices, where B is described with respect to the rotated system obtained via A. Denote by C the matrix, obtained from the composition of the rotations in order A, B, that is,*

$$C := BA.$$

Then the following property holds:

$$C^* = A^* B^*.$$

Proof. Let S be the original system before the application of no rotation; denote by S' the system obtained by applying to S the rotation A, and by S'' the system obtained from S' when applying the rotation B, that is,

$$S \xrightarrow{A} S' \xrightarrow{B} S''.$$

Rigid body kinematics

With the notation used previously and by the previous lemma we have the expressions

$$A^* := A_{S'},$$
$$B^* := B_{S''} = B_{S'}$$
(2.106)

whose product (in the order that follows) is obtained:

$$B^* A^* = B_{S'} A_{S'} = C_{S'}.$$
(2.107)

On the other hand, in view of (2.103) and using (2.107),

$$C_{S''} = B_{S'}^T C_{S'} B_{S'} = B_{S'}^T B_{S'} A_{S'} B_{S'}.$$

But in view of the property $B_{S'}^T B_{S'} = I$, we get

$$C_{S''} = A_{S'} B_{S'}.$$
(2.108)

Since

$$C^* := C_{S''},$$

by (2.106) and (2.108), it follows that

$$C^* = A^* B^*.$$

□

The newly tested result has its natural application in obtaining the matrix that describes the rotation, corresponding to the description in the Euler angles, which is the object of the example in the end of this chapter.

2.6 Rotations and quaternions

As shown below, the so-called *quaternions* represent a very useful tool for the modeling and analysis of rotations.

2.6.1 Quaternions

Definition 2.13. A construction of the type

$$\Lambda := \lambda_0 + \sum_{j=1}^{3} \lambda_j \mathbf{i}_j, \quad \lambda_j \in R, \; j = 1, 2, 3,$$
(2.109)

is called the **Hamiltonian quaternion**, hereinafter simply called **quaternion**. The set of quaternions is denoted by $\mathcal{H}$.

In mathematics, the quaternions are a number system that extends the complex numbers. They were first described and applied to mechanics in three-dimensional space by Irish mathematician William Rowan Hamilton in 1843. A feature of quaternions is that multiplication of two quaternions is non-commutative. Hamilton defined a quaternion as the quotient of two directed lines in a three-dimensional space or equivalently as the quotient of two vectors.

Definition 2.14. Consider the quaternions

$$\Lambda := \lambda_0 + \sum_{j=1}^{3} \lambda_j \mathbf{i}_j,$$

$$\Delta := \delta_0 + \sum_{j=1}^{3} \delta_j \mathbf{i}_j.$$

(i) It is said that $\Lambda = \Delta$ if and only if

$$\lambda_j = \delta_j, \quad j = 0, ..., 3.$$

(ii) The sum $\Lambda + \Delta$ is defined as

$$\Lambda + \Delta = \lambda_0 + \delta_0 + \sum_{j=0}^{3} \left(\lambda_j + \delta_j \right) \mathbf{i}_j.$$

(iii) The product of quaternions $\Lambda \circ \Delta$ is defined as the distributive operation on the sum, where the following relations are satisfied:

$$\left. \begin{array}{l} \mathbf{i}_1 \mathbf{i}_2 = \mathbf{i}_3, \quad \mathbf{i}_2 \mathbf{i}_3 = \mathbf{i}_1, \quad \mathbf{i}_3 \mathbf{i}_1 = \mathbf{i}_2, \\ \mathbf{i}_2 \mathbf{i}_1 = -\mathbf{i}_3, \quad \mathbf{i}_3 \mathbf{i}_2 = -i_1, \quad \mathbf{i}_1 \mathbf{i}_3 = -i_2, \\ \mathbf{i}_j \mathbf{i}_j = -1, \quad j = 1, 2, 3. \end{array} \right\} \quad (2.110)$$

Remark 2.6. It is clear that the set of quaternions with the defined addition operation forms an Abelian group structure whose neutral element is 0, defined by

$$0 := 0 + \sum_{j=1}^{3} 0 \mathbf{i}_j.$$

The algebraic structure formed by the set of quaternions with the operations of sum and product, defined above, is denoted by $\mathcal{H}$.

Remark 2.7. By the given product definition, the quaternion of (2.109) may be represented in the form

$$\Lambda := \lambda_0 + \boldsymbol{\lambda}, \tag{2.111}$$

Rigid body kinematics

where λ_0 and

$$\boldsymbol{\lambda} := \sum_{j=1}^{3} \lambda_j \mathbf{i}_j$$

are called the **scalar** and **vector parts** of $\boldsymbol{\Lambda}$, respectively.

From the product definition property (2.110) and with the representation (2.111) the following formula for the quaternions product is obtained immediately.

Lemma 2.7. *If*

$$\boldsymbol{\Lambda} := \lambda_0 + \boldsymbol{\lambda}, \ \boldsymbol{\Delta} := \delta_0 + \boldsymbol{\delta},$$

then

$$\boldsymbol{\Lambda} \circ \boldsymbol{\Delta} := \lambda_0 \delta_0 - (\boldsymbol{\lambda}, \boldsymbol{\delta}) + \lambda_0 \boldsymbol{\delta} + \delta_0 \boldsymbol{\lambda} + [\boldsymbol{\lambda}, \boldsymbol{\delta}], \quad (2.112)$$

where $(\cdot, \cdot)$ *and* $[\cdot, \cdot]$ *denote the usual scalar and vector products.*

Proof. Using the properties (2.110) it follows that

$$\boldsymbol{\Lambda} \circ \boldsymbol{\Delta} = \left(\lambda_0 + \sum_{j=1}^{3} \lambda_j \mathbf{i}_j \right) \left(\delta_0 + \sum_{j=1}^{3} \delta_j \mathbf{i}_j \right) =$$

$$\lambda_0 \delta_0 + \lambda_0 \boldsymbol{\delta} + \delta_0 \boldsymbol{\lambda} + \sum_{j=1}^{3} \lambda_j \mathbf{i}_j \sum_{j=1}^{3} \delta_j \mathbf{i}_j =$$

$$\lambda_0 \delta_0 + \lambda_0 \boldsymbol{\delta} + \delta_0 \boldsymbol{\lambda} + \sum_{i=1}^{3} \sum_{j=1}^{3} \lambda_i \delta_j \mathbf{i}_j \mathbf{i}_i =$$

$$\lambda_0 \delta_0 + \lambda_0 \boldsymbol{\delta} + \delta_0 \boldsymbol{\lambda} - (\boldsymbol{\lambda}, \boldsymbol{\delta}) + \sum_{i=1}^{3} \sum_{j \neq i}^{3} \lambda_i \delta_j \mathbf{i}_j \mathbf{i}_i =$$

$$\lambda_0 \delta_0 - (\boldsymbol{\lambda}, \boldsymbol{\delta}) + \lambda_0 \boldsymbol{\delta} + \delta_0 \boldsymbol{\lambda} + [\boldsymbol{\lambda}, \boldsymbol{\delta}].$$

$\square$

Definition 2.15. Let

$$\boldsymbol{\Lambda} := \lambda_0 + \boldsymbol{\lambda}.$$

Then the quaternion

$$\boldsymbol{\Lambda}^* := \lambda_0 - \boldsymbol{\lambda}$$

is referred to as the **quaternion conjugated to** $\boldsymbol{\Lambda}$.

For given

$$\Lambda := \lambda_0 + \lambda$$

and

$$\Lambda^* := \delta_0 + \delta, \ \delta_0 = \lambda_0, \ \delta = -\lambda,$$

by the properties of the scalar and vector products, it is verified that

$$\Lambda \circ \Lambda^* = \Lambda^* \circ \Lambda = \lambda_0 \delta_0 - (\lambda, \delta) = \sum_{j=0}^{4} \lambda_j \delta_j \in R.$$

Definition 2.16. The scalar amount

$$\|\Lambda\| := (\Lambda \circ \Lambda^*)^{1/2} = (\Lambda^* \circ \Lambda)^{1/2} \qquad (2.113)$$

it is called the **quaternion norm** of Λ.

In addition to the properties of the product already given, the operation $\Lambda \circ \Delta$ of the product satisfies several more, which are easily verified from the previous developments.

Lemma 2.8 (Additional properties of the quaternion product). *Let Λ, Δ, Γ be quaternions, that is, $\Lambda, \Delta, \Gamma \in \mathcal{H}$. Then the following properties hold:*

1. *No commutativity: We have*

$$\Lambda \circ \Delta \neq \Delta \circ \Lambda.$$

2. *Associativity: We have*

$$(\Lambda \circ \Delta) \circ \Gamma = \Lambda \circ (\Delta \circ \Gamma).$$

3. *Unity: We have*

$$\Lambda \circ 1 = 1 \circ \Lambda = \Lambda, \qquad (2.114)$$

where $1 \in \mathcal{H}$ is given by

$$1 = 1 + \sum_{j=1}^{3} 0 \mathbf{i}_j.$$

4. *Conjugate product: We have*

$$(\Lambda \circ \Delta)^* = \Delta^* \circ \Lambda^*. \qquad (2.115)$$

5. Given that

$$\|\boldsymbol{\Lambda} \circ \boldsymbol{\Delta}\|^2 := \boldsymbol{\Lambda} \circ \boldsymbol{\Delta} \circ (\boldsymbol{\Lambda} \circ \boldsymbol{\Delta})^*,$$

in view of (2.115) and (2.113) it follows that

$$\|\boldsymbol{\Lambda} \circ \boldsymbol{\Delta}\|^2 = \boldsymbol{\Lambda} \circ \boldsymbol{\Delta} \circ \boldsymbol{\Delta}^* \circ \boldsymbol{\Lambda}^* = \|\boldsymbol{\Delta}\|^2 \boldsymbol{\Lambda} \circ \boldsymbol{\Lambda}^* = \|\boldsymbol{\Delta}\|^2 \|\boldsymbol{\Lambda}\|^2,$$

which means that

$$\|\boldsymbol{\Lambda} \circ \boldsymbol{\Delta}\| = \|\boldsymbol{\Lambda}\| \|\boldsymbol{\Delta}\|. \tag{2.116}$$

6. If $\boldsymbol{\Lambda} \neq 0$, we may define the multiplicative inverse of $\boldsymbol{\Lambda}$ as the quaternion, denoted by $\boldsymbol{\Lambda}^{-1}$, with the property

$$\boldsymbol{\Lambda}^{-1} \circ \boldsymbol{\Lambda} = \boldsymbol{\Lambda} \circ \boldsymbol{\Lambda}^{-1} = 1, \tag{2.117}$$

whereby

$$\boldsymbol{\Lambda}^* \circ \boldsymbol{\Lambda} \circ \boldsymbol{\Lambda}^{-1} = \boldsymbol{\Lambda}^*,$$

implying

$$\boldsymbol{\Lambda}^{-1} = \frac{1}{\|\boldsymbol{\Lambda}\|^2} \boldsymbol{\Lambda}^*,$$

where (2.113) has been used.

Remark 2.8. As can be seen immediately, the set $\mathcal{H}$ turns out to be a **ring with division**.

The next exercise illustrates the use of the quaternion set $\mathcal{H}$ in the solution of quadratic equations.

Example 2.10. Consider the following quadratic equation with respect to the variable X defined in the quaternion space $\mathcal{H}$:

$$\left.\begin{array}{r}X^2 + aX + b = 0, \\ \dfrac{a^2}{4} - b < 0, \quad a, b \in R.\end{array}\right\} \tag{2.118}$$

The solutions of (2.118) depend on the algebraic structure to which X belongs.

a) If X is considered as a real number, i.e., $X \in R$, then there are no solutions at all.
b) If X is considered as a complex number, i.e., $X \in \mathbb{C}$, then there exist two solutions:

$$X_{1,2} = -\frac{a}{2} \pm j\sqrt{\left|\frac{a^2}{4} - b\right|}, \quad j^2 = -1.$$

c) If X is considered as a quaternion, i.e., $X \in \mathcal{H}$, then it has the following representation:

$$X = x_0 + \mathbf{x}, \quad \mathbf{x} = \sum_{k=1}^{3} x_k \mathbf{i}_k, \quad x_k \in R, \; k = 0, \ldots, 3,$$

and we need to find the exact expressions for x_k ($k = 0, 1, 2, 3$).

In view of (2.112) we have

$$X^2 := XX = x_0^2 - \langle \mathbf{x}, \mathbf{x} \rangle + 2x_0 \mathbf{x},$$

which after substitution into (2.118) leads to

$$x_0^2 - \langle \mathbf{x}, \mathbf{x} \rangle + b + ax_0 + (2x_0 + a)\mathbf{x} = 0. \tag{2.119}$$

The relation (2.119) is satisfied if and only if the real and vectorial parts are equal to zero simultaneously, namely,

$$x_0^2 - \langle \mathbf{x}, \mathbf{x} \rangle + b + ax_0 = 0, \tag{2.120}$$
$$(2x_0 + a)\mathbf{x} = 0. \tag{2.121}$$

Since it has already been seen that if $\mathbf{x} = \mathbf{0}$ there are no solutions ($x = x_0 \in R$), from (2.121) it follows that

$$x_0 = -\frac{a}{2}.$$

This value, being substituted in (2.120), leads to

$$\langle \mathbf{x}, \mathbf{x} \rangle = -\frac{a^2}{4} + b, \tag{2.122}$$

which must be non-negative and which matches the condition given in (2.118). Expression (2.122) is a condition only on the magnitude of $\mathbf{x}$:

$$\|\mathbf{x}\| = \sqrt{b - \frac{a^2}{4}}.$$

So, all $\mathbf{x}$ are of the form

$$\mathbf{x} = \sqrt{b - \frac{a^2}{4}} \, \mathbf{e},$$

where $\mathbf{e}$ is any unitary vector ($\|\mathbf{e}\| = 1$). Then the set of quaternions that are solutions of (2.118) is given by

$$X = -\frac{a}{2} + \sqrt{b - \frac{a^2}{4}} \, \mathbf{e}.$$

Rigid body kinematics

This means that there are infinitely many solutions of the quadratic equation (2.118) in the quaternion space $\mathcal{H}$.

Quaternions as rotation operators

Below we illustrate one of the possible effective applications of quaternions.

Definition 2.17. It is said that a quaternion $\Lambda = \lambda_0 + \lambda$ is **normalized** if

$$\|\Lambda\|^2 = \lambda_0^2 + \sum_{j=1}^{3} \lambda_j^2 = 1.$$

We can observe that $\lambda_0^2 \leq 1$ and hence there exist ζ such that λ_0 can be represented as

$$\lambda_0 := \cos\frac{\zeta}{2},$$

and

$$\|\lambda\|^2 = \sum_{j=1}^{3} \lambda_j^2 = 1 - \lambda_0^2 = \sin^2\frac{\zeta}{2},$$

$$\bar{\Lambda} = \bar{\mathbf{e}} \cdot \sin\frac{\zeta}{2},$$

so that

$$\lambda = \bar{\mathbf{e}}\sin\frac{\zeta}{2}, \quad \|\bar{\mathbf{e}}\| = 1.$$

This finally gives the general representation of a normalized quaternion Λ_{norm}:

$$\Lambda_{norm} = \cos\frac{\zeta}{2} + \bar{\mathbf{e}}\sin\frac{\zeta}{2}. \tag{2.123}$$

Theorem 2.2. *Consider a vector* $\mathbf{r}$ *that has a common point O with a vector* $\bar{\mathbf{e}}$. *Then the transformation*

$$\left.\begin{array}{l}\mathbf{r}' = \Lambda_{norm} \circ \mathbf{r} \circ \Lambda_{norm}^* = \\ \left(\cos\dfrac{\zeta}{2} + \bar{\mathbf{e}}\sin\dfrac{\zeta}{2}\right) \circ \mathbf{r} \circ \left(\cos\dfrac{\zeta}{2} - \bar{\mathbf{e}}\sin\dfrac{\zeta}{2}\right) = \\ \mathbf{r}\cos\zeta + [\bar{\mathbf{e}}, \mathbf{r}]\sin\zeta + \bar{\mathbf{e}}(\bar{\mathbf{e}}, \mathbf{r})[1 - \cos\zeta]\end{array}\right\} \tag{2.124}$$

*is **a rotation of the vector** $\mathbf{r}$ with respect to the axis $\bar{\mathbf{e}}$ with the angle (see Fig. 2.21)*

$$\zeta = 2\arccos\lambda_0.$$

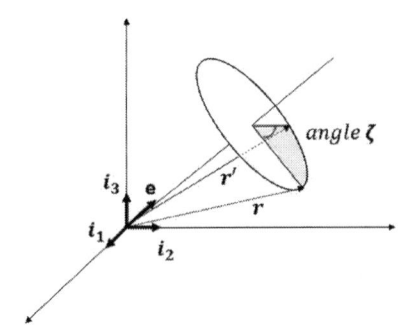

Figure 2.21 Quaternion as a rotation operator.

Proof. 1) First we have to show that (2.124) defines a rotation, that is, its norm is the same both before and after the application of the rotation. Indeed, using the rule (2.116) we get

$$\|\mathbf{r}'\| = \|\Lambda_{norm}\| \, \|\Lambda^*_{norm}\| \, \|\mathbf{r}\| = \|\mathbf{r}\|,$$

since $\|\Lambda_{norm}\|$ and $\|\Lambda^*_{norm}\|$ are normalized quaternions.

2) Using (2.112), let us calculate first

$$\left(\cos\frac{\zeta}{2} + \bar{\mathbf{e}}\sin\frac{\zeta}{2}\right) \circ \mathbf{r} = \mathbf{r}\cos\frac{\zeta}{2} - (\bar{\mathbf{e}}, \mathbf{r})\sin\frac{\zeta}{2} + [\bar{\mathbf{e}}, \mathbf{r}]\sin\frac{\zeta}{2}.$$

Then, by the same rule (2.112), it follows that

$$\Lambda_{norm} \circ \mathbf{r} \circ \Lambda^*_{norm} = \left(\cos\frac{\zeta}{2} + \bar{\mathbf{e}}\sin\frac{\zeta}{2}\right) \circ \mathbf{r} \circ \left(\cos\frac{\zeta}{2} - \bar{\mathbf{e}}\sin\frac{\zeta}{2}\right) =$$

$$\left[-(\bar{\mathbf{e}}, \mathbf{r})\sin\frac{\zeta}{2} + \left(\mathbf{r}\cos\frac{\zeta}{2} + [\bar{\mathbf{e}}, \mathbf{r}]\sin\frac{\zeta}{2}\right)\right] \circ \left[\cos\frac{\zeta}{2} + \left(-\bar{\mathbf{e}}\sin\frac{\zeta}{2}\right)\right]$$

$$= -(\bar{\mathbf{e}}, \mathbf{r})\cos\frac{\zeta}{2}\sin\frac{\zeta}{2} + \cos\frac{\zeta}{2}\left(\mathbf{r}\cos\frac{\zeta}{2} + [\bar{\mathbf{e}}, \mathbf{r}]\sin\frac{\zeta}{2}\right)$$

$$- (\bar{\mathbf{e}}, \mathbf{r})\sin\frac{\zeta}{2}\left[-\bar{\mathbf{e}}\sin\frac{\zeta}{2}\right] - \left(\mathbf{r}\cos\frac{\zeta}{2} + [\bar{\mathbf{e}}, \mathbf{r}]\sin\frac{\zeta}{2}, \left[-\bar{\mathbf{e}}\sin\frac{\zeta}{2}\right]\right)$$

$$+ \left[\left(\mathbf{r}\cos\frac{\zeta}{2} + [\bar{\mathbf{e}}, \mathbf{r}]\sin\frac{\zeta}{2}\right), \left(-\bar{\mathbf{e}}\sin\frac{\zeta}{2}\right)\right].$$

So,

$$\mathbf{r}' = -(\bar{\mathbf{e}}, \mathbf{r})\cos\frac{\zeta}{2}\sin\frac{\zeta}{2} + (\mathbf{r}, \bar{\mathbf{e}})\cos\frac{\zeta}{2}\sin\frac{\zeta}{2}$$

$$\cos\frac{\zeta}{2}\left(\mathbf{r}\cos\frac{\zeta}{2} + [\bar{\mathbf{e}}, \mathbf{r}]\sin\frac{\zeta}{2}\right) - (\bar{\mathbf{e}}, \mathbf{r})\sin\frac{\zeta}{2}\left[-\bar{\mathbf{e}}\sin\frac{\zeta}{2}\right]$$

Rigid body kinematics

$$-[\mathbf{r},\bar{\mathbf{e}}]\cos\frac{\zeta}{2}\sin\frac{\zeta}{2} - [[\bar{\mathbf{e}},\mathbf{r}],\bar{\mathbf{e}}]\sin^2\frac{\zeta}{2}$$

$$= \mathbf{r}\cos^2\frac{\zeta}{2} + [\bar{\mathbf{e}},\mathbf{r}]\cos\frac{\zeta}{2}\sin\frac{\zeta}{2} + \bar{\mathbf{e}}(\bar{\mathbf{e}},\mathbf{r})\sin^2\frac{\zeta}{2}$$

$$-[\mathbf{r},\bar{\mathbf{e}}]\cos\frac{\zeta}{2}\sin\frac{\zeta}{2} + \underbrace{[\bar{\mathbf{e}},[\bar{\mathbf{e}},\mathbf{r}]]}_{\bar{\mathbf{e}}(\bar{\mathbf{e}},r)-\mathbf{r}}\sin^2\frac{\zeta}{2} =$$

$$\mathbf{r}\left(\cos^2\frac{\zeta}{2} - \sin^2\frac{\zeta}{2}\right) + 2[\bar{\mathbf{e}},\mathbf{r}]\cos\frac{\zeta}{2}\sin\frac{\zeta}{2} + 2\bar{\mathbf{e}}(\bar{\mathbf{e}},\mathbf{r})\sin^2\frac{\zeta}{2}$$

$$= \mathbf{r}\cos\zeta + [\bar{\mathbf{e}},\mathbf{r}]\sin\zeta + \bar{\mathbf{e}}(\bar{\mathbf{e}},\mathbf{r})[1-\cos\zeta].$$

Here we have used the following trigonometric relations:

$$\cos^2\frac{\zeta}{2} - \sin^2\frac{\zeta}{2} = \cos\zeta,$$

$$\cos^2\frac{\zeta}{2} + \sin^2\frac{\zeta}{2} = 1,$$

$$2\sin^2\frac{\zeta}{2} = 1 - \cos\zeta.$$

So, finally we have (2.124).

Note that the square of the norm of the right-hand side in (2.124) is equal to $\|\mathbf{r}\|^2 = r^2$. Indeed, defining the angle $\theta := \widehat{(\bar{\mathbf{e}},\mathbf{r})}$, we have

$$\|\mathbf{r}\cos\zeta + [\bar{\mathbf{e}},\mathbf{r}]\sin\zeta + \bar{\mathbf{e}}(\bar{\mathbf{e}},\mathbf{r})[1-\cos\zeta]\|^2 / r^2 =$$

$$\cos^2\zeta + \sin^2\theta\sin^2\zeta + (1-\cos\zeta)^2\cos^2\theta + 2\cos^2\theta\cos\zeta[1-\cos\zeta] =$$

$$\cos^2\zeta + \sin^2\theta\sin^2\zeta + (1-\cos\zeta)^2\cos^2\theta + 2\cos^2\theta\cos\zeta[1-\cos\zeta] =$$

$$\cos^2\zeta + \sin^2\theta\sin^2\zeta + \left(1 - 2\cos\zeta + \cos^2\zeta\right)\cos^2\theta + 2\cos^2\theta\cos\zeta$$

$$-2\cos^2\theta\cos^2\zeta = \cos^2\zeta + \sin^2\theta\sin^2\zeta + \cos^2\theta - \cos^2\theta\cos^2\zeta =$$

$$\cos^2\zeta + \sin^2\theta\sin^2\zeta + \cos^2\theta\sin^2\zeta =$$

$$\cos^2\zeta + \left[\sin^2\theta + \cos^2\theta\right]\sin^2\zeta = 1.$$

3) To prove that this is a rotation around the direction $\bar{\mathbf{e}}$ we need to prove that the projections of $\mathbf{r}'$ and $\mathbf{r}$ to the vector $\bar{\mathbf{e}}$ are equal. In view of (2.124) it follows that

$$(\mathbf{r}',\bar{\mathbf{e}}) = (\bar{\mathbf{e}},r)\cos\zeta + \|\bar{\mathbf{e}}\|^2(\bar{\mathbf{e}},\mathbf{r})[1-\cos\zeta]$$
$$= (\bar{\mathbf{e}},r)\cos\zeta + (\bar{\mathbf{e}},\mathbf{r})[1-\cos\zeta] = (\bar{\mathbf{e}},r).$$

4) Finally we need to prove that the angle ζ' of rotation is ζ. For $\zeta \in [0, \pi/2]$ (see Fig. 2.21) it follows that

$$\cos\zeta' = \frac{(\mathbf{r} - h\bar{\mathbf{e}},r' - h\bar{\mathbf{e}})}{\rho^2} = \rho^{-2}\left[(\mathbf{r},\mathbf{r}') - h(\mathbf{e},\bar{r}+r') + h^2\right] =$$

$$\rho^{-2}\left[(\mathbf{r},\mathbf{r}') - h\left(\mathbf{e}, \bar{2r}\cos\zeta + [\bar{\mathbf{e}},\mathbf{r}]\sin\zeta + \bar{\mathbf{e}}\,(\bar{\mathbf{e}},\mathbf{r})\,[1-\cos\zeta]\right) + h^2\right] =$$
$$\rho^{-2}\left[r^2\cos\zeta - h2r\cos\zeta\cos\theta - hr\cos\theta\,[1-\cos\zeta] + h^2\right] =$$
$$\rho^{-2}\left[r^2\cos\zeta - hr\cos\zeta\cos\theta - hr\cos\theta + h^2\right] =$$
$$\rho^{-2}\left[r^2\cos\zeta - \cos\zeta\cos^2\theta - \cos^2\theta + \cos^2\theta\right] = \underbrace{[\frac{r}{\rho}\sin\theta]^2}_{1}\cos\zeta = \cos\zeta,$$

which means that

$$\zeta' = \zeta + \pi k, \quad k = \ldots, -1, 0, 1, \ldots.$$

For physical reasons only the unique value $k = 0$ makes sense, such that

$$\zeta' = \zeta.$$

□

Summary 2.2. From this fact one can see that the parameters of the normalized quaternions Λ_{norm} **coincide** with the Rodríguez–Hamilton parameters which define a rotation, that is,

$$\lambda_0 = \cos\frac{\zeta}{2}, \quad \lambda_1 = e_x \sin\frac{\zeta}{2}, \quad \lambda_2 = e_y \sin\frac{\zeta}{2}, \quad \lambda_3 = e_z \sin\frac{\zeta}{2},$$
$$e_x^2 + e_y^2 + e_z^2 = 1.$$

2.6.2 Composition or summation of rotations as a quaternion

Suppose that Λ_{norm} and $\mathbf{M}_{norm}$ are two normalized quaternions and that $\mathbf{r} \in R^3$. The notation

$$\mathbf{r} \stackrel{\Lambda_{norm}}{\rightarrow} \mathbf{r}' \stackrel{\mathbf{M}_{norm}}{\rightarrow} \mathbf{r}''$$

indicates that the quaternion given by Λ_{norm} is first applied on $\mathbf{r}$ and then the quaternion given by $\mathbf{M}_{norm}$ is applied to the obtained vector $\mathbf{r}''$. In other words,

$$\mathbf{r}' = \Lambda_{norm} \circ \mathbf{r} \circ \Lambda_{norm}^*,$$
$$\mathbf{r}'' = \mathbf{M}_{norm} \circ \mathbf{r}' \circ \mathbf{M}_{norm}^* = \mathbf{M}_{norm} \circ (\Lambda_{norm} \circ \mathbf{r} \circ \Lambda_{norm}^*) \circ \mathbf{M}_{norm}^*$$
$$= (\mathbf{M}_{norm} \circ \Lambda_{norm}) \circ \mathbf{r} \circ (\Lambda_{norm}^* \circ \mathbf{M}_{norm}^*).$$

In view of the relation

$$\Lambda^* \circ \mathbf{M}^* = (\mathbf{M} \circ \Lambda)^*$$

Rigid body kinematics

and defining

$$\mathbf{N}_{norm} := \mathbf{M}_{norm} \circ \mathbf{\Lambda}_{norm},$$

finally, we obtain

$$\mathbf{r}'' = \mathbf{N}_{norm} \circ \mathbf{r} \circ \mathbf{N}^*_{norm}.$$

Since $\mathbf{\Lambda}_{norm}$ and $\mathbf{M}_{norm}$ are normalized quaternions, it follows that

$$\|\mathbf{N}\| = \|\mathbf{M}\| \cdot \|\mathbf{\Lambda}\| = 1,$$

which means that $\mathbf{N}$ is also normalized, namely, $\mathbf{N} = \mathbf{N}_{norm}$.

If there are K rotations, that is, $\mathbf{\Lambda}_{norm,i}$ ($i = 1, .., K$), then the final or summary effect can be represented as only one rotation, given by

$$\mathbf{N}_{norm} = \mathbf{\Lambda}_{norm,K} \circ \mathbf{\Lambda}_{norm,K-1} \circ \mathbf{\Lambda}_{norm,K-2} \circ \cdots \circ \mathbf{\Lambda}_{norm,1}.$$

It should be mentioned that all quaternions are given in the same initial coordinate system.

Quaternion as a transformation of a coordinate system

Sometimes it occurs that two quaternions are given in different basis systems. This provokes a problem in the analysis of the resulting two-step rotation. We will show now how to avoid this trouble.

Now let us consider each quaternion as a transformation of the coordinate system

$$\left.\begin{aligned}\mathbf{i}'_j &= \mathbf{\Lambda}_{norm} \circ \mathbf{i}_j \circ \mathbf{\Lambda}^*_{norm} \quad (j = 1, 2, 3), \\ \mathbf{i}_j &= \mathbf{\Lambda}^*_{norm} \circ \mathbf{i}'_j \circ \mathbf{\Lambda}_{norm}.\end{aligned}\right\} \quad (2.125)$$

Hence, we can represent the vector $\mathbf{r}$ as

$$\mathbf{r} = \lambda_1 \mathbf{i}_1 + \lambda_2 \mathbf{i}_2 + \lambda_3 \mathbf{i}_3 = \mathbf{\Lambda}^*_{norm} \circ (\lambda_1 \mathbf{i}'_1 + \lambda_2 \mathbf{i}'_2 + \lambda_3 \mathbf{i}'_3) \circ \mathbf{\Lambda}_{norm}.$$

In this reduced form we can represent the expression above as

$$\mathbf{r}_{(i'_1, i'_2, i'_3)} = \mathbf{\Lambda}_{norm} \circ \mathbf{r}_{(i_1, i_2, i_3)} \circ \mathbf{\Lambda}^*_{norm},$$

where $\mathbf{r}_{(i'_1, i'_2, i'_3)}$ is the vector $\mathbf{r} := \mathbf{r}_{(i_1, i_2, i_3)}$ given in a new coordinate system $(\mathbf{i}'_1, \mathbf{i}'_2, \mathbf{i}'_3)$. Let now the quaternion $\mathbf{\Lambda}_{1,norm}$ be given in the basis (i_1, i_2, i_3) and let the quaternion $\mathbf{\Lambda}_{2,norm,(i'_1, i'_2, i'_3)}$ be defined in the basis (i'_1, i'_2, i'_3). To present the second quaternion in the original basis (i_1, i_2, i_3) let us use formula (2.125):

$$\left.\begin{array}{l}\boldsymbol{\Lambda}_{2,norm,(i_1',i_2',i_3')} = \lambda_{2,0} + \sum_{j=1}^{3}\lambda_{2,j}\mathbf{i}_j' = \\ \lambda_{2,0} + \sum_{j=1}^{3}\lambda_{2,j}\boldsymbol{\Lambda}_{1,norm,(i_1,i_2,i_3)} \circ \mathbf{i}_j \circ \boldsymbol{\Lambda}_{1,norm,(i_1,i_2,i_3)}^{*} = \\ \boldsymbol{\Lambda}_{1,norm,(i_1,i_2,i_3)} \circ \left(\lambda_{2,0} + \sum_{j=1}^{3}\lambda_{2,j}\mathbf{i}_j\right) \circ \boldsymbol{\Lambda}_{1,norm,(i_1,i_2,i_3)}^{*} = \\ \boldsymbol{\Lambda}_{1,norm,(i_1,i_2,i_3)} \circ \boldsymbol{\Lambda}_{2,norm,(i_1,i_2,i_3)} \circ \boldsymbol{\Lambda}_{1,norm,(i_1,i_2,i_3)}^{*}.\end{array}\right\} \quad (2.126)$$

Remark 2.9. It is important to note that the coordinates $\lambda_{2,j}$ ($j = 0, 1, .., 3$) of the quaternions $\boldsymbol{\Lambda}_{2,norm,(i_1',i_2',i_3')}$ in the basis (i_1', i_2', i_3') are the same as in the basis i_1, i_2, i_3).

Lemma 2.9. *If*

$$\mathbf{N} = \boldsymbol{\Lambda}_{2,norm,(i_1',i_2',i_3')} \circ \boldsymbol{\Lambda}_{1,norm,(i_1,i_2,i_3)}, \quad (2.127)$$

that is, $\mathbf{N}$ is a composition of two quaternions given in different basic coordinates, then the same quaternion $\mathbf{N}_{(i_1,i_2,i_3)}$ given in the initial basis can be expressed as

$$\mathbf{N}_{(i_1,i_2,i_3)} = \boldsymbol{\Lambda}_{1,norm,(i_1,i_2,i_3)} \circ \boldsymbol{\Lambda}_{2,norm,(i_1,i_2,i_3)}.$$

Proof. Substituting the relation (2.126) into (2.127) gives

$$\mathbf{N} = \boldsymbol{\Lambda}_{2,norm,(i_1',i_2',i_3')} \circ \boldsymbol{\Lambda}_{1,norm,(i_1,i_2,i_3)} = $$
$$\left(\boldsymbol{\Lambda}_{1,norm,(i_1,i_2,i_3)} \circ \boldsymbol{\Lambda}_{2,norm,(i_1,i_2,i_3)} \circ \boldsymbol{\Lambda}_{1,norm,(i_1,i_2,i_3)}^{*}\right) \circ \boldsymbol{\Lambda}_{1,norm,(i_1,i_2,i_3)} = $$
$$\boldsymbol{\Lambda}_{1,norm,(i_1,i_2,i_3)} \circ \boldsymbol{\Lambda}_{2,norm,(i_1,i_2,i_3)} := \mathbf{N}_{(i_1,i_2,i_3)}.$$

□

Definition 2.18. A quaternion $\boldsymbol{\Lambda}_{2,norm,(i_1,i_2,i_3)}$ is called **proper** with respect to the quaternion $\boldsymbol{\Lambda}_{2,norm,(i_1',i_2',i_3')}$.

Summary 2.3. We may conclude that if $\mathbf{N}$ is a composition of several quaternions given in different sequential basic coordinates

$$\mathbf{N} = \boldsymbol{\Lambda}_{K,(i_1^{(K)},i_2^{(K)},i_3^{(K)})} \circ \boldsymbol{\Lambda}_{K-1,(i_1^{(K-1)},i_2^{(K-1)},i_3^{(K-1)})} \circ \ldots \circ \boldsymbol{\Lambda}_{1(i_1,i_2,i_3)},$$

then

$$\mathbf{N}_{(i_1,i_2,i_3)} = \boldsymbol{\Lambda}_{1(i_1,i_2,i_3)} \circ \boldsymbol{\Lambda}_{K-2,(i_1,i_2,i_3)} \circ \ldots \circ \boldsymbol{\Lambda}_{K,(i_1,i_2,i_3)}. \quad (2.128)$$

Rigid body kinematics

Example 2.11. Let two rotations apply to the vector **r**: the first with respect to $\mathbf{i}_2$ on the angle φ, and the second with respect to $\mathbf{i}_3$ on the angle θ. Let us obtain the final position vector $\mathbf{r}''$, using the quaternion composition approach. We have

$$\bar{r} = a\mathbf{i}_1 - a\mathbf{i}_2, \quad a = r\frac{\sqrt{2}}{2}.$$

Our rotations are realized by

$$\Lambda_1 = \cos\frac{\varphi}{2} + \mathbf{i}_2 \sin\frac{\varphi}{2},$$
$$\Lambda_2 = \cos\frac{\theta}{2} + \mathbf{i}'_3 \sin\frac{\theta}{2}.$$

Then, the composition of rotations is given by

$$\mathbf{N} := \Lambda_2 \circ \Lambda_1 = (\cos\frac{\theta}{2} + \mathbf{i}'_3 \sin\frac{\theta}{2}) \circ (\cos\frac{\varphi}{2} + \mathbf{i}_2 \sin\frac{\varphi}{2}) =$$

$$\Lambda_{1,norm,(i_1,i_2,i_3)} \circ \Lambda_{2,norm,(i_1,i_2,i_3)} = (\cos\frac{\varphi}{2} + \mathbf{i}_2 \sin\frac{\varphi}{2}) \circ (\cos\frac{\theta}{2} + \mathbf{i}_3 \sin\frac{\theta}{2}) =$$

$$\cos\frac{\varphi}{2}\cos\frac{\theta}{2} + \mathbf{i}_2 \sin\frac{\varphi}{2}\cos\frac{\theta}{2} + \mathbf{i}_3 \cos\frac{\varphi}{2}\sin\frac{\theta}{2} + \underbrace{\mathbf{i}_2\mathbf{i}_3}_{\mathbf{i}_1} \sin\frac{\varphi}{2}\sin\frac{\theta}{2}.$$

To simplify the obtained expressions let us define

$$n_0 := \cos\frac{\varphi}{2}\cos\frac{\theta}{2}, \quad n_1 := \sin\frac{\varphi}{2}\cos\frac{\theta}{2},$$
$$n_2 := \cos\frac{\varphi}{2}\sin\frac{\theta}{2}, \quad n_3 := \sin\frac{\varphi}{2}\sin\frac{\theta}{2}.$$

Since the coordinates coincide we can eliminate the apostrophe, which finally gives

$$\mathbf{r}'' = \mathbf{N} \circ \mathbf{r} \circ \mathbf{N}^* =$$
$$a(n_0 + n_1\mathbf{i}_1 + n_2\mathbf{i}_2 + n_3\mathbf{i}_3) \circ (\mathbf{i}_1 - \mathbf{i}_2) \circ (n_0 - n_1\mathbf{i}_1 - n_2\mathbf{i}_2 - n_3\mathbf{i}_3) =$$
$$a(n_0 + n_1\mathbf{i}_1 + n_2\mathbf{i}_2 + n_3\mathbf{i}_3) \circ$$
$$([n_1 - n_2] + n_0(\mathbf{i}_1 - \mathbf{i}_2) - n_1\mathbf{i}_3 - n_2\mathbf{i}_3 + n_3\mathbf{i}_2 + n_3\mathbf{i}_1) =$$
$$a(n_0 + n_1\mathbf{i}_1 + n_2\mathbf{i}_2 + n_3\mathbf{i}_3) \circ$$
$$([n_1 - n_2] + [n_0 + n_3]\mathbf{i}_1 + [n_3 - n_0]\mathbf{i}_2 - [n_2 + n_1]\mathbf{i}_3) =$$
$$an_0[n_1 - n_2] + an_0([n_0 + n_3]\mathbf{i}_1 + [n_3 - n_0]\mathbf{i}_2 - [n_2 + n_1]\mathbf{i}_3) +$$
$$a[n_1 - n_2](n_1\mathbf{i}_1 + n_2\mathbf{i}_2 + n_3\mathbf{i}_3) -$$
$$-an_1[n_0 + n_3] - an_2[n_3 - n_0] + an_3[n_2 + n_1] =$$
$$a(n_1[n_1 - n_2] + n_0[n_3 + n_0])\mathbf{i}_1 +$$
$$a(n_0[n_3 - n_0] + n_2[n_1 - n_2])\mathbf{i}_2 +$$
$$a(n_3[n_1 - n_2] - n_0[n_2 + n_1])\mathbf{i}_3.$$

By performing operations and rearranging terms with respect to $(\mathbf{i}_1, \mathbf{i}_2, \mathbf{i}_3)$, we finally obtain

$$\mathbf{r}'' = \lambda_0 + \lambda_1 \mathbf{i}_1 + \lambda_2 \mathbf{i}_2 + \lambda_3 \mathbf{i}_3,$$

where

$$\lambda_0 = 0,$$
$$\lambda_1 = a(n_1[n_1 - n_2] + n_0[n_3 + n_0]),$$
$$\lambda_2 = a(n_0[n_3 - n_0] + n_2[n_1 - n_2]),$$
$$\lambda_3 = a(n_3[n_1 - n_2] - n_0[n_2 + n_1]).$$

2.7 Differential kinematic equations (DKEs)

2.7.1 DKEs in Euler coordinates

If the quaternion $\mathbf{A}$, corresponding to Euler's description of a rotation, is denoted by the sequential rotations Ψ, Θ, and Φ, respectively, then by (2.128) we have

$$\mathbf{A} = \Phi \Theta \Psi, \quad \mathbf{A}^* = \Psi^* \Theta^* \Phi^*,$$

where (in view of (2.71)–(2.73))

$$\Psi^* := \begin{bmatrix} \cos\psi & \sin\psi & 0 \\ -\sin\psi & \cos\psi & 0 \\ 0 & 0 & 1 \end{bmatrix}, \quad \Theta^* := \begin{bmatrix} 1 & 0 & 0 \\ 0 & \cos\theta & \sin\theta \\ 0 & -\sin\theta & \cos\theta \end{bmatrix},$$

$$\Phi^* := \begin{bmatrix} \cos\phi & 0 & -\sin\phi \\ 0 & 1 & 0 \\ \sin\phi & 0 & \cos\phi \end{bmatrix}.$$

This implies

$$\mathbf{A}^* = \begin{bmatrix} \cos\phi\cos\psi + \sin\phi\sin\theta\sin\psi & \cos\theta\sin\psi & -\sin\phi\cos\psi + \cos\phi\sin\theta\sin\psi \\ -\cos\phi\sin\psi + \sin\phi\sin\theta\cos\psi & \cos\theta\cos\psi & \sin\phi\sin\psi + \cos\phi\sin\theta\cos\psi \\ \sin\phi\cos\theta & -\sin\theta & \cos\phi\cos\theta \end{bmatrix}.$$

Let us define the dynamics of the Euler angles in the following way:

$$\dot{\Psi}_{(x,y,z)} = \begin{pmatrix} 0 \\ 0 \\ \dot{\psi} \end{pmatrix}, \quad \dot{\Theta}_{(x',y',z')} = \begin{pmatrix} 0 \\ 0 \\ \dot{\theta} \end{pmatrix}, \quad \dot{\Phi}_{(x'',y'',z'')} = \begin{pmatrix} 0 \\ \dot{\phi} \\ 0 \end{pmatrix}.$$

This gives

$$\bar{\omega} = \Phi^\top \Theta^\top \dot{\Psi}_{(x,y,z)} + \Phi^\top \dot{\Theta}_{(x',y',z')} + \dot{\Phi}_{(x'',y'',z'')}.$$

Define the components of $\bar{\omega}$ as

$$\bar{\omega} = \begin{pmatrix} p \\ q \\ r \end{pmatrix},$$

where

$$\left. \begin{array}{l} p = \dot{\psi}\sin\theta\sin\varphi + \dot{\theta}\cos\varphi, \\ q = \dot{\psi}\sin\theta\cos\varphi + \dot{\theta}\sin\varphi, \\ r = \dot{\psi}\cos\theta + \dot{\varphi}, \end{array} \right\}$$

from which it follows that

$$\left. \begin{array}{l} \dot{\psi} = (p\sin\varphi + q\cos\varphi)\dfrac{1}{\sin\theta}, \\ \dot{\theta} = p\cos\varphi - q\sin\varphi, \\ \dot{\varphi} = -(p\sin\varphi + q\cos\varphi)\dfrac{1}{\cot\theta} + r. \end{array} \right\} \quad (2.129)$$

Eqs. (2.129) are referred to as the *differential kinematic equations (DKEs) in Euler angles*.

2.7.2 DKEs in quaternions: Poisson equation

For a rotation, given by the quaternion $\mathbf{\Lambda}(t)$, define its derivative as

$$\dot{\mathbf{\Lambda}}(t) := \lim_{\Delta t \to 0} \frac{\mathbf{\Lambda}(t + \Delta t) - \mathbf{\Lambda}(t)}{\Delta t}.$$

We can express the term $\mathbf{\Lambda}(t + \Delta t)$ as a composition of rotations, that is, the first rotation $\mathbf{\Lambda}(t) = \lambda_0(t) + \boldsymbol{\lambda}(t)$ is at time t, followed by application of the rotation $\delta\mathbf{\Lambda}(\Delta t)$ with respect to a new basis, realized at time Δt:

$$\mathbf{\Lambda}(t + \Delta t) = \delta\mathbf{\Lambda}(\Delta t) \circ \mathbf{\Lambda}(t).$$

Here

$$\delta\mathbf{\Lambda}(\Delta t) = \cos(\frac{\omega(t)\Delta t}{2}) + \bar{e}_{\bar{\omega}}\sin(\frac{\omega(t)\Delta t}{2})$$
$$= 1 + \bar{e}_{\bar{\omega}}\frac{\omega(t)}{2}\Delta t + o(\Delta t),$$

from which it follows that

$$\mathbf{\Lambda}(t + \Delta t) = (1 + \bar{e}_{\bar{\omega}}\frac{\omega(t)}{2}\Delta t + o(\Delta t)) \circ \mathbf{\Lambda}(t)$$
$$= \mathbf{\Lambda}(t) + \frac{\bar{\omega}(t) \circ \mathbf{\Lambda}(t)}{2}\Delta t + o(\Delta t).$$

Reordering terms and dividing by Δt, we obtain

$$\frac{\Lambda(t+\Delta t)-\Lambda(t)}{\Delta t}=\frac{\bar{\omega}(t)\circ\Lambda(t)}{2}+\frac{o(\Delta t)}{\Delta t}.$$

Taking $\Delta t \to 0$, we finally get

$$\dot{\Lambda}(t)=\frac{1}{2}\bar{\omega}(t)\circ\Lambda(t)=\frac{1}{2}\Lambda(t)\circ\bar{\omega}_{(x',y',z')}(t). \tag{2.130}$$

This differential equation for quaternion dynamics is known as the *Poisson equation*. If

$$\bar{\omega}(t)=p(t)\mathbf{i}_1+q(t)\mathbf{i}_2+r(t)\mathbf{i}_3,$$

then (since the components are the same)

$$\bar{\omega}_{(x',y',z')}(t)=p(t)\mathbf{i}'_1+q(t)\mathbf{i}'_2+r(t)\mathbf{i}'_3.$$

Substitution the last formula in (2.130) gives the representation of the Poisson equation in the "*open format*":

$$\left.\begin{array}{l}2\dot{\lambda}_0(t)=-p(t)\lambda_1(t)-q(t)\lambda_2(t)-r(t)\lambda_3(t),\\ 2\dot{\lambda}_1(t)=p(t)\lambda_0(t)+r(t)\lambda_2(t)-q(t)\lambda_3(t),\\ 2\dot{\lambda}_2(t)=q(t)\lambda_0(t)-r(t)\lambda_1(t)+p(t)\lambda_3(t),\\ 2\dot{\lambda}_3(t)=r(t)\lambda_0(t)+q(t)\lambda_1(t)-p(t)\lambda_2(t).\end{array}\right\} \tag{2.131}$$

Summary 2.4. From the above developments we can conclude that:
- the vector Poisson equation, given in quaternions (Rodriguez–Hamilton parameters), is **linear**, but with time-varying parameters $\bar{\omega}(t)$,
- the dynamics (2.129), given in the Euler angle, is extremely **nonlinear** and may have the non-singularity problem when $\sin\theta = 0$ or $\cot\theta = 0$.

2.8 Exercises

Exercise 2.1. A disk, mounted at right angles to the OC rod, rotates around the OC with a constant angular velocity ω_1 (see Fig. 2.22). The rod, in turn, performs harmonic oscillations in the vertical plane XY according to the law

$$\varphi(t)=\varphi_0\sin(\omega_0 t).$$

Show that the time dependence of the angular velocity ω of the disk and the angular acceleration ε are described by the formulas

$$\omega(t)=\sqrt{\omega_1^2+\varphi_0^2\omega_0^2\cos^2(\omega_0 t)},$$

Rigid body kinematics

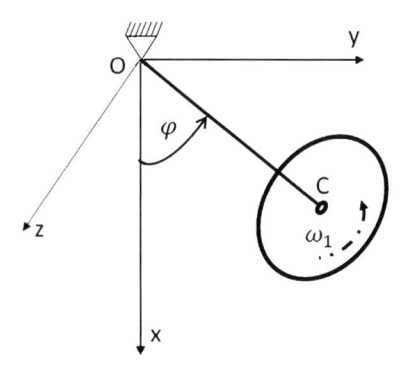

Figure 2.22 The disk, mounted at right angles to the rod.

$$\varepsilon(t) = \varphi_0 \omega_0 \sqrt{\omega_0^2 + (\omega_1^2 - \omega_0^2) \cos^2(\omega_0 t)}.$$

Exercise 2.2. The "Segner wheel" rotates with an angular acceleration of ε, currently having an angular velocity of ω (see Fig. 2.23).

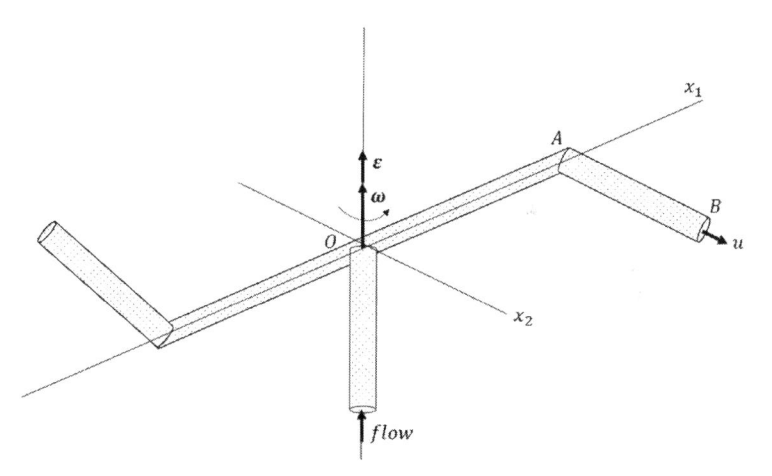

Figure 2.23 The "Segner wheel."

The relative velocity of the outflow of fluid particles is equal to $u = \text{const}$. Show that the absolute velocity and acceleration of fluid particles in the output section B are equal to

$$\mathbf{v} = \begin{bmatrix} u - \omega a \\ \omega a \end{bmatrix}, \quad \mathbf{w} = \begin{bmatrix} -(\omega^2 + \varepsilon) a \\ 2\omega u + \varepsilon a - \omega^2 a \end{bmatrix},$$

assuming that
$$OA = AB = a, \angle OAB = \pi/2.$$

Exercise 2.3. The ends A and B of the rod move along two mutually perpendicular straight lines OX and OY (see Fig. 2.24). The speed of point A is constant. Show that the acceleration of any point of the rod is always orthogonal to the axis OY and varies inversely with the cube of the distance of this point from this axis, that is, show that

$$w_P \sim \frac{1}{x_P^3}.$$

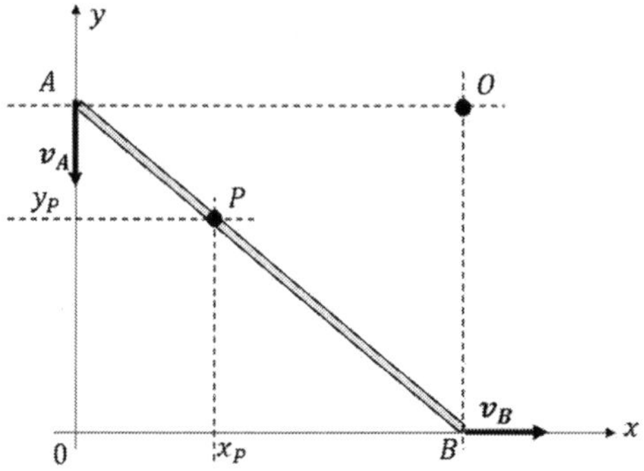

Figure 2.24 The rod moving along two mutually perpendicular straight lines.

Exercise 2.4. Solve the quaternion equation with respect to quaternion X:

a)
$$X \circ \Lambda = M,$$

where the quaternions

$$\Lambda = \lambda_0 + \boldsymbol{\lambda} = (1, 0, 1, -1), \quad M = \mu_0 + \boldsymbol{\mu} = (1, 1, -1, 1)$$

are given;

b)
$$\Lambda \circ X^2 = X \circ \Lambda.$$

Exercise 2.5. A rigid body realizes the rotation (a regular precision) with $\omega_1 = $ const (the rotation with respect to its main axis of symmetry) and $\omega_2 = \overline{\text{const}}$ (the rotation with respect to the axis z), keeping the constant angle $\theta = \widehat{(\omega_1, \omega_2)}$. We need to describe the movement of this body in the Rodriguez–Hamilton parameters using the Poisson equation (2.131).

Dynamics

Contents

3.1 Main dynamics characteristics   90
     3.1.1 System of material points   90
     3.1.2 Three main dynamics characteristics   91
3.2 Axioms or Newton's laws   91
     3.2.1 Newton's axioms   91
     3.2.2 Expression for $\dot{\mathbf{Q}}$ 92
     3.2.3 Expression for $\dot{\mathbf{K}}_A$ 94
3.3 Force work and potential forces   95
     3.3.1 Elementary and total force work   96
     3.3.2 Potential forces   96
     3.3.3 Force power and expression for $\dot{T}$ 98
     3.3.4 Conservative systems   99
3.4 Virial of a system   100
     3.4.1 Main definition of virial   100
     3.4.2 Virial for homogeneous potential energies   101
3.5 Properties of the center of mass   103
     3.5.1 Dynamics of the center of inertia (mass)   103
3.6 "King/König/Rey" theorem   103
     3.6.1 Principle theorem   103
     3.6.2 Moment of inertia and the impulse moment with respect to a pivot   105
     3.6.3 A rigid flat body rotating in the same plane   106
     3.6.4 Calculation of moments of inertia for different rigid bodies   108
     3.6.5 König theorem application   112
     3.6.6 Steiner's theorem on the inertia moment   114
3.7 Movements with friction   123
3.8 Exercises   127

The motivation of this chapter is the study of the existing relationship between the kinematic quantities of a system's points and their causes, that is, the forces. This study leads to the introduction of important concepts, such as kinetic energy, momentum, moment of an impulse, moment of a force, etc., which will allow to obtain very important results discussed below in detail. König's and Steiner's theorems are discussed. Friction effects are considered.

Classical and Analytical Mechanics. https://doi.org/10.1016/B978-0-32-389816-4.00014-4
Copyright © 2021 Elsevier Inc. All rights reserved.

3.1 Main dynamics characteristics

3.1.1 System of material points

Consider a set of mobile material points located in space. Now consider such a space of a coordinate system originating in some point O. The obtained construction, denoted by S, will be called a system of *material points*. For the particle $i \in S$, m_i represents its mass, while $\mathbf{r}_i$ and $\mathbf{v}_i$ denote its position vector and its velocity with respect to O. Fig. 3.1 shows these aspects.

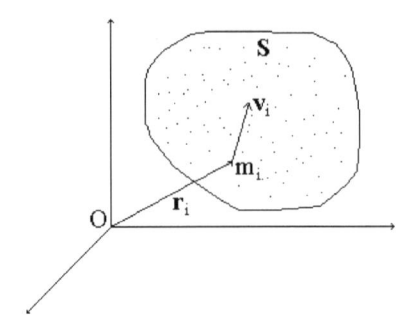

Figure 3.1 A set of material points referring to a coordinate system.

Definition 3.1. A **pole** is a point A in space. If $\mathbf{r}_A$ is the position vector of the point A with respect to O, the vector

$$\mathbf{r}_{i,A} := \mathbf{r}_i - \mathbf{r}_A \tag{3.1}$$

is known as the **position vector** of point $i \in S$ with respect to pole A (see Fig. 3.2).

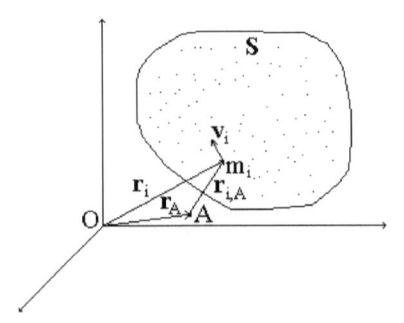

Figure 3.2 Relationship between the pole A and the origin O.

Dynamics

3.1.2 Three main dynamics characteristics

This chapter is heavily based on the concepts of kinetic energy, momentum, and impulse momentum of a system of material points: the first of them is **scalar** and the other two are **vectors**. The definitions of these concepts are presented below.

Definition 3.2. Consider the system of material points S.

1. The scalar quantity

$$T := \frac{1}{2} \sum_{i \in S} m_i v_i^2 \tag{3.2}$$

is called **the kinetic energy** of S. Here $v_i := \|\mathbf{v}_i\|$ is the norm of the velocity $\mathbf{v}_i$ of the point $i \in S$.

2. The vector quantity

$$\mathbf{Q} := \sum_{i \in S} m_i \mathbf{v}_i \tag{3.3}$$

is called the **impulse** of S.

3. The vector quantity

$$\mathbf{K}_A := \sum_{i \in S} \left[\mathbf{r}_{i,A}, m_i \mathbf{v}_i \right] \tag{3.4}$$

is known as the **moment of the impulse** of S with respect to pole A.

In the following paragraphs of this chapter we will obtain the main dynamics law of mechanics, namely, we will try to get the exact expressions for $\dot{\mathbf{Q}}$, $\dot{\mathbf{K}}_A$, and $\dot{T}$.

3.2 Axioms or Newton's laws

In addition to the concepts previously introduced, in the formal construction addressed in this text Newton's laws are presented in the form of the following axioms.

3.2.1 Newton's axioms

Axiom 3.1 (Newton's first law). *Every material particle not subject to any external stimulus, can only move with uniform rectilinear speed or remain at rest.*

Axiom 3.2 (Second law of Newton). *By the definition, the total force* $\mathbf{F}$ *is*

$$\mathbf{F} = \dot{\mathbf{Q}}. \tag{3.5}$$

Denoting by $\mathbf{F}_i$ the force exerted on the particle $i \in S$ one can represent $\mathbf{F}$ as a total force acting in the system S, that is,

$$\mathbf{F} := \sum_{i \in S} \mathbf{F}_i = \sum_{i \in S} m_i \dot{\mathbf{v}}_i. \tag{3.6}$$

Axiom 3.3 (Newton's third law). *Given $i, j \in S$, we denote by $\mathbf{F}_{ij}$ the force exerted on the particle i by the particle $j \neq i$. Then*

$$\mathbf{F}_{ij} = -\mathbf{F}_{ji}.$$

The vector quantity $\mathbf{F}$, defined in (3.6), as we have already mentioned, is the **total force** acting on the system S. It can be represented as

$$\mathbf{F} = \mathbf{F}_{ext} + \mathbf{F}_{int},$$

where $\mathbf{F}_{ext}$ is the force exerted on S by **external** agents and

$$\mathbf{F}_{int} := \sum_{i \in S} \sum_{\substack{j \in S \\ i \neq j}} \mathbf{F}_{ij} \tag{3.7}$$

is the net of **internal forces** that results from the actions of the particles among themselves.

Lemma 3.1. *In any system S of material points*

$$\mathbf{F}_{int} = 0.$$

Proof. By (3.7) we have

$$\mathbf{F}_{int} = (\mathbf{F}_{12} + \mathbf{F}_{13} + \cdots) + (\mathbf{F}_{21} + \mathbf{F}_{23} + \cdots) + \cdots =$$
$$(\mathbf{F}_{12} + \mathbf{F}_{21}) + (\mathbf{F}_{13} + \mathbf{F}_{31}) \cdots + (\mathbf{F}_{ij} + \mathbf{F}_{ji}) + \cdots$$

and in view of Newton's third law we get the desired result. □

3.2.2 Expression for $\dot{\mathbf{Q}}$

By the previous lemma, Eq. (3.5) is reduced to

$$\dot{\mathbf{Q}} = \mathbf{F}_{ext}. \tag{3.8}$$

Definition 3.3. Whereas the mass of a system S of material points is given by

$$M := \sum_{i \in S} m_i, \tag{3.9}$$

the point CI with position vector

$$\mathbf{r}_{CI} := \frac{1}{M} \sum_{i \in S} m_i \mathbf{r}_i \qquad (3.10)$$

is called the **center of mass** or the **inertial center** of S.

Remark 3.1. From the definition of center of mass (3.10), the expression of the velocity of this point with respect to the origin O is immediately obtained: the derivation of (3.10) results in

$$\mathbf{v}_{CI} := \frac{1}{M} \sum_{i \in S} m_i \dot{\mathbf{r}}_i = \frac{1}{M} \sum_{i \in S} m_i \mathbf{v}_i. \qquad (3.11)$$

Definition 3.4. If $\mathbf{F}_i$ represents the action on the material point $i \in S$, the vector quantity

$$\mathbf{M}_{\mathbf{F}_A} := \sum_{i \in S} \left[\mathbf{r}_{i,A}, \mathbf{F}_i \right] \qquad (3.12)$$

is called the **moment of forces** with respect to pole A.

Lemma 3.2. *Only external forces contribute to the moment of the forces (3.12), that is,*

$$\mathbf{M}_{\mathbf{F}_A} = \mathbf{M}_{\mathbf{F}_{ext,A}} := \sum_{i \in S} \left[\mathbf{r}_{i,A}, \mathbf{F}_{i,ext} \right]. \qquad (3.13)$$

Proof. Since

$$\mathbf{F}_i = \mathbf{F}_{i,ext} + \mathbf{F}_{i,int},$$

expression (3.12) can be rewritten as

$$\mathbf{M}_{\mathbf{F}_A} = \mathbf{M}_{\mathbf{F}_{ext,A}} + \mathbf{M}_{\mathbf{F}_{int,A}},$$

where

$$\mathbf{M}_{\mathbf{F}_{int,A}} := \sum_{i \in S} \left[\mathbf{r}_{i,A}, \mathbf{F}_{i,int} \right].$$

But in view of the identities

$$\mathbf{M}_{\mathbf{F}_{int,A}} = \sum_{i \in S} \left[\mathbf{r}_{i,A}, \mathbf{F}_{i,int} \right] = \sum_{i \in S} \left[\mathbf{r}_{i,A}, \sum_{j \in S,\, j \neq i} \mathbf{F}_{ij} \right] = \sum_{i \in S} \sum_{j \in S,\, j \neq i} \left[\mathbf{r}_{i,A}, \mathbf{F}_{ij} \right],$$

$$\mathbf{M}_{\mathbf{F}_{int,A}} = \sum_{j \in S} \left[\mathbf{r}_{j,A}, \mathbf{F}_{j,int} \right] = \sum_{j \in S} \left[\mathbf{r}_{j,A}, \sum_{i \in S,\, i \neq j} \mathbf{F}_{ji} \right] = \sum_{\substack{j \in S \\ i \neq j}} \sum_{i \in S} \left[\mathbf{r}_{j,A}, \mathbf{F}_{ji} \right],$$

summation of both identities and using Newton's third law for $i \neq j$ we get

$$2\mathbf{M_{F_{int,A}}} = \sum_{i \in S} \sum_{\substack{j \in S \\ i \neq j}} [\mathbf{r}_{i,A}, \mathbf{F}_{ij}] + \sum_{j \in S} \sum_{\substack{i \in S \\ i \neq j}} [\mathbf{r}_{j,A}, \mathbf{F}_{ji}] =$$

$$\sum_{i \in S} \sum_{\substack{j \in S \\ i \neq j}} ([\mathbf{r}_{i,A}, \mathbf{F}_{ij}] + [\mathbf{r}_{j,A}, \mathbf{F}_{ji}]) = \sum_{i \in S} \sum_{\substack{j \in S \\ i \neq j}} ([\mathbf{r}_{i,A}, \mathbf{F}_{ij}] + [\mathbf{r}_{j,A}, -\mathbf{F}_{ij}]) =$$

$$\sum_{i \in S} \sum_{\substack{j \in S \\ i \neq j}} [\mathbf{r}_{i,A} - \mathbf{r}_{j,A}, \mathbf{F}_{ij}] = 0.$$

Here we have used the fact that the vector $\mathbf{r}_{i,A} - \mathbf{r}_{j,A}$ is always parallel to the vector $\mathbf{F}_{ij}$. So, $\mathbf{M_{F_{int,A}}} = 0$. The lemma is proven. □

In view of this result, it follows that

$$\mathbf{M_{F_A}} = \mathbf{M_{F_{ext,A}}}. \tag{3.14}$$

3.2.3 Expression for $\dot{\mathbf{K}}_A$

The following is one of the key results in dynamics.

Theorem 3.1 (Rizal's formula). *In a system of material points with constant masses*

$$\dot{\mathbf{K}}_A = \mathbf{M_{F_{ext,A}}} + M[\mathbf{v}_{CI}, \mathbf{v}_A], \tag{3.15}$$

where $\mathbf{v}_{CI}$ and $\mathbf{v}_A$ denote, respectively, the velocities of the center of mass CI and the velocity of the pole A with respect to the origin O.

Proof. By (3.1), the definition (3.4) can be represented as

$$\mathbf{K}_A = \sum_{i \in S} [\mathbf{r}_i - \mathbf{r}_A, m_i \mathbf{v}_i].$$

Deriving this expression under the consideration that m_i is constant for all $i \in S$, we obtain

$$\left. \begin{array}{l} \dot{\mathbf{K}}_A = \sum_{i \in S} \dfrac{d}{dt}[\mathbf{r}_i - \mathbf{r}_A, m_i \mathbf{v}_i] = \sum_{i \in S} [\mathbf{v}_i - \mathbf{v}_A, m_i \mathbf{v}_i] + \\[2mm] \sum_{i \in S} [\mathbf{r}_i - \mathbf{r}_A, m_i \dot{\mathbf{v}}_i] = -\left[\mathbf{v}_A, \sum_{i \in S} m_i \mathbf{v}_i\right] + \sum_{i \in S} [\mathbf{r}_{i,A}, m_i \dot{\mathbf{v}}_i], \end{array} \right\} \tag{3.16}$$

since

$$\sum_{i \in S} [\mathbf{v}_i, m_i \mathbf{v}_i] = \sum_{i \in S} m_i [\mathbf{v}_i, \mathbf{v}_i] = 0.$$

Now, in view of (3.3) we have

$$m_i \dot{\mathbf{v}}_i = \dot{\mathbf{Q}}_i,$$

and by Newton's second law

$$\mathbf{F}_i = \dot{\mathbf{Q}}_i,$$

where $\mathbf{F}_i$ is the force acting on i. Therefore (3.16) becomes

$$\dot{\mathbf{K}}_A = -M \left[\mathbf{v}_A, \frac{1}{M} \sum_{i \in S} m_i \mathbf{v}_i \right] + \sum_{i \in S} [\mathbf{r}_{i,A}, \mathbf{F}_i]. \qquad (3.17)$$

Using (3.11) and (3.12), expression (3.17) can be rewritten as

$$\dot{\mathbf{K}}_A = -M [\mathbf{v}_A, \mathbf{v}_{CI}] + \mathbf{M}_{\mathbf{F}_A} = M [\mathbf{v}_{CI}, \mathbf{v}_A] + \mathbf{M}_{\mathbf{F}_A}.$$

The theorem is proven. $\square$

Remark 3.2. There are some important special cases of the Rizal's formula.
1. If $\mathbf{v}_A = 0$ or $\mathbf{v}_{CI} = 0$ we have

$$\dot{\mathbf{K}}_A = \mathbf{M}_{\mathbf{F}_{ext,A}}.$$

2. If additionally

$$\mathbf{F}_{ext,i} = 0 \quad \forall i \in S,$$

then

$$\mathbf{M}_{\mathbf{F}_{ext,A}} = \mathbf{0},$$

implying

$$\mathbf{K}_A = \underset{t}{\mathrm{const.}}$$

3.3 Force work and potential forces

This section addresses its study, which is predominantly based on the concept of work of a force. Potential forces are a type of forces with very specific and useful characteristics.

3.3.1 Elementary and total force work

Definition 3.5. Consider a force **F** acting on a particle that traverses a path L. Denote by $d\mathbf{r}$ the differential element of L (see Fig. 3.3). The scalar amount

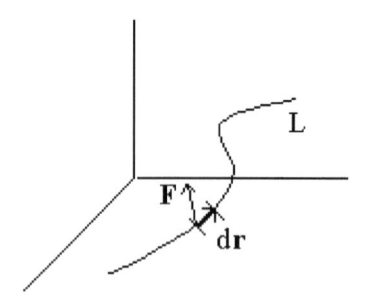

$$\delta A := (\mathbf{F}, d\mathbf{r}) \tag{3.18}$$

Figure 3.3 Relationship between a force and elementary displacement.

is called the **elementary work** of force **F** on path $d\mathbf{r}$, and the value

$$A := \int_L \delta A \tag{3.19}$$

is referred to as the **total force work** of **F** on path L.

Remark 3.3. The amount δA in general does not represent the total differential of some function; that is why we are using δA instead of dA.

Below, we demonstrate some interesting results related with the notions above.

3.3.2 Potential forces

Definition 3.6. Assume that a trajectory L has initial $\mathbf{r}_{ini}$ and terminal $\mathbf{r}_{term}$ positions. If the work of force **F** is such that it does not depend on the form of L but only on its initial and terminal points, that is,

$$A = A(\mathbf{r}_{ini}, \mathbf{r}_{term}), \tag{3.20}$$

then it is said that **F** is a **potential force**.

Lemma 3.3. *Force* **F** *is potential if and only if there is a scalar function*

$$\Pi = \Pi(\mathbf{r}) \tag{3.21}$$

such that

$$\mathbf{F} = -\nabla \Pi(\mathbf{r}). \tag{3.22}$$

Dynamics

Proof. *Necessity.* Since **F** is potential, we have

$$\int_L (\mathbf{F}, d\mathbf{r}) = A(\mathbf{r}_{ini}, \mathbf{r}_{term}), \tag{3.23}$$

and

$$\int_{\mathbf{r}}^{\mathbf{r}+\Delta\mathbf{r}} (\mathbf{F}, d\mathbf{r}) = A(\mathbf{r}_{ini}, \mathbf{r} + \Delta\mathbf{r}) - A(\mathbf{r}_{ini}, \mathbf{r}). \tag{3.24}$$

Now, the condition (3.23) implies that A is sufficiently smooth, and hence, by the Taylor expansion, the relation (3.24) can be expressed as

$$\int_{\mathbf{r}}^{\mathbf{r}+\Delta\mathbf{r}} (\mathbf{F}, d\mathbf{r}) = (\mathbf{F}, d\mathbf{r}) + o(|\Delta\mathbf{r}|) =$$
$$A(\mathbf{r}_{ini}, \mathbf{r} + \Delta\mathbf{r}) - A(\mathbf{r}_{ini}, \mathbf{r}) = (\nabla_{\mathbf{r}} A(\mathbf{r}_{ini}, \mathbf{r}), \Delta\mathbf{r}) + o(|\Delta\mathbf{r}|),$$

or equivalently, as

$$(\mathbf{F}, \Delta\mathbf{r}) + o(|\Delta\mathbf{r}|) = (\nabla_{\mathbf{r}} A(\mathbf{r}_{ini}, \mathbf{r}), \Delta\mathbf{r}) + o(|\Delta\mathbf{r}|).$$

This implies

$$(\mathbf{F} - \nabla_{\mathbf{r}} A(\mathbf{r}_{ini}, \mathbf{r}), \Delta\mathbf{r}) = o(|\Delta\mathbf{r}|). \tag{3.25}$$

Tending to $\Delta\mathbf{r} \to d\mathbf{r}$, the relation (3.25) becomes

$$(\mathbf{F} - \nabla_{\mathbf{r}} A(\mathbf{r}_{ini}, \mathbf{r}), d\mathbf{r}) = 0,$$

and, since it is true for any $d\mathbf{r}$, it is concluded that

$$\mathbf{F} - \nabla_{\mathbf{r}} A(\mathbf{r}_{ini}, \mathbf{r}) = 0,$$

or

$$\mathbf{F} = \nabla_{\mathbf{r}} A(\mathbf{r}_{ini}, \mathbf{r}).$$

Once the point $\mathbf{r}_{ini}$ has been fixed and remains constant, it is possible to represent A in the form

$$A(\mathbf{r}_{ini}, \mathbf{r}) = -\Pi(\mathbf{r}),$$

so that

$$\mathbf{F} = -\nabla_{\mathbf{r}} \Pi(\mathbf{r}).$$

Sufficiency. Since there exists $\Pi : R^3 \to R$,

$$\mathbf{F}(\mathbf{r}) = -\nabla \Pi(\mathbf{r}),$$

and by the definition (3.18) we have

$$\delta A = (\mathbf{F}, d\mathbf{r}) = -(\nabla \Pi(\mathbf{r}), d\mathbf{r}). \tag{3.26}$$

But note that the term $(\nabla \Pi(\mathbf{r}), d\mathbf{r})$ turns out to be the total differential of the function Π:

$$(\nabla \Pi(\mathbf{r}), d\mathbf{r}) = d\Pi(\mathbf{r}),$$

and (3.26) can be represented as

$$\delta A = -d\Pi(\mathbf{r}), \tag{3.27}$$

such that the work of $\mathbf{F}$ along path L with initial and final positions $\mathbf{r}_{ini}$ and $\mathbf{r}_{term}$ is given by

$$A = \int_L \delta A = -\int_L d\Pi(\mathbf{r}) = \Pi(\mathbf{r}_{ini}) - \Pi(\mathbf{r}_{term}) = A(\mathbf{r}_{ini}, \mathbf{r}_{term}).$$

Here we have used that the integral of a total differential along any way L depends only on the initial and final positions. The lemma is proven. □

Remark 3.4. According to (3.27), in the case of potential forces only δA is effectively a total differential.

3.3.3 Force power and expression for $\dot{T}$

Definition 3.7. Let $\mathbf{F}_i$ be the force exerted on the particle $i \in S$. The elementary work carried out by all the forces in S is given by

$$\delta A := \sum_{i \in S} (\mathbf{F}_i, d\mathbf{r}_i), \tag{3.28}$$

which, suppose, is done in the time interval δt. The quantity

$$N := \frac{\delta A}{\delta t}$$

is said to be the **power**, developed by these forces.

Lemma 3.4. *Let S be a system of particles with constant masses subject to the action of external and internal forces. The elementary work of the forces and the variation of the kinetic energy of S keep the relationship*

$$dT = \delta A, \tag{3.29}$$

or equivalently,

$$\dot{T} = N. \tag{3.30}$$

Proof. From (3.18) it follows that for S

$$\delta A = \sum_{i \in S} (\mathbf{F}_i, d\mathbf{r}_i) = \sum_{i \in S} (m_i \dot{\mathbf{v}}_i, d\mathbf{r}_i),$$

where Newton's second law and the fact that m_i ($i \in S$) is constant have been used. Equivalently, using the definitions of derivative and differential and the properties of the internal product, we have

$$\delta A = \sum_{i \in S} \lim_{\Delta t \to 0} \left(m_i \frac{\Delta \mathbf{v}_i}{\Delta t}, \Delta \mathbf{r}_i \right) = \sum_{i \in S} m_i \lim_{\Delta t \to 0} \left(\Delta \mathbf{v}_i, \frac{\Delta \mathbf{r}_i}{\Delta t} \right) = \sum_{i \in S} m_i (d\mathbf{v}_i, \mathbf{v}_i),$$

where, considering the expression for the differential of the internal product and the definition of kinetic energy T of S, we get

$$\delta A = d \left(\frac{1}{2} \sum_{i \in S} m_i (\mathbf{v}_i, \mathbf{v}_i) \right) = dT,$$

which implies (3.30). $\square$

3.3.4 Conservative systems

Definition 3.8. Let S be a system in which all forces are potentials, that is, the force acting on the particle $i \in S$ is given by $\mathbf{F}_i = -\nabla \Pi (\mathbf{r}_i)$, where $\mathbf{r}_i$ is the position of the particle. Such a system is called **conservative**.

Lemma 3.5. *In a system S that is conservative the property*

$$E(t) := T(t) + \sum_{i \in S} \Pi (\mathbf{r}_i(t)) = \underset{t}{\text{const}} \qquad (3.31)$$

is met.

Proof. Using (3.28) and the property of potential forces we have

$$\delta A = \sum_{i \in S} (\mathbf{F}_i, d\mathbf{r}_i) = -\sum_{i \in S} (\nabla \Pi (\mathbf{r}_i), d\mathbf{r}_i) = -\sum_{i \in S} d\Pi (\mathbf{r}_i),$$

whose replacement in (3.29) leads to the relationship

$$d \left[T + \sum_{i \in S} \Pi (\mathbf{r}_i) \right] = 0,$$

or equivalently, to (3.31). $\square$

Remark 3.5. The quantities $\sum_{i \in S} \Pi(\mathbf{r}_i(t))$ and $E(t)$ receive the names of **potential energy** and **mechanical energy** of the system S, and consequently, the result (3.31) of the previous corollary is called **the principle of conservation of mechanical energy**, which justifies the name of conservative systems.

3.4 Virial of a system

In systems of material points certain *average characteristics* have interesting relationships. These aspects are studied in this section.

3.4.1 Main definition of virial

Definition 3.9. Consider a scalar function $\beta(t)$ for $t \geq 0$. The amount

$$\beta_\tau := \frac{1}{\tau} \int_{t=0}^{\tau} \beta(t)\, dt, \quad \tau > 0, \ \beta_0 = \beta(0),$$

is said to be the **average value** of $\beta(t)$ in the interval $[0, \tau]$.

Definition 3.10. Suppose that S is a system of material points subject to the action of the forces $\mathbf{F}_i(t), i \in S$. The amount

$$V_\tau := -\frac{1}{2}\left(\sum_{i \in S}(\mathbf{F}_i(t), \mathbf{r}_i(t))\right)_\tau \tag{3.32}$$

is referred to as the **virial** of S.

In the following the forces and the position vectors are dependent on t, but this dependency by economy in the notation is omitted.

Theorem 3.2 (On the virial of the system). *For the virial of a particle system S we have*

$$V_\tau = T_\tau \tag{3.33}$$

if any of two conditions are met:
a) *trajectories and velocities of system points are bounded and $\tau = \infty$;*
b) *the trajectories of the points of the system are periodic with period τ_{per} and $\tau = \tau_{per}$.*

Proof. Define

$$G := \sum_{i \in S}(m_i \mathbf{v}_i, \mathbf{r}_i), \tag{3.34}$$

Dynamics

whose time derivative is

$$\dot{G} = \sum_{i \in S} [(m_i \dot{\mathbf{v}}_i, \mathbf{r}_i) + (m_i \mathbf{v}_i, \mathbf{v}_i)] = \sum_{i \in S} (\mathbf{F}_i, \mathbf{r}_i) + 2T.$$

Here we have used the Newton's second law as well as the definition (3.2) for the kinetic energy. Integrating this expression in time interval $[0, \tau]$, $\tau > 0$, and dividing by τ, we get

$$\frac{1}{\tau}[G(\tau) - G(0)] = \frac{1}{\tau}\int_0^\tau \sum_{i \in S} (\mathbf{F}_i(t), \mathbf{r}_i(t)) \, dt + \frac{2}{\tau}\int_0^\tau T(t) \, dt$$

or, with the concept of average quantity and definition of virial (3.32) it follows that

$$\frac{1}{\tau}[G(\tau) - G(0)] = 2(-V_\tau + T_\tau). \tag{3.35}$$

a) If the trajectories of the points of S are bounded, there exists

$$G^+ := \max_{t \geq 0} |G(t)| < \infty,$$

which leads to

$$\lim_{t \geq 0} \frac{1}{\tau}[G(\tau) - G(0)] = 0,$$

and in view of (3.35) it follows that $V_\infty = T_\infty$.

b) If the trajectories of the points of S are periodic with period τ_{per}, one has

$$G(\tau_{per}) = G(0),$$

and by (3.35) we get

$$T_{\tau_{per}} = V_{\tau_{per}}.$$

□

3.4.2 Virial for homogeneous potential energies

The following results concern a special case when the potential energy is homogeneous.

Definition 3.11. A function $f : R^n \to R$ with the property

$$f(\lambda \mathbf{r}) = \lambda^s f(\mathbf{r}), \quad \lambda > 0,$$

is said to be **positively homogeneous** (or simply, **homogeneous**) **with homogeneity order** s.

We also have

$$\mathbf{F}_i = -\nabla \Pi_i(\mathbf{r}_i), \quad i \in S. \tag{3.36}$$

Lemma 3.6. *Suppose that Π_i is homogeneous with homogeneity order s and that the trajectories of the points in S comply with the restrictions of the virial theorem, Theorem 3.2. Then*

$$T_\tau = \frac{s}{2}\Pi_\tau, \qquad (3.37)$$

where

$$\Pi_\tau := \left(\sum_{i \in S} \Pi_i\left(\mathbf{r}_i\right)\right)_\tau.$$

Proof. By the virial theorem, Theorem 3.2, with $\tau = \infty$ or $\tau = \tau_{per}$, depending on the case, we have

$$T_\tau = V_\tau. \qquad (3.38)$$

Now, using the definition of virial of the system (3.32) and in view of (3.36),

$$V_\tau = \frac{1}{2\tau}\int_0^\tau \sum_{i \in S}\left(\nabla_\mathbf{r}\Pi_i\left(\mathbf{r}_i\right), \mathbf{r}_i\right) dt. \qquad (3.39)$$

But, we have

$$\left(\nabla_\mathbf{r}\Pi_i\left(\mathbf{r}\right), \mathbf{r}\right) = \left.\frac{\partial \Pi_i\left(\lambda \mathbf{r}\right)}{\partial \lambda}\right|_{\lambda=1}.$$

Hence, (3.39) can be represented as

$$V_\tau = \frac{1}{2}\sum_{i \in S}\frac{1}{\tau}\int_0^\tau \left.\frac{\partial \Pi_i\left(\lambda \mathbf{r}_i\right)}{\partial \lambda}\right|_{\lambda=1} dt,$$

and, in view of homogeneity Π_i,

$$V_\tau = \frac{1}{2}\sum_{i \in S}\frac{1}{\tau}\int_0^\tau \left.\frac{\partial \lambda^s \Pi_i\left(\mathbf{r}_i\right)}{\partial \lambda}\right|_{\lambda=1} dt =$$

$$\frac{s}{2}\lambda^{s-1}\bigg|_{\lambda=1}\sum_{i \in S}\frac{1}{\tau}\int_0^\tau \Pi_i\left(\mathbf{r}_i\right) dt = \frac{s}{2}\left(\sum_{i \in S}\Pi_i\left(\mathbf{r}_i\right)\right)_\tau = \frac{s}{2}\Pi_\tau.$$

The lemma is proven. □

Corollary 3.1. *Given*

$$T_\tau + \Pi_\tau = \frac{1}{\tau}\int_0^\tau \left[\sum_{i \in S}m_i v_i^2(t) + \sum_{i \in S}\Pi_i\left(\mathbf{r}_i(t)\right)\right]dt =$$

Dynamics

$$\frac{1}{\tau} \int_0^\tau \left[T(t) + \sum_{i \in S} \Pi(\mathbf{r}_i(t)) \right] dt,$$

taking into account that all forces are potential with the homogeneous potential energy with the index s, by (3.31) we have

$$T_\tau + \Pi_\tau = E = \operatorname*{const}_t.$$

Combining this equation with (3.37) leads to

$$T_\tau = \frac{s}{s+2} E, \quad \Pi_\tau = \frac{2}{s+2} E. \tag{3.40}$$

3.5 Properties of the center of mass

In this section the *dynamic properties of the center of mass* (or inertia) are established.

3.5.1 Dynamics of the center of inertia (mass)

Lemma 3.7. *In a system S of particles $i \in S$ the dynamics of the center of mass with the coordinate vector $\mathbf{r}_{CI}$ is as follows:*

$$M\ddot{\mathbf{r}}_{CI} = \mathbf{F}_{ext}, \tag{3.41}$$

where M represents the total mass of the system S.

Proof. By Newton's second law,

$$\dot{\mathbf{Q}} = \mathbf{F}_{ext}$$

or, considering m_i constant for all $i \in S$ and using the definition of inertial center (3.10), we have

$$\dot{\mathbf{Q}} = \frac{d}{dt} \sum_{i \in S} m_i \mathbf{v}_i = M \frac{d}{dt} \frac{1}{M} \sum_{i \in S} m_i \mathbf{v}_i = M \ddot{\mathbf{r}}_{CI},$$

which gives (3.41). □

3.6 "King/König/Rey" theorem

3.6.1 Principle theorem

Theorem 3.3 (König theorem). *Suppose that we deal with two reference systems: one absolute with origin O and another relative with origin O' (see Fig. 3.4). The kinetic*

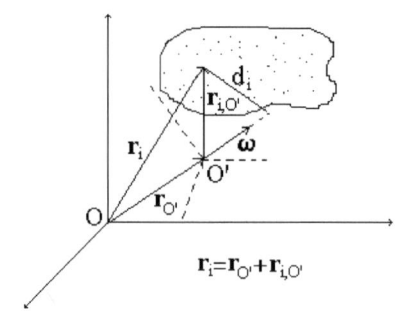

Figure 3.4 Relationship between the coordinates of an absolute and an auxiliary system for calculating kinetic energy.

energy of a system S containing some material particles can be calculated as

$$T = T_{O'} + T_{rel,O'} + M\left(\mathbf{v}_{O'}, \mathbf{v}_{CI,O'}\right), \qquad (3.42)$$

where

- $\mathbf{v}_{O'}$ is the absolute velocity of the pole (origin) O',
- $\mathbf{v}_{CI,O'}$ is the velocity of the center of mass with respect to the pole O',
- $T_{O'} := \frac{1}{2} M v_{O'}^2$ is the kinetic energy of the mass of S if it were concentrated in the pole O',
- $T_{rel,O'} := \frac{1}{2} \sum_{i \in S} m_i v_{i,O'}^2$ is the kinetic energy of S calculated with respect to the pole O',
- $\mathbf{v}_{i,O'}$ is the velocity of the point $i \in S$ relative to O'.

Proof. By the definition of the kinetic energy (3.2)

$$T = \frac{1}{2} \sum_{i \in S} m_i \left(\mathbf{v}_i, \mathbf{v}_i\right),$$

and taking into account that

$$\mathbf{v}_i = \mathbf{v}_{O'} + \mathbf{v}_{i,O'},$$

we have

$$T = \frac{1}{2} \sum_{i \in S} m_i \left[\left(\mathbf{v}_{O'}, \mathbf{v}_{O'}\right) + \left(\mathbf{v}_{i,O'}, \mathbf{v}_{i,O'}\right) + 2\left(\mathbf{v}_{O'}, \mathbf{v}_{i,O'}\right)\right] =$$

$$\frac{1}{2} M v_{O'}^2 + \frac{1}{2} \sum_{i \in S} m_i v_{i,O'}^2 + \left(\mathbf{v}_{O'}, \sum_{i \in S} m_i \mathbf{v}_{i,O'}\right).$$

Finally, multiplying and dividing by M and in view of (3.11) it follows that

$$T = T_{O'} + T_{rel,O'} + M\left(\mathbf{v}_{O'}, \mathbf{v}_{CI,O'}\right).$$

□

Corollary 3.2. *The König theorem has two very important particularizations.*
1. *If the origin O' coincides with the center of mass (inertia) of the system, then $\mathbf{v}_{CI,O'} = 0$ and*

$$T = T_{O'} + T_{rel,O'}. \tag{3.43}$$

2. *If S is a rigid body, whose pivot coincides with O', then there exists a vector $\boldsymbol{\omega}$ relative to O' such that, if $\mathbf{r}_{i,O'}$ denotes the position vector of the point $i \in S$ relative to O', the following relation holds:*

$$\mathbf{v}_{i,O'} := \frac{d}{dt}\mathbf{r}_{i,O'} = \left[\boldsymbol{\omega}, \mathbf{r}_{i,O'}\right]$$

and

$$v_{i,O'}^2 = \omega^2 r_{i,O'}^2 \sin^2\left(\widehat{\boldsymbol{\omega}, \mathbf{r}_{i,O'}}\right),$$

so that

$$T_{rel,O'} = \frac{\omega^2}{2}\sum_{i \in S} m_i r_{i,O'}^2 \sin^2\left(\widehat{\boldsymbol{\omega}, \mathbf{r}_{i,O'}}\right) = \frac{\omega^2}{2}\sum_{i \in S} m_i d_i^2,$$

where

$$d_i := r_{i,O'} \sin\left(\widehat{\boldsymbol{\omega}, \mathbf{r}_{i,O'}}\right)$$

denotes the distance to the line of action of $\boldsymbol{\omega}$ (the axis of rotation). The quantity

$$\mathrm{I}_{\boldsymbol{\omega}} := \sum_{i \in S} m_i d_i^2 \tag{3.44}$$

*is called the **instantaneous moment of inertia** or simply **moment of inertia** of the rigid body S with respect to the axis of $\boldsymbol{\omega}$. That is why*

$$T_{rel,O'} = \frac{1}{2}\mathrm{I}_{\boldsymbol{\omega}}\omega^2. \tag{3.45}$$

3.6.2 Moment of inertia and the impulse moment with respect to a pivot

Definition 3.12. In general terms, for a system S of points and an axis $AA\prime$, the quantity

$$\mathrm{I}_{AA'} := \sum_{i \in S} m_i d_i^2, \tag{3.46}$$

where d_i denotes the distance from $i \in S$ to $AA\prime$, is called **the moment of inertia** of S with respect to the axis $AA\prime$.

The concept of moment of inertia also appears in other developments, as the following exercise shows.

3.6.3 A rigid flat body rotating in the same plane

Let S be a rigid flat body that rotates in its own plane with respect to the pivot O' with angular velocity ω (Fig. 3.5). The impulse moment with respect to the pivot is given by

$$\mathbf{K}_{O'} := \sum_{i \in S} [\mathbf{r}_{i,O'}, m_i \mathbf{v}_i] = \sum_{i \in S} m_i [\mathbf{r}_{i,O'}, \mathbf{v}_i]$$

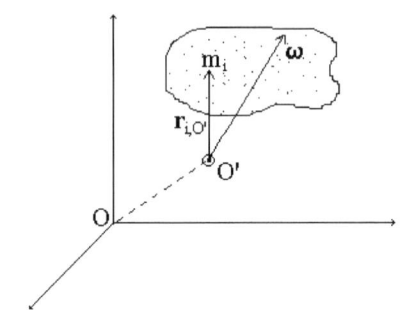

Figure 3.5 Flat body rotating in its plane with respect to a pivot O'.

and, by Euler's theorem (see Chapter 2),

$$\mathbf{v}_i = [\omega, \mathbf{r}_{i,O'}],$$

which implies

$$\begin{aligned}\mathbf{K}_{O'} &= \sum_{i \in S} m_i \left[\mathbf{r}_{i,O'}, [\omega, \mathbf{r}_{i,O'}]\right] = \\ &\sum_{i \in S} m_i \left[(\mathbf{r}_{i,O'}, \mathbf{r}_{i,O'})\omega - (\mathbf{r}_{i,O'}, \omega)\mathbf{r}_{i,O'}\right],\end{aligned} \qquad (3.47)$$

where the alternative formula of the triple vector product has been used (see Chapter 1). Whereas

$$(\mathbf{r}_{i,O'}, \omega) = 0 \ \forall i \in S$$

and

$$(\mathbf{r}_{i,O'}, \mathbf{r}_{i,O'}) = d_i^2$$

Dynamics

(where d_i is the distance from point $i \in S$ to the axis of rotation), one has

$$\mathbf{K}_{O'} = \left(\sum_{i \in S} m_i d_i^2\right) \boldsymbol{\omega},$$

or, by the definition of moment of inertia (3.44), we finally arrive at the presentation

$$\mathbf{K}_{O'} = I_\omega \boldsymbol{\omega}. \tag{3.48}$$

Rotation of a body with geometric and mass symmetry

The result of the previous subsection is valid in a slightly more general situation, namely, when we deal with the rotation of a body with **geometric and mass symmetry** with respect to the axis of rotation. To see this, recall from Chapter 1 that

$$\mathbf{r}_{i,O'} = \frac{(\boldsymbol{\omega}, \mathbf{r}_{i,O'})}{\omega^2} \boldsymbol{\omega} + \frac{[\boldsymbol{\omega}, [\mathbf{r}_{i,O'}, \boldsymbol{\omega}]]}{\omega^2},$$

which, after substitution in (3.47), leads to

$$\mathbf{K}_{O'} = \sum_{i \in S} m_i \left\{ \left((\mathbf{r}_{i,O'}, \mathbf{r}_{i,O'}) - \frac{1}{\omega^2}(\mathbf{r}_{i,O'}, \boldsymbol{\omega})^2\right) \boldsymbol{\omega} - \frac{1}{\omega^2}(\mathbf{r}_{i,O'}, \boldsymbol{\omega})[\boldsymbol{\omega}, [\mathbf{r}_{i,O'}, \boldsymbol{\omega}]]\right\}, \tag{3.49}$$

where the term $[\boldsymbol{\omega}, [\mathbf{r}_{i,O'}, \boldsymbol{\omega}]]$ has radial direction with respect to the axis of rotation. So, under the condition of geometric and mass symmetry with respect to the axis of rotation, for any $i \in S$, there exists $i^* \in S$ (the *mirror image* of $i \in S$) such that

$$m_i = m_{i^*}, \quad r_{i,O'} = r_{i^*,O'}, \mathbf{r}_{i,O'} = -\mathbf{r}_{i^*,O'},$$

and therefore,

$$[\boldsymbol{\omega}, [\mathbf{r}_{i,O'}, \boldsymbol{\omega}]] + [\boldsymbol{\omega}, [\mathbf{r}_{i^*,O'}, \boldsymbol{\omega}]] = 0,$$

where the second term on the right side of (3.49) is canceled, so that

$$\mathbf{K}_{O'} = \sum_{i \in S} m_i \left((\mathbf{r}_{i,O'}, \mathbf{r}_{i,O'}) - \frac{1}{\omega^2}(\mathbf{r}_{i,O'}, \boldsymbol{\omega})^2\right) \boldsymbol{\omega}. \tag{3.50}$$

But

$$(\mathbf{r}_{i,O'}, \mathbf{r}_{i,O'}) = r_{i,O'}^2 \text{ and } \frac{1}{\omega^2}(\mathbf{r}_{i,O'}, \boldsymbol{\omega})^2 = \left(\text{comp}_{\boldsymbol{\omega}}^{\mathbf{r}_{i,O'}}\right)^2$$

and by the Pythagorean theorem

$$d_i^2 = r_{i,O'}^2 - \left(\text{comp}_{\boldsymbol{\omega}}^{\mathbf{r}_{i,O'}}\right)^2.$$

In view of that, (3.50) may be rewritten as

$$\mathbf{K}_{O'} = \left(\sum_{i \in S} m_i d_i^2\right) \omega = \mathbf{I}_\omega \omega, \qquad (3.51)$$

and since, in this case, the location of O' is not important, unless it is on the axis of rotation (we may move ω along the rotation axis), the notation

$$\mathbf{K}_\omega := \mathbf{K}_{O'}$$

can be used in all situations with geometric and mass symmetry with respect to the axis of rotation.

3.6.4 Calculation of moments of inertia for different rigid bodies

The result (3.45) makes use of the moment of inertia, which is a function of the geometry and the distribution of the mass of the rigid body in question. In the following examples this mechanical characteristic is calculated for several geometries.

Example 3.1. Using the definition (3.46), let us calculate the moment of inertia $I_{xx'}$ (with respect to the indicated axis) of the bodies shown in Fig. 3.6. All bodies have mass M of uniform distribution.

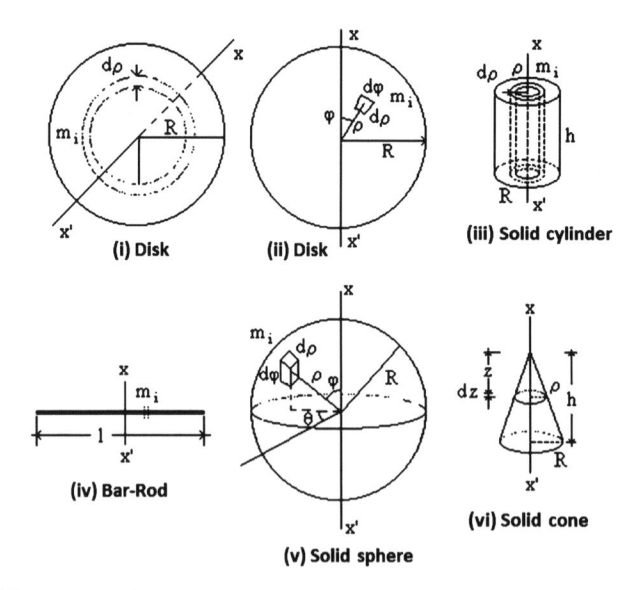

Figure 3.6 Some bodies of simple geometry for the calculation of their moment of inertia.

i) For the disc, with respect to its center, we have

$$I_{xx'} = \sum_{i \in S} m_i d_i^2 = \int_{\rho=0}^{R} \rho^2 dm(\rho),$$

where $dm(\rho)$ is the mass of the elemental ring of radius ρ and width $d\rho$, whose area is

$$dA = 2\pi\rho d\rho,$$

while the total area of the disk is

$$A = \pi R^2.$$

Given the condition of uniform distribution of the mass, it follows that

$$dm = \frac{M}{A} dA = \frac{2M}{R^2} \rho d\rho,$$

so that

$$I_{xx'} = \frac{2M}{R^2} \int_0^R \rho^3 d\rho = \frac{MR^2}{2}.$$

ii) For the disc, regarding a diameter,

$$I_{xx'} = \sum_{i \in S} m_i d_i^2 = 4 \int_{\varphi=0}^{\pi/2} \int_{\rho=0}^{R} \rho^2 \sin^2\varphi\, dm(\rho, \varphi),$$

where $dm(\rho, \varphi)$ is the mass of the differential element located at the point (ρ, φ) and whose area is

$$dA = \rho d\rho d\varphi.$$

Since the mass has uniform distribution and the total area is

$$A = \pi R^2,$$

we have

$$dm = \frac{M}{A} dA = \frac{M}{\pi R^2} \rho d\rho d\varphi.$$

Hence,

$$I_{xx'} = \sum_{i \in S} m_i d_i^2 = \frac{4M}{\pi R^2} \int_{\varphi=0}^{\pi/2} \int_{\rho=0}^{R} \rho^3 \sin^2\varphi\, d\rho d\varphi =$$

$$\frac{MR^2}{\pi} \int_{\varphi=0}^{\pi/2} \sin^2\varphi\, d\varphi = \frac{MR^2}{\pi} \int_{\varphi=0}^{\pi/2} \left(\frac{1-\cos 2\varphi}{2}\right) d\varphi =$$

$$\frac{MR^2}{2\pi}\left[\varphi - \frac{\sin 2\varphi}{2}\right]_0^{\pi/2} = \frac{MR^2}{4}.$$

iii) For the shown solid cylinder the same procedure is followed as for the case disk (i), only here $dm\,(\rho)$ is the mass of the elementary hollow cylinder of radius ρ, width $d\rho$, and height h, whose volume is

$$dV = 2\pi h\rho d\rho.$$

Since the total volume of the cylinder V is

$$V = \pi hR^2,$$

in view of the uniform mass distribution we have

$$dm = \frac{M}{V}dV = \frac{2M}{R^2}\rho d\rho.$$

So,

$$I_{xx'} = \frac{2M}{R^2}\int_0^R \rho^3 d\rho = \frac{MR^2}{2}.$$

iv) For the bar without thickness, with respect to the axis shown, it follows that

$$I_{xx'} = \sum_{i\in S} m_i d_i^2 = 2\int_{s=0}^{l/2} s^2 dm,$$

where dm is the mass of the differential element of length ds. Because of the uniformly distributed mass condition, we have

$$dm = \frac{M}{l}ds,$$

which is why

$$I_{xx'} = 2\frac{M}{l}\int_{s=0}^{l/2} s^2 ds = \frac{Ml^2}{12}.$$

v) For the solid sphere we have

$$I_{xx'} = \sum_{i\in S} m_i d_i^2 = \int_M d^2\,(\rho,\varphi)\,dm\,(\rho,\varphi,\theta),$$

where $dm\,(\rho,\varphi,\theta)$ denotes the mass of the differential element with volume

$$dV = \rho^2 \sin\varphi d\rho d\varphi d\theta$$

Dynamics

and $d(\rho, \varphi)$ is its distance to the axis xx', which is

$$d(\rho, \varphi) = \rho \sin \varphi.$$

Since the total volume of the body is

$$V = \frac{4}{3}\pi R^3$$

and the mass is distributed evenly, we get

$$dm = \frac{M}{V}dV = \frac{3M}{4\pi R^3}\rho^2 \sin\varphi d\theta d\rho d\varphi.$$

Therefore,

$$I_{xx'} = \frac{3M}{4\pi R^3} \int_{\varphi=0}^{\pi} \int_{\rho=0}^{R} \int_{\theta=0}^{2\pi} \rho^4 \sin^3 \varphi d\theta d\rho d\varphi = \frac{3MR^2}{10} \int_0^{\pi} \sin^3 \varphi d\varphi.$$

Taking into account that

$$\int_{\varphi=0}^{\pi} \sin^3 \varphi d\varphi = -\int_{\varphi=0}^{\pi} \sin^2 \varphi d(\cos\varphi) - \sin^2 \varphi \cos\varphi \Big|_0^{\pi} +$$

$$\int_{\varphi=0}^{\pi} \cos\varphi d\left(\sin^2 \varphi\right) = 2\int_{\varphi=0}^{\pi} \sin\varphi \cos^2 \varphi d\varphi =$$

$$-2\int_{\varphi=0}^{\pi} \cos^2 \varphi d(\cos\varphi) - 2\int_{\varphi=0}^{\pi} \cos^2 \varphi d(\cos\varphi) =$$

$$-\frac{2}{3} \cos^3 \varphi \Big|_0^{\pi} = -\frac{2}{3}(-1-1) = \frac{4}{3},$$

we may conclude that

$$I_{xx'} = \frac{2}{5}MR^2.$$

vi) Finally, for the solid cone with respect to its longitudinal axis, the result of part (i) will be used. So,

$$I_{xx'} = \sum_{i \in S} m_i d_i^2 = \int_{z=0}^{h} dI_{xx'}(z),$$

where $dI_{xx'}(z)$ denotes the moment of inertia with respect to its center of the elementary disk of radius ρ and height dz (see Fig. 3.6(vi)). Now, given the geometry we have the relationship

$$\rho = \frac{R}{h}z,$$

so the volume of the differential disk results in

$$dV = \pi\rho^2 dz = \frac{\pi R^2}{h^2} z^2 dz,$$

while that of the whole body is

$$V = \frac{\pi}{3} R^2 h.$$

The elementary disk has mass dm:

$$dm = \frac{M}{V} dV = \frac{3M}{h^3} z^2 dz,$$

and according to the result (i),

$$dI_{xx'} = \frac{\rho^2}{2} dm = \frac{3MR^2}{2h^5} z^4 dz,$$

which gives

$$I_{xx'} = \frac{3MR^2}{2h^5} \int_0^h z^4 dz = \frac{3}{10} MR^2.$$

3.6.5 König theorem application

The following two examples are solved by a direct application of the König theorem, Theorem 3.3.

Example 3.2. The rigid circular chain of Fig. 3.7 has mass M evenly distributed and rolls with constant speed as shown. Determine its kinetic energy.

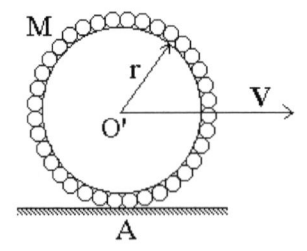

Figure 3.7 Rigid chain rolling with constant speed.

Locate two systems: one absolute fixed to the floor and the other relative fixed to the center of the circle formed by the chain. Since in this case the center of inertia coincides with the origin O' of the relative system, the result (3.43) is applicable, and therefore

$$T = \frac{1}{2}MV^2 + \frac{1}{2}\sum_{i \in S} m_i v_{i,O'}^2.$$

But, applying the results of Chapter 2, point A contacts the floor:

$$\mathbf{v}_A = \mathbf{v}_{O'} + \mathbf{v}_{A,O'} = \mathbf{V} + \mathbf{v}_{A,O'} = 0,$$

which is why

$$\mathbf{v}_{A,O'} = -\mathbf{V}.$$

And since all the points of the chain have velocity relative to O' of equal magnitude, we have

$$v_{i,O'} = v_{A,O'} = V,$$

which implies

$$T = \frac{1}{2}MV^2 + \frac{1}{2}MV^2 = MV^2.$$

Example 3.3. The disk in Fig. 3.8 has mass M evenly distributed and rolls with constant speed, as indicated. Determine its kinetic energy.

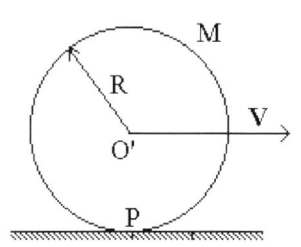

Figure 3.8 Rolling disc.

The solution procedure is very similar to that followed in the preceding example. An absolute coordinate system fixed to the floor and a relative one fixed to the center of the disk are placed. Since the center of mass of the disk coincides with the origin O' of the relative system, the result (3.43) is directly applicable, and in this case the relative kinetic energy is given by (3.45). So,

$$T = \frac{1}{2}MV^2 + \frac{1}{2}I_\omega \omega^2. \tag{3.52}$$

By the results of Chapter 2, we have for the point P, contacting with the floor,

$$\mathbf{v}_P = \mathbf{v}_{O'} + \mathbf{v}_{P,O'} = \mathbf{V} + [\boldsymbol{\omega}, \mathbf{r}_{P,O'}] = 0. \tag{3.53}$$

But since **V** and $[\boldsymbol{\omega}, \mathbf{r}_{P,O'}]$ are of the opposite directions and considering that $r_{P,O'} = R$, from (3.53) we obtain

$$V - \omega R = 0,$$

and as a consequence

$$\omega = \frac{V}{R}. \tag{3.54}$$

On the other hand, the moment of inertia of the disk with respect to the axis of rotation is given by (see the exercise in the end of the chapter)

$$I_\omega = \frac{MR^2}{2}. \tag{3.55}$$

So the substitution of (3.54) and (3.55) in (3.52) finally leads to

$$T = \frac{3}{4}MV^2.$$

3.6.6 Steiner's theorem on the inertia moment

The result that follows is extremely useful in the calculation of moments of inertia.

Theorem 3.4 (Steiner). *Consider the solid of mass M depicted in Fig. 3.9. Denote by $I_{AA'}$ and $I_{OO'}$ the moments of inertia with respect to the axes AA', which does not pass through the center of mass, and OO', which does and which is parallel to AA' and is separated from it at a distance d. We have the following relationship between such moments of inertia:*

$$I_{AA'} = I_{OO'} + Md^2.$$

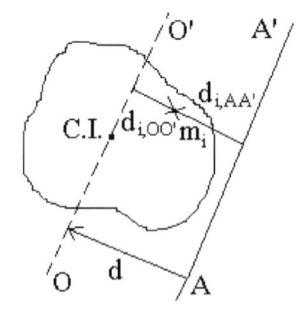

Figure 3.9 Solid body rotating around an axis that does not pass through its inertial center.

Proof. Denote by $d_{i,AA'}$ and $d_{i,OO'}$ the distances of the material point $i \in S$ up to the axes AA' and OO', respectively. By the definition (3.46) we have

$$I_{AA'} = \sum_{i \in S} m_i d_{i,AA'}^2.$$

But, by the law of cosines

$$d_{i,AA'}^2 = d^2 + d_{i,OO'}^2 - 2d d_{i,OO'} \cos \alpha_i$$

(see Fig. 3.10), and therefore

$$\begin{aligned} I_{AA'} &= \sum_{i \in S} m_i \left(d^2 + d_{i,OO'}^2 - 2d d_{i,OO'} \cos \alpha_i \right) = \\ &\quad M d^2 + I_{OO'} - 2d \sum_{i \in S} m_i d_{i,OO'} \cos \alpha_i. \end{aligned} \qquad (3.56)$$

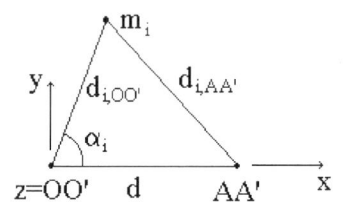

Figure 3.10 Diagram of distances of a point $i \in S$ to the axes OO' and AA'.

Now, note that if CI is chosen as the origin of the coordinate system, we have

$$\mathbf{r}_{CI} := \frac{1}{M} \sum_{i \in S} m_i \mathbf{r}_i = \mathbf{0}.$$

By the component representation $\mathbf{r} = (x, y, z)^T$ it follows that

$$\sum_{i \in S} m_i x = 0, \quad \sum_{i \in S} m_i y = 0, \quad \sum_{i \in S} m_i z = 0. \qquad (3.57)$$

If in Fig. 3.10 the coordinate system is selected in such a way that the axes z and OO' are coincident and the x-axis is directed towards the axis AA', it follows that

$$d_{i,OO'} \cos \alpha_i = x_i,$$

from which by (3.57) we obtain

$$\sum_{i \in S} m_i d_{i,OO'} \cos \alpha_i = 0,$$

and the affirmation follows from (3.56). $\square$

Example 3.4. Calculate the moment of inertia of the solid cylinder (see Fig. 3.11) with respect to the axis shown. The body has a uniformly distributed mass M.

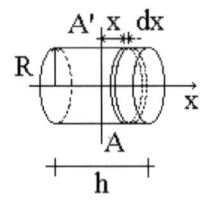

Figure 3.11 Solid cylinder rotating on a transverse axis.

Here the Steiner's theorem, Theorem 3.4, may be directly applicable. To do that let us consider the volume of the elementary disk, which is

$$dV = \pi R^2 dx.$$

Since the total volume of the solid is

$$V = \pi R^2 h,$$

the following mass of the elementary disk results:

$$dM = \frac{M}{V} dV = \frac{M}{h} dx.$$

So, using the result of the aforementioned example (see Exercise (ii)), we have

$$dI_{OO'} = \frac{MR^2}{4h} dx,$$

where OO' represents the axis parallel to AA' passing through the center of inertia of the considered elementary disk. Now, by Steiner's theorem, Theorem 3.4,

$$dI_{AA'} = \frac{M}{h} \left(\frac{R^2}{4} + x^2 \right) dx,$$

from which we get

$$I_{AA'} = 2 \int_{x=0}^{h/2} dI_{AA'}(x) = \frac{2M}{h} \int_{x=0}^{h/2} \left(\frac{R^2}{4} + x^2 \right) dx$$
$$= \frac{M}{4} \left(R^2 + \frac{h^2}{3} \right).$$

The examples that follow illustrate some points studied in this chapter; particularly, the first two make use of the concept of center of velocities (see Chapter 2) and its applicability in the solution of kinematic problems.

Dynamics

Example 3.5. In the articulated bar of Fig. 3.12 each half has length l and mass m, and in addition, the bar is placed in such a way that the articulation is exactly half of the opening in the floor, which has exactly a width l. In the initial moment, to which the situation shown corresponds, the bar begins to fall. Determine the velocity of the joint, denoted by D, at the moment when the ends of the bar are touching the corners of the opening. Consider that the slippage is frictionless.

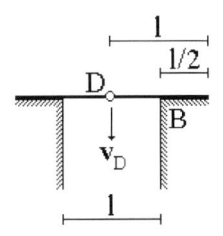

Figure 3.12 Articulated bar about to fall.

The general situation is illustrated in Fig. 3.13. First it will be shown that when the bar falls, the velocities of the contact points of the bar with the corners of the opening have direction towards the articulation D.

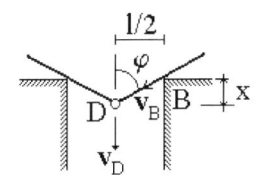

Figure 3.13 Bar falling.

a) For this, locate two coordinate systems: one absolute fixed to the floor and the other fixed relative to D. Given the symmetry conditions of the problem, it is enough to analyze the situation for the right half. So, for point B of contact we have (see Chapter 2)

$$\mathbf{v}_B = \mathbf{v}_D + \left[\boldsymbol{\omega}, \overline{DB}\right], \tag{3.58}$$

where

$$\omega = |\dot{\varphi}|. \tag{3.59}$$

On the other hand, if $\sin\varphi \neq 0$ we have

$$\left|\overline{DB}\right| = \frac{l/2}{\sin\varphi} \tag{3.60}$$

and

$$x = \frac{l}{2}\cot\varphi.$$

The temporary derivation of this last relation leads to

$$\dot{x} = -\frac{l}{2\sin^2\varphi}\dot{\varphi},$$

so that

$$|\dot{\varphi}| = \frac{2}{l}|\dot{x}|\sin^2\varphi. \tag{3.61}$$

Considering that $v_D = |\dot{x}|$, that $\boldsymbol{\omega}$ is orthogonal to $\overline{DB}$, and that the relations (3.59) and (3.60) hold, the following equality is reached:

$$\frac{|[\boldsymbol{\omega}, \overline{DB}]|}{v_D} = \frac{\omega|\overline{DB}|}{v_D} = \frac{l|\dot{\varphi}|}{2|\dot{x}|\sin\varphi},$$

or, with (3.61),

$$\frac{|[\boldsymbol{\omega}, \overline{DB}]|}{v_D} = \sin\varphi. \tag{3.62}$$

Fig. 3.14 shows a vector diagram with the vectors $\mathbf{v}_D$ and $[\boldsymbol{\omega}, \overline{DB}]$. From this diagram and (3.58) we can see that

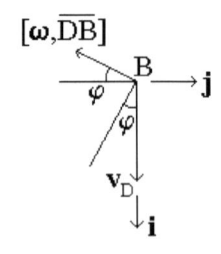

$$\mathbf{v}_B = \left(v_D - |[\boldsymbol{\omega}, \overline{DB}]|\sin\varphi\right)\mathbf{i} - |[\boldsymbol{\omega}, \overline{DB}]|\cos\varphi\mathbf{j}.$$

Figure 3.14 Vectors $\mathbf{v}_D$ and $[\boldsymbol{\omega}, \overline{DB}]$.

But in view of (3.62) the last formula may be represented as

$$\mathbf{v}_B = v_D\cos\varphi\left(\cos\varphi\mathbf{i} - \sin\varphi\mathbf{j}\right),$$

from which it follows that the direction of $\mathbf{v}_B$ is towards D. This means that the point B **is always in the contact with the corner.**

Dynamics

b) The knowledge of the direction of $\mathbf{v}_B$ allows locating the center of velocities C of the right half of the bar; in particular, at the moment of interest, i.e., when the bar is detaching from the floor, it has the configuration of Fig. 3.15. In this situation it is known that with respect to C (the center of velocities) there exists $\mathbf{\Omega}$ such that

$$\mathbf{v}_D = \left[\mathbf{\Omega}, \overline{DC}\right]$$

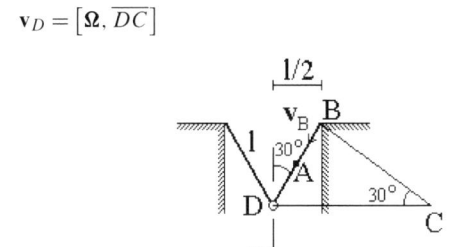

Figure 3.15 Bar detaching from the floor.

with $\mathbf{\Omega}$ orthogonal to $\overline{DC}$, so that

$$v_D = \Omega \left|\overline{DC}\right|. \tag{3.63}$$

From Fig. 3.15 it is also seen that

$$\frac{l}{\left|\overline{DC}\right|} = \sin 30° = \frac{1}{2},$$

so

$$\left|\overline{DC}\right| = 2l.$$

But we deal with a situation with no loss of energy, and therefore

$$T(t) + \sum_{i \in S} \Pi\left(\mathbf{r}_i(t)\right) = \operatorname*{const.}_{t} \tag{3.64}$$

So, taking g as the acceleration of gravity and as a reference level for potential energy the level of the floor, we have for $t = 0$

$$T(0) = 0, \quad \sum_{i \in S} \Pi\left(\mathbf{r}_i(0)\right) = 0, \tag{3.65}$$

whereas for the final moment t_f (see Fig. 3.15)

$$T\left(t_f\right) = \frac{1}{2} I_\Omega \Omega^2,$$
$$\sum_{i \in S} \Pi\left(\mathbf{r}_i\left(t_f\right)\right) = -\frac{1}{2} mgl \cos 30° = -\frac{\sqrt{3}}{4} mgl. \tag{3.66}$$

Now, by Steiner's theorem

$$I_\Omega = I_A + m\,|\overline{AC}|^2,$$

where I_A represents the moment of inertia with respect to the axis perpendicular to the plane of Fig. 3.15 and passing through the center of inertia, denoted by A, of the right half of the bar. Since

$$I_A = \frac{ml^2}{12},$$
$$|\overline{BC}| = |\overline{DC}|\cos 30° = \sqrt{3}l$$

and

$$|\overline{AC}| = \sqrt{|\overline{BC}|^2 + (l/2)^2} = l\sqrt{3 + 1/4} = \sqrt{13}\,l/2.$$

Therefore

$$I_\Omega = I_A + m\,|\overline{AC}|^2 = \left(\frac{1}{12} + \frac{13}{4}\right)ml^2 = \frac{10}{3}ml^2,$$

and as a result

$$T(t_f) = \frac{1}{2}\left(\frac{10}{3}ml^2\right)\Omega^2 = \frac{5}{3}ml^2\Omega^2. \tag{3.67}$$

The results (3.65), (3.66), and (3.67) together with (3.64) give

$$\frac{5}{3}ml^2\Omega^2 = \frac{\sqrt{3}}{4}mgl$$

and

$$\Omega^2 = \frac{3\sqrt{3}}{20}\frac{g}{l}.$$

In view of (3.63), we finally obtain

$$v_D = \Omega\,|\overline{DC}| = \sqrt{\frac{3\sqrt{3}}{20}\frac{g}{l}}\,2l = \sqrt{\frac{3\sqrt{3}}{5}gl}.$$

Example 3.6. The articulated bars in Fig. 3.16 have length l and mass M and the structure is tied with a weightless rope at points B and C. At time $t = 0$ the string is cut and the assembly begins to slide without friction. Determine $\mathbf{v}_B(h)$ considering that the initial height is h_0.

Given the symmetry of the structure, the problem can be reduced to the study of the left half, which is shown in Fig. 3.17. From this figure we have

$$\mathbf{v}_B = [\boldsymbol{\omega}, \overline{AB}],$$

Dynamics

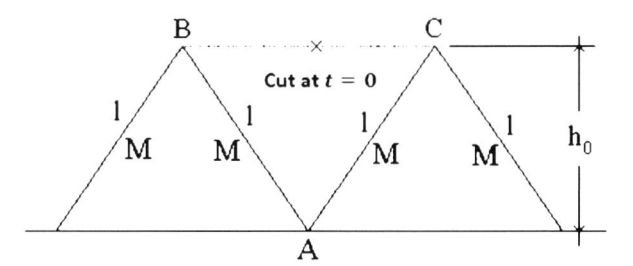

Figure 3.16 Set of articulated bars.

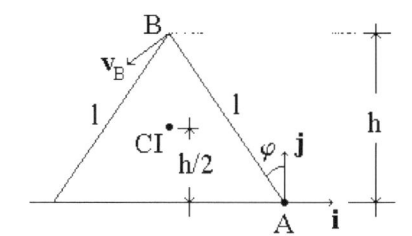

Figure 3.17 Left half of the structure of articulated bars.

where, for the system shown,

$$\boldsymbol{\omega}(t) = \omega(t)\mathbf{k}, \quad \omega(t) = \dot{\varphi}(t),$$

and

$$\overline{AB} = l(-\sin\varphi\mathbf{i} + \cos\varphi\mathbf{j}),$$

so that

$$\mathbf{v}_B = -\omega l(\cos\varphi\mathbf{i} + \sin\varphi\mathbf{j}). \tag{3.68}$$

Now, considering that the only force to be taken into account (weight) is potential, it is true that

$$E(t) := T(t) + \sum_{i \in S} \Pi(\mathbf{r}_i(t)) = \underset{t}{\text{const.}} \tag{3.69}$$

In the initial state ($t = 0$), we have

$$\sum_{i \in S} \Pi(\mathbf{r}_i(0)) = Mgh_0, \quad T(0) = 0,$$

where g denotes the gravity constant. Hence, for all $t \geq 0$

$$E(t) \equiv Mgh_0. \tag{3.70}$$

At the general instant t the potential energy is

$$\sum_{i \in S} \Pi(\mathbf{r}_i(t)) = Mgh(t), \tag{3.71}$$

whereas, taking reference points absolute and relative to A and B, respectively, for the application of the König theorem, we have for the kinetic energy

$$T_A(t) = T_B(t) + T_{rel,B}(t) + 2M\left(\mathbf{v}_B(t), \mathbf{v}_{CI,B}(t)\right), \tag{3.72}$$

where

$$T_B = \frac{1}{2}(2M)v_B^2 = Ml^2\omega^2.$$

Since $v_B = \omega l$, it follows that

$$T_{rel,B} = 2\left(\frac{1}{2}I_\omega \omega^2\right) = I_\omega \omega^2,$$

where I_ω is the moment of inertia of a single bar with respect to the axis of rotation (perpendicular to the plane of Fig. 3.17 and passing B), which is given by

$$I_\omega = \int_0^l x^2 dm = \int_0^l x^2 \frac{M}{l} dx = \frac{1}{3}Ml^2.$$

So,

$$T_{rel,B} = \frac{1}{3}Ml^2\omega^2.$$

To calculate the third term, $\mathbf{v}_{CI,B}$ is required. The position vector of CI with respect to point B, expressed in the absolute system located in A, is

$$\mathbf{r}_{CI,B} = -\frac{l}{2}\cos\varphi \mathbf{j},$$

where

$$\mathbf{v}_{CI,B} = \frac{l}{2}\dot\varphi \sin\varphi \mathbf{j},$$

which is why

$$\left(\mathbf{v}_B, \mathbf{v}_{CI,B}\right) = -\frac{1}{2}l^2\omega^2 \sin^2\varphi.$$

By the previous results and in view of (3.72) we have

$$T_A = Ml^2\omega^2\left(\frac{4}{3} - \sin^2\varphi\right) = M\omega^2\left(\frac{l^2}{3} + h^2\right), \tag{3.73}$$

because

$$\sin^2\varphi = \frac{l^2 - h^2}{l^2}. \tag{3.74}$$

Substitution of (3.70), (3.71), and (3.73) in (3.69) leads to

$$\omega^2\left(\frac{l^2}{3} + h^2\right) + gh = gh_0,$$

and hence

$$\omega = \left(g\frac{h_0 - h}{\frac{l^2}{3} + h^2}\right)^{1/2}. \tag{3.75}$$

Relations (3.74) and (3.75) together with (3.68) finally allow to obtain

$$\mathbf{v}_B(t) = -\left(g\frac{h_0 - h(t)}{\frac{l^2}{3} + h(t)^2}\right)^{1/2}\left(h\mathbf{i} + \sqrt{l^2 - h(t)^2}\mathbf{j}\right).$$

3.7 Movements with friction

Friction appears on the contact surface with bodies when they tend to slide on a surface or they are sliding. In the first case the phenomenon is called *static friction*, and in the second, *dynamic friction*. The first has the effect of producing a force that opposes starting the movement and whose maximum value is reached when the slip is about to begin. This condition allows to derive a simple relation for the calculation of said force. Regarding dynamic friction, its definition is given below.

Axiom 3.4. *Dynamic friction force appears on the contact surface, is proportional to the pressure force, and acts in the direction opposite to the movement (see Fig. 3.18), namely,*

$$\mathbf{F}_{fr} = -f_{fr}N\mathbf{e}_S, \quad |\mathbf{e}_S| = 1, \tag{3.76}$$

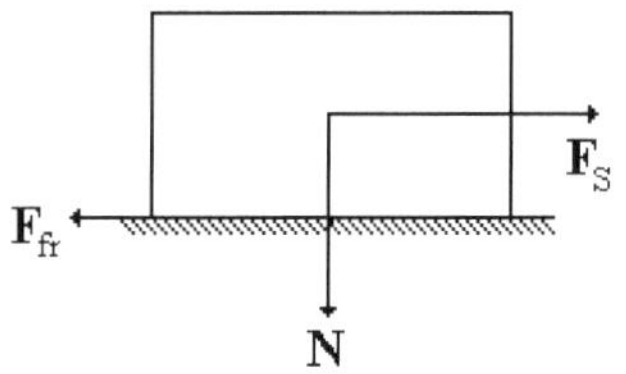

Figure 3.18 Body subject to friction.

where

- f_{fr} is called a **friction coefficient** which depends only on the physical (non-geometric) characteristics of the surfaces in contact,
- N is the **magnitude of the pressure force** perpendicular to the contact surfaces,
- $\mathbf{e}_S$ represents a unitary vector in the direction of the tendency to slide (in the direction of the **rolling force** $\mathbf{F}_S$ in Fig. 3.18).

In Fig. 3.19 it is depicted how the amplitude of a friction force $\|\mathbf{F}_{fr}\|$ depends on the amplitude of the applied external force $\|\mathbf{Q}\|$.

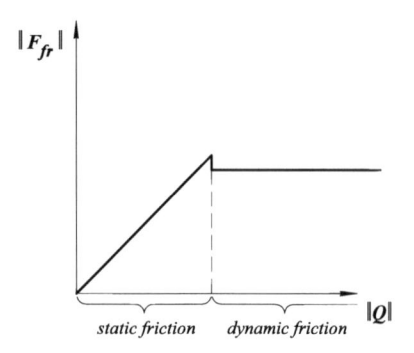

Figure 3.19 How a friction force $\|\mathbf{F}_{fr}\|$ depends on the amplitude of the applied external force $\|\mathbf{Q}\|$.

Example 3.7. Let m, α, and f_{fr} be (see Fig. 3.20) the mass of the body, the angle of the inclined plane, and the coefficient of friction, respectively. If m and f_{fr} are given, calculate α^*, which is the angle of the plane for which motion is initiated.

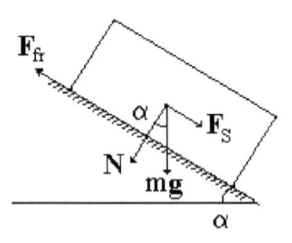

Figure 3.20 Body on an inclined plane.

The sliding condition is

$$F_S \geq F_{fr},$$

that is,

$$mg \sin\alpha \geq f_{fr} N = f_{fr} mg \cos\alpha,$$

which gives

$$\tan\alpha \geq f_{fr},$$

or finally,

$$\alpha^* = \arctan f_{fr}.$$

Example 3.8. Consider the solid sphere on the inclined plane (see Fig. 3.21) with the mass m and the radius ρ. The friction coefficient is assumed to be f_{fr}. Calculate α^*, the critical angle for which the sphere begins to slide, in addition to rolling.

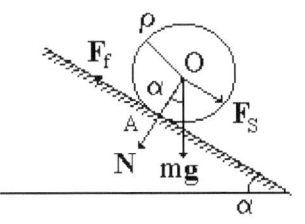

Figure 3.21 Sphere on an inclined plane.

The force responsible for the bearing of the sphere is due to friction, so first, let us determine the relationship between the value of this force and the angle of the plane under conditions of pure bearing. Under these conditions we have by Newton's second law

$$m\dot{v} = F_S - F_f = mg\sin\alpha - F_f, \qquad (3.77)$$

where F_f denotes the force generated by the friction, which under the conditions indicated in Fig. 3.21 gives

$$F_f \leq F_{fr},$$

since it is not necessarily close to starting the slide. On the other hand, the magnitude of momentum of the impulse around the axis of rotation of the sphere is given by (see (3.51))

$$K_\omega = I_\omega \omega,$$

where I_ω is the moment of inertia of the sphere around a diameter and ω is its angular velocity. From the exercise above we have

$$K_\omega = \frac{2}{5}m\rho^2\omega,$$

so that

$$\dot{K}_\omega = \frac{2}{5}m\rho^2\dot{\omega}. \qquad (3.78)$$

Now, by the Rizal's formula

$$\dot{K}_\omega = F_f \rho,$$

which in combination with (3.78) leads to

$$m\rho\dot{\omega} = \frac{5}{2} F_f. \tag{3.79}$$

Additionally, since $v_A = 0$, we have the cinematic relation

$$v = \rho\omega,$$

and hence,

$$\dot{v} = \rho\dot{\omega}. \tag{3.80}$$

This expression being substituted into (3.79) gives

$$m\dot{v} = \frac{5}{2} F_f$$

and, in view of (3.77), this implies

$$\frac{5}{2} F_f = mg \sin\alpha - F_f,$$

or equivalently,

$$F_f = \frac{2}{7} mg \sin\alpha. \tag{3.81}$$

On the other hand, from (3.76), the maximum friction force that can be generated at contact point A is given by

$$F_{fr} = f_{fr} N,$$

with

$$N = mg \cos\alpha,$$

that is,

$$F_{fr} = f_{fr} mg \cos\alpha. \tag{3.82}$$

The value α^* corresponds to the situation when

$$F_f = F_{fr},$$

which using (3.81) and (3.82) gives

$$\alpha^* = \arctan\left(\frac{7}{2} f_{fr}\right).$$

3.8 Exercises

Exercise 3.1. A uniform circular cone is placed on the base of a smooth horizontal table. The cone is given the angular velocity ω_0, so that the speeds of the points of its axis of symmetry are zero. Show that the angular velocity of the cone will be equal to

$$\omega = \frac{3(1+k)}{13+3k}\omega,$$

if a ball is dropped from its top to the base, the mass of which is k times less than the mass of the cone.

Exercise 3.2. Two identical balls can move without friction on the sides of a right angle located in a horizontal plane. Balls carry charges of different signs and begin to move from a state of rest. Show that they will simultaneously be at the top of the corner.

Exercise 3.3. A homogeneous stick AB of length $2a$ is pivotally fixed at point B. From the end of the stick, the material point D begins to move, the mass of which is equal to the mass of the stick. At the initial moment, the stick is in a horizontal position. Having received the push, it begins to rotate clockwise in a vertical plane. Show that the time T, in which the point D reaches the end A of the stick, is equal to

$$T = \frac{1}{\omega}\arcsin\frac{2a\omega^2}{g}(2-\ln 3),$$

if it moves in such a way that the angular velocity ω of the stick remains constant.

Exercise 3.4. The physical pendulum consists of a homogeneous ball of radius r, suspended on a weightless rod to the fulcrum point O. The lower point of the ball describes a circle of radius R. Another same ball is placed in a circular groove of radius R and rolls along it without slipping (see Fig. 3.22). At the initial moment the balls are on the same level and begin to move without initial speed. Show that the ratio of the highest speeds of the centers of the balls is equal to

$$\frac{v_1}{v_2} = \sqrt{\frac{7r^2+5R^2-10Rr}{7(R-r)^2}}.$$

Show also that with the ratio of

$$R = 2r$$

between R and r, these velocities will be the same.

Exercise 3.5. A homogeneous cylinder of radius r and mass m freely rolls from a stationary cylinder of radius R. The cylinder begins to move from a state of rest as a result of a small impulse (see Fig. 3.23). The coefficient of sliding friction is equal to f. Show that:

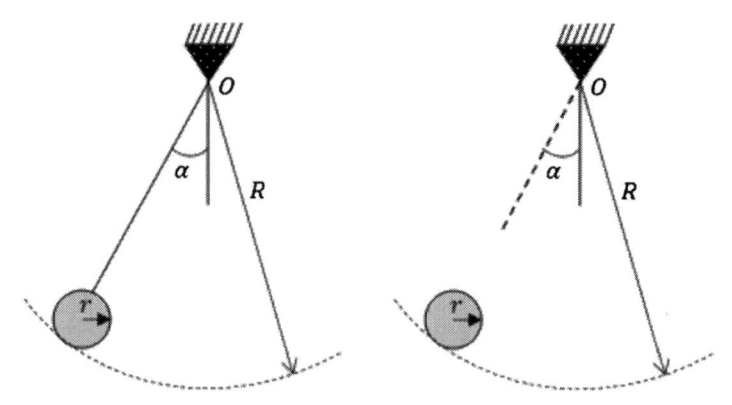

Figure 3.22 The physical pendulum, consisting of a homogeneous ball, is suspended on a weightless rod to the fulcrum point O. Another same ball is in a circular groove and rolls along it.

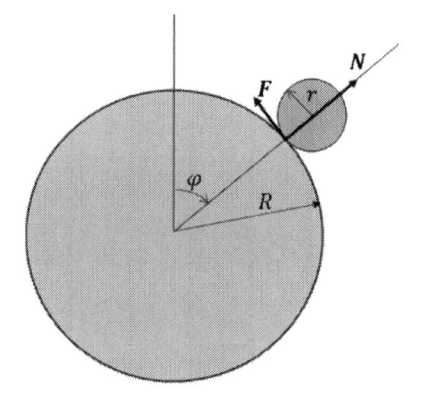

Figure 3.23 A homogeneous cylinder freely rolling from a stationary cylinder.

- all values of the angle φ, at which the rolling occurs without slipping, are given by the inequality

$$\varphi < 2\arctan\frac{\sqrt{1+33f^2}-1}{11f},$$

- the corresponding velocities of the center of the cylinder (as a function of φ) are

$$v(\varphi) = \sqrt{\frac{4}{3}g(R+r)(1-\cos\varphi)},$$

- the normal reaction force $N = N(\varphi)$ is given by

$$N(\varphi) = \frac{mg}{3}(7\cos\varphi - 4),$$

- the friction force $F = F(\varphi)$ is

$$F(\varphi) = \frac{mg}{3}\sin\varphi.$$

Non-inertial and variable-mass systems

Contents

4.1 Non-inertial systems   131
     4.1.1   Newton's second law regarding a relative system   132
     4.1.2   Rizal's theorem in a relative system   134
     4.1.3   Kinetic energy and work in a relative system   137
     4.1.4   Some examples dealing with non-inertial systems   139
4.2 Dynamics of systems with variable mass   142
     4.2.1   Reactive forces and the Meshchersky equation   142
     4.2.2   Tsiolkovsky's rocket formula and other examples   143
4.3 Exercises   151

The relationships obtained in the previous chapters are based on the consideration that the absolute reference system is not accelerated. Systems in which this condition is met are called *inertial*. In this chapter we analyze the dynamics of *non-inertial systems*, that is, systems whose absolute reference undergoes an acceleration. Another consideration made previously is in relation to the mass; this has been assumed constant. The treatment of some cases in which the mass is variable is the other aspect of this chapter. The Meshchersky and Tsiolkovsky's rocket formulas are derived and analyzed.

4.1 Non-inertial systems

Definition 4.1. A coordinate system that does not experience acceleration is called an **inertial system**; otherwise it is called a **non-inertial system**.

Recall that in a system S of material points, referring to an inertial system, the following three relationships, discussed in the previous chapter, are satisfied:

(a) the second law of Newton,

$$\dot{\mathbf{Q}} = \mathbf{F}_{ext} ; \tag{4.1}$$

(b) the Rizal's theorem,

$$\dot{\mathbf{K}}_A = \mathbf{M}_{\mathbf{F}_{ext}, A} + M\left[\mathbf{v}_{CI}, \mathbf{v}_A\right] ; \tag{4.2}$$

(c) the following relationship between the work, developed by the forces acting on S, and the increase of its kinetic energy:

$$dT = \delta A \text{ or } \dot{T} = N. \tag{4.3}$$

4.1.1 Newton's second law regarding a relative system

In this subsection we will solve the problem of rewriting Eqs. (4.1)–(4.3) for the case in which the reference system is non-inertial. To do that, it will be sufficient to rewrite them with respect to a relative reference system.

From what we observed in Chapter 2, it is known that the acceleration of a point $\mathbf{r}_{abs}$ with respect to an absolute coordinate system can be described in terms of the acceleration $\mathbf{w}_O$, the angular velocity $\boldsymbol{\omega}$, and the acceleration $\boldsymbol{\varepsilon}$ of a relative coordinate system with origin O, namely,

$$\mathbf{w}_{abs} = \mathbf{w}_{tr} + \mathbf{w}_{rel} + \mathbf{w}_{cor}, \tag{4.4}$$

where the translation $\mathbf{w}_{tr}$ and Coriolis $\mathbf{w}_{cor}$ accelerations are

$$\mathbf{w}_{tr} := \mathbf{w}_O + [\boldsymbol{\varepsilon}, \mathbf{r}_{rel}] + [\boldsymbol{\omega}, [\boldsymbol{\omega}, \mathbf{r}_{rel}]] \tag{4.5}$$

and

$$\mathbf{w}_{cor} := 2[\boldsymbol{\omega}, \mathbf{v}_{rel}], \tag{4.6}$$

while $\mathbf{r}_{rel}$, $\mathbf{v}_{rel}$, and $\mathbf{w}_{rel}$ represent the relative position, velocity, and acceleration of the point, and are given by

$$\mathbf{r}_{rel} = x\mathbf{i} + y\mathbf{j} + z\mathbf{k}, \ \mathbf{v}_{rel} = \dot{x}\mathbf{i} + \dot{y}\mathbf{j} + \dot{z}\mathbf{k}, \ \mathbf{w}_{rel} = \ddot{x}\mathbf{i} + \ddot{y}\mathbf{j} + \ddot{z}\mathbf{k}$$

(here $\mathbf{i}, \mathbf{j}, \mathbf{k}$ are unitary orths of a non-inertial coordinate system).

Now, it has been seen in Chapter 3 that the inertial center CI of a system S of material points with constant total mass M satisfies the relation

$$\dot{\mathbf{Q}} = M\ddot{\mathbf{r}}_{CI,abs}, \tag{4.7}$$

where $\mathbf{r}_{CI,abs}$ denotes the absolute position of CI. Since

$$\mathbf{w}_{CI,abs} := \ddot{\mathbf{r}}_{CI,abs},$$

expression (4.7) can be represented as

$$\dot{\mathbf{Q}} = M\mathbf{w}_{CI,abs}.$$

This, in combination with Newton's second law (4.1), leads to

$$M\mathbf{w}_{CI,abs} = \mathbf{F}_{ext},$$

or, using (4.4),

$$M\left(\mathbf{w}_{CI,tr} + \mathbf{w}_{CI,rel} + \mathbf{w}_{CI,cor}\right) = \mathbf{F}_{ext}.$$

From the last equation it follows that

$$M\mathbf{w}_{CI,rel} = \mathbf{F}_{ext} - M\mathbf{w}_{CI,tr} - M\mathbf{w}_{CI,cor}. \tag{4.8}$$

Definition 4.2. The vector

$$\mathbf{R}_{CI,tr} := -M\mathbf{w}_{CI,tr} \tag{4.9}$$

is referred to as **inertial translation force**, and the vector

$$\mathbf{R}_{CI,cor} = -M\mathbf{w}_{CI,cor} \tag{4.10}$$

is called **inertial Coriolis force**.

The definitions (4.9) and (4.10) allow to obtain the final expression of (4.8), which is the relative counterpart of the law (4.1) and describes **the dynamics of** CI **with respect to the relative system**:

$$M\mathbf{w}_{CI,rel} = \mathbf{F}_{ext} + \mathbf{R}_{CI,tr} + \mathbf{R}_{CI,cor}. \tag{4.11}$$

The example that appears next illustrates the usefulness of the expression obtained.

Example 4.1. Consider the inclined plane as in Fig. 4.1, which is subject to the acceleration **w** shown. On the plane there is a body of mass m that slides without friction. Determine the magnitude of **w** so that the sliding of the body is towards the highest point of the plane. Look at the plane from the relative coordinate system that is displayed. Since this system is not rotating, in view of (4.5) and (4.6) we have

$$\boldsymbol{\omega} = 0, \quad \boldsymbol{\varepsilon} = 0,$$

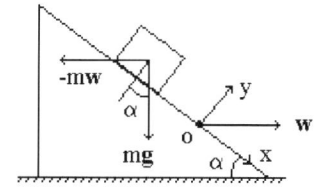

Figure 4.1 Body on an accelerated inclined plane.

implying

$$\mathbf{R}_{CI,tr} = -m\mathbf{w}, \quad \mathbf{R}_{CI,cor} = 0.$$

From (4.11), in order for the body to accelerate in the negative direction of the x-axis, we must have

$$R_{CI,tr} \cos\alpha \geq mg \sin\alpha,$$

or

$$mw \cos\alpha \geq mg \sin\alpha,$$

which finally gives

$$w \geq g \tan\alpha.$$

4.1.2 Rizal's theorem in a relative system

Recall (from Chapter 3) that the moment of the impulse of the system S with respect to pole A is given by

$$\mathbf{K}_A := \sum_{i \in S} \left[\mathbf{r}_{i,A}, m_i \mathbf{v}_{i,abs} \right],$$

where $\mathbf{r}_{i,A}$ denotes the position of the point $i \in S$, with mass m_i, with respect to the pole A. In this context we can introduce the following definition.

Definition 4.3. Let the particle system S be referenced to a relative system with origin O. The relative momentum of the impulse of S with respect to pole A is given by

$$\mathbf{K}_{rel,A} := \sum_{i \in S} \left[\mathbf{r}_{i,A}, m_i \mathbf{v}_{i,rel} \right],$$

where $\mathbf{v}_{i,rel}$ denotes the velocity of the point $i \in S$ relative to O.

In view of this definition we have

$$\dot{\mathbf{K}}_{rel,A} = \sum_{i \in S} \left(\left[\dot{\mathbf{r}}_{i,A}, m_i \mathbf{v}_{i,rel} \right] + \left[\mathbf{r}_{i,A}, m_i \dot{\mathbf{v}}_{i,rel} \right] \right), \tag{4.12}$$

and in particular, if $A = O$, by (2.45) it follows that

$$\mathbf{r}_{i,O} = \mathbf{r}_{i,rel}, \quad \dot{\mathbf{r}}_{i,O} = \mathbf{v}_{i,rel}, \quad \dot{\mathbf{v}}_{i,rel} = \mathbf{w}_{i,rel} + \frac{1}{2}\mathbf{w}_{i,cor},$$

and the relation (4.12) is reduced to

$$\dot{\mathbf{K}}_{rel} := \dot{\mathbf{K}}_{rel,O} = \sum_{i \in S} \left[\mathbf{r}_{i,rel}, m_i \left(\mathbf{w}_{i,rel} + \frac{1}{2}\mathbf{w}_{i,cor} \right) \right] \tag{4.13}$$

since $\left[\mathbf{v}_{i,rel}, m_i \mathbf{v}_{i,rel} \right] = 0$. Using (4.4) we have

$$\mathbf{w}_{i,rel} = \mathbf{w}_{i,abs} - \mathbf{w}_{i,tr} - \mathbf{w}_{i,cor} \tag{4.14}$$

and hence, (4.13) can be expressed as

$$\dot{\mathbf{K}}_{rel} = \sum_{i \in S} \left[\mathbf{r}_{i,rel}, m_i \mathbf{w}_{i,abs} \right] + \sum_{i \in S} \left[\mathbf{r}_{i,rel}, -m_i \mathbf{w}_{i,tr} \right] \\ + \sum_{i \in S} \left[\mathbf{r}_{i,rel}, -\frac{1}{2} m_i \mathbf{w}_{i,cor} \right]. \quad (4.15)$$

Define for the particle $i \in S$ the **inertial translation force** as

$$\mathbf{R}_{i,tr} := -m_i \mathbf{w}_{i,tr} = -m_i \left(\mathbf{w}_O + \left[\boldsymbol{\varepsilon}, \mathbf{r}_{i,rel} \right] + \left[\boldsymbol{\omega}, \left[\boldsymbol{\omega}, \mathbf{r}_{i,rel} \right] \right] \right) \quad (4.16)$$

and the **inertial Coriolis force** as

$$\mathbf{R}_{i,cor} = -\frac{1}{2} m_i \mathbf{w}_{i,cor} = -m_i \left[\boldsymbol{\omega}, \mathbf{v}_{i,rel} \right]. \quad (4.17)$$

Here we have used that $\mathbf{w}_{i,cor} := 2 \left[\boldsymbol{\omega}, \mathbf{v}_{i,rel} \right]$. Since by Newton's second law

$$m_i \mathbf{w}_{i,abs} = \mathbf{F}_i, \quad (4.18)$$

with $\mathbf{F}_i$ as the total force acting on the particle $i \in S$, expression (4.15) takes the form

$$\dot{\mathbf{K}}_{rel} = \sum_{i \in S} \left[\mathbf{r}_{i,rel}, \mathbf{F}_i \right] + \sum_{i \in S} \left[\mathbf{r}_{i,rel}, \mathbf{R}_{i,tr} \right] + \sum_{i \in S} \left[\mathbf{r}_{i,rel}, \mathbf{R}_{i,cor} \right]. \quad (4.19)$$

Recall (also from Chapter 3) that the moment of the external forces with respect to the origin O is given by

$$\mathbf{M}_{\mathbf{F}_{ext},O} := \sum_{i \in S} \left[\mathbf{r}_{i,rel}, \mathbf{F}_{i,ext} \right] = \sum_{i \in S} \left[\mathbf{r}_{i,rel}, \mathbf{F}_i \right].$$

Defining the **moments** $\mathbf{M}_{tr,O}$ and $\mathbf{M}_{cor,O}$ **of the inertial translation forces** and the **inertial Coriolis forces** with respect to the origin O as

$$\mathbf{M}_{tr,O} := \sum_{i \in S} \left[\mathbf{r}_{i,rel}, \mathbf{R}_{i,tr} \right] \quad \text{and} \quad \mathbf{M}_{cor,O} := \sum_{i \in S} \left[\mathbf{r}_{i,rel}, \mathbf{R}_{i,cor} \right],$$

respectively, we arrive at the final expression for the relative counterpart of (4.2), that is, the law that governs **the dynamics of the relative momentum of the impulse**:

$$\dot{\mathbf{K}}_{rel} = \mathbf{M}_{\mathbf{F}_{ext},O} + \mathbf{M}_{tr,O} + \mathbf{M}_{cor,O}. \quad (4.20)$$

Expression (4.20) takes simple forms in some particular cases.

a) For example, if

$$\mathbf{v}_{i,rel} = 0 \quad \forall i \in S,$$

then

$$\mathbf{R}_{i,cor} = -2m_i\left[\boldsymbol{\omega}, \mathbf{v}_{i,rel}\right] = 0, \quad \mathbf{M}_{cor,O} := \sum_{i \in S} m_i\left[\mathbf{r}_{i,rel}, \mathbf{R}_{i,cor}\right] = 0,$$

and hence,

$$\dot{\mathbf{K}}_{rel} = \mathbf{M}_{\mathbf{F}_{ext},O} + \mathbf{M}_{tr,O}. \tag{4.21}$$

b) Another simplification occurs if the relative system does not rotate, that is,

$$\boldsymbol{\omega} \equiv 0.$$

In this case (4.16) has the reduced form

$$\mathbf{R}_{i,tr} = -m_i \mathbf{w}_O,$$

so that

$$\left.\begin{aligned}\mathbf{M}_{tr,O} &= \sum_{i \in S}\left[\mathbf{r}_{i,rel}, -m_i \mathbf{w}_O\right] = \\ &\left[-\sum_{i \in S} m_i \mathbf{r}_{i,rel}, \mathbf{w}_O\right] = \left[\mathbf{r}_{CI,rel}, -M\mathbf{w}_O\right],\end{aligned}\right\} \tag{4.22}$$

where the definition of the inertial center has been used.

The example below uses the newly obtained results, in particular expression (4.22).

Example 4.2. Fig. 4.2 shows a sphere, of radius ρ and mass m uniformly distributed, resting against a step of height $h < \rho$. Determine the magnitude of the horizontal acceleration **w** that the "floor" must have for the sphere to climb the step. Look at the sphere in a relative coordinate system with origin O at the point around which the rotation of the sphere is to be verified. Since the coordinate system moves with the solid, it is true that $\mathbf{v}_{i,rel} = 0 \ \forall i \in S$. So, formula (4.22) is applicable. In view of the fact that in the limit case we have

$$\boldsymbol{\varepsilon} = 0, \quad \boldsymbol{\omega} = 0,$$

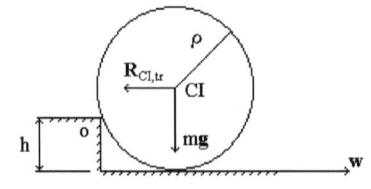

Figure 4.2 Sphere against a step.

by (4.22) we have

$$\mathbf{M}_{tr,O} = [\mathbf{r}_{CI,rel}, -m\mathbf{w}].$$

Moreover, the reaction of the top point of the step on the sphere has null moment. Therefore,

$$\mathbf{M}_{\mathbf{F}_{ext},O} = \sum_{i \in S}[\mathbf{r}_{i,rel}, m_i \mathbf{g}] = \left[\frac{1}{m}\sum_{i \in S} m_i \mathbf{r}_{i,rel}, m\mathbf{g}\right] = [\mathbf{r}_{CI,rel}, m\mathbf{g}].$$

The limit condition is given by

$$M_{tr,O} = M_{\mathbf{F}_{ext},O},$$

that is to say,

$$w(\rho - h) = g\sqrt{\rho^2 - (\rho - h)^2},$$

and the sphere goes up when

$$w > g\sqrt{\left(\frac{\rho}{\rho - h}\right)^2 - 1}.$$

4.1.3 Kinetic energy and work in a relative system

From what we observed in Chapter 3, the kinetic energy of a system of material points S referenced to an absolute system is given by

$$T := \frac{1}{2}\sum_{i \in S} m_i \left(\mathbf{v}_{i,abs}, \mathbf{v}_{i,abs}\right),$$

where $\mathbf{v}_{i,abs}$ represents the velocity of the point $i \in S$ with mass m_i with respect to the absolute system. Likewise, in the same Chapter 3 the amount

$$T_{rel,O} := \frac{1}{2}\sum_{i \in S} m_i \left(\mathbf{v}_{i,O}, \mathbf{v}_{i,O}\right), \qquad (4.23)$$

where O is the origin of a relative system and $\mathbf{v}_{i,O}$ is the velocity of $i \in S$ relative to O, was called the kinetic energy of S relative to O. For clarity of the notation, in what follows $T_{rel,O}$ will be denoted simply T_{rel} and will be called **relative kinetic energy**, while $\mathbf{v}_{i,O}$ will be denoted $\mathbf{v}_{i,rel}$, whereupon (4.23) adopts the expression

$$T_{rel} := \frac{1}{2}\sum_{i \in S} m_i \left(\mathbf{v}_{i,rel}, \mathbf{v}_{i,rel}\right). \qquad (4.24)$$

Taking into account the expression for the differential of an internal product in (4.24) and the consideration that m_i is constant we have

$$dT_{rel} = \sum_{i \in S} m_i \left(\dot{\mathbf{v}}_{i,rel}dt, \mathbf{v}_{i,rel}\right) = \sum_{i \in S} m_i \left(\dot{\mathbf{v}}_{i,rel}, \mathbf{v}_{i,rel}dt\right),$$

where using the relations (2.45)

$$\dot{\mathbf{v}}_{i,rel} = \mathbf{w}_{i,rel} + \frac{1}{2}\mathbf{w}_{i,cor}, \quad \mathbf{v}_{i,rel}dt = d\mathbf{r}_{i,rel},$$

with $\mathbf{r}_{i,rel}$ denoting the position of the point $i \in S$ with respect to the relative system, we arrive at

$$dT_{rel} = \sum_{i \in S} \left(m_i \mathbf{w}_{i,rel}, d\mathbf{r}_{i,rel}\right) + \frac{1}{2}\sum_{i \in S} \left(m_i \mathbf{w}_{i,cor}, d\mathbf{r}_{i,rel}\right).$$

Now using the alternative expression (4.5) of $\mathbf{w}_{i,rel}$, we then have

$$dT_{rel} = \sum_{i \in S} \left(m_i \mathbf{w}_{i,abs}, d\mathbf{r}_{i,rel}\right) + \sum_{i \in S} \left(-m_i \mathbf{w}_{i,tr}, d\mathbf{r}_{i,rel}\right) +$$

$$\sum_{i \in S} \left(-m_i \mathbf{w}_{i,cor}, d\mathbf{r}_{i,rel}\right) + \frac{1}{2}\sum_{i \in S} \left(m_i \mathbf{w}_{i,cor}, d\mathbf{r}_{i,rel}\right) =$$

$$\sum_{i \in S} \left(m_i \mathbf{w}_{i,abs}, d\mathbf{r}_{i,rel}\right) - \sum_{i \in S} \left(m_i \mathbf{w}_{i,tr}, d\mathbf{r}_{i,rel}\right) - \frac{1}{2}\sum_{i \in S} \left(m_i \mathbf{w}_{i,cor}, d\mathbf{r}_{i,rel}\right).$$

And considering now the definitions of the translatory and inertial Coriolis forces as well as Newton's second law (4.16)–(4.18) we find

$$dT_{rel} = \sum_{i \in S} \left(\mathbf{F}_i, d\mathbf{r}_{i,rel}\right) + \sum_{i \in S} \left(\mathbf{R}_{i,tr}, d\mathbf{r}_{i,rel}\right) + \sum_{i \in S} \left(\mathbf{R}_{i,cor}, d\mathbf{r}_{i,rel}\right),$$

where

$$\mathbf{F}_i = m_i \mathbf{w}_{i,abs}, \quad \mathbf{R}_{i,tr} = -m_i \mathbf{w}_{i,tr}, \quad \mathbf{R}_{i,cor} = -\frac{1}{2}m_i \mathbf{w}_{i,cor} = -\left[\boldsymbol{\omega}, \mathbf{v}_{i,rel}\right].$$

Finally, remembering the concept of elementary work of a force on a trajectory, we have as the final expression for the relative counterpart of (4.3)

$$dT_{rel} = \delta A + \delta A_{tr} + \delta A_{cor}, \tag{4.25}$$

with

$$\left.\begin{array}{l} \delta A := \sum_{i \in S} \left(\mathbf{F}_i, d\mathbf{r}_{i,rel}\right), \quad \delta A_{tr} := \sum_{i \in S} \left(\mathbf{R}_{i,tr}, d\mathbf{r}_{i,rel}\right), \\ \delta A_{cor} := \sum_{i \in S} \left(\mathbf{R}_{i,cor}, d\mathbf{r}_{i,rel}\right). \end{array}\right\} \tag{4.26}$$

However, there is an interesting simplification of (4.26).

Lemma 4.1. *We have*

$$\delta A_{cor} = 0, \quad \forall t \geq 0,$$

*which means that the **Coriolis forces** (normal ones as well as inertial) **do not produce any work**.*

Proof. In view of (4.26) it follows that

$$\delta A_{cor} = -\sum_{i \in S} \left(m_i \left[\boldsymbol{\omega}, \mathbf{v}_{i,rel} \right], d\mathbf{r}_{i,rel} \right),$$

and taking into account that

$$d\mathbf{r}_{i,rel} = \mathbf{v}_{i,rel} dt,$$

we get

$$\delta A_{cor} = -\sum_{i \in S} \left(m_i \left[\boldsymbol{\omega}, \mathbf{v}_{i,rel} \right], \mathbf{v}_{i,rel} \right) dt = 0$$

by the existing orthogonality of $\left[\boldsymbol{\omega}, \mathbf{v}_{i,rel} \right]$ and $\mathbf{v}_{i,rel}$. $\square$

In view of this fact the final expression (4.25) looks as

$$dT_{rel} = \delta A + \delta A_{tr}. \tag{4.27}$$

4.1.4 Some examples dealing with non-inertial systems

Some aspects of the results obtained in the previous developments are illustrated in the exercises that are now presented.

Example 4.3. Assume that a ring with the radius ρ, shown in Fig. 4.3, is rotating with angular velocity $\boldsymbol{\omega}$ about the indicated axis. The distance between the axis of rotation and the center of the ring is $a > \rho$. A small ring of mass m surrounds the wire that forms the ring, being able to slide on it without friction. We need to find the value of ω such that the moving small ring maintains the angle φ_0 with respect to the vertical line. The answer can be achieved by expression (4.11). To do this, the relative coordinate system shown is fixed to the ring. In these conditions, since the ring is small, we have

$$\mathbf{r}_{CI,rel} = \mathbf{v}_{CI,rel} = \mathbf{w}_{CI,rel} = 0.$$

Moreover

$$\mathbf{w}_O = \mathbf{w}_{CI,abs}.$$

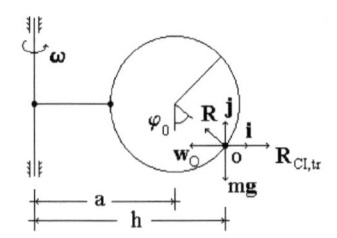

Figure 4.3 Ring in rotation with a small ring.

whereby

$$\mathbf{R}_{CI,tr} = -m\mathbf{w}_{CI,abs}, \quad \mathbf{R}_{CI,cor} = 0.$$

So, the expression to be considered is reduced to

$$\mathbf{F}_{ext} - m\mathbf{w}_{CI,abs} = 0. \tag{4.28}$$

We also have

$$\mathbf{F}_{ext} = \mathbf{R} + m\mathbf{g},$$

where $\mathbf{R}$ denotes the action of the small ring on the ring, which, in view of the absence of friction and considering that the small ring does not move with respect to the main ring, has a radial direction. It is also known (Chapter 2) that the speed of the ring has magnitude

$$v = \omega h \text{ with } h := a + \rho \sin \varphi_0. \tag{4.29}$$

Then, from the observations in (4.29),

$$\mathbf{w}_{CI,abs} = -\frac{v^2}{h}\mathbf{i} = -\omega^2 h \mathbf{i}.$$

So, from (4.28)

$$m\omega^2 h - R \sin \varphi_0 = 0, \quad -mg + R \cos \varphi_0 = 0,$$

which gives

$$\frac{\omega^2 h}{g} = \tan \varphi_0,$$

and with the value of h, given in (4.29) and which is positive, it follows that

$$\omega = \sqrt{\frac{g \tan \varphi_0}{a + \rho \sin \varphi_0}}.$$

Non-inertial and variable-mass systems   141

Example 4.4. Consider the series of n articulated bars in Fig. 4.4. Each bar has length l and mass m. The upper fixed point experiences an acceleration w. Calculate the angles that the bars form with the vertical one when all the bars have reached the acceleration w. The force diagram of the k-th bar in the state in which all the bars have the acceleration w, and therefore do not rotate, is shown in Fig. 4.5. In this diagram, a relative system originating in O is fixed to the bar. The force $\mathbf{R}_k$ represents the action exerted by the previous bar, while the forces $-(n-k)m\mathbf{w}$ and $(n-k)m\mathbf{g}$ denote the components of the traction exerted by the next bar. Since the relative system has $\boldsymbol{\varepsilon} = \boldsymbol{\omega} = 0$, the expression to be used is (4.21) with $\dot{\mathbf{K}}_{rel} = 0$. In this case

$$\mathbf{M}_{\mathbf{F}_{ext},O} + \mathbf{M}_{tr,O} = 0. \tag{4.30}$$

Figure 4.4 Accelerated series of articulated bars.

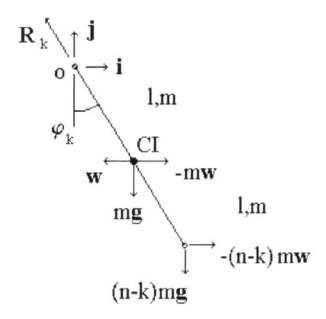

Figure 4.5 Diagram of forces on a bar.

Proceeding to calculate the terms of (4.30), we get

$$\mathbf{M}_{\mathbf{F}_{ext},O} := \sum_{i \in S} \left[\mathbf{r}_{i,rel}, \mathbf{F}_{i,ext}\right] =$$

$$\sum_{i \in S} \left[\mathbf{r}_{i,rel}, m_i \mathbf{g}\right] + (n-k)ml\left(-g\sin\varphi_k + w\cos\varphi_k\right)\mathbf{k},$$

where

$$\sum_{i \in S} \left[\mathbf{r}_{i,rel}, m_i \mathbf{g}\right] = \left[\mathbf{r}_{CI,rel}, m\mathbf{g}\right] = -\frac{1}{2}mgl\sin\varphi_k \mathbf{k}.$$

Therefore

$$\mathbf{M}_{F_{ext},O} = ml\left[-\frac{1}{2}g\sin\varphi_k + (n-k)(-g\sin\varphi_k + w\cos\varphi_k)\right]\mathbf{k} \qquad (4.31)$$

and

$$\mathbf{M}_{tr,O} = \sum_{i\in S}[\mathbf{r}_{i,rel}, -m_i\mathbf{w}] = [\mathbf{r}_{CI,rel}, -m\mathbf{w}] = \frac{1}{2}mwl\cos\varphi_k\mathbf{k}. \qquad (4.32)$$

Substitution of (4.31) and (4.32) into (4.30) leads, finally, to the following conclusion:

$$\tan\varphi_k = \frac{w}{g}, \quad k = 1, \ldots, n,$$

which means that all angles are equal.

4.2 Dynamics of systems with variable mass

This section deals with a problem different from that of the previous section: it is considered that the mass of the systems is not constant, which will generalize some dynamic relationships obtained in Chapter 3. In this section no more relative systems are considered, so the quantities are with respect to an absolute system.

4.2.1 Reactive forces and the Meshchersky equation

Recall that in a system S of material points, Newton's second law has the expression

$$\dot{\mathbf{Q}} = \mathbf{F}_{ext}, \qquad (4.33)$$

where the impulse $\mathbf{Q}$ of the system S is calculated as

$$\mathbf{Q} = \sum_{i\in S} m_i \mathbf{v}_i,$$

so that

$$\dot{\mathbf{Q}} = \frac{d}{dt}\sum_{i\in S} m_i \mathbf{v}_i$$

or, using the fact that the velocity of the center of inertia is given by

$$\mathbf{v}_{CI} = \frac{1}{M}\sum_{i\in S} m_i \mathbf{v}_i,$$

where M is the total mass of S. In the case when M is allowed to be variable we arrive at the following dynamic equation:

$$\dot{\mathbf{Q}} = \frac{d}{dt}(M\mathbf{v}_{CI}) = \dot{M}\mathbf{v}_{CI} + M\dot{\mathbf{v}}_{CI}. \tag{4.34}$$

If a **reactive force** is defined as

$$\mathbf{F}_{reac} := -\dot{M}\mathbf{v}_{CI} = -\operatorname{sign}(\dot{M})\mu\mathbf{v}_{CI}, \tag{4.35}$$

where

$$\mu := |\dot{M}|$$

is referred to as **expenditure**, the Newton's second law (4.33) allows to rewrite (4.34) as

$$M\dot{\mathbf{v}}_{CI} = \mathbf{F}_{ext} + \mathbf{F}_{reac}. \tag{4.36}$$

The relationship (4.36) is called the **Meshchersky equation**.[1]

The examples and exercises that follow illustrate several interesting cases of the dynamics of variable-mass systems.

4.2.2 Tsiolkovsky's rocket formula and other examples

Example 4.5. Consider the mobile tank shown in Fig. 4.6. The container has a hole through which its contents leak and the system is not subject to external forces. Considering that the expenditure is constant over time, we will try to determine the law that follows the speed of the mobile. In view of the fact that the mass of the system is decreasing, the law (4.36) adopts the simplified form

$$M(t)\dot{v}_{CI}(t) = \mu v_{CI}(t). \tag{4.37}$$

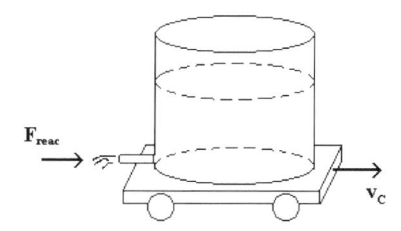

Figure 4.6 Tank with drain.

Since the expense is constant, we have

$$M(t) = M(0) - \mu t,$$

---
[1] It was obtained by I.V. Meshchersky in 1897 for a variable-mass body of the material points.

so (4.37) is expressed as

$$\frac{dv_{CI}}{v_{CI}} = \frac{\mu}{M(0) - \mu t} dt,$$

whose integration with respect to time leads to

$$\ln \frac{v_{CI}(t)}{v_{CI}(0)} = -\ln \frac{M(0) - \mu t}{M(0)}$$

or, equivalently, to

$$v_{CI}(t) = v_{CI}(0) \frac{1}{1 - \frac{\mu}{M(0)} t},$$

valid for any $0 \leq t \leq \frac{M(0)}{\mu}$.

Example 4.6. Fig. 4.7 shows a rocket from which mass with a relative velocity **u** emerges. Let us try to find the expression for the speed of the rocket. Recall that the general impulse of the system at time t is given by

$$\mathbf{Q}(t) = M(t) \mathbf{v}(t),$$

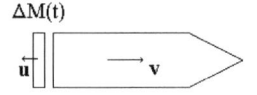

Figure 4.7 Rocket shedding mass.

whereas at time $t + \Delta t$, at which the mass quantity $\Delta M(t)$ has the additional speed $\mathbf{u}(t)$, it is

$$\mathbf{Q}(t + \Delta t) = M(t)[\mathbf{v}(t) + \Delta \mathbf{v}(t)] + \Delta M(t) \mathbf{u}(t),$$

from which it follows that

$$\frac{\mathbf{Q}(t + \Delta t) - \mathbf{Q}(t)}{\Delta t} = M(t) \frac{\Delta \mathbf{v}(t)}{\Delta t} + \frac{\Delta M(t)}{\Delta t} \mathbf{u}(t).$$

If $\Delta t \to 0$, by the definition of derivative, we obtain

$$\dot{\mathbf{Q}} = M(t) \dot{\mathbf{v}}(t) + \dot{M}(t) \mathbf{u}(t),$$

where, by the Newton's second law (4.33) $\dot{\mathbf{Q}} = \mathbf{F}_{ext}$, the following relationship follows:

$$M(t) \dot{\mathbf{v}}(t) + \dot{M}(t) \mathbf{u}(t) = \mathbf{F}_{ext}. \tag{4.38}$$

Non-inertial and variable-mass systems 145

In the particular case where $\mathbf{F}_{ext} \equiv 0$ and $\mathbf{u}$ is collinear with $\mathbf{v}(0)$, Eq. (4.38) is reduced to

$$\dot{v}(t) = -u(t) \frac{\dot{M}(t)}{M(t)}. \tag{4.39}$$

Note that here $\dot{M}(t)$ is negative, since the mass of the rocket is decreasing in time. If in addition the magnitude of $\mathbf{u}$ is constant, the integration of (4.39) leads to the expression

$$v(t) = v(0) + u \ln \frac{M(0)}{M(t)}, \tag{4.40}$$

which is known as the **Tsiolkovsky rocket formula**.[2]

The Tsiolkovsky rocket equation (4.39), classical rocket equation, or ideal rocket equation is a mathematical equation that describes the motion of vehicles that follow the basic principle of a rocket: a device that can apply acceleration to itself using thrust by expelling part of its mass with high velocity can thereby move due to the conservation of momentum.[3]

Example 4.7. In Fig. 4.8 a rocket is shown with n fuel tanks, each of which contains a mass m_0, in addition to a capsule of mass m. Assuming that u is the relative speed with which the combustion gases are released and supposing that the rocket starts from a given level with the velocity $v(0) = 0$ and that v^* is the speed that is required to be reached by the capsule, we need to calculate the number n of tanks required. This problem can be resolved with the formula of Tsiolkovsky (4.40). To do that, note that

$$v(0) = 0, \quad M(0) = nm_0 + m.$$

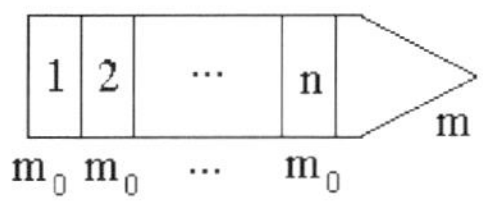

Figure 4.8 Rocket with n fuel tanks.

---

[2] Konstantin Eduardovich Tsiolkovsky (September 17, 1857–September 19, 1935) was a Russian rocket scientist and pioneer of the astronautic theory. Along with the French Robert Esnault-Pelterie, the German Hermann Oberth, and the American Robert H. Goddard, he is considered to be one of the founding fathers of modern rocketry and astronautics. His works later inspired leading Soviet rocket engineers such as Sergei Korolev and Valentin Glushko and contributed to the success of the Soviet space program.

[3] The equation is named after Russian scientist Konstantin Tsiolkovsky, who independently derived it and published it in his 1903 work. The equation had been derived earlier by the British mathematician William Moore in 1810, and later published in a separate book in 1813. The minister William Leitch, who was a capable scientist, also independently derived the fundamentals of rocketry in 1861.

While the derivation of the rocket equation is a straightforward calculus exercise, Tsiolkovsky is honored as being the first to apply it to the question of whether rockets could achieve speeds necessary for space travel.

Robert Goddard in America independently developed the equation in 1912 when he began his research to improve rocket engines for possible space flight. Hermann Oberth in Europe independently derived the equation about 1920 as he studied the feasibility of space travel.

If we denoted by $\bar{M}(t)$ the amount of fuel mass, consumed up to time t, we obtain the expression

$$M(t) = nm_0 + m - \bar{M}(t).$$

So, from formula (4.40) it follows that

$$v(t) = u \ln \frac{nm_0 + m}{nm_0 + m - \bar{M}(t)}. \tag{4.41}$$

Now, at the moment t^* when $v(t) = v^*$, the mass consumed should be equal to

$$\bar{M}(t^*) = nm_0.$$

Hence by (4.41) we get

$$v^* = u \ln \frac{nm_0 + m}{m},$$

from which it follows that

$$n = \frac{m}{m_0} \left[\exp\left(\frac{v^*}{u}\right) - 1 \right].$$

Since n can only take integer values we may conclude that

$$n = \text{int} \left\{ \frac{m}{m_0} \left[\exp\left(\frac{v^*}{u}\right) - 1 \right] \right\} + 1,$$

where $\text{int}\{\cdot\}$ denotes the "integer part" function.

Example 4.8. The container in Fig. 4.9 is a hollow cylinder of radius ρ that can rotate about the vertical axis. In addition to the possible external forces, the content of the cylinder is leaking tangentially with a relative speed of magnitude $u(t)$. Consequently, an action that rotates the container is exerted. Let us determine the expression that governs the dynamics of angular velocity.

From Section 3.2 of Chapter 3, it is known that for the considered configuration of the problem the impulse moment $K_\omega(t)$ at time t is given by

$$K_\omega(t) = I_\omega(t) \omega(t), \tag{4.42}$$

whereas for the instant $t + \Delta t$, with $\Delta t > 0$, in which the mass $\Delta M(t)$ has come out, the impulse moment $K_\omega(t + \Delta t)$ is

$$K_\omega(t + \Delta t) = I_\omega(t) [\omega(t) + \Delta \omega(t)] + \rho \Delta M(t) u(t), \tag{4.43}$$

so, by (4.42) and (4.43) it follows that

$$\frac{K_\omega(t + \Delta t) - K_\omega(t)}{\Delta t} = I_\omega(t) \frac{\Delta \omega(t)}{\Delta t} + \rho u(t) \frac{\Delta M}{\Delta t}.$$

Non-inertial and variable-mass systems 147

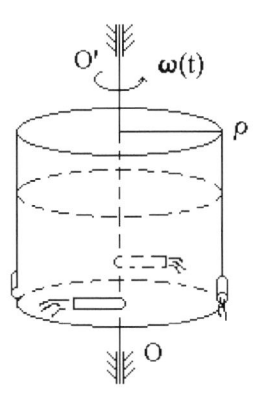

Figure 4.9 Cylinder driven by the tangential leakage of its contents.

If Δt tends to zero, by the definition of the corresponding derivatives,

$$\dot{M}(t) := \lim_{\Delta t \to 0} \frac{\Delta M(t)}{\Delta t}$$

implies

$$\dot{K}_\omega(t) = I_\omega(t)\dot{\omega}(t) + \rho u(t)\dot{M}. \tag{4.44}$$

Now, considering the relationship

$$\dot{K}_\omega(t) = \left(\mathbf{M}_{\mathbf{F}_{ext,O}}, \mathbf{e}_{OO'}\right),$$

where $\mathbf{M}_{\mathbf{F}_{ext,O}}$ represents the moment of the external forces with respect to the point O and $\mathbf{e}_{OO'}$ denotes a unit vector in the direction O to O', expression (4.44) adopts the final expression

$$I_\omega(t)\dot{\omega}(t) = \left(\mathbf{M}_{\mathbf{F}_{ext,O}}, \mathbf{e}_{OO'}\right) - \rho u \dot{M}. \tag{4.45}$$

In our case the problem on the container in Fig. 4.9 has as conditions

$$\omega(0) = 0, \quad u = \underset{t}{\text{const}}, \quad \mathbf{F}_{ext} \equiv \mathbf{0}, \quad h(0) = h_0, \quad |\dot{M}| = \mu = \underset{t}{\text{const}}, \tag{4.46}$$

where $h(t)$ represents the height of the content at time t. Let us obtain the expression of

$$\omega = \omega(t) \text{ and } \omega = \omega(h).$$

From (4.45) and considering the conditions (4.46) we have

$$I_\omega(t)\dot{\omega}(t) = -\rho u \dot{M},$$

whose integration is reached:

$$\omega(t) = -\rho u \int_0^t \frac{dM}{I_\omega(t)}.$$

But recall from Chapter 3 that

$$I_\omega(t) = \frac{1}{2} M(t) \rho^2,$$

which leads to

$$\omega(t) = 2\frac{u}{\rho} \ln \frac{M(0)}{M(t)}. \tag{4.47}$$

If we denote by δ the density of the content of the container, we have the following relationships:

$$M(0) = \delta \pi \rho^2 h_0, \quad M(t) = \pi \delta \rho^2 h(t). \tag{4.48}$$

This allows to rewrite (4.47) in the form

$$\omega(t) = 2\frac{u}{\rho} \ln \frac{h_0}{h(t)}, \tag{4.49}$$

which represents one of the requested dependences:

$$\omega(h) = 2\frac{u}{\rho} \ln \frac{h_0}{h}.$$

Also from (4.48) we have

$$\dot{M}(t) = \pi \delta \rho^2 \dot{h}(t),$$

so that

$$\dot{h}(t) = \frac{\dot{M}(t)}{\pi \delta \rho^2}, \tag{4.50}$$

and considering that the expenditure $\mu := |\dot{M}(t)|$ is constant and $\dot{M}(t) = -\mu \leq 0$, integration of (4.50) gives

$$h(t) = h_0 - \frac{\mu}{\pi \delta \rho^2} t. \tag{4.51}$$

Finally, substituting (4.51) in (4.49) yields the other sought expression:

$$\omega(t) = 2\frac{u}{\rho} \ln \frac{h_0}{h_0 - \frac{\mu}{\pi \delta \rho^2} t}.$$

The example that follows is interesting because a very important differential equation (namely, *Bernoulli's equation*) appears in the solution process.

Example 4.9 (Kelly problem). Suppose that one end of the chain in Fig. 4.10 is falling. We will try to determine the law that governs the length $x(t)$ of the segment that is in the vacuum (no friction appears), considering as initial conditions

$$x(0) = \dot{x}(0) = 0.$$

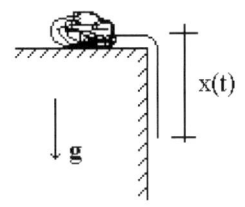

Figure 4.10 Chain with one end falling.

If the constant γ represents the mass per unit length of the chain, by Newton's second law (4.33), applied to the segment, we have

$$\frac{d}{dt}(\gamma x \dot{x}) = \gamma x g,$$

from which it follows that

$$\dot{x}^2 + x\ddot{x} = xg. \tag{4.52}$$

The nonlinear differential equation that is obtained is called the **Bernoulli equation**. To solve Eq. (4.52), note that the velocity $\dot{x}$ is a function of x, and then, with the notation

$$\dot{x} = v(x), \tag{4.53}$$

by temporary derivation we get

$$\ddot{x} = \frac{d}{dt}v(x) = \frac{d}{dx}v(x)\dot{x} = v'(x)v(x), \tag{4.54}$$

where v' denotes the derivative of v with respect to its argument x. Note now that

$$v'(x)v(x) = \frac{1}{2}\frac{d}{dx}v^2(x). \tag{4.55}$$

So, the relationships (4.53)–(4.55) allow to rewrite (4.52) as

$$v^2(x) + \frac{1}{2}x\frac{d}{dx}v^2(x) = xg,$$

or, with the change of variable

$$u(x) = v^2(x),$$

as

$$u(x) + \frac{1}{2}xu'(x) = xg. \qquad (4.56)$$

If for $u(x)$ we propose the form

$$u(x) = kx, \quad k = \underset{x}{\text{const}}, \quad x \geq 0,$$

its derivative with respect to x is

$$u'(x) = k,$$

and consequently (4.56) is represented as

$$kx + \frac{1}{2}xk = xg,$$

from which it follows that

$$k = \frac{2}{3}g,$$

and therefore

$$v^2(x) = u(x) = \frac{2}{3}gx,$$

or, by (4.53),

$$\dot{x}^2 = \frac{2}{3}gx,$$

implying

$$\dot{x} = \sqrt{\frac{2}{3}g}\sqrt{x}.$$

Separating variables, it can be represented as

$$\frac{dx}{\sqrt{x}} = \sqrt{\frac{2}{3}g}\,dt,$$

whose integration, given that $x(0) = 0$, leads to

$$2\sqrt{x} = \sqrt{\frac{2}{3}g}\,t,$$

or finally,

$$x(t) = \frac{gt^2}{6}. \qquad (4.57)$$

Expression (4.57) is referred to as the **Kelly formula**.

4.3 Exercises

Exercise 4.1. A homogeneous disk can roll without sliding along the horizontal directional axis Ox, rotating with a constant angular velocity ω around the vertical axis of Oy. Show that the law of relative motion of the disk is described by the equation

$$3\ddot{x}(t) = 2\omega x(t),$$

and its dynamics may be expressed as

$$x(t) = C_1 e^{\lambda t} + C_2 e^{-\lambda t},$$

where C_1, C_2 are constants depending on the initial values $x(0)$, $\dot{x}(0)$ and

$$\lambda = \frac{\sqrt{6}}{3}\omega.$$

Exercise 4.2. A mine has been dug in the Earth, the direction of which at each point coincides with the direction of the plumb at this point. Show that the shape of the mine is given by the formula

$$x = ay^\gamma, \ 0 < \gamma \le 1,$$

if the Earth is considered a homogeneous rotating ball.

Exercise 4.3. A thin flexible inextensible ABC thread is laid in a vertical smooth pipe of a sufficiently small diameter. The ends A and C are fixed, and point B occupies the lowest position, with $BC = s_0$ (see Fig. 4.11). At some moment, the end C is released without initial velocity. Show that the dependence of the speed v of the moving part of the thread on its length s is described by the formula

$$v(s) = \frac{2}{3s}\sqrt{3g\left(s_0^3 - s^3\right)}.$$

Find also the time it takes for the thread to fully straighten.

Exercise 4.4. The jet vessel is driven by a pump that draws water of density ρ through the inlet of the horizontal channel and throws it in the opposite direction (see Fig. 4.12). The relative velocity of the water at the inlet is u. The area of the inlet and

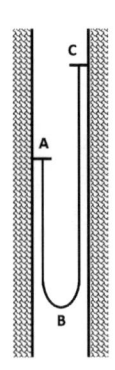

Figure 4.11 A thin flexible inextensible thread in a vertical smooth pipe.

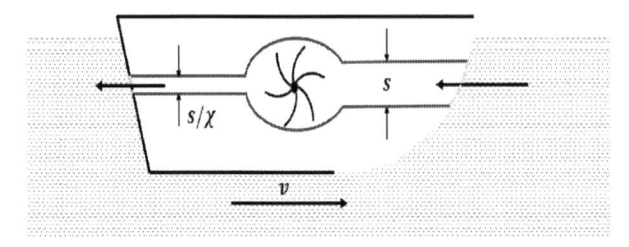

Figure 4.12 A jet vessel driven by a pump.

outlet is S and $S/\varkappa$ ($\varkappa > 1$), respectively. The mass of the vessel with the water in it is equal to m. Show that the acceleration time of the vessel from speed v_0 to v_1 is

$$t_{accel} = \frac{mp}{2k} \ln\left(\frac{1 + pv_1}{1 - pv_1} \cdot \frac{1 - pv_0}{1 + pv_0}\right),$$

where

$$p = \frac{1}{u}\sqrt{\frac{k}{\rho S (\varkappa - 1)}},$$

assuming that it is affected by the force F_{res} of water resistance proportional to the square of the speed, namely,

$$F_{res} = -kv^2.$$

Exercise 4.5. Show that the impulse of a rocket, which moves linearly in the absence of external forces, reaches its maximum value at a time when the rocket velocity v becomes equal to the velocity of the gas u, which is assumed to be constant.

Euler's dynamic equations

Contents

5.1 Tensor of inertia 153
5.2 Relative kinetic energy and impulse momentum 156
    5.2.1 Relative kinetic energy 156
    5.2.2 Relative impulse momentum 157
5.3 Some properties of inertial tensors 158
    5.3.1 Tensor of inertia as a non-negative symmetric matrix 158
    5.3.2 Eigenvalues and eigenvectors of inertial tensors 159
    5.3.3 Examples using tensors of inertia 163
5.4 Euler's dynamic equations 174
    5.4.1 Special cases of Euler's equations 175
5.5 Dynamic reactions caused by the gyroscopic moment 182
5.6 Exercises 185

In this chapter we continue with the study of the dynamics of solids. The dynamic equations corresponding to the rotation of bodies are obtained. They are called *Euler's dynamic equations*. For this purpose, a fundamental concept of the geometry of solids is introduced, namely, the inertia tensor, which is key in the description of the equations sought. The inertial tensor will allow calculating fundamental quantities such as kinetic energy and impulse moment with reduced expressions. In the central part of the chapter the proposed objective is achieved, once some main properties of the inertia tensor have been stated. The chapter concludes with the application (not trivial, but very productive) of Euler's dynamic equations to the study of special systems such as the gyroscope and considers in detail arising dynamic reactions. Several examples and exercises illustrate the presented theory.

5.1 Tensor of inertia

Consider a rigid body shown in Fig. 5.1. It is desired to calculate the moment of inertia of that body with respect to the axis with direction given by the unit vector **e**. With reference to Fig. 5.1 the moment of inertia is given by

$$I_\mathbf{e} := \sum_{i \in S} m_i h_i^2, \tag{5.1}$$

where

$$h_i^2 = r_i^2 - (OP_i)^2 \tag{5.2}$$

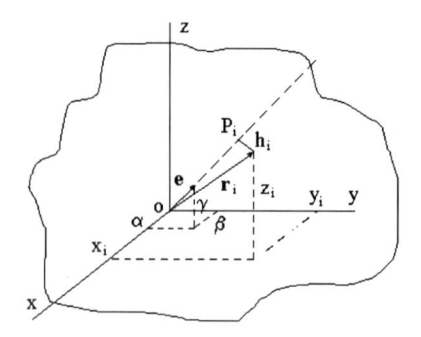

Figure 5.1 Solid body and a generic axis.

with

$$OP_i = (\mathbf{e}, \mathbf{r}_i). \tag{5.3}$$

Considering that

$$\mathbf{r}_i = \begin{pmatrix} x_i & y_i & z_i \end{pmatrix}^\mathsf{T}$$

and

$$\mathbf{e} = \begin{pmatrix} \alpha & \beta & \gamma \end{pmatrix}^\mathsf{T},$$

(5.3) can be represented as

$$OP_i = \alpha x_i + \beta y_i + \gamma z_i.$$

Therefore expression (5.2) becomes

$$h_i^2 = x_i^2 + y_i^2 + z_i^2 - (\alpha x_i + \beta y_i + \gamma z_i)^2 = \\ x_i^2 + y_i^2 + z_i^2 - \alpha^2 x_i^2 - \beta^2 y_i^2 - \gamma^2 z_i^2 - 2\alpha\beta x_i y_i - 2\alpha\gamma x_i z_i - 2\beta\gamma y_i z_i.$$

Taking into account that

$$\alpha^2 + \beta^2 + \gamma^2 = 1,$$

it follows that

$$h_i^2 = \alpha^2(y_i^2 + z_i^2) + \beta^2(x_i^2 + z_i^2) + \gamma^2(x_i^2 + y_i^2) \\ - 2\alpha\beta x_i y_i - 2\alpha\gamma x_i z_i - 2\beta\gamma y_i z_i.$$

Euler's dynamic equations

Substitution of this last expression into (5.1) gives

$$\begin{aligned}\mathbb{I}_e &= \alpha^2 \sum_{i \in S} m_i(y_i^2 + z_i^2) + \beta^2 \sum_{i \in S} m_i(x_i^2 + z_i^2) + \gamma^2 \sum_{i \in S} m_i(x_i^2 + y_i^2) \\ &- 2\alpha\beta \sum_{i \in S} m_i x_i y_i - 2\alpha\gamma \sum_{i \in S} m_i x_i z_i - 2\beta\gamma \sum_{i \in S} m_i y_i z_i.\end{aligned} \quad (5.4)$$

Definition 5.1. The amounts

$$\left.\begin{aligned} I_{xx} &:= \sum_{i \in S} m_i(y_i^2 + z_i^2), \\ I_{yy} &:= \sum_{i \in S} m_i(x_i^2 + z_i^2), \\ I_{zz} &:= \sum_{i \in S} m_i(x_i^2 + y_i^2) \end{aligned}\right\} \quad (5.5)$$

are called **principal moments of inertia**, while

$$\left.\begin{aligned} I_{xy} &:= \sum_{i \in S} m_i x_i y_i, \\ I_{xz} &:= \sum_{i \in S} m_i x_i z_i, \\ I_{yz} &:= \sum_{i \in S} m_i y_i z_i \end{aligned}\right\} \quad (5.6)$$

are called **centrifugal moments**, both with respect to the corresponding axes.

The previous definitions (5.5) and (5.6) allow expressing (5.4) as

$$\mathbb{I}_e = \alpha^2 I_{xx} + \beta^2 I_{yy} + \gamma^2 I_{zz} - 2\alpha\beta I_{xy} - 2\alpha\gamma I_{xz} - 2\beta\gamma I_{yz},$$

or, as can easily be verified,

$$\mathbb{I}_e = (\mathbf{e}, \mathbb{I}\mathbf{e}), \quad (5.7)$$

where the matrix $\mathbb{I}$ is defined as follows:

$$\mathbb{I} := \begin{bmatrix} I_{xx} & -I_{xy} & -I_{xz} \\ -I_{yx} & I_{yy} & -I_{yz} \\ -I_{zx} & -I_{zy} & I_{zz} \end{bmatrix}. \quad (5.8)$$

Remark 5.1. Because

$$I_{xz} = I_{xz}, \quad I_{xy} = I_{yx}, \quad I_{yz} = I_{zy},$$

the matrix $\mathbb{I}$ is symmetric, i.e.,

$$\mathbb{I} = \mathbb{I}^T.$$

Definition 5.2. Matrix $\mathbb{I}$ is called the **inertial tensor** (or **tensor of inertia**) of the solid body given with respect to the coordinate system considered.

The *inertia tensor* is a very useful element to describe the most important mechanical concepts of solid bodies, such as kinetic energy and momentum, which are developed in what follows for certain common cases.

5.2 Relative kinetic energy and impulse momentum

5.2.1 Relative kinetic energy

This section uses the nomenclature and the results obtained in Section 3.6 of Chapter 3. This section is based on the König theorem, namely, the total kinetic energy T_{total} of a rigid body is calculated as

$$T_{total} = T_{O'} + T_{rel,O'} + M\left(\mathbf{v}_{O'}, \mathbf{v}_{CI,O'}\right),$$

where, if it is a rigid body rotating with angular velocity $\boldsymbol{\omega}$ relative to the pivot O', we have

$$T_{rel,O'} = \frac{1}{2} \mathbf{I}_\omega \omega^2, \qquad (5.9)$$

with $\mathbf{I}_\omega$ denoting the moment of inertia of the body with respect to the line of action of $\boldsymbol{\omega}$. Then, using (5.7), we get

$$T_{rel,O'} = \frac{1}{2} \left(\mathbf{e}_\omega, \mathbb{I}\mathbf{e}_\omega\right) \omega^2,$$

where $\mathbb{I}$ refers to a coordinate system with origin O'. So, we have the following result.

Lemma 5.1. *In the conditions stated*

$$T_{rel,O'} = \frac{1}{2} \boldsymbol{\omega}^T \mathbb{I} \boldsymbol{\omega}. \qquad (5.10)$$

Proof. Since

$$\boldsymbol{\omega} = \omega \mathbf{e}_\omega$$

it follows that

$$T_{rel,O'} = \frac{1}{2} \left(\mathbf{e}_\omega, \mathbb{I}\mathbf{e}_\omega\right) \omega^2 = \frac{1}{2} \left((\omega \mathbf{e}_\omega), \mathbb{I}(\omega \mathbf{e}_\omega)\right) = \frac{1}{2} \boldsymbol{\omega}^T \mathbb{I} \boldsymbol{\omega}.$$

□

Remark 5.2. Note that the tensor of inertia $\mathbb{I}$ does not depend on the direction $\mathbf{e}_\omega$ of the angular velocity vector $\boldsymbol{\omega}$. This fact simplifies the expression for $\dot{T}_{rel,O'}$, namely,

a) using (5.9) we have

$$\dot{T}_{rel,O'} = \mathbb{I}_\omega \omega \dot{\omega} + \frac{\omega^2}{2} \frac{d}{dt} \mathbb{I}_\omega,$$

b) and using (5.10) we have

$$T_{rel,O'} = \omega^T \mathbb{I} \dot{\omega},$$

which does not require to calculate $\dfrac{d}{dt} \mathbb{I}_\omega$.

5.2.2 Relative impulse momentum

On the other hand, by the Euler's theorem for any generic point i in the rigid body S its relative velocity is equal to

$$\mathbf{v}_{i,O'} = [\boldsymbol{\omega}, \mathbf{r}_{i,O'}],$$

where $\mathbf{r}_{i,O'}$ and $\mathbf{v}_{i,O'}$ denote the position and velocity of $i \in S$ with respect to the origin O'. Since the impulse moment relative to O' is defined as

$$\mathbf{K}_{rel,O'} := \sum_{i \in S} [\mathbf{r}_{i,O'}, m_i \mathbf{v}_{i,O'}]$$

we may conclude that

$$\mathbf{K}_{rel,O'} = \sum_{i \in S} [\mathbf{r}_{i,O'}, m_i [\boldsymbol{\omega}, \mathbf{r}_{i,O'}]].$$

Using the relation

$$[\mathbf{a}, [\mathbf{b}, \mathbf{c}]] = (\mathbf{a}, \mathbf{c}) \mathbf{b} - (\mathbf{a}, \mathbf{b}) \mathbf{c},$$

we get

$$\mathbf{K}_{rel,O'} = \sum_{i \in S} m_i \left[(\mathbf{r}_{i,O'}, \mathbf{r}_{i,O'}) \boldsymbol{\omega} - (\mathbf{r}_{i,O'}, \boldsymbol{\omega}) \mathbf{r}_{i,O'} \right] =$$

$$\sum_{i \in S} m_i (x_i^2 + y_i^2 + z_i^2) \boldsymbol{\omega} - \sum_{i \in S} m_i (x_i \omega_x + y_i \omega_y + z_i \omega_z) \mathbf{r}_i,$$

from which it is easily verified that

$$\left. \begin{array}{l} K_{rel,O'x} = \omega_x I_{xx} - \omega_y I_{xy} - \omega_z I_{xz}, \\ K_{rel,O'y} = -\omega_x I_{xy} + \omega_y I_{yy} - \omega_z I_{yz}, \\ K_{rel,O'z} = -\omega_x I_{zx} - \omega_y I_{zy} + \omega_z I_{zz}. \end{array} \right\} \qquad (5.11)$$

From these relations and from (5.8) we have the result that follows.

Lemma 5.2. *The following representations hold:*

$$\mathbf{K}_{rel,O'} = \mathbb{I}\boldsymbol{\omega} \qquad (5.12)$$

and

$$T_{rel,O'} = \frac{1}{2}\boldsymbol{\omega}^T \mathbf{K}_{rel,O'}. \qquad (5.13)$$

Proof. This lemma directly follows from (5.12) and (5.10). □

Remark 5.3. When the origins O' (relative and possibly mobile) and O (absolute and fixed) coincide, the quantities $\mathbf{K}_{rel,O'}$ and $T_{rel,O'}$ will be denoted simply as $\mathbf{K}_O$ and T_O.

5.3 Some properties of inertial tensors

Several properties characterize $\mathbb{I}$ and its revision is the subject of this section.

5.3.1 Tensor of inertia as a non-negative symmetric matrix

Definition 5.3. A symmetric matrix $M \in \Re^{n \times n}$ with the property

$$\mathbf{x}^T M \mathbf{x} \geq 0, \quad \forall \mathbf{x} \in R^n$$

it is said to be a **positive semi-definite matrix**, which is denoted as

$$M \geq 0.$$

Proposition 5.1. *Since $I_e \geq 0$, from (5.7) it follows that a symmetric matrix $\mathbb{I}$ can be an inertia tensor if and only if*

$$\mathbb{I} \geq 0.$$

The following criterion allows to establish if a matrix is positive semi-definite.

Lemma 5.3 (Silvester). *A symmetric matrix is positive semi-definite if and only if its main minors are non-negative.*

The proof can be found, for example, in (Poznyak, 2008, Theorem 7.2).
The previous criterion trivially leads to the statement that follows.

Lemma 5.4. *For the symmetric matrix*

$$\mathbb{I} = \begin{bmatrix} I_{xx} & -I_{xy} & -I_{xz} \\ -I_{yx} & I_{yy} & -I_{yz} \\ -I_{zx} & -I_{zy} & I_{zz} \end{bmatrix}$$

to be a tensor of inertia, it must meet the following properties:

1.
$$I_{xx} \geq 0,$$

2.
$$I_{xx}I_{yy} \geq I_{xy}^2,$$

from where it follows that
$$I_{yy} \geq 0,$$

3.
$$I_{xx}I_{yy}I_{zz} \geq 2I_{xy}I_{yz}I_{zx} + I_{yy}I_{xz}^2 + I_{xx}I_{yz}^2 + I_{zz}I_{xy}^2.$$

Example 5.1. Let the matrix G be given as

$$G = \begin{bmatrix} 1 & -d & 0 \\ -d & 2 & 0 \\ 0 & 0 & 3 \end{bmatrix}.$$

What should be the value of d such that the matrix G is an inertial tensor? Clearly the first major minor is positive. The non-negativity condition for the second one gives

$$2 \geq d^2.$$

The condition on the third major minor does not provide any extra relationship. So, the solution is

$$-\sqrt{2} \leq d \leq \sqrt{2}.$$

5.3.2 Eigenvalues and eigenvectors of inertial tensors

The following problem allows to obtain important conclusions concerning $\mathbb{I}$.

Problem 5.1. Let us consider a solid body referring to a coordinate system as shown in Fig. 5.1. Calculate the minimum and maximum moments of inertia and obtain the corresponding directions, that is, determine the extremes of the function

$$I_e = e^T \mathbb{I} e \to \operatorname*{extr}_{e \in R^3} \tag{5.14}$$

with the restriction

$$\|e\| = 1. \tag{5.15}$$

Before addressing the solution to the proposed problem, recall the Lagrange multipliers approach to deal with this type of optimization problems.

Let $f : R^n \to R$ and $\mathbf{g} : R^n \to R^m$, $m \leq n$, be continuously differentiable functions. Suppose we want to find the points $\mathbf{x} \in R^n$ in which the function f reaches its extreme values and also meeting the condition

$$\mathbf{g}(\mathbf{x}) = 0.$$

Definition 5.4. *If the point $\mathbf{x}_0 \in R^n$ is such that the Jacobian matrix $\dfrac{\partial \mathbf{g}}{\partial \mathbf{x}}(\mathbf{x}_0)$ has a range m, it is said to be a regular point of $\mathbf{g}$.*

From the theory of constraint optimization (see for example Poznyak (2008)), if $\mathbf{x}_0 \in R^n$ is a regular point of $\mathbf{g}$ and at that point f has an extremum (maximum or minimum), then the function $\mathcal{L}(\mathbf{x}, \lambda)$, called Lagrangian and defined as

$$\mathcal{L}(\mathbf{x}, \lambda) := f(\mathbf{x}) - (\lambda, \mathbf{g}(\mathbf{x})),$$

also has a global (non-constraint) extreme at $\mathbf{x}_0$ for some vector $\lambda = \lambda_0 \in R^m$ which is called the *Lagrange multipliers vector*. Both variables $\mathbf{x}_0$ and λ_0 satisfy the extremum condition

$$\nabla_{\mathbf{x}} \mathcal{L}(\mathbf{x}_0, \lambda_0) = 0, \tag{5.16}$$

from which we get

$$\mathbf{x}_0 = \mathbf{x}_0(\lambda_0),$$

and λ_0 is calculated via the restriction

$$\mathbf{g}(\mathbf{x}_0(\lambda_0)) = 0.$$

Now we can enunciate the following result.

Lemma 5.5. *Extremal directions $\mathbf{e}$ satisfy the relationship*

$$\mathbb{I}\mathbf{e} = \lambda \mathbf{e} \tag{5.17}$$

while the corresponding extremal moments of inertia are given by the corresponding values of the parameter λ_i ($i = 1, 2, 3$).

Proof. Applying the described method to the optimization problem (5.14)–(5.15) with

$$\mathbf{x} = \mathbf{e}, \quad f(\mathbf{e}) = \mathbf{e}^T \mathbb{I} \mathbf{e}, \quad \mathbf{g}(\mathbf{e}) = \|\mathbf{e}\|^2 - 1 = 0$$

yields (with $\lambda \in R$)

$$\mathcal{L}(\mathbf{e}, \lambda) = \mathbf{e}^T \mathbb{I} \mathbf{e} - \lambda (\mathbf{e}^T \mathbf{e} - 1),$$

so that in view of (5.16) it follows that

$$\nabla_{\mathbf{e}}\mathcal{L}(\mathbf{e}, \lambda) = 2\left(\mathbb{I}\mathbf{e} - \lambda\mathbf{e}\right) = 0,$$

which leads to (5.17). With this result it follows that in these same directions

$$\mathbf{I}_{\mathbf{e}} = \mathbf{e}^T\mathbb{I}\mathbf{e} = \lambda\mathbf{e}^T\mathbf{e} = \lambda, \text{ since } \mathbf{e}^T\mathbf{e} = 1.$$

□

Remark 5.4. From the previous lemma it is concluded that the extreme directions are determined by the **eigenvectors** $\mathbf{e}_i$ of the inertia tensor, while the corresponding values of the moments of inertia are given by the associated **eigenvalues** λ_i ($i = 1, 2, 3$).

Example 5.2. Find the extreme moments of inertia I_{max} and I_{min} and the extreme directions $\mathbf{e}_{max}$ and $\mathbf{e}_{min}$ corresponding to the inertial tensor

$$\mathbb{I} = \begin{bmatrix} 3 & -1 & 0 \\ -1 & 2 & 0 \\ 0 & 0 & 1 \end{bmatrix}.$$

The eigenvalues of $\mathbb{I}$ are given by the roots of the associated characteristic polynomial of $\mathbb{I}$ with respect to λ, i.e.,

$$\det\left(\mathbb{I} - \lambda I_{3\times 3}\right) = 0, \qquad (5.18)$$

where $I_{3\times 3}$ denotes the identity matrix of order 3. Substituting $\mathbb{I}$ in (5.18) gives

$$\det \begin{bmatrix} 3-\lambda & -1 & 0 \\ -1 & 2-\lambda & 0 \\ 0 & 0 & 1-\lambda \end{bmatrix} = (1-\lambda)\left[(3-\lambda)(2-\lambda) - 1\right] = 0,$$

whose roots are

$$\lambda_{min} = I_{min} = 1, \quad \lambda_2 = \frac{1}{2}\left(5 - \sqrt{5}\right), \quad \lambda_{max} = I_{max} = \frac{1}{2}\left(5 + \sqrt{5}\right).$$

Now, the extreme directions may be determined from the relation

$$(\mathbb{I} - \lambda I)\mathbf{e} = 0.$$

Consider first $\lambda_{min} = 1$. We must have

$$\begin{bmatrix} 2 & -1 & 0 \\ -1 & 1 & 0 \\ 0 & 0 & 0 \end{bmatrix} \begin{bmatrix} e_x \\ e_y \\ e_z \end{bmatrix} = 0,$$

or equivalently,

$$\left.\begin{array}{l} 2e_x - e_y = 0, \\ -e_x + e_y = 0, \\ e_z \text{ any}, \end{array}\right\}$$

from which it follows (since $\|\mathbf{e}_{\max}\| = 1$) that

$$\mathbf{e}_{\min} = \begin{bmatrix} 0 & 0 & 1 \end{bmatrix}^\mathsf{T}.$$

For $\lambda_{\max} = \frac{1}{2}\left(5 + \sqrt{5}\right)$ it follows that

$$\begin{bmatrix} 3 - \lambda_{\max} & -1 & 0 \\ -1 & 2 - \lambda_{\max} & 0 \\ 0 & 0 & 1 - \lambda_{\max} \end{bmatrix} \begin{bmatrix} e_x \\ e_y \\ e_z \end{bmatrix} = 0,$$

and hence,

$$\left.\begin{array}{l} e_y = (3 - \lambda_{\max})\, e_x, \\ e_x = (2 - \lambda_{\max})\, e_y, \\ e_z = 0. \end{array}\right\} \tag{5.19}$$

The first two equations in (5.19) are linearly independent, but by the condition $\|\mathbf{e}_{\max}\| = 1$ we have

$$e_x^2 + e_y^2 = \left[1 + (3 - \lambda_{\max})^2\right] e_x^2 = 1,$$

so that

$$\mathbf{e}_{\max} = \frac{1}{\sqrt{1 + (3 - \lambda_{\max})^2}} \begin{bmatrix} 1 & (3 - \lambda_{\max}) & 0 \end{bmatrix}^\mathsf{T}.$$

Definition 5.5. If the origin of the reference coordinate system coincides with the center of inertia, the main moments of inertia are called **central moments of inertia**.

Remark 5.5. If, in addition to the condition of the previous definition, the axes of the system are selected according to the extreme directions, one in the direction of $\mathbf{e}_{\max}$, another in the direction of $\mathbf{e}_{\min}$, and the third orthogonal to both, then the tensor of inertia adopts the reduced form

$$\mathbb{I} = \begin{bmatrix} I_{xx} & 0 & 0 \\ 0 & I_{yy} & 0 \\ 0 & 0 & I_{zz} \end{bmatrix}. \tag{5.20}$$

5.3.3 Examples using tensors of inertia

The examples and exercises that make up this section illustrate and complement the theory of the chapter presented so far.

Example 5.3. Consider the cube and the coordinate system shown in Fig. 5.2. The cube has side a and mass M, which is uniformly distributed. The coordinate system is located in the center of the solid and is parallel to its edges. Calculate the tensor of inertia.

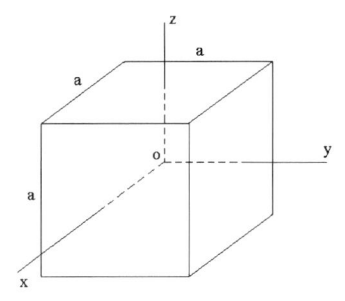

Figure 5.2 A cube with side a and the reference system.

Given the symmetry of the figure with respect to the coordinate system, we have

$$I_{xx} = I_{yy} = I_{zz}, \quad I_{xy} = I_{xz} = I_{yz} = 0.$$

By the definition

$$I_{xx} = \sum_{i \in S} m_i h_i^2, \tag{5.21}$$

where h_i denotes the distance of the element of mass m_i from the coordinate axis x. In our case (5.21) it translates into

$$I_{xx} = \int_{x=-\frac{a}{2}}^{\frac{a}{2}} \int_{y=-\frac{a}{2}}^{\frac{a}{2}} \int_{z=-\frac{a}{2}}^{\frac{a}{2}} h^2(y,z)\, dm.$$

By the uniform distribution of the mass

$$dm = \frac{M}{a^3} dx\,dy\,dz,$$

and it follows from Fig. 5.3 that

$$h^2(y,z) = y^2 + z^2,$$

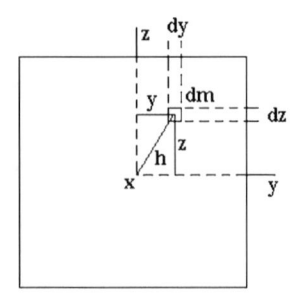

Figure 5.3 A differential element of mass and its position with respect to the x-axis.

and hence,

$$I_{xx} = \frac{M}{a^3} \int_{x=-\frac{a}{2}}^{\frac{a}{2}} \int_{z=-\frac{a}{2}}^{\frac{a}{2}} \int_{y=-\frac{a}{2}}^{\frac{a}{2}} \left(y^2 + z^2\right) dy\, dz\, dx =$$

$$\frac{M}{a^3} a \left(\int_{z=-\frac{a}{2}}^{\frac{a}{2}} \int_{y=-\frac{a}{2}}^{\frac{a}{2}} \left(y^2 + z^2\right) dy\, dz \right) =$$

$$\frac{M}{a^3} a \left(a \frac{y^3}{3} \Big|_{-\frac{a}{2}}^{\frac{a}{2}} + a \frac{z^3}{3} \Big|_{-\frac{a}{2}}^{\frac{a}{2}} \right) = M \frac{a^2}{6}.$$

So, finally,

$$\mathbb{I} = M \frac{a^2}{6} I_{3 \times 3},$$

where $I_{3 \times 3}$ is the unitary matrix of the third order.

Example 5.4.

a) Let us calculate the tensor of inertia for the cylinder of mass M uniformly distributed with respect to the coordinate system shown in Fig. 5.4. Recall that in Section 3.6 of Chapter 3 the main moments have already been calculated:

$$I_{xx} = I_{yy} = \frac{M}{4}\left(\rho^2 + \frac{h^2}{3}\right), \quad I_{zz} = \frac{M}{2}\rho^2. \tag{5.22}$$

In addition, by the symmetry with respect to the coordinate system, it is concluded that

$$I_{xy} = I_{xz} = I_{yz} = 0,$$

Euler's dynamic equations

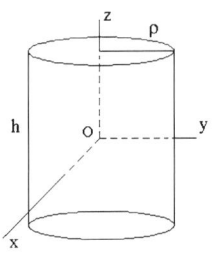

Figure 5.4 Cylinder and reference system.

whereby

$$\mathbb{I} = \begin{bmatrix} \frac{M}{4}\left(\rho^2 + \frac{h^2}{3}\right) & 0 & 0 \\ 0 & \frac{M}{4}\left(\rho^2 + \frac{h^2}{3}\right) & 0 \\ 0 & 0 & \frac{M}{2}\rho^2 \end{bmatrix}.$$

b) Fig. 5.5 shows two solids of uniformly distributed mass referring to two coordinate systems located in their centers of inertia. Determine their inertia tensors. The center of inertia is determined first. It is clear that it is on the axis of symmetry. Since the body has volume

$$V = \frac{\pi R^2 h}{3},$$

(i) Solid cone (ii) Hollow cylinder

Figure 5.5 Solid cone and hollow cylinder referring to two coordinate systems.

the following density of mass results:

$$\mu = \frac{3M}{\pi R^2 h}.$$

On the other hand, the expression for the calculation of the position of the center of inertia has to be found at a height of the base given by

$$h_{CI} = \frac{1}{M} \int_0^h w \, dm,$$

where w denotes the position with respect to the base of the cone, which is the elementary disk of radius ρ and differential height dw, whose mass results in

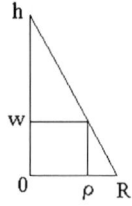

$$dm = \mu\pi\rho^2 dw.$$

Figure 5.6 Relationship between the radius ρ of the elementary disk and its height w.

From Fig. 5.6 we have

$$\rho = \frac{R}{h}(h-w).$$

So,

$$h_{CI} = \frac{\mu\pi R^2}{Mh^2}\int_0^h w(h-w)^2 dw = \frac{\mu\pi R^2}{Mh^2}\frac{h^4}{12} = \frac{h}{4}.$$

We can now calculate the moments of inertia. To determine I_{yy} consider the elementary disk of radius ρ and height dz (see Fig. 5.7), which has the mass

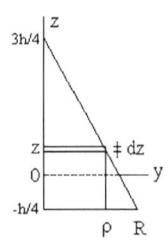

$$dm = \mu\pi\rho^2 dz,$$

Figure 5.7 Elementary disk of radius ρ and height dz.

where now

$$\rho = R\left(\frac{3}{4} - \frac{z}{h}\right). \tag{5.23}$$

This gives

$$dm = \mu\pi R^2\left(\frac{3}{4} - \frac{z}{h}\right)^2 dz = \frac{3M}{h}\left(\frac{3}{4} - \frac{z}{h}\right)^2 dz. \tag{5.24}$$

Recall from Section 3.6 of Chapter 3 that the moment of inertia of the disk in question with respect to the axis, passing through its diameter, has the expression

$$d\mathrm{I}_{yy} = \frac{1}{4}\rho^2 dm.$$

This result together with the Steiner's theorem allows us to obtain the formula

$$\mathrm{I}_{yy} = \int_{z=-h/4}^{3h/4} \left(\frac{1}{4}\rho^2 + z^2\right) dm,$$

where with relationships (5.23)–(5.24) we get

$$\mathrm{I}_{yy} = \frac{3M}{h} \int_{-h/4}^{3h/4} \left(\frac{3}{4} - \frac{z}{h}\right)^2 \left[\frac{1}{4}R^2\left(\frac{3}{4} - \frac{z}{h}\right)^2 + z^2\right] dz = \frac{3}{80}M\left(h^2 + 4R^2\right).$$

Recalling that

$$\mathrm{I}_{xx} = \mathrm{I}_{yy}$$

and

$$\mathrm{I}_{zz} = \frac{3}{10}MR^2$$

(see Section 3.6 of Chapter 3). In addition, by the symmetry it is verified that

$$\mathrm{I}_{xy} = \mathrm{I}_{xz} = \mathrm{I}_{yz} = 0.$$

So finally,

$$\mathbb{I} = \begin{bmatrix} \frac{3}{80}M\left(h^2 + 4R^2\right) & 0 & 0 \\ 0 & \frac{3}{80}M\left(h^2 + 4R^2\right) & 0 \\ 0 & 0 & \frac{3}{10}MR^2 \end{bmatrix}.$$

c) Assuming that the cylinder is solid with mass M_s, its density is given as

$$\mu = \frac{M_s}{\pi R^2 h},$$

and since for the hollow cylinder we have

$$M = \pi \mu h \left(R^2 - \rho^2\right),$$

the following relationship is satisfied:

$$M_s = \frac{MR^2}{R^2 - \rho^2}.$$

Now, using expressions (5.22), it is concluded that

$$I_{xx} = I_{yy} = \frac{M_s}{4}\left(R^2 + \frac{h^2}{3}\right) - \frac{(M_s - M)}{4}\left(\rho^2 + \frac{h^2}{3}\right) =$$

$$\frac{M_s}{4}\left(R^2 - \rho^2\right) + \frac{M}{4}\left(\rho^2 + \frac{h^2}{3}\right) = \frac{M}{4}\left(R^2 + \rho^2 + \frac{h^2}{3}\right)$$

and

$$I_{zz} = \frac{M_s}{2}R^2 - \frac{(M_s - M)}{2}\rho^2 =$$

$$\frac{M_s}{2}\left(R^2 - \rho^2\right) + \frac{M}{2}\rho^2 = \frac{M}{2}\left(R^2 + \rho^2\right),$$

whereas, by symmetry, the centrifugal moments are equal to zero:

$$I_{xy} = I_{xz} = I_{yz} = 0,$$

which is why

$$\mathbb{I} = \begin{bmatrix} \frac{M}{4}\left(R^2 + \rho^2 + \frac{h^2}{3}\right) & 0 & 0 \\ 0 & \frac{M}{4}\left(R^2 + \rho^2 + \frac{h^2}{3}\right) & 0 \\ 0 & 0 & \frac{M}{2}\left(R^2 + \rho^2\right) \end{bmatrix}.$$

In the examples that follow, the obtained expressions are applied.

Example 5.5. Find $\mathbf{K}_O$ and T for the body of mass M shown in Fig. 5.8. The reference system is located in the center of inertia of the solid and the mass has a uniform distribution. For the present case the quantities sought are given by the relations (5.12)–(5.13), that is,

$$\mathbf{K}_O = \mathbb{I}\boldsymbol{\omega}, \quad T = \frac{1}{2}\boldsymbol{\omega}^T \mathbf{K}_O.$$

From Fig. 5.8 we have

$$\boldsymbol{\omega} = \begin{bmatrix} 0 \\ \omega \cos \alpha \\ \omega \sin \alpha \end{bmatrix},$$

Euler's dynamic equations

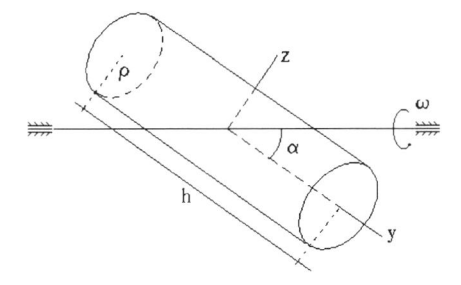

Figure 5.8 Solid cylinder rotating obliquely.

and by the result of the second exercise of this section it is easily verified that

$$\mathbf{K}_O = \begin{bmatrix} 0 \\ \frac{M}{2}\rho^2 \omega \cos\alpha \\ \frac{M}{4}\left(\rho^2 + \frac{h^2}{3}\right)\omega \sin\alpha \end{bmatrix}$$

and

$$T = \frac{M}{4}\omega^2 \left[\rho^2 \cos^2\alpha + \frac{1}{2}\left(\rho^2 + \frac{h^2}{3}\right)\sin^2\alpha\right].$$

Example 5.6. Fig. 5.9 shows a pendulum of length R formed by a disk of radius ρ and uniformly distributed mass M. It is considered that the mass of the pendulum arm is zero. Calculate the speed with which the center of the disk passes through the lower position if the pendulum is released with zero velocity from the horizontal position. As it is a conservative system, we have for the lower point of the trajectory

$$T = Mg(R - \rho). \tag{5.25}$$

Figure 5.9 Pendulum.

On the other side, by (5.10) we have

$$T = \frac{1}{2}\boldsymbol{\omega}^T \mathbb{I}\boldsymbol{\omega}, \tag{5.26}$$

where by the results of Section 3.6 of Chapter 3 in conjunction with Steiner's theorem

$$I_{xx} = \frac{1}{4}M\rho^2, \quad I_{yy} = M\left(\frac{1}{4}\rho^2 + (R-\rho)^2\right), \quad I_{zz} = M\left(\frac{1}{2}\rho^2 + (R-\rho)^2\right),$$

while by the symmetry of the disk

$$I_{xy} = I_{xz} = I_{yz} = 0,$$

which finally gives

$$\mathbb{I} = \begin{bmatrix} \frac{1}{4}M\rho^2 & 0 & 0 \\ 0 & M\left(\frac{1}{4}\rho^2 + (R-\rho)^2\right) & 0 \\ 0 & 0 & M\left(\frac{1}{2}\rho^2 + (R-\rho)^2\right) \end{bmatrix}.$$

Now, the rotation velocity in the lower point is given by

$$\boldsymbol{\omega} = \begin{pmatrix} 0 & 0 & \omega \end{pmatrix}^T,$$

so that expression (5.26) becomes

$$T = \frac{1}{4}M\omega^2\left(\rho^2 + 2(R-\rho)^2\right)$$

and by (5.25)

$$\frac{1}{2}M\omega^2\left(\frac{1}{2}\rho^2 + (R-\rho)^2\right) = Mg(R-\rho),$$

from which we derive

$$\omega = \sqrt{\frac{2g(R-\rho)}{\frac{1}{2}\rho^2 + (R-\rho)^2}},$$

which allows to obtain the speed sought via the expression

$$v = \omega(R-\rho) = (R-\rho)\sqrt{\frac{4g(R-\rho)}{\rho^2 + 2(R-\rho)^2}}.$$

Example 5.7. Fig. 5.10 shows a solid cylinder of uniformly distributed mass M and rotating eccentrically with respect to the axis shown. Regarding the reference system shown, let us calculate $\mathbf{K}_O$ and T. By the relations (5.12)–(5.13) we have

$$\mathbf{K}_O = \mathbb{I}\boldsymbol{\omega}, \quad T = \frac{1}{2}\boldsymbol{\omega}^T\mathbf{K}_O,$$

Euler's dynamic equations

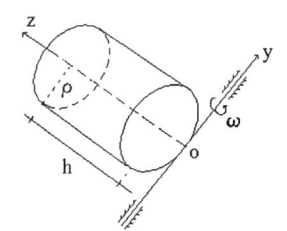

Figure 5.10 Solid cylinder in eccentric rotation.

and (see Fig. 5.10)

$$\boldsymbol{\omega} = \begin{bmatrix} 0 & \omega & 0 \end{bmatrix}^T.$$

In view of the results of the second exercise of this section and by the Steiner's theorem it follows that

$$I_{xx} = \frac{M}{4}\left(\rho^2 + \frac{4h^2}{3}\right), \quad I_{yy} = \frac{M}{4}\left(5\rho^2 + \frac{4h^2}{3}\right), \quad I_{zz} = \frac{3M}{2}\rho^2.$$

Body symmetricity gives

$$I_{xy} = I_{yz} = 0$$

since

$$I_{xz} = \sum_{i \in S} m_i x_i z_i = \sum_{i \in S} m_i x_i (z_i - \frac{1}{2}h + \frac{1}{2}h)$$
$$= \sum_{i \in S} m_i x_i (z_i - \frac{1}{2}h) + \frac{Mh}{2}\frac{1}{M}\sum_{i \in S} m_i x_i.$$

But

$$\sum_{i \in S} m_i x_i (z_i - \frac{1}{2}h) = 0, \quad \frac{1}{M}\sum_{i \in S} m_i x_i = \rho,$$

which leads to

$$I_{xz} = \frac{Mh\rho}{2}.$$

So, finally,

$$\mathbb{I} = \begin{bmatrix} \frac{M}{4}\left(\rho^2 + \frac{4h^2}{3}\right) & 0 & -\frac{Mh\rho}{2} \\ 0 & \frac{M}{4}\left(5\rho^2 + \frac{4h^2}{3}\right) & 0 \\ -\frac{Mh\rho}{2} & 0 & \frac{3M}{2}\rho^2 \end{bmatrix}.$$

Hence

$$\mathbf{K}_O = \begin{bmatrix} 0 \\ \frac{M}{4}\omega\left(5\rho^2 + \frac{4h^2}{3}\right) \\ 0 \end{bmatrix}$$

and

$$T = \frac{M}{4}\omega^2\left(5\rho^2 + \frac{4h^2}{3}\right).$$

Example 5.8. Fig. 5.11 shows the profile of a disk rotating eccentrically and obliquely. The disk has a radius ρ and its mass M is evenly distributed. Calculate $\mathbf{K}_O$ and T. Again, the amounts sought are given by the relationships (5.12)–(5.13):

$$\mathbf{K}_O = \mathbb{I}\boldsymbol{\omega}, \quad T = \frac{1}{2}\boldsymbol{\omega}^T \mathbf{K}_O.$$

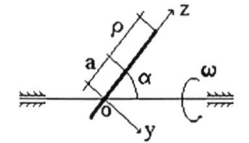

Figure 5.11 Disk in eccentric and oblique rotation.

From Fig. 5.11 we can see that

$$\boldsymbol{\omega} = \begin{bmatrix} 0 & \omega\sin\alpha & \omega\cos\alpha \end{bmatrix}^T.$$

As above, the moments of inertia can be calculated using the results obtained in Section 3.6 of Chapter 3 and the Steiner's theorem:

$$\mathrm{I}_{xx} = M\left(\frac{1}{4}\rho^2 + a^2\right), \quad \mathrm{I}_{yy} = M\left(\frac{1}{2}\rho^2 + a^2\right), \quad \mathrm{I}_{zz} = \frac{1}{4}M\rho^2,$$

whereas by the symmetry with respect to x

$$\mathrm{I}_{xz} = 0$$

and taking into account that $y = 0$ we have

$$I_{xy} = I_{yz} = 0.$$

Therefore

$$\mathbb{I} = \begin{bmatrix} M\left(\frac{1}{4}\rho^2 + a^2\right) & 0 & 0 \\ 0 & M\left(\frac{1}{2}\rho^2 + a^2\right) & 0 \\ 0 & 0 & \frac{1}{4}M\rho^2 \end{bmatrix},$$

which finally leads to

$$\mathbf{K}_O = \begin{bmatrix} 0 \\ M\omega\left(\frac{1}{2}\rho^2 + a^2\right)\sin\alpha \\ \frac{1}{4}M\omega\rho^2\cos\alpha \end{bmatrix}$$

and

$$T = \frac{1}{2}M\omega^2\left[\left(\frac{1}{2}\rho^2 + a^2\right)\sin^2\alpha + \frac{1}{4}\rho^2\cos^2\alpha\right].$$

Example 5.9. In Fig. 5.12 a profile disk is shown. It is subjected to the two rotations shown. Determine $\mathbf{K}_O$ considering that the disk has a uniformly distributed mass M. By (5.12)

$$\mathbf{K}_O = \mathbb{I}\boldsymbol{\omega}.$$

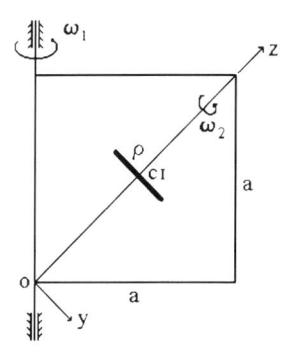

Figure 5.12 Disk subjected to two rotations.

Now, by Section 3.6 of Chapter 3 and the Steiner's theorem it follows that

$$I_{xx} = I_{yy} = M\left(\frac{1}{4}\rho^2 + \frac{1}{2}a^2\right), \quad I_{zz} = \frac{1}{2}M\rho^2.$$

In view of symmetry we also have

$$I_{xz} = I_{xy} = I_{yz} = 0.$$

Hence

$$\mathbb{I} = \begin{bmatrix} \frac{1}{2}M\left(\frac{1}{2}\rho^2 + a^2\right) & 0 & 0 \\ 0 & \frac{1}{2}M\left(\frac{1}{2}\rho^2 + a^2\right) & 0 \\ 0 & 0 & \frac{1}{2}M\rho^2 \end{bmatrix}.$$

From Fig. 5.12 we have

$$\omega = \begin{bmatrix} 0 & -\frac{\sqrt{2}}{2}\omega_1 & \frac{\sqrt{2}}{2}\omega_1 + \omega_2 \end{bmatrix}^T,$$

and therefore

$$\mathbf{K}_O = \begin{bmatrix} 0 \\ -\frac{\sqrt{2}}{4}M\omega_1\left(\frac{1}{2}\rho^2 + a^2\right) \\ \frac{1}{2}M\rho^2\left(\frac{\sqrt{2}}{2}\omega_1 + \omega_2\right) \end{bmatrix}.$$

5.4 Euler's dynamic equations

Recall that if ω is the angular velocity of the solid with respect to the pivot O', which is fixed at the origin O of the coordinate system, then by (5.12), (5.13), and the Rizal's theorem

$$\mathbf{K}_O = \mathbb{I}\omega, \quad T = \frac{1}{2}\omega^T \mathbf{K}_O, \quad \dot{\mathbf{K}}_O = \mathbf{M}_{\mathbf{F}_{ext},O}. \tag{5.27}$$

Recall also that if the main moments of inertia are central (origin O is located in the inertial center of the solid) and the coordinate system has been chosen according to the directions in which the moments of inertia are extreme, then the inertial tensor has the form

$$\mathbb{I} = \begin{bmatrix} A & 0 & 0 \\ 0 & B & 0 \\ 0 & 0 & C \end{bmatrix}, \tag{5.28}$$

where

$$A := I_{xx}, \quad B := I_{yy}, \quad C := I_{zz},$$

Euler's dynamic equations

which gives
$$\mathbf{K}_O = Ap\mathbf{i} + Bq\mathbf{j} + Cr\mathbf{k}, \tag{5.29}$$
if
$$\boldsymbol{\omega} = p\mathbf{i} + q\mathbf{j} + r\mathbf{k}. \tag{5.30}$$

Deriving (5.29) with respect to time under the consideration that the origin O remains in the inertial center of the solid and the reference system rotates with the body, we get

$$\dot{\mathbf{K}}_O = A\dot{p}\mathbf{i} + B\dot{q}\mathbf{j} + C\dot{r}\mathbf{k} + Ap\frac{d}{dt}\mathbf{i} + Bq\frac{d}{dt}\mathbf{j} + Cr\frac{d}{dt}\mathbf{k}.$$

By Euler's theorem
$$\left.\begin{aligned}\frac{d}{dt}\mathbf{i} &= [\boldsymbol{\omega}, \mathbf{i}] = r\mathbf{j} - q\mathbf{k}, \\ \frac{d}{dt}\mathbf{j} &= [\boldsymbol{\omega}, \mathbf{j}] = -r\mathbf{i} + p\mathbf{k}, \\ \frac{d}{dt}\mathbf{k} &= [\boldsymbol{\omega}, \mathbf{k}] = q\mathbf{i} - p\mathbf{j},\end{aligned}\right\} \tag{5.31}$$

which is why
$$\left.\begin{aligned}\dot{\mathbf{K}}_O = [A\dot{p} + (-B + C)\,qr]\mathbf{i} + [B\dot{q} + (A - C)\,pr]\mathbf{j} + \\ [C\dot{r} + (-A + B)\,pq]\mathbf{k}.\end{aligned}\right\} \tag{5.32}$$

From the third relation of (5.27) and considering (5.32), we come to the so-called the *Euler dynamic equations*, which are formed by the group of nonlinear differential equations

$$\left.\begin{aligned}A\dot{p} + (C - B)\,qr &= \left(\mathbf{M}_{\mathbf{F}_{ext},O}\right)_x, \\ B\dot{q} + (A - C)\,pr &= \left(\mathbf{M}_{\mathbf{F}_{ext},O}\right)_y, \\ C\dot{r} + (B - A)\,pq &= \left(\mathbf{M}_{\mathbf{F}_{ext},O}\right)_z.\end{aligned}\right\} \tag{5.33}$$

These equations describe the dynamics of the components p, q, and r of the rotation vector $\boldsymbol{\omega}$. There are several important particular cases of the previous equations, which are discussed in what follows.

5.4.1 Special cases of Euler's equations

1. The Euler case

In this case the moment of the external forces is null, namely,
$$\mathbf{M}_{\mathbf{F}_{ext},O} = \mathbf{0}. \tag{5.34}$$

It follows that

$$\dot{\mathbf{K}}_O = \mathbf{0}, \tag{5.35}$$

and therefore the Euler dynamic equations take the form

$$\left.\begin{array}{l} A\dot{p} + (C - B)qr = 0, \\ B\dot{q} + (A - C)pr = 0, \\ C\dot{r} + (B - A)pq = 0. \end{array}\right\} \tag{5.36}$$

If the first equation of (5.36) is multiplied by p, the second by q, and the third by r and then the resulting expressions are added, the following relationship is reached:

$$Ap\dot{p} + Bq\dot{q} + Cr\dot{r} = 0, \tag{5.37}$$

which corresponds to

$$\frac{1}{2}\frac{d}{dt}[Ap^2 + Bq^2 + Cr^2] = 0. \tag{5.38}$$

Now, note that from the second equation of (5.27)

$$T = \frac{1}{2}\boldsymbol{\omega}^T \mathbf{K}_O = \frac{1}{2}\begin{bmatrix} p & q & r \end{bmatrix}\begin{bmatrix} Ap \\ Bq \\ Cr \end{bmatrix} = \frac{1}{2}(Ap^2 + Bq^2 + Cr^2), \tag{5.39}$$

that is, in view of (5.38),

$$\frac{d}{dt}T = 0.$$

Then the following result has been demonstrated.

Lemma 5.6. *In a rotating solid body, if* $\mathbf{M}_{\mathbf{F}_{ext},O} = \mathbf{0}$, *then*

$$T \underset{t}{=} \text{const}. \tag{5.40}$$

To state the following result, remember that the angular acceleration has been defined as

$$\boldsymbol{\varepsilon} = \dot{\boldsymbol{\omega}}.$$

Lemma 5.7. *In the same conditions of the previous lemma*

$$\boldsymbol{\varepsilon} \perp \mathbf{K}_O.$$

Proof. From the expression of $\boldsymbol{\omega}$ given (5.30) we have

$$\boldsymbol{\varepsilon} = \frac{d}{dt}(p\mathbf{i} + q\mathbf{j} + r\mathbf{k}) = \dot{p}\mathbf{i} + \dot{q}\mathbf{j} + \dot{r}\mathbf{k} + p\frac{d}{dt}\mathbf{i} + q\frac{d}{dt}\mathbf{j} + r\frac{d}{dt}\mathbf{k},$$

Euler's dynamic equations

but by the Euler's theorem

$$p\frac{d}{dt}\mathbf{i} + q\frac{d}{dt}\mathbf{j} + r\frac{d}{dt}\mathbf{k} =$$
$$p\,[\boldsymbol{\omega}, \mathbf{i}] + q\,[\boldsymbol{\omega}, \mathbf{j}] + r\,[\boldsymbol{\omega}, \mathbf{k}] = [\boldsymbol{\omega}, \boldsymbol{\omega}] = 0.$$

Therefore

$$\boldsymbol{\varepsilon} = \dot{p}\mathbf{i} + \dot{q}\mathbf{j} + \dot{r}\mathbf{k}.$$

Note now that

$$(\boldsymbol{\varepsilon}, \mathbf{K}_O) = \begin{bmatrix} \dot{p} & \dot{q} & \dot{r} \end{bmatrix} \begin{bmatrix} Ap \\ Bq \\ Cr \end{bmatrix} = Ap\dot{p} + Bq\dot{q} + Cr\dot{r}$$

and in view of (5.37) the lemma is proven. $\square$

Remark 5.6. The case of Euler entails two important conditions on the dynamics of the components of $\boldsymbol{\omega}$. The first one is obtained from (5.39) and (5.40):

$$Ap^2 + Bq^2 + Cr^2 = 2T = \underset{t}{\text{const.}} \qquad (5.41)$$

For the second, note that as a consequence of (5.35)

$$\mathbf{K}_O = \underset{t}{\text{const}},$$

whereby

$$K_O^2 = \underset{t}{\text{const}},$$

or, in view of (5.29),

$$A^2 p^2 + B^2 q^2 + C^2 r^2 = K_O^2 = \underset{t}{\text{const.}} \qquad (5.42)$$

Additional conditions on the geometry of the solid allow obtaining more results.

First case: $A \neq B$ (restriction to the symmetry of the solid). This extra condition allows, starting from (5.41) and (5.42), to arrive at explicit expressions for the determination of p, q, and r. To see it, note that multiplying (5.41) by B and subtracting the resulting expression from (5.42) yields

$$A(A - B)p^2 + C(C - B)r^2 = K_O^2 - 2BT.$$

Define

$$(A - B)p^2 = f_1\left(r^2\right), \qquad (5.43)$$

which leads to
$$f_1\left(r^2\right) := \frac{1}{A}\left[K_O^2 - 2BT + C(B-C)r^2\right].$$

On the other hand, multiplying (5.41) by A and subtracting the resulting expression from (5.42), we get
$$B(B-A)q^2 + C(C-A)r^2 = K_O^2 - 2AT,$$

which implies
$$(A-B)q^2 = f_2\left(r^2\right) \tag{5.44}$$

with
$$f_2\left(r^2\right) := -\frac{1}{B}\left[K_O^2 - 2AT + C(A-C)r^2\right].$$

The product of (5.43) and (5.44) leads to
$$(A-B)^2 q^2 p^2 = f_1\left(r^2\right) f_2\left(r^2\right),$$

or equivalently,
$$(A-B)qp = \pm\sqrt{f_1\left(r^2\right) f_2\left(r^2\right)}.$$

But given that by the third equation of (5.36)
$$C\dot{r} = (A-B)qp,$$

we get the differential expression for r:
$$C\dot{r} \mp \sqrt{f_1\left(r^2\right) f_2\left(r^2\right)} = 0,$$

which can be solved in view of the fact that it only depends on one variable r. The achievement of the expression for r will in turn allow the corresponding expressions to be found,
$$p = p(r), \quad q = q(r),$$

in view of (5.41) and (5.42).

Second case: $A = B$. The solid exhibits a certain symmetry and is called the *Lagrange case*. Based on this condition, by the third equation of (5.36) we have
$$\dot{r} = 0,$$

that is,
$$r(t) = \operatorname*{const}_{t}, \tag{5.45}$$
and as a result from (5.42) we get
$$p^2 + q^2 = \operatorname*{const}_{t}. \tag{5.46}$$
Immediately, the results (5.45) and (5.46) allow to formulate the following lemma.

Lemma 5.8. *In the Lagrange case we have*
$$\omega = \|\boldsymbol{\omega}\| = \operatorname*{const}_{t}. \tag{5.47}$$

2. Gyroscope

In Fig. 5.13 the device called *gyroscope* is shown. It is formed by a body subjected to the two rotations shown. The gyroscope has special importance, mainly to determine the presence of a certain condition of movement called *regular precession*, which possesses interesting properties.

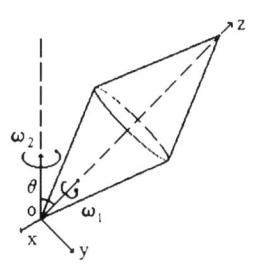

Figure 5.13 Gyroscope.

The following definition is related to Fig. 5.13.

Definition 5.6. A movement produced by a gyroscope with the condition of symmetry $A = B$ is called **regular precession** if the following conditions are met:

(i)
$$\omega_1 = \|\boldsymbol{\omega}_1\| = \operatorname*{const}_{t},$$

(ii)
$$\boldsymbol{\omega}_2 = \overline{\operatorname*{const}_{t}},$$

(iii)
$$\theta := \widehat{\boldsymbol{\omega}_1, \boldsymbol{\omega}_2} = \operatorname*{const}_{t}.$$

The requirements (i)–(iii) allow to formulate an explicit condition for the existence of the regular precession movement. In Fig. 5.13 it is seen that the angular velocity $\boldsymbol{\omega}$, with which the solid moves, is given by

$$\boldsymbol{\omega} = \boldsymbol{\omega}_1 + \boldsymbol{\omega}_2$$

since $\boldsymbol{\omega}_1$ and $\boldsymbol{\omega}_2$ have a common pivot. Note also that p, q, and r are the components of $\boldsymbol{\omega}$ with respect to the coordinate system that appears and is fixed to the solid. On the other hand, in view of the conditions (i)–(iii) that the movement in question must meet, it is concluded that

$$K_O = \underset{t}{\text{const.}} \qquad (5.48)$$

So, if we rewrite $\mathbf{K}_O$ in the form

$$\mathbf{K}_O = K_O \mathbf{e}_{\mathbf{K}_O}$$

with $\mathbf{e}_{\mathbf{K}_O}$ as a unit vector in the direction of $\mathbf{K}$, from (5.48) we have

$$\dot{\mathbf{K}}_O = K_O \dot{\mathbf{e}}_{\mathbf{K}_O},$$

and by the Euler's theorem, the configuration of the movement, and conditions (i)–(iii) it follows that

$$\dot{\mathbf{e}}_{\mathbf{K}_O} = [\boldsymbol{\omega}_2, \mathbf{e}_{\mathbf{K}_O}],$$

which gives

$$\dot{\mathbf{K}}_O = [\boldsymbol{\omega}_2, \mathbf{K}_O],$$

and considering that

$$\dot{\mathbf{K}}_O = \mathbf{M}_{\mathbf{F}_{ext},O},$$

we finally get

$$[\boldsymbol{\omega}_2, \mathbf{K}_O] = \mathbf{M}_{\mathbf{F}_{ext},O}. \qquad (5.49)$$

This last relation will allow to prove the following result.

Theorem 5.1 (N.E. Zhukovski, 1916). *To achieve a regular precession movement with the parameters $\boldsymbol{\omega}_1$, $\boldsymbol{\omega}_2$, and θ, it is necessary and sufficient to apply the following external force moment:*

$$\mathbf{M}_{\mathbf{F}_{ext},O} = [\boldsymbol{\omega}_2, \boldsymbol{\omega}_1]\left[C + (C - A)\frac{\omega_2}{\omega_1}\cos\theta\right]. \qquad (5.50)$$

Proof. Note that for the configuration shown in Fig. 5.13, we have

$$p = 0, \quad q = -\omega_2 \sin\theta, \quad r = \omega_1 + \omega_2 \cos\theta,$$

and therefore, formula (5.29) becomes

$$\mathbf{K}_O = -A(\omega_2 \sin\theta)\mathbf{j} + C(\omega_1 + \omega_2 \cos\theta)\mathbf{k}, \tag{5.51}$$

taking into account that $A = B$. The relationship (5.51) and the fact that

$$\boldsymbol{\omega}_2 = \begin{bmatrix} 0 & -\omega_2 \sin\theta & \omega_2 \cos\theta \end{bmatrix}^\mathsf{T}$$

allow to obtain the left part of (5.49). Indeed,

$$[\boldsymbol{\omega}_2, \mathbf{K}_O] = \begin{bmatrix} \mathbf{i} & \mathbf{j} & \mathbf{k} \\ 0 & -\omega_2 \sin\theta & \omega_2 \cos\theta \\ 0 & -A\omega_2 \sin\theta & C(\omega_1 + \omega_2 \cos\theta) \end{bmatrix}$$

$$= \left[-C\omega_2 \sin\theta (\omega_1 + \omega_2 \cos\theta) + A\omega_2^2 \sin\theta \cos\theta \right] \mathbf{i}$$

$$= \left[-(C-A)\omega_2^2 \sin\theta \cos\theta - C\omega_1 \omega_2 \sin\theta \right] \mathbf{i}$$

$$= -\omega_1 \omega_2 \sin\theta \left[(C-A)\frac{\omega_2}{\omega_1} \cos\theta + C \right] \mathbf{i}$$

$$= [\boldsymbol{\omega}_2, \boldsymbol{\omega}_1] \left[(C-A)\frac{\omega_2}{\omega_1} \cos\theta + C \right],$$

which proves (5.50). □

Example 5.10. A gyroscope is in space; given the magnitudes ω_1, ω_2 let us calculate the angle θ, formed when the regular precession is presented.

In space we have

$$\mathbf{M}_{\mathbf{F}_{ext},O} = 0,$$

so that expression (5.50) leads to two possibilities:

a)

$$[\boldsymbol{\omega}_2, \boldsymbol{\omega}_1] = 0,$$

so

$$\theta \text{ is any},$$

and

b)

$$(C-A)\frac{\omega_2}{\omega_1} \cos\theta + C = 0,$$

giving

$$\theta = \arccos\left(\frac{C}{A-C}\frac{\omega_1}{\omega_2}\right).$$

The Euler's equations (5.33) allow to obtain other properties of the regular precession movement, in addition to those given by the Zhukovski's theorem. From Fig. 5.13 and by the configuration of the movement it is concluded that the projection of ω on the z-axis remains constant, that is,

$$r = \operatorname*{const}_{t}, \qquad (5.52)$$

implying

$$\dot{r} \equiv 0.$$

This fact and the condition of symmetry $A = B$, to be substituted in the third of Euler's equations, lead to a necessary condition of the movement of regular precession:

$$\left(\mathbf{M}_{\mathbf{F}_{ext}.O}\right)_z = 0,$$

which coincides with what was predicted by the previous theorem. In addition, from the fact that K_O is constant and by (5.29) it follows that

$$A^2 p^2 + B^2 q^2 + C^2 r^2 = \operatorname*{const}_{t},$$

which, in view of (5.52) and the premise $A = B$, is reduced to the following condition on the components p and q of ω with respect to the fixed coordinate system to the body shown in Fig. 5.13:

$$p^2 + q^2 = \operatorname*{const}_{t}.$$

5.5 Dynamic reactions caused by the gyroscopic moment

When the solids rotate eccentrically, or when the axis of rotation does not coincide with any of the directions in which the moments of inertia are extreme or is perpendicular to them, additional forces are generated in the supports of the axis of rotation. These forces are called *dynamic reactions* and are very important especially for the design of said supports. To understand the above, the two mentioned cases are analyzed through two very illustrative examples.

Example 5.11. Suppose a disk of mass M rotates on an axis that does not pass through its center of inertia, as illustrated in Fig. 5.14. Clearly, if there were no rotation (static case)

$$\mathbf{F}_A^{stat} + \mathbf{F}_B^{stat} = -m\mathbf{g}.$$

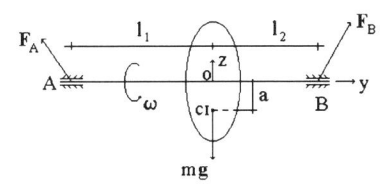

Figure 5.14 Disk rotating eccentrically.

However, in the presence of rotation, an additional force due to the dynamic effect appears, namely,

$$\mathbf{F}^{din} = -M\omega^2 \mathbf{a},$$

where **a** represents the position vector of the center of inertia of the disk. So now we have

$$\mathbf{F}_B = \mathbf{F}_B^{stat} + \mathbf{F}_B^{din}, \quad \mathbf{F}_A = \mathbf{F}_A^{stat} + \mathbf{F}_A^{din},$$

where the superscripts *stat* and *din* denote that the force in question is due to the static part and to the dynamic part, respectively. Clearly

$$F_A^{stat} + F_B^{stat} = mg,$$

while

$$F_A^{din} + F_B^{din} = m\omega^2 a.$$

To determine the values of F_A^{din} and F_B^{din} the condition is used that, for the type of support shown in Fig. 5.14, the moment of forces is zero, that is,

$$M_A^{din} = 0, \quad M_B^{din} = 0.$$

Therefore

$$M_A^{din} = M\omega^2 a l_1 - F_B^{din}(l_1 + l_2) = 0,$$
$$M_B^{din} = M\omega^2 a l_2 - F_A^{din}(l_1 + l_2) = 0,$$

from which we finally get

$$F_A^{din} = M\omega^2 a \frac{l_2}{(l_1 + l_2)}, \quad F_B^{din} = M\omega^2 a \frac{l_1}{(l_1 + l_2)}.$$

So, the situation becomes dangerous if the condition

$$F_{A,B}^{din} \geq F_{crit}^{din}$$

is violated, where F_{crit}^{din} represents the permitted design value of the supports.

Example 5.12. Assume that the solid shown in Fig. 5.15 is in space. The illustrated coordinate system has its origin in the center of inertia of the body and its directions follow the directions in which the moments of inertia are extreme. The axis of rotation of the solid is in the plane yz, and it also passes through the center of inertia, but it keeps an inclination α with respect to the axis y. Because of the condition that the body is in space, it does not present the weight force. Therefore we have the first relationship

$$F_A^{din} = F_B^{din}. \tag{5.53}$$

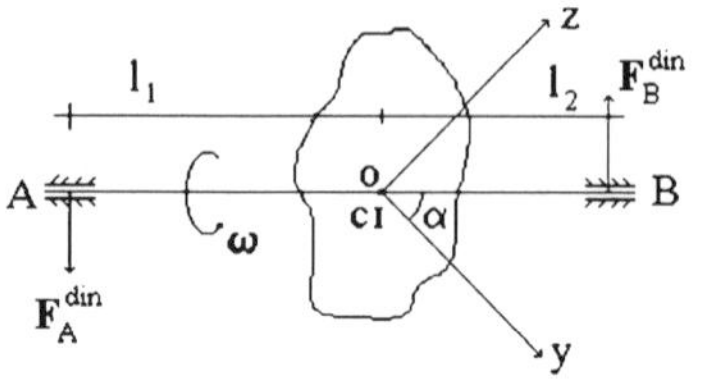

Figure 5.15 Solid rotating around a given direction.

The second relation is given by the first of the Euler's equations (5.33), namely,

$$A\dot{p} + (C - B)qr = F_A^{din}l_1 + F_B^{din}l_2. \tag{5.54}$$

But from Fig. 5.15 one can see that

$$p \equiv 0,$$

so

$$\dot{p} = 0,$$

which, in view of (5.53) and (5.54), gives

$$F_A^{din} = F_B^{din} = \frac{C-B}{l_1+l_2}qr = \frac{C-B}{2(l_1+l_2)}\omega^2 \sin 2\alpha, \tag{5.55}$$

since here

$$q := \omega \cos\alpha, \quad r := \omega \sin\alpha.$$

The following example is a direct application of the newly obtained result and illustrates the magnitude of the forces that can be generated.

Example 5.13. The solid cylinder of uniformly distributed mass M of Fig. 5.16 satisfies the conditions of the previous exercise. Calculate F_A^{din} and F_B^{din} for the values

$$\alpha = 30°, \ \omega = 1000 \text{ rad/s}, \ M = 10 \text{ g}, \ l = 2 \text{ m}, \ r = 0.1 \text{ m}, \ h = 1 \text{ m}.$$

Euler's dynamic equations 185

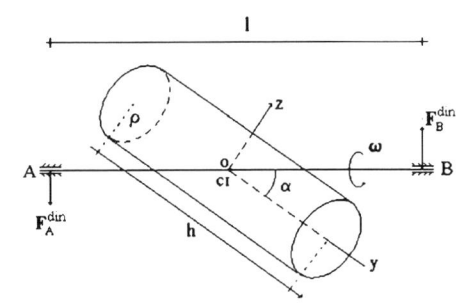

Figure 5.16 Solid cylinder rotating around a given direction.

From the second example of Section 5.4 we have

$$B := I_{yy} = \frac{1}{2}Mr^2, \quad C := I_{zz} = \frac{1}{4}M\left(r^2 + \frac{h^2}{3}\right),$$

so (5.55) leads to

$$F_A^{din} = F_B^{din} = \frac{\frac{h^2}{3} - r^2}{8l} M\omega^2 \sin 2\alpha,$$

and for the given values

$$F_A^{din} = F_B^{din} = 175.01 \text{ N}.$$

5.6 Exercises

Exercise 5.1. It is required to find the main axes of inertia at point A of a homogeneous circular cylinder of mass m, height H, and base radius R (see Fig. 5.17). Show that one of the main axes is perpendicular to the plane passing through the axis of the cylinder and point A, and the other two lie in this plane and make the angles α and $\pi/2 - \alpha$ with the generatrix of the cylinder. For the case $H = \sqrt{3}R$, show that the inertia tensor in the principal axes for point A is

$$\mathbb{I} = \frac{mR^2}{4} \begin{bmatrix} 9 & 0 & 0 \\ 0 & 9 & 0 \\ 0 & 0 & 2 \end{bmatrix}.$$

Exercise 5.2. A biaxial gyro platform carries two identical gyroscopes rotating with a constant angular velocity ω. A special device holds the axis of the first gyroscope in the plane of the platform, and the second perpendicular to it. The centers of inertia of the gyroscopes C_1 and C_2 are located in the plane of the platform at a distance a from its center (see Fig. 5.18).

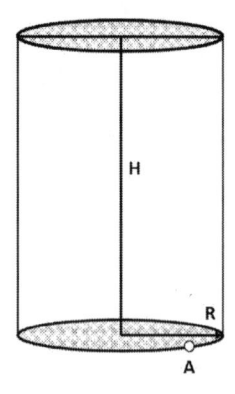

Figure 5.17 A homogeneous circular cylinder of mass m, height H, and base radius R.

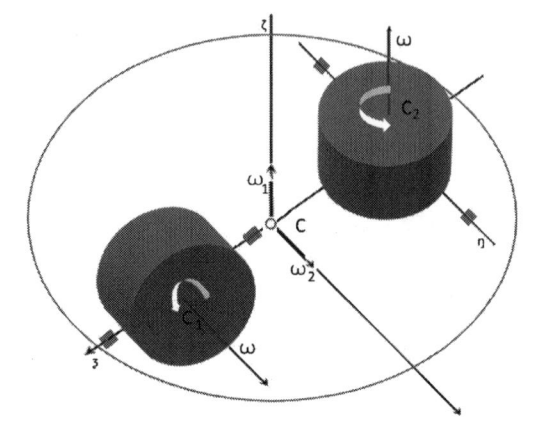

Figure 5.18 A biaxial gyro platform carries two identical gyroscopes rotating with a constant angular velocity.

Considering gyroscopes to be thin homogeneous dikes of mass m and radius r, show that:

1) In the case when the platform rotates with an angular velocity ω_1 around the axis of $C\zeta$, perpendicular to the plane of the platform, the kinetic energy of the system is equal to

$$T_1 = \frac{mr^2}{2}\omega(\omega+\omega_1) + m\left(a^2 + \frac{3}{8}r^2\right)\omega_1^2.$$

2) In the case when the platform rotates with an angular velocity ω_2 around the axis of $C\eta$, parallel to the axis of the first gyroscope, the kinetic energy of the system

is equal to
$$T_2 = \frac{mr^2}{2}\omega(\omega+\omega_2) + m\left(a^2 + \frac{3}{8}r^2\right)\omega_2^2.$$

Exercise 5.3. A constant (in value and in direction) moment of external forces $\mathbf{M}_0$ is applied to a symmetric ($A = B \neq C$) solid body with a fixed point O. Show that, if at the initial moment the angular velocity of the body was equal to zero, then its dependence on time will be given by the formulas

$$\left.\begin{aligned} p(t) &= \frac{t}{A}\left(M_0^{(1)}\cos\frac{\Psi t^2}{2} - M_0^{(2)}\sin\frac{\Psi t^2}{2}\right), \\ q(t) &= \frac{t}{A}\left(M_0^{(1)}\sin\frac{\Psi t^2}{2} + M_0^{(2)}\cos\frac{\Psi t^2}{2}\right), \\ r(t) &= \frac{M_0^{(3)}}{C}t, \end{aligned}\right\}$$

where
$$\Psi = M_0^{(3)}\frac{C-A}{AC}$$

and $M_0^{(i)}$ ($i = 1, 2, 3$) are the projections of the moment of external forces $\mathbf{M}_0$ to the main axis of inertia in the initial body position.

Exercise 5.4. To stabilize the angle of various objects, gyroscopes are used, applying a moment to the object which compensates for the external disturbing effect. The servo-gyroscope drive circuit is shown in Fig. 5.19. Supposing $A = B \neq C$, show that the angular velocities ω_1 and ω_2, at which the disturbing periodic moment $M = M_0\cos\omega t$, applied to the outer frame and directed along the axis 1-1, are

$$\omega_1 = \frac{M_0}{C\omega}, \quad \omega_2 = \omega.$$

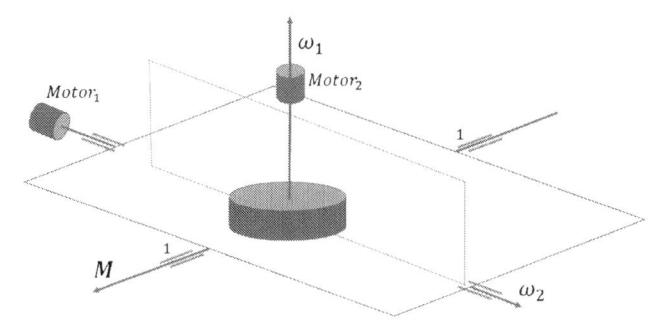

Figure 5.19 The servo-gyroscope drive circuit.

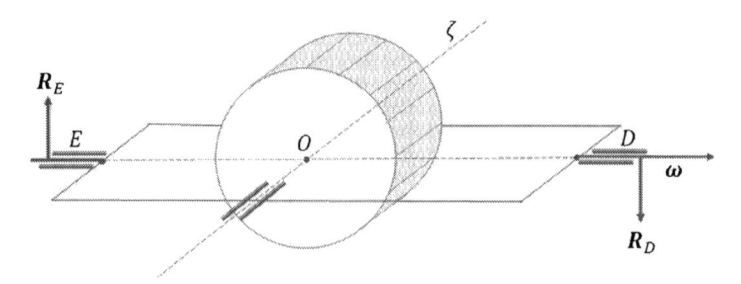

Figure 5.20 The frame of the balancing gyroscope mounted on a fixed base using bearings D and E.

Exercise 5.5. The frame of the balancing gyroscope is mounted on a fixed base using bearings D and E. The gyro rotor performs n revolutions per second around the axis $O\zeta$. Distance $DE = l$ (see Fig. 5.20). The moment of inertia of the rotor relative to the axis of symmetry is C. Having neglected the mass of the frame, show that the strengths of the dynamic reactions to the frame bearings D and E, caused by the gyroscopic moment, are

$$F_D^{din} = F_E^{din} = 2\pi n \omega \frac{C}{l}.$$

Dynamic Lagrange equations

Contents

6.1 Mechanical connections 189
6.2 Generalized forces 192
6.3 Dynamic Lagrange equations 195
6.4 Normal form of Lagrange equations 204
6.5 Electrical and electromechanical models 207
    6.5.1 Some physical relations 208
    6.5.2 Table of electromechanical analogies 210
6.6 Exercises 217

Newton's second law and Euler's dynamic equations are the formalism that allows to obtain the equations of movement in mechanical systems. However its application is usually complicated if the geometry of the movement is not simple and/or by the presence of restrictions to it. The Lagrange equations, whose study is addressed in this chapter, are an essential tool for these cases, since they naturally include the constraints, in addition to being based on the concept of generalized coordinates, which allow describing the dynamics in terms of the variables, associated with the degrees of freedom of the system. This particularity also makes it possible to apply the same formalism to electrical and even electromechanical systems. Fundamental parts of the Lagrange equations are generalized forces, which are defined and characterized before obtaining said equations.

6.1 Mechanical connections

In general, the different material points of the mechanical systems keep connections to each other, called connections or *mechanical constraints*. These are relationships that define the movement and may be described by mathematical expressions. In the examples that follow, several cases are illustrated.

Example 6.1.

(a) Consider the simple pendulum in Fig. 6.1. Assume that the arm is rigid and has length l. It is clear that the dynamic evolution of the distal point of the pendulum is restricted to the variety

$$x^2(t) + y^2(t) = l^2, \ z(t) = 0.$$

These two expressions represent the mechanical connections of this system.

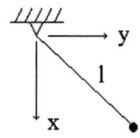

Figure 6.1 Simple pendulum with rigid arm.

(b) Consider a simple pendulum again, but suppose now that the massless arm is extensible and that its length follows the temporal law $l(t)$ (Fig. 6.2). In these circumstances the dynamics of the distal point of the pendulum is conditioned by the mechanical constraints

$$x^2(t) + y^2(t) = l^2(t), \quad z(t) = 0.$$

Figure 6.2 Simple pendulum with extendable arm.

(c) Let the disk of radius r roll without sliding as shown in Fig. 6.3. The point of the disk in contact with the x-axis obeys the differential equation

$$\dot{x}(t) = r\dot{\phi}(t),$$

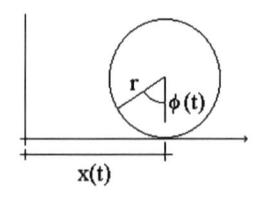

Figure 6.3 Disc rolling without sliding.

so the mechanical connections that restrict the dynamics of this point are given by relationships

$$x(t) - r\phi(t) = \underset{t}{\text{const}}, \quad z(t) = 0.$$

(d) Suppose that the point shown in Fig. 6.4 moves with velocity v of constant magnitude. In this case the movement is given in such a way that the following relationship must be satisfied:

$$\dot{x}^2(t) + \dot{y}^2(t) + \dot{z}^2(t) = v^2 = \underset{t}{\text{const}}.$$

Dynamic Lagrange equations

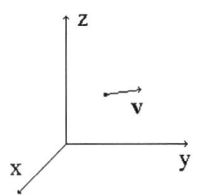

Figure 6.4 Point moving with constant magnitude speed.

(e) Consider a disk that moves on a horizontal plane, possibly with sliding, as shown in Fig. 6.5. According to the coordinate system used, the movement of the points of the disk obeys the restriction

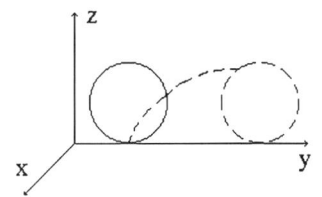

$$z(t) \geq 0.$$

Figure 6.5 Disk with movement on the xy-plane.

Let S be a system of N material points. Quite generally, the mechanical connections between the points of S can be expressed by relations of the type

$$f_k(t, \mathbf{R}(t), \dot{\mathbf{R}}(t)) = 0, \quad k = 1, \dots, m, \tag{6.1}$$

where m is the number of mechanical constraints, while $\mathbf{R}(t)$ is the vector, formed by the position vectors of all the points, i.e.,

$$\mathbf{R}(t) := \begin{bmatrix} \mathbf{r}_1(t) \\ \vdots \\ \mathbf{r}_N(t) \end{bmatrix} \in R^{3N}.$$

A restriction

$$f_k(\mathbf{R}(t), \dot{\mathbf{R}}(t)) = 0,$$

which is not explicitly dependent on time, is called *stationary*.

Definition 6.1. A vector $\mathbf{R}$ that satisfies the mechanical constraints (6.1) is said to be in a **possible position** of the points of S.

Definition 6.2. If the relation of a **mechanical connection**, for example f_k, is integrable, then it can be represented in the form

$$\tilde{f}_k(t, \mathbf{R}(t)) = 0,$$

and it is called **holonomic**. Otherwise, which is the general case, it is called **non-holonomic**. A system whose mechanical constraints are holonomic receives the name of **holonomic**.

Clearly, if S is a holonomic system with m mechanical connections, it is possible to enter the parameter vector

$$\mathbf{q} := \begin{bmatrix} q_1 & q_2 & \cdots & q_n \end{bmatrix}^T, \tag{6.2}$$

where

$$n = 3N - m,$$

such that the possible positions of the material points of the mechanical system can be expressed as

$$\mathbf{R}(t) = \mathbf{R}(t, \mathbf{q}). \tag{6.3}$$

The number n is known as the **number of degrees of freedom** of system S.

Definition 6.3. The set of n parameters that make up $\mathbf{q}$ in (6.2) is said to be **independent** if the expression

$$\sum_{i=1}^{n} \lambda_i \frac{\partial \mathbf{R}(t)}{\partial q_i} = 0, \quad \lambda_i \in R,$$

is satisfied if and only if

$$\lambda_i = 0, \ i = 1, ..., n.$$

Definition 6.4. The n parameters q_i, $i = 1, ..., n$, in (6.2) are called **generalized coordinates** if the following conditions are satisfied:
- for each instant of time, function (6.3) is uniquely defined between the set of possible positions $\mathbf{R}(t)$ of the material points and a certain region of the space of $\mathbf{q}$,
- the components of $\mathbf{q}$ are independent.

6.2 Generalized forces

Definition 6.5. For the possible position

$$\mathbf{r}_k = \mathbf{r}_k(t, \mathbf{q}), \quad k = 1, ..., N,$$

the infinitesimal displacement corresponding to this position is given by

$$d\mathbf{r}_k := \frac{\partial \mathbf{r}_k}{\partial t} dt + \sum_{i=1}^{n} \frac{\partial \mathbf{r}_k}{\partial q_i} \delta q_i, \quad k = 1, ..., N,$$

and is called a **possible transfer**. If in particular this displacement does not depend on time, that is, $\frac{\partial \mathbf{r}_k}{\partial t} = 0$, we obtain

$$\delta \mathbf{r}_k = \sum_{i=1}^{n} \frac{\partial \mathbf{r}_k}{\partial q_i} \delta q_i, \quad k = 1, ..., N.$$

This is referred to as a **virtual possible transfer**. Henceforth, only possible translations are considered; that is why we will use the term "virtual transfer," omitting the word "possible."

If the translations of all the material points of S are virtual, the work developed by the forces in the system is given by

$$\left. \begin{aligned} \delta A|_{\delta \mathbf{R}} &:= \sum_{k=1}^{N} (\mathbf{F}_k, \delta \mathbf{r}_k) = \sum_{k=1}^{N} \left(\mathbf{F}_k, \sum_{i=1}^{n} \frac{\partial \mathbf{r}_k}{\partial q_i} \delta q_i \right) \\ &= \sum_{i=1}^{n} \left(\sum_{k=1}^{N} \left(\mathbf{F}_k, \frac{\partial \mathbf{r}_k}{\partial q_i} \right) \right) \delta q_i = \sum_{i=1}^{n} Q_i \delta q_i, \end{aligned} \right\} \quad (6.4)$$

where

$$Q_i := \sum_{k=1}^{N} \left(\mathbf{F}_k, \frac{\partial \mathbf{r}_k}{\partial q_i} \right) \quad (6.5)$$

and $\mathbf{F}_k$, $k = 1, ...N$, denotes the total force on the point $k \in S$.

Definition 6.6. The vector

$$\mathbf{Q} := \begin{bmatrix} Q_1 & Q_2 & \cdots & Q_n \end{bmatrix}^T,$$

where the component Q_i, $i = 1, ...n$, is given by (6.5), is called vector of **generalized forces**. The component Q_i is the generalized force corresponding to the coordinate q_i.

Remark 6.1. Note that (6.4) can be used to calculate the generalized force Q_i. Indeed, from this equation it is seen that if we fix the generalized coordinates q_j, $j \neq i$, that is, $\delta q_j = 0$, $j \neq i$, we have

$$Q_i = \frac{\delta A|_{\delta \mathbf{R}}}{\delta q_i}. \quad (6.6)$$

Mechanical restrictions are translated into forces on material points. These forces are called *reaction forces*. The total reaction force exerted on the k-th material point of the mechanical system will be denoted by $\mathbf{F}_{k,reac}$. Then we have

$$\mathbf{F}_k = \mathbf{F}_{k,ext} + \mathbf{F}_{k,reac}, \quad k = 1, ..., N,$$

where $\mathbf{F}_{k,ext}$ is called *external forces* and encompasses the external and internal actions exerted on the k-th particle of S.

Remark 6.2. In view of the definitions

$$Q_{i,ext} := \sum_{k=1}^{N} \left(\mathbf{F}_{k,ext}, \frac{\partial \mathbf{r}_k}{\partial q_i} \right), \quad Q_{i,reac} := \sum_{k=1}^{N} \left(\mathbf{F}_{k,reac}, \frac{\partial \mathbf{r}_k}{\partial q_i} \right), \quad i = 1, ..., n, \tag{6.7}$$

where $Q_{i,ext}$ and $Q_{i,reac}$ are called **generalized external** and **reaction forces** corresponding to q_i, respectively, we get

$$Q_i = Q_{i,ext} + Q_{i,reac},$$

or, in vector form,

$$\mathbf{Q} = \mathbf{Q}_{ext} + \mathbf{Q}_{reac}.$$

Definition 6.7. The mechanical constraints of a system are called **ideal** if the work, developed by the reaction forces on any virtual translation $\delta \mathbf{R}$, is zero, i.e.,

$$\delta A_{reac}|_{\delta \mathbf{R}} = \sum_{k=1}^{N} \left(\mathbf{F}_{k,reac}, \delta \mathbf{r}_k \right) = \sum_{i=1}^{n} Q_{i,reac} \delta q_i = 0,$$

for all δq_i.

The following result is obvious.

Lemma 6.1. *In systems whose mechanical connections are ideal, we have*

$$Q_{i,reac} = 0 \ \forall i = 1, ...n.$$

In the following example it is shown that the systems that appear have ideal mechanical connections.

Example 6.2.

a) Fig. 6.6 shows a simple pendulum. This system is composed of a material point, and therefore

$$\delta A_{reac}|_{\delta \mathbf{R}} = (\mathbf{F}_{reac}, \delta \mathbf{r}) = (\mathbf{F}_{reac}, \delta \mathbf{r}) = 0$$

Dynamic Lagrange equations

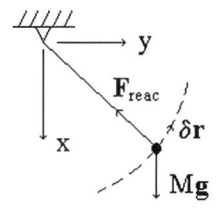

Figure 6.6 A simple pendulum.

since $\mathbf{F}_{reac}$ and $\delta\mathbf{r}$ are orthogonal. Therefore, in this system the mechanical restriction is ideal.

b) A disc rolls without sliding as shown in Fig. 6.7. For this system $\mathbf{F}_{reac}$ is given by the sum of the friction force and the reaction of the floor, applied on the material point A; therefore

$$\delta A_{reac}|_{\delta\mathbf{R}} = (\mathbf{F}_{reac}, \delta\mathbf{r}_A) = \left(\mathbf{F}_{reac}, \frac{\delta\mathbf{r}_A}{\delta t}\right)\delta t = (\mathbf{F}_{reac}, \mathbf{v}_A)\delta t = 0$$

since $\mathbf{v}_A = 0$. So, the mechanical restriction of this system is also ideal.

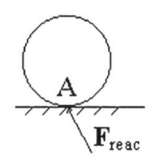

Figure 6.7 Disk in movement.

6.3 Dynamic Lagrange equations

Lemma 6.2 (Lagrange, 1750). *Let S be a holonomic system of N material points whose masses do not depend on the velocity $\dot{\mathbf{q}}$ or the position $\mathbf{q}$. In such a system the dynamics is governed by the following differential equation:*

$$\frac{d}{dt}\frac{\partial T}{\partial \dot{q}_i} - \frac{\partial T}{\partial q_i} = Q_i, \quad i = 1, ..., n, \tag{6.8}$$

where T is the total kinetic energy of S.

Proof. Let m_k be the mass of the k-th particle of S. By the Newton's second law

$$\frac{d}{dt}(m_k\mathbf{v}_k) = \mathbf{F}_k, \quad k = 1, ..., N. \tag{6.9}$$

Multiplying scalarly (6.9) by $\dfrac{\partial \mathbf{r}_k}{\partial q_i}$ and adding on $k = 1, ..., N$, we get

$$\sum_{k=1}^{N} \left(\frac{d}{dt}(m_k \mathbf{v}_k), \frac{\partial \mathbf{r}_k}{\partial q_i} \right) = \sum_{k=1}^{N} \left(\mathbf{F}_k, \frac{\partial \mathbf{r}_k}{\partial q_i} \right) = Q_i. \tag{6.10}$$

Given the relation

$$\left(\frac{d}{dt}(m_k \mathbf{v}_k), \frac{\partial \mathbf{r}_k}{\partial q_i} \right) = \frac{d}{dt}\left(m_k \mathbf{v}_k, \frac{\partial \mathbf{r}_k}{\partial q_i} \right) - \left(m_k \mathbf{v}_k, \frac{d}{dt}\frac{\partial \mathbf{r}_k}{\partial q_i} \right)$$

the left member of (6.10) can be represented as

$$\sum_{k=1}^{N} \left(\frac{d}{dt}(m_k \mathbf{v}_k), \frac{\partial \mathbf{r}_k}{\partial q_i} \right) = \sum_{k=1}^{N} \left[\frac{d}{dt}\left(m_k \mathbf{v}_k, \frac{\partial \mathbf{r}_k}{\partial q_i} \right) - \left(m_k \mathbf{v}_k, \frac{d}{dt}\frac{\partial \mathbf{r}_k}{\partial q_i} \right) \right].$$

But, following a procedure similar to that used in Section 1.3 of Chapter 1, it is easily verified that

$$\frac{\partial \mathbf{r}_k}{\partial q_i} = \frac{\partial \mathbf{v}_k}{\partial \dot{q}_i}, \quad \frac{d}{dt}\frac{\partial \mathbf{r}_k}{\partial q_i} = \frac{\partial \mathbf{v}_k}{\partial q_i}.$$

Therefore

$$\sum_{k=1}^{N} \left(\frac{d}{dt}(m_k \mathbf{v}_k), \frac{\partial \mathbf{r}_k}{\partial q_i} \right) = \sum_{k=1}^{N} \left[\frac{d}{dt}\left(m_k \left(\mathbf{v}_k, \frac{\partial \mathbf{v}_k}{\partial \dot{q}_i} \right) \right) - m_k \left(\mathbf{v}_k, \frac{\partial \mathbf{v}_k}{\partial q_i} \right) \right],$$

and since

$$\left(\mathbf{v}_k, \frac{\partial \mathbf{v}_k}{\partial \dot{q}_i} \right) = \frac{\partial}{\partial \dot{q}_i} \left(\frac{1}{2}(\mathbf{v}_k, \mathbf{v}_k) \right), \quad \left(\mathbf{v}_k, \frac{\partial \mathbf{v}_k}{\partial q_i} \right) = \frac{\partial}{\partial q_i} \left(\frac{1}{2}(\mathbf{v}_k, \mathbf{v}_k) \right),$$

we arrive at

$$\sum_{k=1}^{N} \left(\frac{d}{dt}(m_k \mathbf{v}_k), \frac{\partial \mathbf{r}_k}{\partial q_i} \right) = \frac{d}{dt}\frac{\partial T}{\partial \dot{q}_i} - \frac{\partial T}{\partial q_i},$$

where

$$T := \frac{1}{2} \sum_{k=1}^{N} m_k (\mathbf{v}_k, \mathbf{v}_k).$$

$\square$

Lemma 6.3. *If the external forces are potential, then the generalized external forces are also potential, that is, if there are N scalar functions such that*

$$\mathbf{F}_{k,ext} = -\nabla \Pi_k (\mathbf{r}_k), \quad k = 1, ..., N,$$

Dynamic Lagrange equations

then

$$\mathbf{Q}_{ext} = -\nabla_{\mathbf{q}} V(t, \mathbf{q}),$$

where

$$V(t, \mathbf{q}) := \Pi(\mathbf{R}(t, \mathbf{q})) := \sum_{k=1}^{n} \Pi_k (\mathbf{r}_k(t, \mathbf{q})). \quad (6.11)$$

Proof. Remember that

$$Q_{i,ext} := \sum_{k=1}^{N} \left(\mathbf{F}_{k,ext}, \frac{\partial \mathbf{r}_k}{\partial q_i} \right),$$

and, since external forces are potential, we have

$$Q_{i,ext} = -\sum_{k=1}^{N} \left(\nabla \Pi_k(\mathbf{r}_k), \frac{\partial \mathbf{r}_k}{\partial q_i} \right).$$

But in view of the relation

$$\left(\nabla \Pi_k(\mathbf{r}_k), \frac{\partial \mathbf{r}_k}{\partial q_i} \right) = \frac{\partial}{\partial q_i} \Pi_k(\mathbf{r}_k(t, \mathbf{q})),$$

we may conclude that

$$Q_{i,ext} = -\frac{\partial}{\partial q_i} \sum_{k=1}^{N} \Pi_k(\mathbf{r}_k(t, \mathbf{q})).$$

$\square$

Remark 6.3. Since the external forces can be separated into a potential and a non-potential part, it follows that

$$\mathbf{Q}_{ext} = \mathbf{Q}_{pot} + \mathbf{Q}_{non\text{-}pot}.$$

Therefore Eq. (6.8) can be written in vector form as

$$\frac{d}{dt} \nabla_{\dot{\mathbf{q}}} T - \nabla_{\mathbf{q}} T = \mathbf{Q}_{pot} + \mathbf{Q}_{non\text{-}pot} + \mathbf{Q}_{reac}, \quad (6.12)$$

where

$$\mathbf{Q}_{pot} := -\nabla_{\mathbf{q}} V(t, \mathbf{q})$$

and $V(t, \mathbf{q})$ is given by (6.11).

Definition 6.8. Function

$$L(t, \mathbf{q}, \dot{\mathbf{q}}) := T(t, \mathbf{q}, \dot{\mathbf{q}}) - V(t, \mathbf{q})$$

is referred to as the **Lagrange function** for system S.

The definitions of $\mathbf{Q}_{pot}$ and L allow to express Eq. (6.12) in the form

$$\frac{d}{dt} \nabla_{\dot{\mathbf{q}}} L(t, \mathbf{q}, \dot{\mathbf{q}}) - \nabla_{\mathbf{q}} L(t, \mathbf{q}, \dot{\mathbf{q}}) = \mathbf{Q}_{non\text{-}pot} + \mathbf{Q}_{reac},$$

which is called the **Lagrange equation**. A more common form of this expression is obtained if one considers that the mechanical constraints are *ideal*. In such case

$$\mathbf{Q}_{reac} = \mathbf{0},$$

and the Lagrange equation is reduced to

$$\frac{d}{dt} \nabla_{\dot{\mathbf{q}}} L(t, \mathbf{q}, \dot{\mathbf{q}}) - \nabla_{\mathbf{q}} L(t, \mathbf{q}, \dot{\mathbf{q}}) = \mathbf{Q}_{non\text{-}pot}. \tag{6.13}$$

The Lagrange equation is a very powerful tool in determining the equations of motion of the material points of a mechanical system. This point is illustrated with some examples.

Example 6.3. In Fig. 6.8 a simple pendulum is shown whose arm is a spring of stiffness k. Obtain the equations of motion for the point of mass m. The mechanical constraint in this system is given by the relationship

$$z \equiv 0, \tag{6.14}$$

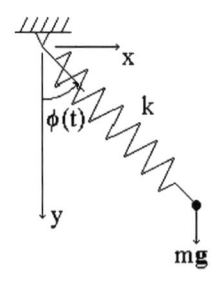

Figure 6.8 Elastic arm pendulum.

which is obviously ideal. Then two additional parameters are required, for example l and ϕ, to fix the system in a space. So, the generalized coordinates may be

$$q_1 := l, \quad q_2 := \phi.$$

Clearly we have

$$x = q_1 \sin q_2, \quad y = q_1 \cos q_2,$$

so that

$$\dot{x} = \dot{q}_1 \sin q_2 + q_1 \dot{q}_2 \cos q_2,$$
$$\dot{y} = \dot{q}_1 \cos q_2 - q_1 \dot{q}_2 \sin q_2.$$

That is why the kinetic energy has the expression

$$T = \frac{1}{2}m(\dot{x}^2 + \dot{y}^2) = \frac{1}{2}m(\dot{q}_1^2 + q_1^2 \dot{q}_2^2).$$

It is now possible to calculate the generalized forces via expression (6.6):

$$Q_1 = \frac{(mg \cos q_2 - k(q_1 - l_0))(\delta q_1)}{\delta q_1} = mg \cos q_2 - k(q_1 - l_0),$$

$$Q_2 = -\frac{mg(\sin q_2)(q_1 \delta q_2)}{\delta q_2} = -mg q_1 \sin q_2,$$

where l_0 represents the nominal length of the spring (without deformation). It is now possible to calculate the Lagrange equations, using (6.8). For the first one we have

$$\frac{d}{dt}\frac{\partial}{\partial \dot{q}_1}T - \frac{\partial}{\partial q_1}T = Q_1,$$

so that

$$\ddot{q}_1 - q_1 \dot{q}_2^2 = g \cos q_2 - \frac{k}{m}(q_1 - l_0), \tag{6.15}$$

while for the second one,

$$\frac{d}{dt}\frac{\partial}{\partial \dot{q}_2}T - \frac{\partial}{\partial q_2}T = Q_2,$$

we get

$$2 q_1 \dot{q}_1 \dot{q}_2 + q_1^2 \ddot{q}_2 = -g q_1 \sin q_2. \tag{6.16}$$

Alternatively, expression (6.13) can be used to arrive at the same results. For this note that if the origin of the coordinate system is chosen as a gravitational potential reference level, it is easily verified that

$$V(t, \mathbf{q}) = \Pi(\mathbf{R}(t, \mathbf{q})) = -mg q_1 \cos q_2 + \frac{1}{2}k(q_1 - l_0)^2.$$

Then the function of Lagrange results:

$$L = T - V = \frac{1}{2}m(\dot{q}_1^2 + q_1^2 \dot{q}_2^2) + mg q_1 \cos q_2 - \frac{1}{2}k(q_1 - l_0)^2.$$

Given that the constraints are ideal and that there are no non-potential forces, for the first equation (6.13) we have

$$\frac{d}{dt}\frac{\partial}{\partial \dot{q}_1}L - \frac{\partial}{\partial q_1}L = 0,$$

implying

$$\frac{d}{dt}(m\dot{q}_1) - \left(mq_1\dot{q}_2^2 + mg\cos q_2 - k(q_1 - l_0)\right) = 0,$$

from which (6.15) follows. For the second Lagrange equation,

$$\frac{d}{dt}\frac{\partial}{\partial \dot{q}_2}L - \frac{\partial}{\partial q_2}L = 0,$$

we have

$$\frac{d}{dt}(mq_1^2\dot{q}_2) + mgq_1 \sin q_2 = 0,$$

which leads to (6.16).

Example 6.4 (Elliptic pendulum). In Fig. 6.9 there is a block of mass M that can slide without friction on a horizontal surface. The movement of the block is due to the pendulum of mass m that is attached as shown.

(a) Obtain Lagrange equations for the system.
(b) Determine approximately the movement for small pendulum angles.

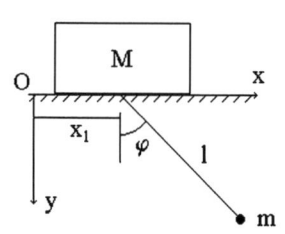

Figure 6.9 Pendulum in a block that slides without friction on a horizontal surface.

a) Denote the masses M and m by 1 and 2, respectively. Since the motion is planar, we have the following two ideal mechanical constraints:

$$y_1 \equiv 0, \quad (x_2 - x_1)^2 + y_2^2 \equiv l^2 = \underset{t}{\text{const}}.$$

So, two generalized coordinates are required. They are chosen as follows:

$$q_1 = x_1, \quad q_2 = \varphi.$$

Dynamic Lagrange equations

The kinetic energy is first determined, which, by the König's theorem, results in

$$T = \frac{1}{2}(M+m)\dot{q}_1^2 + \frac{1}{2}ml^2\dot{q}_2^2 + ml\dot{q}_1\dot{q}_2\cos q_2.$$

If the sliding surface is chosen as the reference level of the potential energy, we have

$$V = -mgl\cos q_2.$$

The last two results allow to obtain the function of Lagrange:

$$L := T - V = \frac{1}{2}(M+m)\dot{q}_1^2 + \frac{1}{2}ml^2\dot{q}_2^2 + ml(\dot{q}_1\dot{q}_2 + g)\cos q_2.$$

Note that there are no non-potential forces. The first Lagrange equation is obtained with

$$\frac{d}{dt}\frac{\partial L}{\partial \dot{q}_1} - \frac{\partial L}{\partial q_1} = 0,$$

where

$$\frac{d}{dt}[(M+m)\dot{q}_1 + ml\dot{q}_2\cos q_2] = 0,$$

or equivalently,

$$(M+m)\ddot{q}_1 + ml\left(\ddot{q}_2\cos q_2 - \dot{q}_2^2\sin q_2\right) = 0. \qquad (6.17)$$

For the second equation we have

$$\frac{d}{dt}\frac{\partial L}{\partial \dot{q}_2} - \frac{\partial L}{\partial q_2} = 0,$$

so

$$\frac{d}{dt}m\left(l^2\dot{q}_2 + l\dot{q}_1\cos q_2\right) + ml(\dot{q}_1\dot{q}_2 + g)\sin q_2 = 0,$$

where finally

$$l\ddot{q}_2 + \ddot{q}_1\cos q_2 + g\sin q_2 = 0. \qquad (6.18)$$

b) If $q_2 \approx 0$ and $\dot{q}_2 \approx 0$, then

$$\sin q_2 \simeq q_2, \quad \cos q_2 \simeq 1, \quad \dot{q}_2^2 \simeq 0, \quad \dot{q}_2^2 \simeq 0.$$

Under these conditions the Lagrange equations (6.17) and (6.18) are reduced to

$$\left.\begin{array}{l}(M+m)\ddot{q}_1 + ml\ddot{q}_2 \simeq 0, \\ l\ddot{q}_2 + \ddot{q}_1 + gq_2 \simeq 0,\end{array}\right\}$$

and from the latter we get

$$\ddot{q}_2 + \omega^2 q_2 \simeq 0,$$

with

$$\omega^2 := \frac{g}{l}\left(1 + \frac{m}{M}\right).$$

From this expression it is deduced that q_2 presents an oscillatory movement of angular frequency ω and period τ with values

$$\omega = \sqrt{\frac{g}{l}\left(1 + \frac{m}{M}\right)}, \quad \tau = \frac{2\pi}{\omega} = \frac{2\pi}{\sqrt{\frac{g}{l}\left(1 + \frac{m}{M}\right)}}.$$

In particular if $M \gg m$, then we get the Huygens formula[1]

$$\omega \simeq \sqrt{\frac{g}{l}}, \quad \tau \simeq 2\pi \sqrt{\frac{l}{g}}.$$

Example 6.5. For the mechanical system of Fig. 6.10 determine the Lagrange equations. Consider that the restoring forces of the springs are of the type

$$F_{res} = k\delta,$$

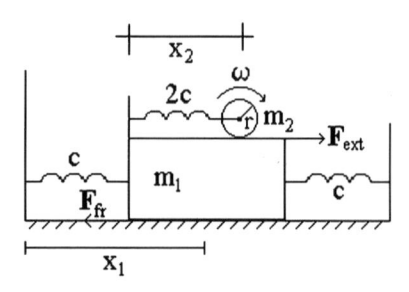

Figure 6.10 Mechanical system of two masses.

where k is the constant called constant of stiffness of the springs and δ is the deformation experienced. Assume that the cylinder is solid and rolls without sliding, while the block experiences viscous friction on its lower surface and is subject to external action as shown. The mechanical system in question consists of two elements subject to the following ideal mechanical restrictions:

$$z_1 = z_2 = y_1 = y_2 \equiv 0.$$

---

[1] Christiaan Huygens (1629–1695).

Dynamic Lagrange equations

Therefore two general coordinates are required to carry out the analysis. The most suitable are

$$q_1 := x_1, \quad q_2 := x_2,$$

where x_1 and x_2 denote, respectively, the displacements of the block and the cylinder. The first step to follow is obtaining the expression for kinetic energy, which is given by

$$T = T_1 + T_2,$$

where the subscripts 1 and 2 denote the block and the cylinder, respectively. Directly we have

$$T_1 = \frac{1}{2} m_1 \dot{q}_1^2,$$

whereas to determine T_2 one must resort to Section 3.5 in Chapter 3. It follows that

$$T_2 = \frac{1}{2} m_2 \dot{q}_1^2 + \frac{1}{2} m_2 \dot{q}_2^2 + m_2 \dot{q}_1 \dot{q}_2 + \frac{1}{2} I_\omega \omega^2,$$

where the moment of inertia I_ω and the angular velocity ω have the expressions

$$I_\omega = \frac{1}{2} m_2 r^2, \quad \omega = \frac{\dot{q}_2}{r},$$

with r as the radius of the cylinder. So, we have

$$T = \frac{1}{2} (m_1 + m_2) \dot{q}_1^2 + m_2 \dot{q}_1 \dot{q}_2 + \frac{3}{4} m_2 \dot{q}_2^2.$$

The next step is to obtain the expression for the potential energy. For a spring of the type of this system, the potential energy is given by

$$\Pi_{res}(\delta) = \frac{k}{2} \delta^2,$$

which is why

$$V = 2 \left(\frac{1}{2} c (q_1 - q_{0,1})^2 \right) + \frac{1}{2} (2c) (q_2 - q_{0,2})^2 =$$
$$c \left[(q_1 - q_{0,1})^2 + (q_2 - q_{0,2})^2 \right],$$

where $q_{0,i}$, $i = 1, 2$, represents the position without deformation of the springs. So, the function of Lagrange results:

$$L := T - V = \frac{1}{2} (m_1 + m_2) \dot{q}_1^2 + m_2 \dot{q}_1 \dot{q}_2 + \frac{3}{4} m_2 \dot{q}_2^2 -$$

$$c\left[(q_1 - q_{0,1})^2 + (q_2 - q_{0,2})^2\right].$$

Now it is time to determine the generalized non-potential forces. By expression (6.6) it immediately follows that

$$Q_{1,non\text{-}pot} = F_{ext} - \beta \dot{q}_1, \quad Q_{2,non\text{-}pot} = 0,$$

where $\beta \dot{q}_1$ is the viscous friction force on the block. Now it is possible to determine the Lagrange equations by (6.13). For the first coordinate equation we have

$$\frac{d}{dt}\frac{\partial}{\partial \dot{q}_1} L - \frac{\partial}{\partial q_1} L = Q_{1,non\text{-}pot},$$

implying

$$\frac{d}{dt}\left[(m_1 + m_2)\dot{q}_1 + m_2 \dot{q}_2\right] + 2c(q_1 - q_{0,1}) = F_{ext} - \beta \dot{q}_1,$$

or finally,

$$(m_1 + m_2)\ddot{q}_1 + m_2 \ddot{q}_2 + \beta \dot{q}_1 + 2c(q_1 - q_{0,1}) = F_{ext}.$$

The second equation is given by

$$\frac{d}{dt}\frac{\partial}{\partial \dot{q}_2} L - \frac{\partial}{\partial q_2} L = Q_{2,non\text{-}pot},$$

which gives

$$\frac{d}{dt}\left(m_2 \dot{q}_1 + \frac{3}{2} m_2 \dot{q}_2\right) + 2c(q_2 - q_{0,2}) = 0,$$

or

$$m_2\left(\ddot{q}_1 + \frac{3}{2}\ddot{q}_2\right) + 2c(q_2 - q_{0,2}) = 0.$$

6.4 Normal form of Lagrange equations

Definition 6.9. It is said that the differential equation

$$\mathbf{F}\left(t, \mathbf{x}, \dot{\mathbf{x}}, \ddot{\mathbf{x}}, \cdots, \mathbf{x}^{(n)}\right) = \mathbf{0}$$

can be presented in the **normal form** if it is algebraically equivalent to

$$\mathbf{x}^{(n)} = \mathbf{G}\left(t, \mathbf{x}, \dot{\mathbf{x}}, \ddot{\mathbf{x}}, \cdots, \mathbf{x}^{(n-1)}\right).$$

Dynamic Lagrange equations

In the generalized coordinates $\mathbf{q} \in \mathbb{R}^n$, the kinetic energy presents a specific expression. Remember that

$$T := \frac{1}{2} \sum_{i \in S} m_i (\mathbf{v}_i, \mathbf{v}_i), \qquad (6.19)$$

but

$$\mathbf{v}_i := \frac{d\mathbf{r}_i}{dt}(t, \mathbf{q}) = \frac{\partial \mathbf{r}_i}{\partial t} + \sum_{j=1}^{n} \frac{\partial \mathbf{r}_i}{\partial q_j} \dot{q}_j,$$

and therefore (6.19) results in

$$T = T_0 + T_1 + T_2,$$

where

$$T_0 := \frac{1}{2} \sum_{i \in S} m_i \left(\frac{\partial \mathbf{r}_i}{\partial t}, \frac{\partial \mathbf{r}_i}{\partial t} \right) = \frac{1}{2} \sum_{i \in S} m_i \left\| \frac{\partial \mathbf{r}_i}{\partial t} \right\|^2,$$

$$T_1 := \sum_{i \in S} m_i \left(\frac{\partial \mathbf{r}_i}{\partial t}, \sum_{j=1}^{n} \frac{\partial \mathbf{r}_i}{\partial q_j} \dot{q}_j \right), \qquad (6.20)$$

$$T_2 := \frac{1}{2} \sum_{i \in S} m_i \left(\sum_{j=1}^{n} \frac{\partial \mathbf{r}_i}{\partial q_j} \dot{q}_j, \sum_{j=1}^{n} \frac{\partial \mathbf{r}_i}{\partial q_j} \dot{q}_j \right) = \frac{1}{2} \sum_{i \in S} m_i \left\| \sum_{j=1}^{n} \frac{\partial \mathbf{r}_i}{\partial q_j} \dot{q}_j \right\|^2. \qquad (6.21)$$

The property of the internal product allows to obtain more compact expressions for T_1 and T_2. From (6.20),

$$T_1 = \sum_{j=1}^{n} \sum_{i \in S} m_i \left(\frac{\partial \mathbf{r}_i}{\partial t}, \frac{\partial \mathbf{r}_i}{\partial q_j} \right) \dot{q}_j = (\mathbf{b}(t, \mathbf{q}), \dot{\mathbf{q}}),$$

with

$$b_j := \sum_{i \in S} m_i \left(\frac{\partial \mathbf{r}_i}{\partial t}, \frac{\partial \mathbf{r}_i}{\partial q_j} \right), \quad j = 1, ..., n,$$

and (6.21) becomes

$$T_2 = \frac{1}{2} \sum_{j=1}^{n} \sum_{k=1}^{n} \left(\sum_{i \in S} m_i \left(\frac{\partial \mathbf{r}_i}{\partial q_j}, \frac{\partial \mathbf{r}_i}{\partial q_k} \right) \right) \dot{q}_j \dot{q}_k = \frac{1}{2} \dot{\mathbf{q}}^T A(t, \mathbf{q}) \dot{\mathbf{q}},$$

where the symmetric matrix $A = (a_{jk}) \in \mathbb{R}^{n \times n}$ has as a generic component

$$a_{jk} = \sum_{i \in S} m_i \left(\frac{\partial \mathbf{r}_i}{\partial q_j}, \frac{\partial \mathbf{r}_i}{\partial q_k} \right). \qquad (6.22)$$

Theorem 6.1 (The basic mechanic theorem). *The newly defined matrix $A(t, \mathbf{q})$ is always non-singular, namely, it has the property*

$$\det A(t, \mathbf{q}) \neq 0 \quad \forall t, \mathbf{q}.$$

Proof. Suppose that $\det A(t, q) = 0$. Then there are numbers $\lambda_k, k = 1, \ldots, n$, not all nulls such that

$$\sum_{k=1}^{n} a_{jk} \lambda_k = 0 \quad \forall j = 1, \ldots, n.$$

Multiplication by λ_j leads to

$$\lambda_j \sum_{k=1}^{n} a_{jk} \lambda_k = 0 \quad \forall j = 1, \ldots, n,$$

and summing over j gives

$$\sum_{j=1}^{n} \sum_{k=1}^{n} a_{jk} \lambda_j \lambda_k = 0.$$

From here and in view of (6.22) it follows that

$$\sum_{j=1}^{n} \sum_{k=1}^{n} \left(\sum_{i \in S} m_i \left(\frac{\partial \mathbf{r}_i}{\partial q_j}, \frac{\partial \mathbf{r}_i}{\partial q_k} \right) \right) \lambda_j \lambda_k = 0,$$

or

$$\sum_{i \in S} m_i \left(\sum_{j=1}^{n} \lambda_j \frac{\partial \mathbf{r}_i}{\partial q_j}, \sum_{k=1}^{n} \lambda_k \frac{\partial \mathbf{r}_i}{\partial q_k} \right) = \sum_{i \in S} m_i \left\| \sum_{k=1}^{n} \lambda_k \frac{\partial \mathbf{r}_i}{\partial q_k} \right\|^2 = 0,$$

which gives

$$\sum_{k=1}^{n} \lambda_k \frac{\partial \mathbf{r}_i}{\partial q_k} = 0 \quad \forall i \in S,$$

or in vector form,

$$\sum_{k=1}^{n} \lambda_k \frac{\partial \mathbf{R}}{\partial q_k} = 0,$$

which contradicts the fact that the parameters $\mathbf{q}$ are independent. □

Corollary 6.1. *The Lagrange equations (6.8) can always be presented in the normal form, i.e.,*

$$\ddot{\mathbf{q}} = \mathcal{F}(t, \mathbf{q}, \dot{\mathbf{q}}, \mathbf{Q}). \tag{6.23}$$

Proof. It has been shown that in generalized coordinates, kinetic energy has the expression

$$T(t, \mathbf{q}, \dot{\mathbf{q}}) = T_0(t, \mathbf{q}) + \mathbf{b}^T(t, \mathbf{q})\dot{\mathbf{q}} + \frac{1}{2}\dot{\mathbf{q}}^T A(t, \mathbf{q})\dot{\mathbf{q}}. \tag{6.24}$$

In addition, from (6.12) the Lagrange equations can be represented as

$$\frac{d}{dt}\nabla_{\dot{\mathbf{q}}}T - \nabla_{\mathbf{q}}T = \mathbf{Q},$$

which is equal to

$$\frac{d}{dt}[\mathbf{b}(t, \mathbf{q}) + A(t, \mathbf{q})\dot{\mathbf{q}}] - \nabla_{\mathbf{q}}T(t, \mathbf{q}, \dot{\mathbf{q}}) = \mathbf{Q},$$

or in the extended form,

$$\frac{\partial \mathbf{b}}{\partial t}(t, \mathbf{q}) + \sum_{j=1}^{n}\frac{\partial \mathbf{b}}{\partial q_j}(t, \mathbf{q})\dot{q}_j + \frac{dA}{dt}(t, \mathbf{q})\dot{\mathbf{q}} + A(t, \mathbf{q})\ddot{\mathbf{q}} - \nabla_{\mathbf{q}}T(t, \mathbf{q}, \dot{\mathbf{q}}) = \mathbf{Q}. \tag{6.25}$$

From here, since $A(t, \mathbf{q})$ is invertible, the result follows. $\square$

Remark 6.4. If the mechanical constraints of the system are stationary, then (6.25) is reduced to

$$A(t, \mathbf{q})\ddot{\mathbf{q}} + \sum_{j=1}^{n}\frac{\partial \mathbf{b}}{\partial q_j}(t, \mathbf{q})\dot{q}_j - \nabla_{\mathbf{q}}T(t, \mathbf{q}, \dot{\mathbf{q}}) = \mathbf{Q},$$

where

$$T(t, \mathbf{q}, \dot{\mathbf{q}}) = \frac{1}{2}\dot{\mathbf{q}}^T A(t, \mathbf{q})\dot{\mathbf{q}},$$

and, if also $m_k \underset{t}{=} \text{const}$, $k = 1, \ldots, N$, and the matrix A, the kinetic energy T and $\mathbf{b}$ do not depend on t, then the equation is further reduced:

$$A(\mathbf{q})\ddot{\mathbf{q}} + \sum_{j=1}^{n}\frac{\partial \mathbf{b}}{\partial q_j}(\mathbf{q})\dot{q}_j - \nabla_{\mathbf{q}}T(\mathbf{q}, \dot{\mathbf{q}}) = \mathbf{Q}.$$

In this last case, the system is called **stationary**.

6.5 Electrical and electromechanical models

Mechanical systems have many similarities with electrical systems. This circumstance allows Lagrange equations, obtained considering systems of the first type, to be applicable to systems of the second class. To see this, a compilation of some of the main relationships of electricity, electromagnetism, and electrical circuits is made below.

6.5.1 Some physical relations

1. The *electromagnetic flux* Φ and the current i that produces it have the relationship

$$\Phi = Li, \qquad (6.26)$$

where L is a constant that depends on geometric factors and environment called **inductance**. An electrical component that behaves according to (6.26) is also called inductance.

2. Electromagnetic flux changes result in electrical potentials. Both are related by the **Faraday's law**

$$u_L = -\frac{d\Phi}{dt},$$

where u_L denotes the voltage at the terminals of the inductance due to the change in flow. Considering (6.26), the Faraday's law can be expressed in the form

$$u_L = -L\frac{di}{dt}. \qquad (6.27)$$

3. In resistive elements the voltage u_R between the terminals of the component and the current i flowing through obey the **Ohm's law**, which is enunciated as

$$u_R = Ri, \qquad (6.28)$$

where R is a constant that depends on the component properties and is called **resistance**.

4. The voltage u_C between the terminals of a capacitance and the charge q that is in its plates follow the relationship

$$u_C = \frac{q}{C} = \frac{1}{C}\int i\,dt, \qquad (6.29)$$

where C is a constant that depends on the geometry and the environment called **capacitance**. If it is considered that the current is defined as the temporal variation of load, that is,

$$i := \frac{dq}{dt}, \qquad (6.30)$$

then, alternatively, (6.29) may be expressed as

$$u_C = \frac{1}{C}\int i\,dt. \qquad (6.31)$$

5. *Kirchhoff's laws* establish two fundamental mathematical relationships in the analysis of electrical circuits:

   The mesh law: *"The sum of the voltages in every closed loop of an electric circuit is zero."*

Dynamic Lagrange equations

The node law: *"The sum of the currents in every point of an electric circuit is zero."*

The laws of Kirchhoff in combination with the relations (6.27), (6.28), and (6.31) allow to establish the dynamic equations that currents and voltages in electrical circuits follow: the mesh law for the first and the node law for the second.

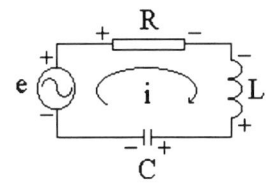

Figure 6.11 Electric circuit in series.

For example, consider the circuit shown in Fig. 6.11. By the mesh law we have

$$u_R + u_C = e + u_L,$$

where by the relations (6.27), (6.28), and (6.31),

$$Ri + L\frac{di}{dt} + \frac{1}{C}\int i\,dt = e. \tag{6.32}$$

In terms of the charge q expression (6.30) results in

$$L\ddot{q} + R\dot{q} + \frac{1}{C}q = e. \tag{6.33}$$

Let us consider another example. Consider now the circuit shown in Fig. 6.12. By the node law,

$$i_R + i_C = i + i_L.$$

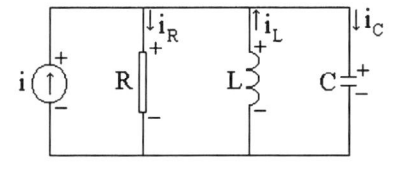

Figure 6.12 Electrical circuit in parallel.

Since the voltage across the terminals of all the electrical components is the same, denoted u, it follows from (6.27), (6.28), and (6.31) that

$$\frac{u}{R} + \frac{1}{L}\int u\,dt + C\dot{u} = i. \tag{6.34}$$

The derivation of (6.34) leads to

$$C\ddot{u} + \frac{1}{R}\dot{u} + \frac{1}{L}u = \frac{di}{dt}. \tag{6.35}$$

6.5.2 Table of electromechanical analogies

Consider a material particle that moves in a line under the conditions

$$T = \frac{a}{2}\dot{q}^2, \quad V = \frac{c}{2}q^2, \quad Q_{non\text{-}pot} = \hat{Q} - b\dot{q}, \quad Q_{reac} = 0,$$

where $\hat{Q}$ is a preassigned force acting on the particle, $b\dot{q}$ denotes viscous friction, and q is its position on the line. Immediately, the corresponding Lagrange equation results in

$$a\ddot{q} + cq = \hat{Q} - b\dot{q}. \tag{6.36}$$

The analysis and comparison of Eq. (6.36) on the one hand and Eqs. (6.33) and (6.35) on the other hand allow to establish in a clear way a series of analogies between mechanical and electrical concepts. Table 6.1 shows these analogies in condensed form.

Table 6.1 Table of electromechanical analogies.

System (coordinates)				T	V	$Q_{non\text{-}pot}$
Mechanics (position q)	a	b	c	$\frac{1}{2}a\dot{q}^2$	$\frac{1}{2}cq^2$	$\hat{Q} - b\dot{q}$
1. Electrical (charge q)	L	R	$\frac{1}{C}$	$\frac{1}{2}L\dot{q}^2$	$\frac{1}{2C}q^2$	$e - R\dot{q}$
2. Electrical (voltage u)	C	$\frac{1}{R}$	$\frac{1}{L}$	$\frac{1}{2}C\dot{u}^2$	$\frac{1}{2L}u^2$	$\frac{di}{dt} - \frac{1}{R}u$

The usefulness of the analogies obtained is evidenced by some examples.

Example 6.6. Let us obtain the dynamic equations that govern the behavior of the charges in the circuit of Fig. 6.13. From Fig. 6.13 and from Table 6.1 we have

$$i_1 := \frac{dq_1}{dt}, \quad i_2 := \frac{dq_2}{dt},$$
$$Q_{1,non\text{-}pot} = e - R_1\dot{q}_1, \quad Q_{2,non\text{-}pot} = -R_2\dot{q}_2,$$

and in addition

$$T = \frac{L_1\dot{q}_1^2}{2} + \frac{L_2\dot{q}_2^2}{2}, \quad V = \frac{q_1^2}{2C_1} + \frac{(q_1 - q_2)^2}{2C_2} + \frac{q_2^2}{2C_3},$$

so that

$$L = T - V = \frac{L_1\dot{q}_1^2}{2} + \frac{L_2\dot{q}_2^2}{2} - \frac{q_1^2}{2C_1} - \frac{(q_1 - q_2)^2}{2C_2} - \frac{q_2^2}{2C_3}.$$

Dynamic Lagrange equations

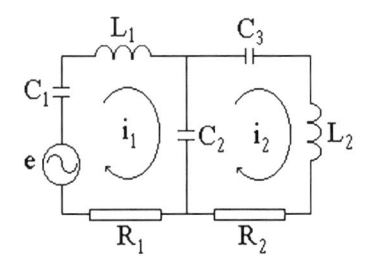

Figure 6.13 Circuit of two loops.

From the first Lagrange equation,

$$\frac{d}{dt}\frac{\partial L}{\partial \dot{q}_1} - \frac{\partial L}{\partial q_1} = Q_{1,non\text{-}pot},$$

it follows then that the dynamic equation for the first loop results in

$$L_1\ddot{q}_1 + R_1\dot{q}_1 + \frac{q_1}{C_1} + \frac{q_1 - q_2}{C_2} = e.$$

On the other hand, the second Lagrange equation,

$$\frac{d}{dt}\frac{\partial L}{\partial \dot{q}_2} - \frac{\partial L}{\partial q_2} = Q_{2,non\text{-}pot},$$

results in the equation for the charge of the second loop:

$$L_2\ddot{q}_2 + R_2\dot{q}_2 + \frac{q_2}{C_3} + \frac{q_2 - q_1}{C_2} = 0.$$

Example 6.7. Fig. 6.14 shows the electrical circuit of a transformer with resistive charge. Obtain the dynamic equations of the charge. Note first that the electromagnetic flux on one branch of the transformer due to the current on the other is given by

$$\Phi_{12} = \mu i_1, \quad \Phi_{21} = \mu i_2, \tag{6.37}$$

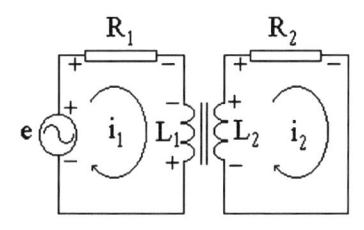

Figure 6.14 Electric transformer.

where μ is a constant of the transformer called **mutual inductance**. From Fig. 6.14 and from Table 6.1 we have

$$i_i := \frac{dq_i}{dt}, \quad i = 1, 2,$$

$$Q_{1,non\text{-}pot} = e - R_1\dot{q}_1, \quad Q_{2,non\text{-}pot} = -R_2\dot{q}_2,$$

and

$$T = \frac{L_1\dot{q}_1^2}{2} + \frac{L_2\dot{q}_2^2}{2} \mp \mu\dot{q}_1\dot{q}_2, \quad V = 0,$$

where the double sign of the third term of T takes into account the effect of mutual inductance. The Lagrange equations are given by

$$\frac{d}{dt}\frac{\partial T}{\partial \dot{q}_i} - \frac{\partial T}{\partial q_i} = Q_{i,non\text{-}pot}, \quad i = 1, 2.$$

Then it follows that the dynamic equation for the first loop results in

$$L_1\ddot{q}_1 \mp \mu\ddot{q}_2 + R_1\dot{q}_1 = e, \tag{6.38}$$

and for the second one

$$L_2\ddot{q}_2 \mp \mu\ddot{q}_1 + R_2\dot{q}_2 = 0. \tag{6.39}$$

As a verification of the technique given by the Lagrange equations, the same equations will be obtained by the alternative method of the mesh Law. By expressions (6.37), the electromagnetic flows on the branches of the transformer are given by

$$\Phi_1 = L_1 i_1 - \Phi_{21} = L_1 i_1 \mp \mu i_2,$$
$$\Phi_2 = L_2 i_2 - \Phi_{12} = L_2 i_2 \mp \mu i_1.$$

By Faraday's law we have

$$U_{L_1} = -\frac{d\Phi_1}{dt}, \quad U_{L_2} = -\frac{d\Phi_2}{dt}$$

and by the mesh law

$$R_1 i_1 = e + U_{L_1}, \quad r_2 i_2 = U_{L_2}$$

or

$$R_1 i_1 + L_1 \frac{di_1}{dt} \mp \mu \frac{di_2}{dt} = e,$$
$$R_2 i_2 + L_2 \frac{di_2}{dt} \mp \mu \frac{di_1}{dt} = 0,$$

where Eqs. (6.38) and (6.39) are followed.

Dynamic Lagrange equations

Example 6.8. In the circuit shown in Fig. 6.15 it is desired to determine the values of the mutual inductance μ of the transformer and capacitor C_{12} such that there is no influence between the loops. From the previous example and from Table 6.1 it follows that

$$T = \frac{1}{2}L_1\dot{q}_1^2 + \frac{1}{2}L_2\dot{q}_2^2 - \mu\dot{q}_1\dot{q}_2$$

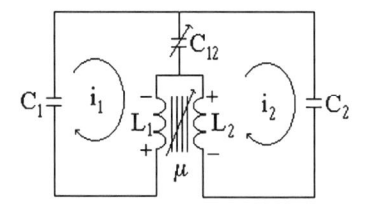

Figure 6.15 Transformer and variable-capacitance circuit.

and

$$V = \frac{q_1^2}{2C_1} + \frac{q_2^2}{2C_2} + \frac{(q_1-q_2)^2}{2C_{12}}.$$

Since

$$Q_{1,non\text{-}pot} = Q_{2,non\text{-}pot} = 0,$$

by the technique of the Lagrange equations we get

$$\left.\begin{array}{l} L_1\ddot{q}_1 - \mu\ddot{q}_2 + \dfrac{q_1}{C_1} + \dfrac{q_1-q_2}{C_{12}} = 0, \\[2mm] L_2\ddot{q}_2 - \mu\ddot{q}_1 + \dfrac{q_2}{C_2} - \dfrac{q_1-q_2}{C_{12}} = 0. \end{array}\right\} \qquad (6.40)$$

From the second equation in (6.40) it follows that

$$\ddot{q}_2 = L_2^{-1}\mu\ddot{q}_1 - L_2^{-1}\frac{q_2}{C_2} + L_2^{-1}\frac{q_1-q_2}{C_{12}}.$$

Substitution the last expression in the first equation of (6.40) gives

$$L_1\ddot{q}_1 - \mu\left(L_2^{-1}\mu\ddot{q}_1 - L_2^{-1}\frac{q_2}{C_2} + L_2^{-1}\frac{q_1-q_2}{C_{12}}\right) + \frac{q_1}{C_1} + \frac{q_1-q_2}{C_{12}} = 0,$$

or

$$\left(L_1L_2 - \mu^2\right)\ddot{q}_1 + \left(\frac{L_2}{C_1} - \frac{\mu}{C_{12}} + \frac{L_2}{C_{12}}\right)q_1 + \left(-\frac{L_2}{C_{12}} + \frac{\mu}{C_2} + \frac{\mu}{C_{12}}\right)q_2 = 0.$$

Analogously, representing $\ddot{q}_1$ from the first equation of (6.40) and substituting it into the second one leads to

$$\left(L_1 L_2 - \mu^2\right) \ddot{q}_2 + \left(\frac{\mu}{C_1} + \frac{\mu}{C_{12}} - \frac{L_1}{C_{12}}\right) q_1 + \left(-\frac{\mu}{C_{12}} + \frac{L_1}{C_2} + \frac{L_1}{C_{12}}\right) q_2 = 0.$$

If you want the dynamics of q_1 and q_2 to be independent of each other, we should satisfy

$$-\frac{L_2}{C_{12}} + \frac{\mu}{C_2} + \frac{\mu}{C_{12}} = 0,$$

$$-\frac{\mu}{C_{12}} + \frac{L_1}{C_2} + \frac{L_1}{C_{12}} = 0,$$

or equivalently,

$$\mu \frac{C_{12}}{C_2} + \mu = L_2, \quad \mu \frac{C_{12}}{C_1} + \mu = L_1,$$

which gives

$$\mu = \frac{L_1 C_1 - L_2 C_2}{C_1 - C_2},$$

$$C_{12} = \frac{L_2 - L_1}{L_1 C_1 - L_2 C_2} C_1 C_2.$$

The following example shows that electromechanical systems can also be approached with the studied technique.

Example 6.9. In Fig. 6.16 an electromechanical system is presented. It is an electrical circuit in which the capacitor is formed by a fixed plate and a mobile plate suspended from a spring. An external force $F(t)$ in addition to the electrical attraction, exerted by the other plate, acts on this plate. So the separation between plates $d(t)$ is a temporary function. Let us obtain the dynamic equations of the electric charge and the position of the moving plate of the capacitor considering that if the distance between plates is a, then the capacity is C_a. Recall that the capacitance of the parallel-plate capacitor shown is obtained by

$$C(t) = \frac{\varepsilon S}{4\pi d(t)}, \tag{6.41}$$

where ε is a constant that depends on the environment and S denotes the area of the plates. Since it is known that if $d = a$, then $C = C_a$, one has

$$C_a = \frac{\varepsilon S}{4\pi a}.$$

With this relation, (6.41) can be expressed as

$$C(t) = C_a \frac{a}{d(t)}.$$

Dynamic Lagrange equations

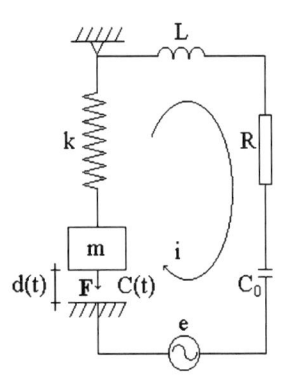

Figure 6.16 Electromechanical variable-capacitor system.

Also if $x := a - d$ denotes the residual separation with respect to a, then

$$C = C_a \frac{a}{a-x}.$$

If the generalized coordinates are selected as

$$q_1 := \int i\, dt, \quad q_2 = x,$$

then by Table 6.1 and in view of Fig. 6.16 we get

$$T = \frac{L}{2}\dot{q}_1^2 + \frac{m}{2}\dot{q}_2^2,$$

$$V = \frac{q_1^2}{2C_0} + \frac{q_1^2}{2C_a a}(a - q_2) + \frac{k}{2}q_2^2 - mgq_2,$$

where k is the stiffness coefficient of the spring. Clearly we also have

$$Q_{1,non\text{-}pot} = e - R\dot{q}_1, \quad Q_{2,non\text{-}pot} = F.$$

We can now get the function of Lagrange $L = T - V$, and since

$$\frac{d}{dt}\frac{\partial L}{\partial \dot{q}_i} - \frac{\partial L}{\partial q_i} = Q_{i,non\text{-}pot}, \quad i = 1, 2,$$

it is verified that the Lagrange equations are given by

$$L\ddot{q}_1 + \frac{q_1}{C_0} + \frac{q_1}{C_a a}(a - q_2) + R\dot{q}_1 = e,$$

$$m\ddot{q}_2 - \frac{q_1^2}{2C_a a} + kq_2 - mg = F(t).$$

Example 6.10. The electromechanical system shown in Fig. 6.17 consists of an electromagnet whose function is to attract the metallic body of mass m shown. The value of the inductance $\hat{L}$ of the electromagnet is a known function of the separation x between the mass m and the core of the electromagnet. Let us:

a) obtain the Lagrange equations for the electric charge in the circuit and for the distance x.
b) define x_{eq} and i_{eq}, considering that the electric voltage source is DC, and determine the equilibrium value for x.

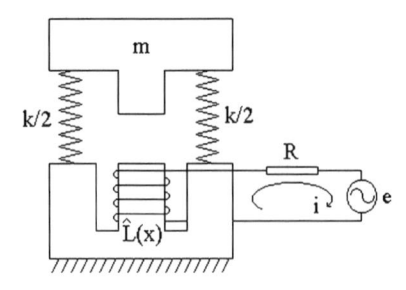

Figure 6.17 Electromechanical system of variable inductance.

a) The general coordinates are

$$q_1 := \int i(t)dt, \quad q_2 := x.$$

Then from Table 6.1 and from Fig. 6.17 it follows that

$$T = \frac{1}{2}\hat{L}(q_2)\dot{q}_1^2 + \frac{1}{2}m\dot{q}_2^2,$$
$$V = \frac{1}{2}k(q_2 - q_{2,0})^2 + mgq_2,$$

where $q_{2,0}$ denotes the value of q_2 for which the springs are not stressed and $k/2$ is the constant of stiffness. Evidently,

$$Q_{1,non\text{-}pot} = e - R\dot{q}_1, \quad Q_{2,non\text{-}pot} = 0.$$

We can now get the function of Lagrange $L = T - V$ which results:

$$L = \frac{1}{2}\left[\hat{L}(q_2)\dot{q}_1^2 + m\dot{q}_2^2 - k(q_2 - q_{2,0})^2\right] - mgq_2.$$

The Lagrange equations

$$\frac{d}{dt}\frac{\partial L}{\partial \dot{q}_i} - \frac{\partial L}{\partial q_i} = Q_{i,non\text{-}pot}, \quad i = 1, 2,$$

become

$$\frac{d}{dt}\left[\hat{L}(q_2)\dot{q}_1\right] + R\dot{q}_1 = e,$$
$$\frac{d}{dt}(m\dot{q}_2) - \hat{L}'(q_2)\dot{q}_1^2 + k(q_2 - q_{2,0}) + mg = 0,$$

or equivalently,

$$\left.\begin{array}{l}\hat{L}(q_2)\ddot{q}_1 + \left[\hat{L}'(q_2)\dot{q}_2 + R\right]\dot{q}_1 = e,\\ m\ddot{q}_2 - \hat{L}'(q_2)\dot{q}_1^2 + k(q_2 - q_{2,0}) + mg = 0.\end{array}\right\} \quad (6.42)$$

b) In equilibrium we have

$$e \equiv E = \underset{t}{\text{const}}, \quad \ddot{q}_1 = 0, \quad \dot{q}_2 = \ddot{q}_2 = 0,$$

which in view of (6.42) gives

$$\dot{q}_1 = i_{eq} = \frac{E}{R}$$

and the value $q_{2,eq} = x_{eq}$ may be found from the nonlinear equation

$$x_{eq} = q_{2,0} - \frac{m}{k}g + \frac{1}{k}\hat{L}'(x_{eq})\left(\frac{E}{R}\right)^2.$$

6.6 Exercises

Exercise 6.1. The Lagrange function of a free relativistic particle with a rest mass m_0 has the form

$$L = -m_0 c\sqrt{1 - c^{-2}(\dot{x}_1^2 + \dot{x}_2^2 + \dot{x}_3^2)},$$

where c is the speed of light. Show that its motion $x_i = x_i(t)$ is described by the relations

$$x_i(t) = \alpha_i t + \beta_i \quad (i = 1, 2, 3).$$

Exercise 6.2. A heavy point can move without friction in the vertical plane of Oxz along the curve

$$z = f(x).$$

Show that the Lagrange equation, describing this movement, has the form

$$\ddot{x}\left[1 + \left(\frac{d}{dx}f\right)^2\right] + \left(\frac{d^2}{dx^2}f\right)\left(\frac{d}{dx}f\right)\dot{x}^2 + \left(\frac{d}{dx}f\right)g = 0,$$

and try to find its first integral.

Exercise 6.3. Using the Lagrange equations, show that the centers of mass of cylinders 1, 2, and 4 (see Fig. 6.18) move vertically with constant accelerations,

$$w_1 = \frac{72}{79}g, \quad w_2 = \frac{58}{79}g, \quad w_4 = \frac{58}{79}g,$$

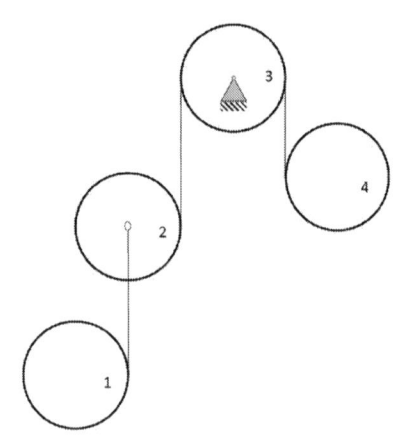

Figure 6.18 Homogeneous cylinders interconnected by inextensible and weightless threads.

and the angular acceleration of cylinder 3 is $\varepsilon_3 = \frac{2}{79}\frac{g}{r}$, assuming that identical cylinders of radius r are homogeneous and interconnected by inextensible and weightless threads that do not slide on the surface of the cylinders.

Exercise 6.4. Compose the Lagrange equations for the two electrical circuits shown in Fig. 6.19.

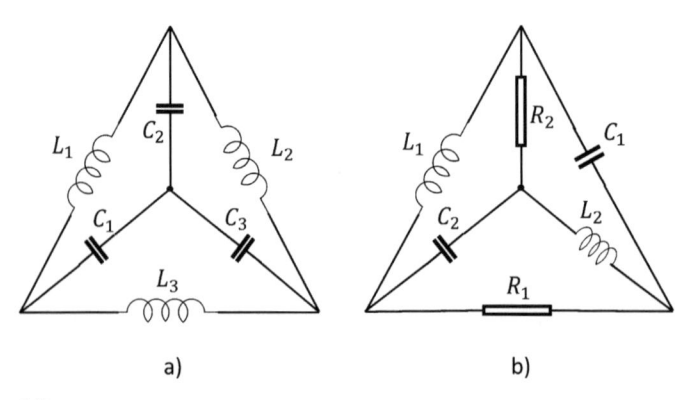

Figure 6.19 Electrical circuits.

Dynamic Lagrange equations   219

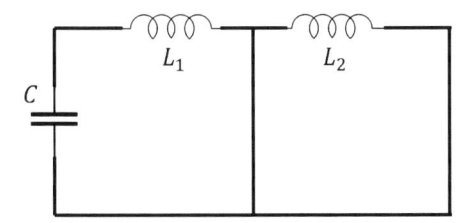

Figure 6.20 Electrical circuit modeling the Lagrange system.

Exercise 6.5. The mechanical system has a Lagrange function

$$L = \frac{m_1}{2}\dot{q}_1^2 + \frac{m_2}{2}\dot{q}_2^2 - \frac{k}{2}q_1^2.$$

Show that the electrical circuit modeling this system has the form as in Fig. 6.20.

Equilibrium and stability

Contents

7.1 Definition of equilibrium   **221**
7.2 Equilibrium in conservative systems   **223**
7.3 Stability of equilibrium   **229**
     7.3.1 Definition of local stability   229
     7.3.2 Stability of equilibrium in conservative systems   232
7.4 Unstable equilibria in conservative systems   **236**
7.5 Exercises   **242**

In dynamic systems in general and in mechanics in particular, the determination of equilibrium positions and their quality of stability are traditional problems of fundamental importance, which to date have been partially solved. In this chapter, with the support of the concepts and results studied up to this point, such as coordinates and generalized forces, the study of these topics is addressed and the most important results are reported. As will be seen, the most developed theory is that dealing with conservative systems, which occupies most of the chapter.

7.1 Definition of equilibrium

Consider the mechanical system S with N material particles and m stationary mechanical constraints. In these circumstances, the expression of the possible positions in generalized coordinates $\mathbf{R}(t) = \mathbf{R}(t, \mathbf{q})$, introduced in the previous chapter, is stationary, that is, $\mathbf{R} = \mathbf{R}(\mathbf{q})$.

Recall from the same chapter that in the normal form the Lagrange equations for the system S have the expression

$$\ddot{\mathbf{q}} = \mathcal{F}(t, \mathbf{q}, \dot{\mathbf{q}}, \mathbf{Q}). \tag{7.1}$$

Definition 7.1. Given $\mathbf{Q}$, it is said that $\mathbf{q}^*$ is an **equilibrium position** of system S if the following condition is met:

$$\mathcal{F}(t, \mathbf{q}^*, \mathbf{0}, \mathbf{Q}) \equiv 0. \tag{7.2}$$

Remark 7.1. In other words, $\mathbf{q}^*$ is an equilibrium position if, when the system is in that position and the velocity is zero, it remains there indefinitely. The condition that the transformation $\mathbf{R} = \mathbf{R}(\mathbf{q})$ is stationary ensures that any equilibrium in space $\mathbf{q}$ defines a balance in space $\mathbf{R}$ and vice versa. For simplicity, equilibrium positions will be called simply equilibria.

Classical and Analytical Mechanics. https://doi.org/10.1016/B978-0-32-389816-4.00018-1
Copyright © 2021 Elsevier Inc. All rights reserved.

Remark 7.2. By Newton's second law, the position $\mathbf{q}^*$ is an equilibrium if and only if in that position the total force acting on each particle of S remains zero, i.e.,

$$\mathbf{F}_i \equiv 0, \quad i \in S.$$

The previous immediate observation allows to formulate the following important result.

Lemma 7.1 (Principle of virtual displacements). *The position $\mathbf{q}^*$ is an equilibrium if and only if in that position the elementary work done by the forces on S along any virtual translation, with respect to $\mathbf{R}^* := \mathbf{R}(\mathbf{q}^*)$, is zero.*

Proof. Let $\mathbf{R}^* := \mathbf{R}(\mathbf{q}^*)$ be the position vector corresponding to the equilibrium $\mathbf{q}^*$. By the previous observation $\mathbf{R}^*$ is an equilibrium only if and only if in that position

$$\mathbf{F}_i \equiv \mathbf{0}, \ i = 1, ..., N,$$

which in turn means that the work done by these forces along any virtual translation with respect to $\mathbf{R}^*$ is null, that is,

$$\delta A|_{\delta \mathbf{R}} := \sum_{k=1}^{N} (\mathbf{F}_k, \delta \mathbf{r}_k) = 0.$$

□

The newly established property results in two very useful criteria.

Corollary 7.1. *The position $\mathbf{q}^*$ is an equilibrium if and only if in that position $\mathbf{Q} \equiv \mathbf{0}$.*

Proof. Recall that in generalized coordinates

$$\delta A|_{\delta \mathbf{R}} := \sum_{i=1}^{n} Q_i \delta q_i.$$

By the previous lemma, $\mathbf{q}^*$ is an equilibrium if and only if in that position

$$\delta A|_{\delta \mathbf{R}} = 0 \ \forall \delta \mathbf{q},$$

and the result follows.

□

Corollary 7.2. *If the system S has ideal constraints, then $\mathbf{q}^*$ is an equilibrium if and only if in that position*

$$\mathbf{Q}_{ext} \equiv \mathbf{0}.$$

Proof. The result is obtained from the previous corollary and the definition

$$\mathbf{Q} := \mathbf{Q}_{ext} + \mathbf{Q}_{reac},$$

since in systems with ideal constraints

$$\mathbf{Q}_{reac} \equiv \mathbf{0}.$$

□

Remark 7.3. The previous results are also valid for the case where the transformation $\mathbf{R}(t) = \mathbf{R}(t, \mathbf{q})$ is non-stationary. To show that it is enough to define the equilibria in the space $\mathbf{R}$ instead of in the space $\mathbf{q}$. It is said that the position $\mathbf{R}^*$ is an equilibrium if, when the system is in that position and the velocities at the initial instant are zero, the system remains at $\mathbf{R}^*$ indefinitely. Note that if the transformation $\mathbf{R}(t) = \mathbf{R}(t, \mathbf{q})$ is non-stationary, the equilibria in $\mathbf{q}$ and in $\mathbf{R}$ do not necessarily coincide. Now, the same procedure of the test of the previous lemma can be used to verify that, in this case, the principle of virtual displacements and the conclusions of two previous immediate corollaries are also valid.

For the needs of this chapter the important result is that given by the first corollary, so it is not important in what space the equilibria are defined.

7.2 Equilibrium in conservative systems

In Chapter 6 we have shown that if all effective forces are potential, the generalized effective forces are also potential and

$$\mathbf{Q}_{ef} = -\nabla_\mathbf{q} V(t, \mathbf{q}), \tag{7.3}$$

where $V(t, \mathbf{q})$ is the potential energy of the system and is obtained as indicated in that chapter. The relation (7.3), in combination with the criterion given by the second previous corollary, allows to establish that in the equilibrium positions the gradient of the potential energy is equal to zero, that is, if the position $\mathbf{q}^*$ is an equilibrium, then

$$\nabla_\mathbf{q} V(t, \mathbf{q}) \equiv 0. \tag{7.4}$$

In particular, the positions where the potential energy reaches its extremes are equilibria, i.e., $\mathbf{q}^*$ is an equilibrium if

$$\mathbf{q}^* := \arg\underset{\mathbf{q}}{\mathrm{ext}}\, V(t, \mathbf{q}).$$

In the following examples, the usefulness of the condition (7.4) is shown.

Example 7.1. In Fig. 7.1 a capacitor with charge q and whose plates are held by springs is shown. The separation d between the plates is variable and each plate has a mass m. Consider that when the springs are not loaded the separation between plates is Δ and the capacity has the nominal value C_Δ. The springs have a constant of stiffness k. Determine the equilibrium position of the plates. Clearly the forces in this

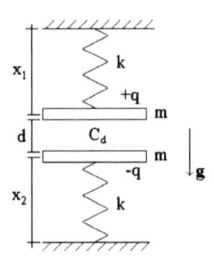

Figure 7.1 Capacitor whose plates are held by springs.

system are potential. It is a system with two masses that move on a straight line; therefore two generalized coordinates are required. Let them be

$$q_1 := x_1 - x_0, \quad q_2 := x_2 - x_0,$$

where x_1 and x_2 denote the current lengths of the springs of the upper and lower masses, respectively, while x_0 represents the length corresponding to the springs without elongation.

Recall from Chapter 6 that the capacity of the capacitor when the separation between the plates is d can be determined by the expression

$$C_d = \frac{\Delta}{d} C_\Delta. \tag{7.5}$$

Recall from the table of electromechanical analogies of the same chapter that the potential energy of a capacity capacitor C_d and charge q is given by

$$V_C(d) = \frac{q^2}{2C_d}.$$

By (7.5) and considering that $d = \Delta - q_1 - q_2$ it follows that

$$V_C(\mathbf{q}) = \frac{q^2}{2\Delta C_\Delta} (\Delta - q_1 - q_2).$$

Then the joint potential energy of the system is given by

$$V = \frac{1}{2} k \left(q_1^2 + q_2^2 \right) - mg \left(q_1 + x_0 + e_0/2 \right)$$
$$- mg(x_0 + \Delta + e_0 - q_2) + \frac{q^2}{2\Delta C_\Delta} (\Delta - q_1 - q_2),$$

where e_0 denotes the thickness of the plates. In equilibrium we have

$$\nabla_\mathbf{q} V(\mathbf{q}) \equiv 0,$$

so that

$$\frac{\partial V}{\partial q_1} = kq_1 - mg - \frac{q^2}{2\Delta C_\Delta} = 0,$$

$$\frac{\partial V}{\partial q_2} = kq_2 + mg - \frac{q^2}{2\Delta C_\Delta} = 0,$$

which leads to

$$q_1^* = \frac{1}{k}\left(\frac{q^2}{2\Delta C_\Delta} + mg\right),$$

$$q_2^* = \frac{1}{k}\left(\frac{q^2}{2\Delta C_\Delta} - mg\right).$$

Example 7.2. A series of masses m_i, $i = 1, \ldots n$, is connected by n springs of stiffness k_i, $i = 1, \ldots n$, as shown in Fig. 7.2. Calculate the equilibrium position. Note that the effective forces on the system are potential. In order to determine the solution, two methods can be followed:

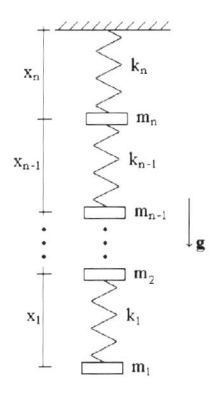

Figure 7.2 Series of masses connected by springs.

- the first one uses the result established in the second previous corollary, without considering that the system is conservative;
- in the second, this fact is exploited via the gradient of the potential energy.

Since the masses move on a straight line, n generalized coordinates are required. Let these be the elongations of the springs, that is,

$$q_i := x_i - x_i^0, \quad i = 1, \ldots n, \tag{7.6}$$

where x_i denotes the current distance between the masses i and $i-1$, and x_i^0 represents the same distance with the spring without elongation.

Method 1. In the balances, and only in them, the following condition is met:

$$Q_i = 0, \quad i = 1, ...n,$$

where

$$Q_i = \sum_{j=1}^{n} F_j \frac{\partial x_j}{\partial q_i} = F_i, \quad i = 1, ...n,$$

with

$$\left.\begin{aligned} F_1 &= m_1 g - k_1 q_1, \\ F_2 &= m_2 g + k_1 q_1 - k_2 q_2, \\ &\vdots \\ F_n &= m_n g + k_{n-1} q_{n-1} - k_n q_n. \end{aligned}\right\}$$

From the condition $F_i = 0$ it follows that

$$\left.\begin{aligned} q_1^* &= \frac{g}{k_1} m_1, \\ q_2^* &= \frac{1}{k_2} \left(m_2 g + k_1 q_1^* \right), \\ &\vdots \\ q_n^* &= \frac{1}{k_n} \left(m_n g + k_{n-1} q_{n-1}^* \right), \end{aligned}\right\}$$

or by replacement,

$$q_i^* = \frac{g}{k_i} \sum_{j=1}^{i} m_i, \quad i = 1, ...n. \tag{7.7}$$

Finally, we obtain

$$x_i^* = \frac{g}{k_i} \sum_{j=1}^{i} m_j + \sum_{j=1}^{i} x_j^0, \quad i = 1, ...n. \tag{7.8}$$

Method 2. With generalized coordinates (7.6) for potential energy we have

$$V(\mathbf{q}) = \frac{1}{2} \sum_{j=1}^{n} k_j q_j^2 - g \sum_{j=1}^{n} m_j \sum_{s=j}^{n} x_s.$$

In equilibrium we must have

$$\nabla_\mathbf{q} V(\mathbf{q}^*) = 0.$$

Equilibrium and stability

But

$$\frac{\partial V}{\partial q_i}(\mathbf{q}) = k_i q_i - g \sum_{j=1}^{n} m_j \sum_{s=j}^{n} \frac{\partial x_s}{\partial q_i},$$

and since

$$\frac{\partial x_s}{\partial q_i} = \delta_{i,s},$$

where the Kronecker symbol $\delta_{i,s}$ is defined as

$$\delta_{i,s} = \begin{cases} 1, & \text{if } i = s, \\ 0, & \text{if } i \neq s, \end{cases}$$

it follows that

$$\frac{\partial V}{\partial q_i}(\mathbf{q}) = k_i q_i - g \sum_{j=1}^{i} m_j.$$

And since we must have $\dfrac{\partial V}{\partial q_i}(\mathbf{q}^*) = 0$, expressions (7.7) and (7.8) are concluded.

The solution of the following problem requires remembering the fundamental result of the convex programming problem on the Euclidean space R^n.

Theorem 7.1 (See, for example, Chapter 21 in (Poznyak, 2008)). *Let Ω be a convex subset of R^n and let $f : \Omega \to R$ and $\mathbf{g} : \Omega \to R^m$ be convex functions. Suppose there is a point $\mathbf{x}_1 \in \Omega$ for which $\mathbf{g}(\mathbf{x}_1) < 0$ (the **Slater condition**). Let*

$$\mu_0 := \inf f(\mathbf{x}) \text{ subject to } \mathbf{x} \in \Omega \text{ and } \mathbf{g}(\mathbf{x}) \leq 0, \tag{7.9}$$

and suppose that μ_0 is finite. Then there is a vector $0 \leq \lambda_0 \in R^m$ such that

$$\mu_0 = \inf_{\mathbf{x} \in \Omega} \mathcal{L}(\mathbf{x}, \lambda_0), \tag{7.10}$$

where

$$\mathcal{L}(\mathbf{x}, \lambda) = f(\mathbf{x}) + (\lambda, \mathbf{g}(\mathbf{x})).$$

*In addition, if the minimum is reached in (7.9) by $\mathbf{x}_0 \in \Omega$, $\mathbf{g}(\mathbf{x}_0) \leq 0$, it is also reached by the same $\mathbf{x}_0$ in (7.10) and the following **complementary slackness condition** holds:*

$$(\lambda_0, \mathbf{g}(\mathbf{x}_0)) = 0. \tag{7.11}$$

*The function $\mathcal{L}$ is referred to as the **Lagrangian function** and the vector $\lambda \in R^m$ is known as the **Lagrange multipliers vector**.*

Example 7.3. Consider a system with n generalized coordinates subject to the following restriction:

$$\sum_{k=1}^{n} q_k^2 \leq 1.$$

Suppose the system is conservative and that the potential energy is given by

$$V(\mathbf{q}) = \sum_{k=1}^{n} \alpha_k q_k. \tag{7.12}$$

We need to calculate equilibrium positions.

Considering that the positions where the potential energy reaches its extremes are equilibria, the present problem can be reformulated in the following terms:

Find $\arg \underset{\mathbf{q} \in R^n}{ext} V(\mathbf{q})$ subject to $\sum_{k=1}^{n} q_k^2 - 1 \leq 0$.

Since $\Omega = R^n$ and $V(\mathbf{q})$ and $g(\mathbf{q}) := \|\mathbf{q}\|^2 - 1$ are convex, the solution of this problem may be found as it is described above. The Lagrangian function (with $m = 1$) is given by

$$\mathcal{L}(\mathbf{q}, \lambda) = \sum_{k=1}^{n} \alpha_k q_k + \lambda \left(\sum_{k=1}^{n} q_k^2 - 1 \right), \quad \lambda \geq 0,$$

which is a quadratic convex function with a global minimum $\mathbf{q}^*$, which may be found from the condition

$$\nabla_\mathbf{q} \mathcal{L}(\mathbf{q}^*, \lambda_0) = 0,$$

or, equivalently,

$$\frac{\partial \mathcal{L}_1}{\partial q_k}(\mathbf{q}^*, \lambda_0) = \alpha_k + 2\lambda_0 q_k^* = 0, \quad k = 1, ..., n, \quad \lambda_0 \geq 0,$$

implying

$$q_k^* = -\frac{\alpha_k}{2\lambda_0}, \quad k = 1, ..., n, \quad \lambda_0 \geq 0. \tag{7.13}$$

To obtain λ_0 the condition (7.11) is used:

$$\lambda_0 g(\mathbf{q}^*) = \lambda_0 \left[\sum_{k=1}^{n} \left(-\frac{\alpha_k}{2\lambda_0} \right)^2 - 1 \right] =$$

$$\lambda_0 \left[\frac{1}{4\lambda_0^2} \sum_{k=1}^{n} \alpha_k^2 - 1 \right] = 0, \quad \lambda_0 \geq 0.$$

Equilibrium and stability

So, the non-trivial solution $\lambda_0 > 0$ is

$$\lambda_0 = \pm \frac{1}{2}\sqrt{\sum_{k=1}^{n} \alpha_k^2}, \qquad (7.14)$$

which, when replaced in (7.13), leads to

$$q_k^* = \mp \frac{\alpha_k}{\sqrt{\sum_{i=1}^{n} \alpha_i^2}}, \quad k = 1, ..., n.$$

Substituting this expression into (7.12) gives

$$V(\mathbf{q}) = \sum_{k=1}^{n} \alpha_k q_k = \mp \sum_{k=1}^{n} \frac{\alpha_k^2}{\sqrt{\sum_{i=1}^{n} \alpha_i^2}} = \mp \sqrt{\sum_{i=1}^{n} \alpha_i^2},$$

which shows that the point $\mathbf{q}^*$, minimizing $V(\mathbf{q})$, is

$$q_k^* = -\frac{\alpha_k}{\sqrt{\sum_{i=1}^{n} \alpha_i^2}}, \quad k = 1, ..., n.$$

7.3 Stability of equilibrium

7.3.1 Definition of local stability

A fundamental problem in mechanics is the qualification of the behavior of systems in the vicinity of equilibria. This qualification can be carried out via the characterization of the behavior of the position and speed variables of the material points. By this reason, this set $\mathbf{p} := (\mathbf{q}, \dot{\mathbf{q}})$ of variables is known as **system states**.

Definition 7.2. An equilibrium position $\mathbf{q}^*$ of a system S is said to be **locally stable equilibrium position** (or **stable in the Lyapunov sense**) if for any $\varepsilon > 0$ there exists $\delta = \delta(\varepsilon) > 0$ such that if the initial state meets

$$\|\mathbf{q}(t_0) - \mathbf{q}^*\| \leq \delta, \quad \|\dot{\mathbf{q}}(t_0)\| \leq \delta, \quad t_0 \geq 0,$$

then for all $t \geq t_0$

$$\|\mathbf{q}(t)\| \leq \varepsilon, \quad \|\dot{\mathbf{q}}(t)\| \leq \varepsilon.$$

In Fig. 7.3 the statement of the previous definition is explained graphically, using a phase diagram. Here it is pointed out that if the equilibrium $\mathbf{q}^*$ is stable locally, then for any $\varepsilon > 0$ we may choose $\delta > 0$ such that the path corresponding to any initial state $\mathbf{p}_{ini}$, contained in the 2δ side hypercube and centered on $\mathbf{C} := (\mathbf{q}^*, \mathbf{0})$, remains indefinitely in the hypercube of side 2ε and with the same center.

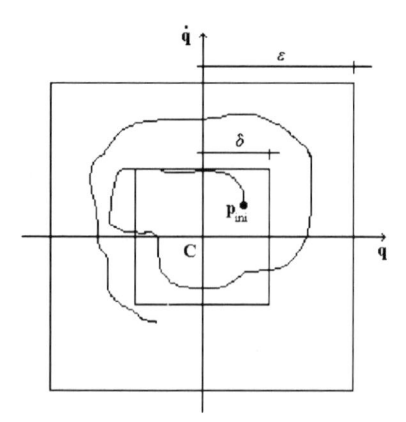

Figure 7.3 Concept of equilibrium local stability.

Definition 7.3. An equilibrium position q^* of a system S is said to be **unstable**, if it is not stable.

The example below illustrates the concept of stability introduced in the preceding definition.

Example 7.4. The system shown in Fig. 7.4 is known as a **linear oscillator**. It consists of a mass m that can move horizontally without friction and is held by a stiffening spring with rigidity k. Determine the stability of the equilibrium states. A generalized coordinate is

$$q := x - x_0,$$

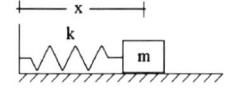

Figure 7.4 Linear oscillator.

where x_0 is the nominal (non-stiffened) length of the spring. Since it is a conservative system with potential energy

$$V(q) = \frac{1}{2}kq^2,$$

the equilibrium points are given by

$$\nabla_q V(q^*) = kq = 0,$$

and therefore there is only one equilibrium point,

$$q^* = 0. \tag{7.15}$$

Equilibrium and stability

To determine the stability of the equilibrium obtained, it is necessary to know the trajectories of the state $(q(t), \dot{q}(t))$; it is easily verified that the kinetic energy is

$$T = \frac{1}{2}m\dot{q}^2,$$

which is why the Lagrange function is given by

$$L(q,\dot{q}) = \frac{1}{2}m\dot{q}^2 - \frac{1}{2}kq^2.$$

Since there are no non-potential forces, it is obtained as a Lagrange equation for the system,

$$m\ddot{q} + kq = 0, \qquad (7.16)$$

and if the angular frequency is defined as

$$\omega^2 := \frac{k}{m} \geq 0,$$

Eq. (7.16) is reduced to the form

$$\ddot{q} + \omega^2 q = 0. \qquad (7.17)$$

From the theory of linear ordinary differential equations it is known that the solution of (7.17) has the following expression:

$$q(t) = q(t_0)\cos\omega(t-t_0) + \frac{\dot{q}(t_0)}{\omega}\sin\omega(t-t_0), \qquad (7.18)$$

with its derivative

$$\dot{q}(t) = -\omega q(t_0)\sin\omega(t-t_0) + \dot{q}(t_0)\cos\omega(t-t_0).$$

These two relationships imply

$$|q(t)| \leq |q(t_0)| + \frac{1}{\omega}|\dot{q}(t_0)|,$$
$$|\dot{q}(t)| \leq \omega|q(t_0)| + |\dot{q}(t_0)|,$$

from which, if $|q(t_0)| \leq \delta$ and $|\dot{q}(t_0)| \leq \delta$, we get

$$|q(t)| \leq \delta\left(1 + \frac{1}{\omega}\right), \quad |\dot{q}(t)| \leq \delta(1+\omega).$$

Defining

$$\varepsilon := \delta \cdot \max\left\{1+\omega, 1+\frac{1}{\omega}\right\},$$

it follows that

$$\delta(\varepsilon) := \frac{\varepsilon}{\max\left\{1+\omega, 1+\frac{1}{\omega}\right\}}. \tag{7.19}$$

The relationship (7.19) allows to ensure the local stability of the equilibrium (7.15).

7.3.2 Stability of equilibrium in conservative systems

Lagrange–Dirichlet theorem

In the previous example, to determine the stability of the equilibrium obtained, the solution (7.18) of the system was used. Since the explicit expression of the solution is not always available in the general case, it is important to have indirect criteria to determine the stability of the balances without resorting to the solution. This is possible and particularly simple in the case of conservative systems.

Theorem 7.2 (Lagrange–Dirichlet).[1] *Let $\mathbf{q}^*$ be an equilibrium point of a conservative system S with potential energy $V(\mathbf{q})$ which is a continuous function. If*

$$\mathbf{q}^* = \arg\min_{\mathbf{q}} V(\mathbf{q})$$

*and there is a neighborhood $\mathcal{A}$ of $\mathbf{q}^*$ such that this **minimum is strict**, namely,*

$$V(\mathbf{q}^*) < V(\mathbf{q}) \; \forall \mathbf{q} \neq \mathbf{q}^*, \; \mathbf{q} \in \mathcal{A},$$

*then $\mathbf{q}^*$ is a **local equilibrium**. In other words, the equilibria, where the potential energy reaches its strict minima, are **locally stable**.*

Proof. Define $\boldsymbol{\Delta} := \mathbf{q} - \mathbf{q}^*$, so that $\dot{\boldsymbol{\Delta}} := \dot{\mathbf{q}}$. Without loss of generality we can accept that $V(\mathbf{q}^*)$ and take $\mathbf{q}^* = 0$. Define the ε-neighborhood of the point $(\mathbf{0}, \mathbf{0})$ as

$$\Omega_\varepsilon := \left\{ (\boldsymbol{\Delta}, \dot{\boldsymbol{\Delta}}) = (\mathbf{q}, \dot{\mathbf{q}}) \mid \|\mathbf{q}\| \leq \varepsilon \text{ and } \|\dot{\mathbf{q}}\| \leq \varepsilon \right\}.$$

In view of (6.24) the kinetic energy $T(t, \mathbf{q}, \dot{\mathbf{q}})$ is a quadratic function and by the continuity property of $V(\mathbf{q})$ we may conclude that the complete mechanical energy

$$E(t, \mathbf{q}, \dot{\mathbf{q}}) = T(t, \mathbf{q}, \dot{\mathbf{q}}) + V(\mathbf{q})$$

---

[1] Johann Peter Gustav Lejeune Dirichlet (February 13, 1805–May 5, 1859) was a German mathematician who made deep contributions to number theory (including creating the field of analytic number theory) and to the theory of Fourier series and other topics in mathematical analysis; he is credited with being one of the first mathematicians to give the modern formal definition of a function. Although his official surname is Lejeune Dirichlet, he is commonly referred to as just Dirichlet, particularly for the eponym. He improved on Lagrange's work on conservative systems by showing that the condition for equilibrium is that the potential energy is minimal.

is also a continuous function vanishing in the point $(\mathbf{0}, \mathbf{0})$ (see Fig. 7.5). Any continuous function attains its minimal $E_* > 0$ and maximal E^* values on the boundary $\bar{\Omega}_\varepsilon$ of the set Ω_ε. Thus, on the boundary $\bar{\Omega}_\varepsilon$ we have

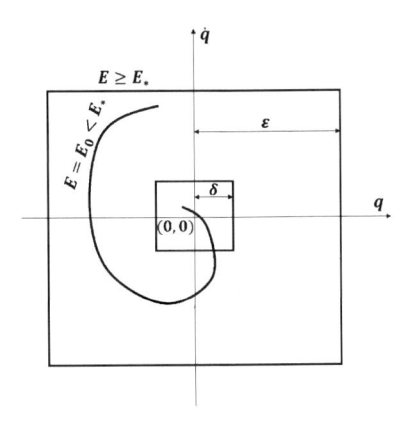

$$E \geq E_* > 0.$$

Figure 7.5 The (ε, δ)-illustration of the local stability of an equilibrium point.

But on the other hand, since the continuous function $E(t, \mathbf{q}, \dot{\mathbf{q}})$ vanishes at $(\mathbf{0}, \mathbf{0})$, there necessarily exists a δ-neighborhood of this point such that

$$E(t, \mathbf{q}, \dot{\mathbf{q}}) < E_*.$$

Hence, if $\mathbf{q}(t_0)$ and $\dot{\mathbf{q}}(t_0)$ are inside of this δ-neighborhood of the point $(0, 0)$, that is,

$$\|\mathbf{q}(t_0)\| \leq \delta \text{ and } \|\dot{\mathbf{q}}(t_0)\| \leq \delta,$$

then

$$E(t, \mathbf{q}(t_0), \dot{\mathbf{q}}(t_0)) < E_*.$$

But for the conservative systems the complete mechanical energy $E(t, \mathbf{q}, \dot{\mathbf{q}})$ remains to be constant, that is,

$$E(t, \mathbf{q}, \dot{\mathbf{q}}) = E_0 = \underset{t}{\text{const}}.$$

Therefore, we may conclude that during the whole time of motion

$$E(t, \mathbf{q}(t), \dot{\mathbf{q}}(t)) < E_*$$

and, as a result, during the motion the trajectories $(\mathbf{q}(t), \dot{\mathbf{q}}(t))$ cannot reach the boundary $\bar{\Omega}_\varepsilon$, where $E(t, \mathbf{q}(t), \dot{\mathbf{q}}(t)) \geq E_*$, keeping the relation $(\mathbf{q}(t), \dot{\mathbf{q}}(t)) \in \Omega_\varepsilon$. □

Under the additional condition that the second derivative of the potential energy exists and is positively defined in an equilibrium, the previous result allows obtaining a sufficient condition to ensure its stability.

Corollary 7.3. *Under the conditions of the previous theorem, if the **Hessian matrix** $\nabla^2 V(\mathbf{q})$ of the potential energy exists and*

$$\nabla^2 V(\mathbf{q}) > 0 \;\forall\; \mathbf{q} \in \mathcal{A} \;-\; \text{a neighborhood of the point } \mathbf{q}^*,$$

then $\mathbf{q}^*$ is locally stable.

Proof. The strict minimum condition for smooth $V(\mathbf{q})$ is as follows:

$$\left.\begin{array}{l} 0 < V(\mathbf{q}) - V(\mathbf{q}^*) \;\forall \mathbf{q} \neq \mathbf{q}^*,\; \mathbf{q} \in \mathcal{A}, \\ \nabla V(\mathbf{q}^*) = 0. \end{array}\right\} \qquad (7.20)$$

Here we have taken into account that $\mathbf{q}^*$ is an equilibrium point, and hence, $\nabla V(\mathbf{q}^*) = 0$. On the other hand, in the small neighborhood of the point $\mathbf{q}^*$ the potential energy can be expressed (using the Taylor expansion) as

$$V(\mathbf{q}) = V(\mathbf{q}^*) + (\nabla V(\mathbf{q}^*), \mathbf{q} - \mathbf{q}^*) + \frac{1}{2}(\mathbf{q} - \mathbf{q}^*)^T \nabla^2 V(\mathbf{q}^*)(\mathbf{q} - \mathbf{q}^*) + o\left\|\mathbf{q} - \mathbf{q}^*\right\|^2 =$$
$$V(\mathbf{q}^*) + \frac{1}{2}(\mathbf{q} - \mathbf{q}^*)^T \nabla^2 V(\mathbf{q}^*)(\mathbf{q} - \mathbf{q}^*) + o\left\|\mathbf{q} - \mathbf{q}^*\right\|^2$$

for any $\mathbf{q} \in \mathcal{A}$, which in combination with (7.20) gives

$$0 < V(\mathbf{q}) - V(\mathbf{q}^*) = (\mathbf{q} - \mathbf{q}^*)^T \nabla^2 V(\mathbf{q}^*)(\mathbf{q} - \mathbf{q}^*) + o\left\|\mathbf{q} - \mathbf{q}^*\right\|^2 \quad \forall \mathbf{q} \in \mathcal{A},$$

which is true if and only if

$$\nabla^2 V(\mathbf{q}^*) > 0.$$

$\square$

Illustrating examples

Two examples illustrate the application of the corollary result.

Example 7.5. Consider a conservative system with potential energy

$$V(\mathbf{q}) = \frac{1}{2}\mathbf{q}^T C \mathbf{q}, \quad C = C^T.$$

Let us determine the equilibrium points and their stability. We will try to carry out the analysis for the matrices

$$C_1 = \begin{bmatrix} 5 & 2.5 \\ 2.5 & 1 \end{bmatrix}, \quad C_2 = \begin{bmatrix} 5 & 2 \\ 2 & 1 \end{bmatrix}.$$

The equilibria are given by the equation $\nabla V(\mathbf{q}^*) = \mathbf{0}$, that is,

$$C\mathbf{q}^* = \mathbf{0},$$

and if $\det C \neq 0$, it follows that the only equilibrium is

$$\mathbf{q}^* = \mathbf{0}.$$

By the previous corollary, the condition $\nabla^2 V(\mathbf{q}^*) > 0$ guarantees $\mathbf{0}$ is stable (globally, as it is a single equilibrium), which translates into the condition

$$C > 0.$$

The matrix C_1 turns out to be not positive definite, which follows from the Sylvester criterion and that $\det C_1 = -1.25 < 0$, so nothing can be ensured about the stability of $\mathbf{0}$. The matrix C_2 is positive definite and consequently $\mathbf{0}$ is stable.

Example 7.6. Suppose a conservative system has potential energy

$$V(\mathbf{q}) = V_0 + (\mathbf{a}, \mathbf{q}) + \frac{1}{2}\mathbf{q}^T C \mathbf{q},$$

where $V_0 \in R$ and $\mathbf{a} \in R^n$ are constants and $\det C \neq 0$. Let us determine the equilibria and their stability if

$$C = \begin{bmatrix} 3 & 0 & \alpha \\ 0 & 2 & 0 \\ \alpha & 0 & 1 \end{bmatrix}.$$

What is the range of values for α that ensure equilibrium stability? In equilibria, the relation $\nabla V(\mathbf{q}^*) = \mathbf{0}$ is satisfied, and therefore

$$\mathbf{a} + C\mathbf{q}^* = \mathbf{0},$$

and the only equilibrium is

$$\mathbf{q}^* = -C^{-1}\mathbf{a}. \tag{7.21}$$

From the previous corollary, for $\mathbf{q}^*$ to be stable (globally, for uniqueness), the condition $\nabla^2 V(\mathbf{q}^*) > 0$ is sufficient to be met, that is,

$$C > 0.$$

In the given matrix C, the first two major minors are positive and

$$\det C = 2\left(3 - \alpha^2\right).$$

It is positive if and only if

$$|\alpha| < \sqrt{3},$$

which, by the Sylvester criteria, turns out to be the stability interval

$$-\sqrt{3} < \alpha < \sqrt{3}$$

for the point $\mathbf{q}^*$ (see (7.21)).

Remark 7.4. The Lagrange–Dirichlet theorem, Theorem 7.2, remains also valid for the wide class of non-conservative systems containing gyroscopic and dissipative forces satisfying locally

$$\sum_{i=1}^{n} \tilde{Q}_i(\mathbf{q}, \dot{\mathbf{q}}) \dot{q}_i \leq 0 \qquad (7.22)$$

for any $(\mathbf{q}, \dot{\mathbf{q}}) \in \Omega_\varepsilon$. Indeed, suppose that at least one $\tilde{Q}_\alpha(\mathbf{0}, \mathbf{0}) \neq 0$ in the equilibrium point $(\mathbf{0}, \mathbf{0})$. But, due to continuity $\tilde{Q}_\alpha(\mathbf{q}, \dot{\mathbf{q}})$, we have $\tilde{Q}_\alpha(\mathbf{q}, \dot{\mathbf{q}}) \neq 0$ in some neighborhood Ω_ε of the origin. So,

$$\sum_{i=1}^{n} \tilde{Q}_i(\mathbf{q}, \dot{\mathbf{q}}) \dot{q}_i = \tilde{Q}_\alpha(\mathbf{q}, \dot{\mathbf{q}}) \dot{q}_\alpha > 0 \qquad (7.23)$$

if we take $\dot{q}_\alpha = \text{sign}\left(\tilde{Q}_\alpha(\mathbf{q}, \dot{\mathbf{q}})\right)$. We are able to realize this selection since all generalized coordinates q_i and $\dot{q}_i$ are independent. So, we obtain the following contradiction: (7.23) contradicts (7.22). This means that for all $i = 1, ..., n$,

$$\tilde{Q}_i(\mathbf{0}, \mathbf{0}) = 0,$$

which indicates that the presence of gyroscopic and dissipative forces does not violate the property of the equilibrium local stability.

7.4 Unstable equilibria in conservative systems

The general problem of determining if an equilibrium is stable has led to more results than determining whether, on the contrary, it is unstable. In the case that the systems are conservative, again there are some conditions that ensure instability.

Remember that in a conservative system, $\mathbf{q}^*$ is an equilibrium if and only if $\nabla V(\mathbf{q}^*) = 0$. Therefore, the serial development of potential energy around $\mathbf{q}^*$ may be written as

$$V(\mathbf{q}) = V(\mathbf{q}^*) + \sum_{j=2}^{\infty} V_j(\mathbf{q}^*, \mathbf{q}), \qquad (7.24)$$

where $V_j(\mathbf{q}^*, \mathbf{q})$ denotes the upper term of j-th order, in particular,

$$V_2(\mathbf{q}^*, \mathbf{q}) := \frac{1}{2}(\mathbf{q} - \mathbf{q}^*)^T \nabla^2 V(\mathbf{q}^*)(\mathbf{q} - \mathbf{q}^*). \qquad (7.25)$$

Equilibrium and stability

The theorems following below are given without their proofs, which can be found in (Malkin, 1952) and (Chetayev, 1965).

Theorem 7.3 (The first Lyapunov theorem on instability). *If in the equilibrium $\mathbf{q}^*$ of a conservative system, the potential energy $V(\mathbf{q})$ does not have a minimum and this condition can be seen from the second order term (7.25), then $\mathbf{q}^*$ is unstable.*

The condition in the previous theorem is satisfied in the following simple case.

Corollary 7.4. *If the Hessian matrix $\nabla^2 V(\mathbf{q}^*)$ has at least one negative eigenvalue, then $\mathbf{q}^*$ is unstable.*

Theorem 7.4 (The second Lyapunov theorem on instability). *Consider a conservative system with potential energy $V(\mathbf{q})$. If in the equilibrium $\mathbf{q}^*$ the function $V(\mathbf{q})$ has a strict maximum and this condition can be determined from the terms $V_j(\mathbf{q}^*, \mathbf{q})$ of lower order ($j \geq 2$) of the development (7.24), then $\mathbf{q}^*$ is unstable.*

The previous theorem has its most useful form in the following corollary.

Corollary 7.5. *As an immediate consequence of the Lyapunov theorem, for (7.25), if the Hessian matrix is negative definite, namely,*

$$\nabla^2 V(\mathbf{q}^*) < 0,$$

then $\mathbf{q}^*$ is unstable.

Theorem 7.5 (The Chetayev theorem on instability). *If the potential energy $V(\mathbf{q}^*)$ of a conservative system is a homogeneous function, that is,*

$$V(\lambda \mathbf{q}) = \lambda^m V(\mathbf{q}) \quad \forall \lambda \in R, \ \forall \mathbf{q} \in R^n,$$

with m some integer, and if in equilibrium $\mathbf{q}^*$ the function $V(\mathbf{q})$ does not have a minimum, then $\mathbf{q}^*$ is unstable.

Example 7.7. Let a conservative system have potential energy

$$V(q) = a(1 - \cos(\alpha q)), \quad a, \alpha \neq 0.$$

Let us rate the stability of its equilibria. The equilibria q^* satisfy

$$\frac{dV}{dq}(q^*) = \alpha a \sin(\alpha q^*) = 0,$$

which is why

$$q_k^* = \frac{k}{\alpha}\pi, \quad k = 0, \pm 1, \pm 2, \dots.$$

The evaluation of the second derivative in q^* leads to

$$\frac{d^2V}{dq^2}(q_k^*) = \alpha^2 a \cos(\alpha q^*) = \alpha^2 a \cos(k\pi).$$

So,

$$\frac{d^2 V}{dq^2}(q_k^*) > 0, \text{ if } k = 0, \pm 2, \pm 4, ...,$$

$$\frac{d^2 V}{dq^2}(q_k^*) < 0, \text{ if } k = \pm 1, \pm 3, ...,$$

and by the Lagrange–Dirichlet theorem and the corollary to the second Lyapunov theorem we may conclude that

q_k^* is locally stable for $k = 0, \pm 2, \pm 4, ...,$

q_k^* is locally unstable for $k = \pm 1, \pm 3, ...$

Example 7.8. We need to determine equilibria and their corresponding stability in a conservative system with the potential energy

$$V(\mathbf{q}) = a \prod_{i=1}^{n} q_i, \ a \neq 0.$$

Solution. The equilibria $\mathbf{q}^*$ are given by the equation $\nabla_\mathbf{q} V(\mathbf{q}^*) = 0$, that is,

$$\frac{\partial V}{\partial q_i}(\mathbf{q}^*) = a \prod_{j=1,\ j \neq i}^{n} q_j^* = 0, \quad i = 1, ...n.$$

That is why

$$\mathbf{q}^* = \mathbf{0}.$$

The application of the corollary to the second theorem of Lyapunov is not possible in view of the fact that $\nabla_\mathbf{q}^2 V(\mathbf{q}^*) = 0$. However, as easily verified, $V(\mathbf{q})$ is homogeneous and in 0 it has no minimum. Therefore, by the Chetayev theorem, the equilibrium found is unstable. □

Example 7.9. Fig. 7.6 shows a conservative system consisting of two spheres with opposite electric charges e_1 and e_2. Both have mass m and are subject to the action of the gravitational attraction force, but one of them is fixed to the origin of the coordinate system. We will determine the set of equilibria and its stability. Clearly, the system has three degrees of freedom. If we denote

$$\mathbf{q} := \begin{bmatrix} x & y & z \end{bmatrix}^T,$$

and if the xy-plane is considered as the reference level of the gravitational potential energy, the joint potential energy results in

$$V(\mathbf{q}^*) = mgz - k\frac{e_1 e_2}{r},$$

Equilibrium and stability

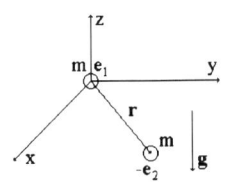

Figure 7.6 A potential electromechanical system.

where $k > 0$ is a constant and $r = \sqrt{x^2 + y^2 + z^2}$. The equilibria can now be found by the expression $\nabla_{\mathbf{q}} V(\mathbf{q}^*) = 0$. Therefore

$$\frac{\partial V}{\partial x}(\mathbf{q}^*) = k \frac{e_1 e_2}{(r^*)^3} x^* = 0, \text{ implying } x^* = 0, \tag{7.26}$$

$$\frac{\partial V}{\partial y}(\mathbf{q}^*) = k \frac{e_1 e_2}{(r^*)^3} y^* = 0, \text{ implying } y^* = 0, \tag{7.27}$$

and

$$\frac{\partial V}{\partial z}(\mathbf{q}^*) = mg + k \frac{e_1 e_2}{(r^*)^3} z^* = 0, \text{ implying } \frac{z^*}{(r^*)^3} = -\frac{mg}{k e_1 e_2}.$$

But, by (7.26) and (7.27),

$$r^* = |z^*|,$$

and hence,

$$\frac{\operatorname{sign} z^*}{|z^*|^2} = -\frac{mg}{k e_1 e_2}.$$

Since $\dfrac{mg}{k e_1 e_2} > 0$, we have

$$\operatorname{sign} z^* = -1$$

and

$$|z^*| = \sqrt{\frac{k e_1 e_2}{mg}},$$

so that

$$z^* = -\sqrt{\frac{k e_1 e_2}{mg}}.$$

The Hessian matrix $\nabla_{\mathbf{q}}^2 V(\mathbf{q}^*)$ may now be calculated:

$$\nabla_{\mathbf{q}}^2 V(\mathbf{q}^*) = \begin{bmatrix} \dfrac{\partial^2 V}{\partial x^2}(\mathbf{q}^*) & \dfrac{\partial^2 V}{\partial x \partial y}(\mathbf{q}^*) & \dfrac{\partial^2 V}{\partial x \partial z}(\mathbf{q}^*) \\ \dfrac{\partial^2 V}{\partial x \partial y}(\mathbf{q}^*) & \dfrac{\partial^2 V}{\partial y^2}(\mathbf{q}^*) & \dfrac{\partial^2 V}{\partial y \partial z}(\mathbf{q}^*) \\ \dfrac{\partial^2 V}{\partial x \partial z}(\mathbf{q}^*) & \dfrac{\partial^2 V}{\partial y \partial z}(\mathbf{q}^*) & \dfrac{\partial^2 V}{\partial z^2}(\mathbf{q}^*) \end{bmatrix},$$

with the elements

$\dfrac{\partial^2 V}{\partial x^2}(\mathbf{q}) = k \dfrac{e_1 e_2}{r^3} \left(1 - \dfrac{3x^2}{r^2}\right)$, implying $\dfrac{\partial^2 V}{\partial x^2}(\mathbf{q}^*) = k \dfrac{e_1 e_2}{|z^*|^3}$,

$\dfrac{\partial^2 V}{\partial x \partial y}(\mathbf{q}) = -3k \dfrac{e_1 e_2}{r^5} xy$, implying $\dfrac{\partial^2 V}{\partial x \partial y}(\mathbf{q}^*) = 0$,

$\dfrac{\partial^2 V}{\partial x \partial z}(\mathbf{q}) = -3k \dfrac{e_1 e_2}{r^5} xz$, implying $\dfrac{\partial^2 V}{\partial x \partial z}(\mathbf{q}^*) = 0$,

$\dfrac{\partial^2 V}{\partial y^2}(\mathbf{q}) = k \dfrac{e_1 e_2}{r^3} \left(1 - \dfrac{3y^2}{r^2}\right)$, implying $\dfrac{\partial^2 V}{\partial y^2}(\mathbf{q}^*) = k \dfrac{e_1 e_2}{|z^*|^3}$,

$\dfrac{\partial^2 V}{\partial y \partial z}(\mathbf{q}) = -3k \dfrac{e_1 e_2}{r^5} yz$, implying $\dfrac{\partial^2 V}{\partial y \partial z}(\mathbf{q}^*) = 0$,

$\dfrac{\partial^2 V}{\partial z^2}(\mathbf{q}) = k \dfrac{e_1 e_2}{r^3} \left(1 - \dfrac{3z^2}{r^2}\right)$, implying $\dfrac{\partial^2 V}{\partial z^2}(\mathbf{q}^*) = -2k \dfrac{e_1 e_2}{|z^*|^3}$.

So,

$$\nabla_{\mathbf{q}}^2 V(\mathbf{q}^*) = k \dfrac{e_1 e_2}{|z^*|^3} \begin{bmatrix} 1 & 0 & 0 \\ 0 & 1 & 0 \\ 0 & 0 & -2 \end{bmatrix},$$

which has a negative eigenvalue, and because of the corollary to the Lyapunov's first theorem, the equilibrium found is unstable.

Example 7.10. Two fixed bars rotate with constant angular velocity ω as shown in Fig. 7.7. On the bars, two dough rings m slide without friction. If the rings are subject to the gravitational field shown and the force of attraction between them, determine their equilibrium positions on the bars and find the stability conditions. The angle α is fixed. The system has two degrees of freedom. The positions of the rings on their respective bar and with respect to their union are chosen as generalized coordinates, as indicated in Fig. 7.7. On the other hand, to show that the system is conservative, note that the force of inertia on the bar 1 ring due to rotation is

$F_{rot} = m \omega^2 q_1 \sin \alpha$,

Equilibrium and stability

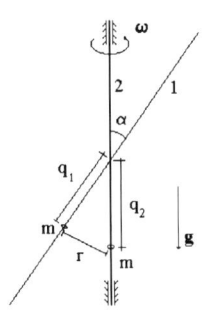

Figure 7.7 Sliding rings on rotating bars.

and it can be represented as

$$F_{rot} = -\frac{d}{dq_1}\left(-\frac{1}{2}m\omega^2 q_1^2 \sin\alpha\right),$$

whence it follows that this force is potential with the corresponding function

$$V_{rot}(\mathbf{q}) = -\frac{1}{2}m\omega^2 q_1^2 \sin\alpha.$$

As is known, the power energy due to the attraction between the masses of the rings is given by

$$V_{at}(\mathbf{q}) = -k\frac{m^2}{r},$$

where $k > 0$ is a constant and

$$r := \sqrt{q_1^2 + q_2^2 - 2q_1 q_2 \cos\alpha}.$$

So, the total potential energy of the system is

$$V(\mathbf{q}) = -mg(q_1 \cos\alpha + q_2) - k\frac{m^2}{r} - \frac{1}{2}m\omega^2 q_1^2 \sin\alpha,$$

whereby equilibria can be determined from the equation $\nabla_{\mathbf{q}} V(\mathbf{q}^*) = \mathbf{0}$. Then the following pair of simultaneous nonlinear equations is obtained:

$$\left.\begin{array}{l} g\cos\alpha + k\dfrac{m}{(r^*)^3}\left(q_2^* \cos\alpha - q_1^*\right) + \omega^2 q_1^* \sin\alpha = 0, \\[2mm] g + k\dfrac{m}{(r^*)^3}\left(q_1^* \cos\alpha - q_2^*\right) = 0. \end{array}\right\} \qquad (7.28)$$

They must be satisfied in equilibrium positions. Given the relations

$$\frac{\partial^2 V}{\partial q_1^2}(\mathbf{q}) = k\frac{m^2}{r^3} - m\omega^2 \sin\alpha,$$

$$\frac{\partial^2 V}{\partial q_2^2}(\mathbf{q}) = k\frac{m^2}{r^3},$$

$$\frac{\partial V}{\partial q_1 q_2}(\mathbf{q}) = -k\frac{m^2}{r^3}\cos\alpha,$$

we can see that the Hessian matrix is given by

$$\nabla_{\mathbf{q}}^2 V(\mathbf{q}^*) = \begin{bmatrix} k\dfrac{m^2}{(r^*)^3} - m\omega^2 \sin\alpha & -k\dfrac{m^2}{(r^*)^3}\cos\alpha \\ -k\dfrac{m^2}{(r^*)^3}\cos\alpha & k\dfrac{m^2}{(r^*)^3} \end{bmatrix} =$$

$$k\frac{m^2}{(r^*)^3}\begin{bmatrix} 1 - \dfrac{\omega^2}{km}(r^*)^3 \sin\alpha & -\cos\alpha \\ -\cos\alpha & 1 \end{bmatrix}.$$

The Sylvester criterion applied to this matrix allows the following conclusions to be established. By the Lagrange–Dirichlet theorem it follows that the equilibrium $\mathbf{q}^*$ is stable if the following inequalities both hold:

$$1 - \frac{\omega^2}{km}(r^*)^3 \sin\alpha > 0,$$

$$1 - \frac{\omega^2}{km}(r^*)^3 \sin\alpha - \cos^2\alpha > 0.$$

By the corollary to Lyapunov's second theorem, instability of $\mathbf{q}^*$ is ensured if

$$1 - \frac{\omega^2}{km}(r^*)^3 \sin\alpha < 0,$$

or

$$1 - \frac{\omega^2}{km}(r^*)^3 \sin\alpha - \cos^2\alpha < 0.$$

7.5 Exercises

Exercise 7.1. A particle of mass m, carrying a charge e, is located in an electric field of a fixed charge q. Show that the equilibrium position of a particle in a uniform gravitational field is displaced vertically from a fixed charge by a distance of $\sqrt{\dfrac{|eq|}{mg}}$ and this equilibrium is unstable.

Equilibrium and stability

Exercise 7.2. The material point is in the gravity field on the surface
$$z = x^2 - xy + y^2.$$
Find the equilibrium positions and investigate their stability if there is no friction in the system and the Oz axis is directed upwards.

Exercise 7.3. A ball suspended on a weightless rod of length l can oscillate in a vertical plane that rotates around the vertical axis, passing through the point of suspension of the pendulum, with a constant angular velocity ω. Show that the angle φ of deviation of the rod from the vertical in the relative equilibrium position is equal to:

- $\varphi^* = 0$ (stable equilibrium position) if $\omega^2 l \leq g$,
- $\varphi^* = \pi$ (unstable equilibrium position) if $\omega^2 l \leq g$,
- $\varphi^* = \pm \arccos\left(\dfrac{g}{\omega^2 l}\right)$ (stable equilibrium positions) if $\omega^2 l > g$,
- $\varphi^* = \{0, \pi\}$ (unstable equilibrium positions) if $\omega^2 l > g$.

Exercise 7.4. Prove the **Earnshaw theorem**: a collection of point charges cannot be maintained in a stable stationary equilibrium configuration solely by the electrostatic interaction of the charges, or in other words, any static configuration of electric charges is unstable.

Exercise 7.5. Two identical balls, connected by a spring of stiffness k, can slide without friction on the sides of a right angle lying in the horizontal plane. The length of the spring in its undeformed state is equal to l_0. Show that the system has a continuum
$$x^2 + y^2 = l_0^2$$
of unstable equilibria.

Oscillations analysis

Contents

8.1 Movements in the vicinity of equilibrium points   245
     8.1.1   Small oscillations   245
     8.1.2   Characteristic polynomial   247
     8.1.3   General solution of the characteristic equation   248
8.2 Oscillations in conservative systems   249
     8.2.1   Some properties of the characteristic equation   249
     8.2.2   Normal coordinates   251
8.3 Several examples of oscillation analysis   255
     8.3.1   Three masses joined by springs in circular dynamics   255
     8.3.2   Three masses joined by springs with dynamics on a straight line   258
     8.3.3   Four spring-bound masses with restricted linear dynamics   260
     8.3.4   Three identical pendula held by springs   262
     8.3.5   Four-loop LC circuits   264
     8.3.6   Finding one polynomial root using other known roots   266
     8.3.7   Hint: how to resolve analytically cubic equations   268
8.4 Exercises   270

Lagrange's equations are an invaluable tool in determining the important properties of mechanical systems. The application of these equations and the study of the consequences derived have been the object of the two preceding chapters. In the one that now begins, one more application is presented to the study of the important problem of oscillations of systems around equilibrium points. Using the usual technique of linearization around an equilibrium point, Lagrange's equations can be approximated by a linear expression that describes in detail the dynamics of the system in a neighborhood sufficiently close to the point of interest. To this approximate expression, all known techniques for linear dynamic systems can be applied, leading to useful conclusions. In addition, if the system in question is restricted to being of the conservative type, then the expression is reduced, which allows to characterize and calculate in a very simple way its solutions, in particular those of interest in this chapter: *oscillations*.

8.1 Movements in the vicinity of equilibrium points

8.1.1 Small oscillations

Recall some facts from Chapter 6, where the Lagrange equations for holonomic systems with ideal constraints are discussed:

$$\frac{d}{dt}\nabla_{\dot{\mathbf{q}}}T(t,\mathbf{q},\dot{\mathbf{q}}) - \nabla_{\mathbf{q}}T(t,\mathbf{q},\dot{\mathbf{q}}) = \mathbf{Q}_{pot} + \mathbf{Q}_{non\text{-}pot}. \tag{8.1}$$

In the vicinity of an equilibrium $\mathbf{q}^*$, position and velocity can be expressed as

$$\mathbf{q}(t) = \mathbf{q}^* + \Delta\mathbf{q}(t), \quad \dot{\mathbf{q}}(t) = \Delta\dot{\mathbf{q}}(t).$$

In this chapter we will consider that both $\Delta\mathbf{q}(t)$ and $\Delta\dot{\mathbf{q}}(t)$ are small, that is,

$$\|\Delta\mathbf{q}(t)\| \ll 1, \quad \|\Delta\dot{\mathbf{q}}(t)\| \ll 1.$$

Suppose the system in question is stationary (Chapter 6, Section 6.4) and that the potential energy is so as well, that is, $V(t,\mathbf{q})$ does not explicitly depend on t. Therefore, in the neighborhood of $\mathbf{q}^*$ we have

$$T(\mathbf{q},\dot{\mathbf{q}}) = \frac{1}{2}\dot{\mathbf{q}}^T A(\mathbf{q})\dot{\mathbf{q}} = \frac{1}{2}\Delta\dot{\mathbf{q}}^T A(\mathbf{q}^*+\Delta\mathbf{q})\Delta\dot{\mathbf{q}}, \quad V(\mathbf{q}) = V(\mathbf{q}^*+\Delta\mathbf{q}),$$

with $A(\mathbf{q}) = A^T(\mathbf{q}) > 0$. By the Taylor expansion

$$A(\mathbf{q}^*+\Delta\mathbf{q}) = A(\mathbf{q}^*) + o(\|\Delta\mathbf{q}\|).$$

Therefore,

$$T(\mathbf{q},\dot{\mathbf{q}}) = \frac{1}{2}\Delta\dot{\mathbf{q}}^T A(\mathbf{q}^*)\Delta\dot{\mathbf{q}} + o\left(\|\Delta\mathbf{q}\|\cdot\|\Delta\dot{\mathbf{q}}\|^2\right) \tag{8.2}$$

and

$$V(\mathbf{q}^*+\Delta\mathbf{q}) = V(\mathbf{q}^*) + \frac{1}{2}\Delta\mathbf{q}^T \nabla^2 V(\mathbf{q}^*)\Delta\mathbf{q} + o\left(\|\Delta\mathbf{q}\|^2\right).$$

Since $\mathbf{q}^*$ is an equilibrium we have $\nabla V(\mathbf{q}^*) = 0$, and therefore the generalized potential force $\mathbf{Q}_{pot} := -\nabla_{\mathbf{q}}V(\mathbf{q})$ may be represented as

$$\mathbf{Q}_{pot} = -\nabla_{\mathbf{q}}\left[V(\mathbf{q}^*) + \frac{1}{2}\Delta\mathbf{q}^T \nabla^2 V(\mathbf{q}^*)\Delta\mathbf{q} + o\left(\|\Delta\mathbf{q}\|^2\right)\right]$$
$$= -\nabla^2 V(\mathbf{q}^*)\Delta\mathbf{q} + o(\|\Delta\mathbf{q}\|).$$

Here we have used the properties

$$\nabla_{\mathbf{q}}V(\mathbf{q}^*) = 0 \text{ and } \nabla_{\mathbf{q}}(\cdot) = \nabla_{\Delta\mathbf{q}}(\cdot).$$

On the other hand, if one considers that the non-potential generalized forces $\mathbf{Q}_{non\text{-}pot}(t,\mathbf{q},\dot{\mathbf{q}})$ do not explicitly depend on t, an analogous development leads to

$$\mathbf{Q}_{non\text{-}pot}(\mathbf{q}^*+\Delta\mathbf{q},\Delta\dot{\mathbf{q}}) = \mathbf{Q}_{non\text{-}pot}(\mathbf{q}^*,\mathbf{0}) + \nabla_{\Delta\mathbf{q}}\mathbf{Q}_{non\text{-}pot}(\mathbf{q}^*,\mathbf{0})\Delta\mathbf{q} +$$
$$\nabla_{\Delta\dot{\mathbf{q}}}\mathbf{Q}_{non\text{-}pot}(\mathbf{q}^*,\mathbf{0})\Delta\dot{\mathbf{q}} + o(\|\Delta\mathbf{q}\|\cdot\|\Delta\dot{\mathbf{q}}\|).$$

Oscillations analysis

In view of the property

$$\mathbf{Q}_{non\text{-}pot}\left(\mathbf{q}^*, \mathbf{0}\right) = \mathbf{0},$$

it follows that

$$\left.\begin{aligned}\mathbf{Q}_{non\text{-}pot}\left(\mathbf{q}^* + \Delta\mathbf{q}, \Delta\dot{\mathbf{q}}\right) = \nabla_{\Delta\mathbf{q}}\mathbf{Q}_{non\text{-}pot}\left(\mathbf{q}^*, \mathbf{0}\right)\Delta\mathbf{q} + \\ \nabla_{\Delta\dot{\mathbf{q}}}\mathbf{Q}_{non\text{-}pot}\left(\mathbf{q}^*, \mathbf{0}\right)\Delta\dot{\mathbf{q}} + o\left(\|\Delta\mathbf{q}\| \cdot \|\Delta\dot{\mathbf{q}}\|\right).\end{aligned}\right\} \quad (8.3)$$

In this context, the Lagrange equation (8.1) can be expressed as (using (8.2)–(8.3))

$$A\left(\mathbf{q}^*\right)\Delta\ddot{\mathbf{q}} = \left[-\nabla^2 V\left(\mathbf{q}^*\right) + \nabla_{\Delta\mathbf{q}}\mathbf{Q}_{non\text{-}pot}\left(\mathbf{q}^*, \mathbf{0}\right)\right]\Delta\mathbf{q} +$$
$$\nabla_{\Delta\dot{\mathbf{q}}}\mathbf{Q}_{non\text{-}pot}\left(\mathbf{q}^*, \mathbf{0}\right)\Delta\dot{\mathbf{q}} + o\left(\|\Delta\mathbf{q}\|, \|\Delta\dot{\mathbf{q}}\|\right),$$

from which the following approximation of the Lagrange equation for small movements in the vicinity of equilibrium $\mathbf{q}^*$ is obtained:

$$A\left(\mathbf{q}^*\right)\Delta\ddot{\mathbf{q}} + B\left(\mathbf{q}^*\right)\Delta\dot{\mathbf{q}} + C\left(\mathbf{q}^*\right)\Delta\mathbf{q} = \mathbf{0}, \quad (8.4)$$

where

$$\left.\begin{aligned}A\left(\mathbf{q}^*\right) &= \nabla^2_{\Delta\dot{\mathbf{q}}}T\left(\mathbf{q}^*, \mathbf{0}\right) = \nabla^2_{\dot{\mathbf{q}}}T\left(\mathbf{q}^*, \mathbf{0}\right),\\ B\left(\mathbf{q}^*\right) &:= -\nabla_{\Delta\dot{\mathbf{q}}}\mathbf{Q}_{non\text{-}pot}\left(\mathbf{q}^*, \mathbf{0}\right),\\ C\left(\mathbf{q}^*\right) &:= \nabla^2 V\left(\mathbf{q}^*\right) - \nabla_{\Delta\mathbf{q}}\mathbf{Q}_{non\text{-}pot}\left(\mathbf{q}^*, \mathbf{0}\right),\end{aligned}\right\} \quad (8.5)$$

and since $\Delta\dot{\mathbf{q}} = \dot{\mathbf{q}}$, $\Delta\ddot{\mathbf{q}} = \ddot{\mathbf{q}}$ (if a shift is made from the origin of the coordinate system so that $\mathbf{q}^* = \mathbf{0}$), whereby $\Delta\mathbf{q} = \mathbf{q}$, then (8.4) can be expressed in the standard form

$$A\ddot{\mathbf{q}}(t) + B\dot{\mathbf{q}}(t) + C\mathbf{q}(t) = \mathbf{0}, \quad A = A\left(\mathbf{q}^*\right), \ B = B\left(\mathbf{q}^*\right), \ C = C\left(\mathbf{q}^*\right) \in R^{n \times n}, \quad (8.6)$$

with some initial conditions

$$\mathbf{q}(0) = \mathbf{q}_0, \quad \dot{\mathbf{q}}(0) = \dot{\mathbf{q}}_0.$$

8.1.2 Characteristic polynomial

To calculate the solutions of (8.6) let us try to find it as

$$\mathbf{q} = e^{\lambda t}\mathbf{u}, \quad (8.7)$$

with λ, a scalar called **frequency**, and a vector $\mathbf{u} \neq \mathbf{0}$, called **amplitude**, both constants which should be determined. Substitution of (8.7) in (8.6) gives

$$\lambda^2 e^{\lambda t} A\mathbf{u} + \lambda e^{\lambda t} B\mathbf{u} + e^{\lambda t} C\mathbf{u} = \mathbf{0},$$

where the so-called **characteristic equation**, associated with (8.6), is obtained:

$$\left(\lambda^2 A + \lambda B + C\right)\mathbf{u} = \mathbf{0}. \tag{8.8}$$

The non-trivial condition $\mathbf{u} \neq \mathbf{0}$ is satisfied if and only if

$$p(\lambda) := \det\left(\lambda^2 A + \lambda B + C\right) = 0. \tag{8.9}$$

The function $p(\lambda)$ is called the **characteristic polynomial**, associated with (8.6), and Eq. (8.9) is referred to as the **characteristic equation**. Opening (8.9), we may represent it in the form

$$p(\lambda) = \lambda^{2n}\rho_0 + \lambda^{2n-1}\rho_1 + \cdots + \rho_{2n} = 0, \quad \rho_i \in R. \tag{8.10}$$

8.1.3 General solution of the characteristic equation

From polynomial theory it follows that there are K different roots of (8.10), each with a certain algebraic multiplicity, so that if m_k is the corresponding multiplicity of the root λ_k, $k = 1, ..., K$, we have

$$\sum_{k=1}^{K} m_k = 2n.$$

Suppose that n_k is the geometric multiplicity of the root λ_k ($k = 1, ..., K$), that is,

$$n_k := \dim \operatorname{Ker}\left(\lambda_k^2 A + \lambda_k B + C\right),$$

with $n_k \leq m_k$, and denote as

$$U_k := \{\mathbf{u}_{k,i}\}_{i=1}^{n_k}$$

the set of amplitude vectors, corresponding to λ_k, obtained from the characteristic equation (8.8), which (as is well known) are *linearly independent*. Suppose also that the vector $\mathbf{u}_{k,i} \in U_k$ corresponds to $n_{k,i}$ repetitions of the root λ_k, where

$$\sum_{i=1}^{n_k} n_{k,i} = m_k.$$

Therefore, by the linearity of (8.6), it is verified that the general form of the **partial basic solution**, corresponding to the root λ_k, is given by

$$\mathbf{q}_k(t) = e^{\lambda_k t} \sum_{i=1}^{n_k} \sum_{j=1}^{n_{k,i}} \gamma_{k,i,j} t^{j-1} \mathbf{u}_{k,i}, \tag{8.11}$$

Oscillations analysis

where

$$\gamma_{k,i,j}, \ k=1,...,K, \ i=1,...,n_k, \ j=1,...n_{k,i},$$

are constants that depend on the initial conditions. Consequently, the general solution $\mathbf{q}(t)$ of the characteristic equation (8.9) is the sum of partial basic solutions $\mathbf{q}_k(t)$, resulting in

$$\mathbf{q}(t) = \sum_{k=1}^{K} \mathbf{q}_k(t). \tag{8.12}$$

8.2 Oscillations in conservative systems

The condition $\mathbf{Q}_{non\text{-}pot} = 0$, typical for conservative systems, allows to obtain interesting simplifications of the results, obtained in the previous section. The most important simplification has to do with matrices B and C in (8.5). It is verified that these are reduced to

$$B = 0, \quad C = \nabla^2 V(\mathbf{q}^*),$$

so (8.6) takes the form

$$A\ddot{\mathbf{q}} + C\mathbf{q} = \mathbf{0}, \quad 0 < A = A^T, \quad C = C^T, \quad A, C \in R^{n \times n}. \tag{8.13}$$

The special properties of (8.13) allow to determine certain important aspects of the concepts previously introduced. Note that for this case, the associated characteristic equation results in

$$\left(\lambda^2 A + C\right) \mathbf{u} = \mathbf{0}, \tag{8.14}$$

with λ^2 satisfying

$$\det\left(\lambda^2 A + C\right) = 0. \tag{8.15}$$

8.2.1 Some properties of the characteristic equation

Resolving (8.15), the solutions obtained allow to determine from (8.14) the corresponding amplitude vectors. But when one of these vectors is known and if the root λ_j^2 to which it belongs is unknown, one can proceed in the following way. Let $\mathbf{u}_j$ be the known vector. By the scalar multiplication of (8.14) by $\mathbf{u}_j$ it follows that

$$\lambda_j^2 \left(\mathbf{u}_j, A\mathbf{u}_j\right) = -\left(\mathbf{u}_j, C\mathbf{u}_j\right),$$

and, in view of the property $A > 0$, we have

$$\lambda_j^2 = -\frac{(\mathbf{u}_j, C\mathbf{u}_j)}{(\mathbf{u}_j, A\mathbf{u}_j)}. \tag{8.16}$$

Proposition 8.1. *Let λ_i^2, λ_j^2 be two solutions of (8.15) and let $\mathbf{u}_i$, $\mathbf{u}_j$ be two corresponding amplitude vectors, obtained from (8.14). If $\lambda_i^2 \neq \lambda_j^2$, then $\mathbf{u}_i$, $\mathbf{u}_j$ are orthogonal,*

$$\mathbf{u}_i \underset{A}{\perp} \mathbf{u}_j,$$

with respect to A, that is,

$$(\mathbf{u}_j, A\mathbf{u}_i) = 0. \tag{8.17}$$

Proof. From (8.14) it follows that

$$\lambda_i^2 A\mathbf{u}_i = -C\mathbf{u}_i, \tag{8.18}$$

$$\lambda_j^2 A\mathbf{u}_j = -C\mathbf{u}_j, \tag{8.19}$$

from where, by scalarly multiplying (8.18) by $\mathbf{u}_j$ and (8.19) by $\mathbf{u}_i$, we get

$$\lambda_i^2 (\mathbf{u}_j, A\mathbf{u}_i) = -(\mathbf{u}_j, C\mathbf{u}_i),$$
$$\lambda_j^2 (\mathbf{u}_i, A\mathbf{u}_j) = -(\mathbf{u}_i, C\mathbf{u}_j),$$

whose difference, under the consideration that the matrices A and C are symmetric, results in

$$\left(\lambda_i^2 - \lambda_j^2\right)(\mathbf{u}_j, A\mathbf{u}_i) = 0,$$

and the conclusion (8.17) follows. □

The immediate result above serves as a tool to prove the following two propositions.

Proposition 8.2. *Let $\sigma := \{\lambda_k^2\}_{k=1}^n$ be the set of solutions of (8.15). Then all elements of σ are real, i.e., $\sigma \subset R$.*

Proof. Suppose there is a solution $\lambda_i^2 \in \sigma$ which is complex, that is,

$$\lambda_i^2 = \mu_i' + j\mu_i'', \quad \mu_i' \in R, \quad 0 \neq \mu_i'' \in R, \quad j := \sqrt{-1}.$$

Therefore, since the coefficients of $\det\left(\lambda^2 A + C\right)$ are real,

$$\overline{\lambda_i^2} = \mu_i' - j\mu_i''$$

also belongs to σ. If

$$\mathbf{u}_i = \mathbf{v}_i' + j\mathbf{v}_i'', \quad \mathbf{v}_i' \in R^n, \quad 0 \neq \mathbf{v}_i'' \in R^n,$$

is the amplitude vector corresponding to λ_i^2, then

$$\overline{\mathbf{u}_i} = \mathbf{v}_i' - j\mathbf{v}_i''$$

corresponds to $\overline{\lambda_i^2}$. Since $\lambda_i^2 \neq \overline{\lambda_i^2}$, from Proposition 8.2, it follows that

$$(\overline{\mathbf{u}_i}, A\mathbf{u}_i) = 0. \tag{8.20}$$

But taking into account that $A = A^T > 0$, we get

$$\begin{aligned}
&\left(\mathbf{v}_i' - j\mathbf{v}_i'', A\left(\mathbf{v}_i' + j\mathbf{v}_i''\right)\right) = \\
&\left(\mathbf{v}_i', A\mathbf{v}_i'\right) + j\left(\mathbf{v}_i', A\mathbf{v}_i''\right) - j\left(\mathbf{v}_i'', A\mathbf{v}_i'\right) + \left(\mathbf{v}_i'', A\mathbf{v}_i''\right) = \\
&\left(\mathbf{v}_i', A\mathbf{v}_i'\right) + \left(\mathbf{v}_i'', A\mathbf{v}_i''\right) > 0.
\end{aligned}$$

So, $\mathbf{v}_i'' \neq \mathbf{0}$ contradicts (8.20). □

Proposition 8.3. *In the context of Proposition 8.2, if $\lambda_i^2 \neq \lambda_j^2$, then the corresponding vectors $\mathbf{u}_i$ and $\mathbf{u}_j$ are linearly independent, which means that the relation*

$$c_i \mathbf{u}_i + c_j \mathbf{u}_j = 0 \tag{8.21}$$

is valid only when

$$c_i = c_j = 0.$$

Proof. Let c_i and c_j be two scalars that make the relationship (8.21) true. Multiply it scalarly by $A\mathbf{u}_i$. Then, by Proposition 8.2

$$\left(c_i\mathbf{u}_i + c_j\mathbf{u}_j, A\mathbf{u}_i\right) = c_i\left(\mathbf{u}_i, A\mathbf{u}_i\right) + c_j\underbrace{\left(\mathbf{u}_j, A\mathbf{u}_i\right)}_{0} = c_i\left(\mathbf{u}_i, A\mathbf{u}_i\right)$$

and in view of the fact that $A > 0$, in order to satisfy (8.21) we must have $c_i = 0$. But if this is so, since $\mathbf{u}_j \neq \mathbf{0}$, from (8.21) we conclude that $c_j = 0$. □

8.2.2 Normal coordinates

In addition to the properties already obtained, we can know more about the form of the solutions of (8.13). We need the following result from linear algebra theory (see, for example, (Poznyak, 2008, Theorem 7.3)).

Theorem 8.1. *For any two quadratic forms*

$$\begin{aligned}
f_A(x) &= (x, Ax), \quad A = A^\mathsf{T}, \\
f_B(x) &= (x, Bx), \quad B = B^\mathsf{T},
\end{aligned}$$

when one quadratic form is strictly positive, i.e., $(x, Ax) > 0$ for any $x \neq 0$, $x \in R^n$, there exists a non-singular transformation

$$S = (\mathbf{s}_1, ..., \mathbf{s}_n)^\mathsf{T}, \quad \mathbf{s}_i \in R^{1 \times n},$$

such that in new variables z, defined as

$$z = S^{-1}x, \quad x = Sz,$$

the given quadratic forms have the following expressions:

$$\left.\begin{aligned}f_A(x) &= (x, Ax) = \left(z, S^\mathsf{T} A S z\right) = (z, z) = \sum_{i=1}^{n} z_i^2, \\ f_B(x) &= (x, Bx) = \left(z, S^\mathsf{T} B S z\right) = \sum_{i=1}^{n} \beta_i z_i^2,\end{aligned}\right\} \quad (8.22)$$

or equivalently,

$$S^\mathsf{T} A S = I_{n \times n}, \quad S^\mathsf{T} B S = \operatorname{diag}(\beta_1, \beta_2, ..., \beta_n). \tag{8.23}$$

Proof. Let T_A transform A to the diagonal form, namely,

$$T_A^\mathsf{T} A T_A = \operatorname{diag}(\alpha_1, \alpha_2, ..., \alpha_n) := \Lambda_A,$$

with $\alpha_i > 0$ ($i = 1, ..., n$). Note that this transformation exists by the spectral theorem and is orthogonal, i.e.,

$$T_A^\mathsf{T} = T_A^{-1}.$$

Then, defining $\Lambda_A^{1/2}$ such that

$$\Lambda_A = \Lambda_A^{1/2} \Lambda_A^{1/2}, \quad \Lambda_A^{1/2} = \operatorname{diag}\left(\sqrt{\alpha_1}, \sqrt{\alpha_2}, ..., \sqrt{\alpha_n}\right),$$

one has

$$\left[\Lambda_A^{-1/2} T_1^\mathsf{T}\right] A \left[T_1 \Lambda_A^{-1/2}\right] = I_{n \times n}.$$

Hence,

$$\tilde{B} := \left[\Lambda_A^{-1/2} T_1^\mathsf{T}\right] B \left[T_1 \Lambda_A^{-1/2}\right]$$

is a symmetric matrix, i.e., $\tilde{B} = \tilde{B}^\mathsf{T}$. Let $T_{\tilde{B}}$ be a unitary matrix transforming $\tilde{B}$ to the diagonal form, that is,

$$T_{\tilde{B}}^\mathsf{T} \tilde{B} T_{\tilde{B}} = \operatorname{diag}\left(\tilde{\beta}_1, \tilde{\beta}_2, ..., \tilde{\beta}_n\right) := \Lambda_{\tilde{B}}.$$

Then the transformation S defined by

$$S := \left[T_A \Lambda_A^{-1/2} \right] T_{\tilde{B}}$$

exactly realizes (8.22) since

$$T^\mathsf{T} A T = T_{\tilde{B}}^\mathsf{T} \left(\left[\Lambda_A^{-1/2} T_1^\mathsf{T} \right] A \left[T_1 \Lambda_A^{-1/2} \right] \right) T_{\tilde{B}} = T_{\tilde{B}}^\mathsf{T} T_{\tilde{B}} = I_{n \times n}.$$

□

Define **normal** or **main coordinates** by

$$\tilde{\mathbf{q}} := S^{-1} \mathbf{q}, \quad \mathbf{q} = S \tilde{\mathbf{q}}. \tag{8.24}$$

Applying the transformation S to (8.13), we get

$$A S \ddot{\tilde{\mathbf{q}}} + C S \tilde{\mathbf{q}} = \mathbf{0}, \tag{8.25}$$

which, when premultiplied by S^T, results in

$$S^T A S \ddot{\tilde{\mathbf{q}}} + S^T C S \tilde{\mathbf{q}} = 0,$$

with $S^T A S = I_{n \times n}$ and $S^T C S = R$, that is, the dynamics of $\tilde{\mathbf{q}}$ is governed by

$$\ddot{\tilde{\mathbf{q}}} + \begin{bmatrix} r_1 & 0 & \cdots & 0 \\ 0 & r_2 & \cdots & 0 \\ \vdots & \vdots & \ddots & \vdots \\ 0 & 0 & \cdots & r_n \end{bmatrix} \tilde{\mathbf{q}} = \mathbf{0}, \ r_i = \tilde{\beta}_i \ (i = 1, ..., n),$$

which represents the following decoupled system of n ordinary second order differential equations:

$$\ddot{\tilde{q}}_i + r_i \tilde{q}_i = 0, \quad i = 1, ..., n. \tag{8.26}$$

Propose

$$\tilde{q}_i = u_i e^{\lambda_i t} \tag{8.27}$$

as a solution for (8.26) with $\lambda_i \in R$ and $0 \neq u_i \in R$ (constants to be determined), which leads to the relationship

$$u_i \left(\lambda_i^2 + r_i \right) e^{\lambda_i t} = 0,$$

from where it follows that λ_i must comply with

$$\lambda_i^2 + r_i = 0,$$

that is,

$$\lambda_i^{(1)} = -\sqrt{-r_i}, \quad \lambda_i^{(2)} = \sqrt{-r_i}.$$

This, together with (8.27), leads to the general solution of (8.26) in the form

$$\tilde{q}_i(t) = u_i \left(\gamma_i^{(1)} e^{-\sqrt{-r_i}t} + \gamma_i^{(2)} e^{\sqrt{-r_i}t} \right) \tag{8.28}$$

with constants $\gamma_i^{(1)}$ and $\gamma_i^{(2)}$ that depend on the initial conditions.

The dynamics of $\tilde{q}_i$ depends on the sign of r_i, as shown in the following.

(a) If $r_i < 0$, then (8.28) takes the form

$$\tilde{q}_i(t) = u_i \left(\gamma_i^{(1)} e^{-\sqrt{|r_i|}t} + \gamma_i^{(2)} e^{\sqrt{|r_i|}t} \right),$$

and it is concluded that the corresponding solution is *not bounded, exponentially increasing*.

(b) If $r_i = 0$, then (8.28) takes the form

$$\tilde{q}_i(t) = u_i \left(\gamma_i^{(1)} + \gamma_i^{(2)} t \right), \tag{8.29}$$

and again it is concluded that the solution obtained is *not bounded, linearly increasing*.

(c) If $r_i > 0$, then (8.28) takes the form

$$\tilde{q}_i(t) = u_i \left(\gamma_i^{(1)} e^{-j\sqrt{r_i}t} + \gamma_i^{(2)} e^{j\sqrt{r_i}t} \right), \quad j := \sqrt{-1},$$

or, in view of the relationship

$$e^{\pm j\alpha} = \cos\alpha \pm j \sin\alpha,$$

it follows that

$$\tilde{q}_i(t) = u_i \gamma_i \sin\left(\sqrt{r_i}t + \varphi_i\right),$$

from which it is concluded that the corresponding solution is a bounded oscillation with the frequency $\sqrt{r_i}$.

Conclusion 8.1. *From the previous discussion on the form of the solution (8.28) two facts emerge:*

(a) *From the substitution of (8.28) in (8.24) it can be verified that, even if a certain root λ_i^2 of (8.15) is present with multiplicity n_k greater than one, the solution* **q** *does not contain polynomials of t, except in the case in which the root $\lambda^2 = 0$ appears, in which the first degree polynomial (8.29) is presented.*[1]

---

[1] This note is related with the so-called "*Lagrange error.*" Lagrange thought that oscillations may increase as polynomials of the order $n_k - 1$.

Oscillations analysis

(b) To have oscillatory movements, we should have $\lambda^2 = -\mu^2$, with $0 < \mu \in R$. So, for this case the characteristic equation and polynomial associated with the solution are

$$\left(-\mu^2 A + C\right) \mathbf{u} = \mathbf{0}, \quad \det\left(-\mu^2 A + C\right) = 0, \quad 0 < \mu \in R. \tag{8.30}$$

Note that if $0 \leq \mu \in R$ is considered, these equations serve for both negative λ^2 and null roots.

8.3 Several examples of oscillation analysis

In this section several examples are developed in order to show some aspects of the results obtained.

8.3.1 Three masses joined by springs in circular dynamics

In Fig. 8.1 a ring is shown on which three mass bodies m are connected in series by stiffness springs k. The movement is frictionless and the masses are not subject to any gravitational field. Determine the general solution for displacements as a function of initial conditions. Since the system is conservative, the theory developed in the previous section can be applied. Clearly, three generalized coordinates are required. They may be selected as the displacements of the masses with respect to a certain position in which the springs are not deformed. The potential energy is expressed by

$$V(\mathbf{q}) = \frac{k}{2}\left[(q_1 - q_3)^2 + (q_2 - q_1)^2 + (q_3 - q_2)^2\right],$$

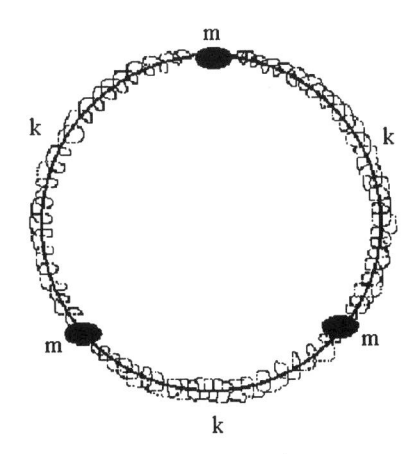

Figure 8.1 Three masses joined by springs in circular dynamics.

from which it follows that

$$\nabla_{\mathbf{q}} V(\mathbf{q}) = k \begin{bmatrix} 2q_1 - q_2 - q_3 \\ -q_1 + 2q_2 - q_3 \\ -q_1 - q_2 + 2q_3 \end{bmatrix}, \quad (8.31)$$

and hence, by the condition $\nabla_{\mathbf{q}} V(\mathbf{q}^*) = 0$,

$$\mathbf{q}^* = \mathbf{0}.$$

This equilibrium may be unstable, because for the initial conditions

$$\mathbf{q}(0) = \mathbf{0}, \dot{\mathbf{q}}(0) = \delta \begin{bmatrix} 1 & 1 & 1 \end{bmatrix}^T, \quad \delta > 0,$$

we have $\|\mathbf{q}(t)\| \to \infty$ when $t \to \infty$. From (8.31) we have

$$C := \nabla_{\mathbf{q}}^2 V(\mathbf{0}) = k \begin{bmatrix} 2 & -1 & -1 \\ -1 & 2 & -1 \\ -1 & -1 & 2 \end{bmatrix}.$$

On the other hand, the kinetic energy is

$$T = \frac{1}{2} m \left(\dot{q}_1^2 + \dot{q}_2^2 + \dot{q}_3^2 \right) = \frac{1}{2} m \dot{\mathbf{q}}^T I_{n \times n} \dot{\mathbf{q}},$$

where $I_{n \times n}$ is the identity matrix of order 3. Then by (8.5) it follows that

$$A := m I_{n \times n}.$$

By (8.15), the $\lambda^2 \in R$ values of the system in question satisfy the following equation:

$$\det \left(\lambda^2 A + C \right) = 0,$$

that is,

$$\det \begin{bmatrix} 2k + \lambda^2 m & -k & -k \\ -k & 2k + \lambda^2 m & -k \\ -k & -k & 2k + \lambda^2 m \end{bmatrix} = 0,$$

or equivalently,

$$\lambda^2 \left[9 \left(\frac{k}{m} \right)^2 + 6 \left(\frac{k}{m} \right) \lambda^2 + \lambda^4 \right] = 0,$$

where the following solutions are obtained:

$$\lambda_1^2 = 0, \quad \lambda_2^2 = -3 \frac{k}{m},$$

Oscillations analysis

with multiplicities 1 and 2, respectively. The fact that $\lambda_1^2 = 0$ shows that $\mathbf{q}^* = \mathbf{0}$ is unstable. From (8.14) the corresponding amplitude vectors are obtained. The one of λ_1^2 is given by

$$C\mathbf{u}_1 = k \begin{bmatrix} 2 & -1 & -1 \\ -1 & 2 & -1 \\ -1 & -1 & 2 \end{bmatrix} \mathbf{u}_1 = 0,$$

or by the following system of two simultaneous equations:

$$2u_{1,1} - u_{1,2} - u_{1,3} = 0,$$
$$-u_{1,1} + 2u_{1,2} - u_{1,3} = 0.$$

If $u_{1,3}$ is taken as a free parameter, the above equations are equivalent to

$$\begin{bmatrix} 2 & -1 \\ -1 & 2 \end{bmatrix} \begin{bmatrix} u_{1,1} \\ u_{1,2} \end{bmatrix} = u_{1,3} \begin{bmatrix} 1 \\ 1 \end{bmatrix},$$

from which

$$\begin{bmatrix} u_{1,1} \\ u_{1,2} \end{bmatrix} = u_{1,3} \begin{bmatrix} 2 & -1 \\ -1 & 2 \end{bmatrix}^{-1} \begin{bmatrix} 1 \\ 1 \end{bmatrix} = u_{1,3} \begin{bmatrix} 1 \\ 1 \end{bmatrix},$$

and if $u_{1,3} = 1$ is chosen, we get

$$\begin{bmatrix} u_{1,1} \\ u_{1,2} \end{bmatrix} = \begin{bmatrix} 1 \\ 1 \end{bmatrix}.$$

For the determination of the amplitude vectors of $\mathbf{u}_2^{(1)}$ and $\mathbf{u}_2^{(2)}$, associated with $\lambda_2^2 = -3\frac{k}{m}$, the results of the previous section can be used. Since $A = mI_{n \times n}$, it follows that these vectors must meet (see (8.32)) the following conditions:

$$\left(\mathbf{u}_1, \mathbf{u}_2^{(1)} \right) = 0, \quad \left(\mathbf{u}_1, \mathbf{u}_2^{(2)} \right) = 0.$$

Moreover, $\mathbf{u}_2^{(1)}$ and $\mathbf{u}_2^{(2)}$ must be independent, which is achieved if they are orthogonal:

$$\left(\mathbf{u}_1^{(1)}, \mathbf{u}_2^{(2)} \right) = 0.$$

These conditions translate into the following system of simultaneous equations:

$$\left. \begin{array}{l} u_{1,1} u_{2,1}^{(1)} + u_{1,2} u_{2,2}^{(1)} + u_{1,3} u_{2,3}^{(1)} = 0, \\ u_{1,1} u_{2,1}^{(2)} + u_{1,2} u_{2,2}^{(2)} + u_{1,3} u_{2,3}^{(2)} = 0, \\ u_{2,1}^{(1)} u_{2,1}^{(2)} + u_{2,2}^{(1)} u_{2,2}^{(2)} + u_{2,3}^{(1)} u_{2,3}^{(2)} = 0, \end{array} \right\}$$

which, in view of how $\mathbf{u}_1$ was chosen, are satisfied if

$$\mathbf{u}_2^{(1)} = \begin{bmatrix} 1 & 0 & -1 \end{bmatrix}^T, \quad \mathbf{u}_2^{(2)} = \begin{bmatrix} 1 & -2 & 1 \end{bmatrix}^T,$$

and consequently, the general solution for the mass system considered is given by

$$\mathbf{q}(t) = \left(\gamma_1^{(1)} + \gamma_1^{(2)} t\right) \begin{bmatrix} 1 \\ 1 \\ 1 \end{bmatrix} +$$

$$\gamma_2 \sin\left(\sqrt{3\frac{k}{m}} t + \varphi_2\right) \begin{bmatrix} 1 \\ 0 \\ -1 \end{bmatrix} + \gamma_3 \sin\left(\sqrt{3\frac{k}{m}} t + \varphi_2\right) \begin{bmatrix} 1 \\ -2 \\ 1 \end{bmatrix},$$

where $\gamma_1^{(1)}, \gamma_1^{(2)}, \gamma_2, \gamma_3, \varphi_1, \varphi_2 \in R$ are constants that depend on the initial conditions.

Remark 8.1. The components of the different vectors $\mathbf{u}$, obtained in the previous example, represent all the directions of movement that the masses of the system can present:

- $\mathbf{u}_1$ corresponds to the movement in which all the masses move in the same direction;
- in $\mathbf{u}_2^{(1)}$ one mass is stopped and the other two move in opposite directions;
- finally, $\mathbf{u}_2^{(2)}$ denotes two masses moving in the same direction and the other in contradiction and with double magnitude.

When the set of masses, as in the preceding example, has similarity in movement, vectors can be determined by inspection, without resolving to Eq. (8.14).

8.3.2 Three masses joined by springs with dynamics on a straight line

A three-mass system is joined by stiffness springs k as shown in Fig. 8.2. Suppose that gravity forces do not act on this system and that the movement occurs without restrictions on the straight line indicated. Determine (if possible by the inspection without resolving equations (8.30)) the law that the movements of the masses follow. The kinetic energy of this system is

$$T = \frac{1}{2} m \left(\dot{q}_1^2 + 2\dot{q}_2^2 + \dot{q}_3^2\right) = \frac{1}{2} m \dot{\mathbf{q}}^T \begin{bmatrix} 1 & 0 & 0 \\ 0 & 2 & 0 \\ 0 & 0 & 1 \end{bmatrix} \dot{\mathbf{q}},$$

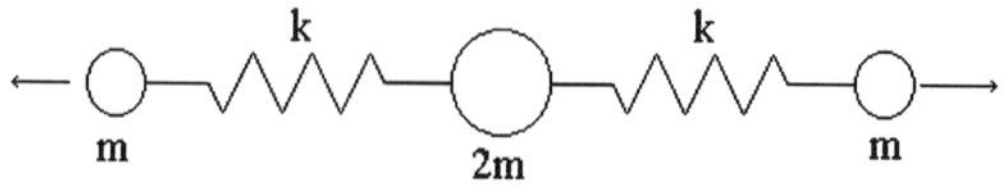

Figure 8.2 Three masses joined by springs with dynamics on a line.

Oscillations analysis

from which it follows that

$$A := m \begin{bmatrix} 1 & 0 & 0 \\ 0 & 2 & 0 \\ 0 & 0 & 1 \end{bmatrix}, \qquad (8.32)$$

while the potential energy has as the expression

$$V(\mathbf{q}) = \frac{k}{2}\left[(q_2 - q_1)^2 + (q_3 - q_2)^2\right],$$

such that

$$\nabla_\mathbf{q} V(\mathbf{q}) = k \begin{bmatrix} q_1 - q_2 \\ -q_1 + 2q_2 - q_3 \\ -q_1 - q_2 + 2q_3 \end{bmatrix} = k \begin{bmatrix} 1 & -1 & 0 \\ -1 & 2 & -1 \\ 0 & -1 & 1 \end{bmatrix} \mathbf{q},$$

implying that

$$\mathbf{q}^* = \mathbf{0}.$$

It is an equilibrium, which is not stable because for the initial conditions

$$\mathbf{q}(0) = \mathbf{0}, \ \dot{\mathbf{q}}(0) = \delta \begin{bmatrix} 1 & 1 & 1 \end{bmatrix}^T, \ \delta > 0.$$

Indeed, in this case $\|\mathbf{q}(t)\| \to \infty$ when $t \to \infty$. We also have

$$C := \nabla_\mathbf{q}^2 V(\mathbf{0}) = k \begin{bmatrix} 1 & -1 & 0 \\ -1 & 2 & -1 \\ 0 & -1 & 1 \end{bmatrix}.$$

Upon inspection it is concluded that two of the amplitude vectors are

$$\mathbf{u}_1 = \begin{bmatrix} 1 \\ 1 \\ 1 \end{bmatrix}, \ \mathbf{u}_2 = \begin{bmatrix} 1 \\ 0 \\ -1 \end{bmatrix}, \qquad (8.33)$$

which are orthogonal with respect to the matrix A that appears in (8.32) and must be orthogonal to $\mathbf{u}_3$ with respect to the same matrix A. Let it be $\mathbf{u}_3 = (\alpha \ \beta \ \gamma)^T$. Then the following simultaneous equations must be satisfied:

$$\left.\begin{array}{l}(\mathbf{u}_1, A\mathbf{u}_3) = m(\alpha + 2\beta + \gamma) = 0, \\ (\mathbf{u}_2, A\mathbf{u}_3) = m(\alpha - \gamma) = 0,\end{array}\right\}$$

whose solution results in

$$\alpha = \gamma, \ \beta = -\gamma,$$

and if $\gamma = 1$ is chosen, then

$$\mathbf{u}_3 = \begin{pmatrix} 1 & -1 & 1 \end{pmatrix}^\mathsf{T}.$$

Using (8.16), we get

$$\lambda_j^2 = -\frac{(\mathbf{u}_j, C\mathbf{u}_j)}{(\mathbf{u}_j, A\mathbf{u}_j)}, \quad i = 1, 2, 3.$$

So,

$$\lambda_1^2 = 0, \quad \lambda_2^2 = -\frac{k}{m}, \quad \lambda_3^2 = -\frac{2k}{m},$$

and therefore, the general solution (oscillation) is

$$\mathbf{q}(t) = \left(\gamma_1^{(1)} + \gamma_1^{(2)} t\right) \begin{bmatrix} 1 \\ 1 \\ 1 \end{bmatrix} +$$

$$\frac{k}{m} \gamma_2 \sin\left(\sqrt{\frac{k}{m}} t + \varphi_2\right) \begin{bmatrix} 1 \\ 0 \\ -1 \end{bmatrix} + \gamma_3 \sin\left(\sqrt{\frac{2k}{m}} t + \varphi_3\right) \begin{bmatrix} 1 \\ -1 \\ 1 \end{bmatrix}.$$

8.3.3 Four spring-bound masses with restricted linear dynamics

A system of four masses m, joined by stiffness springs k, is shown in Fig. 8.3. Determine the expression of mass displacements, considering that they are not subject to gravitational forces.

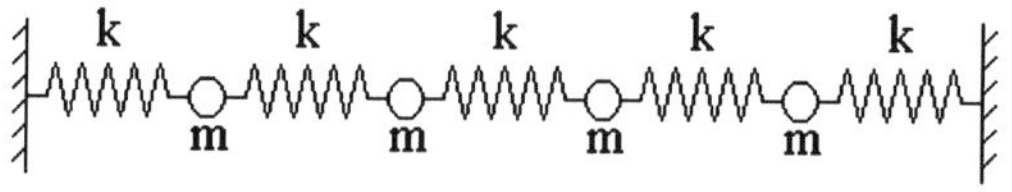

Figure 8.3 Four spring-bound masses with restricted linear dynamics.

Let the generalized coordinates be the displacements of the masses with respect to their position in which the springs are relaxed. The kinetic energy is given by

$$T(\dot{\mathbf{q}}) = \frac{1}{2} m \left(\dot{q}_1^2 + \dot{q}_2^2 + \dot{q}_3^2 + \dot{q}_4^2\right) = \frac{1}{2} m \dot{\mathbf{q}}^\mathsf{T} I_{n \times n} \dot{\mathbf{q}},$$

which is why

$$\nabla^2 T(\dot{\mathbf{q}}) = A := m I_{n \times n}. \tag{8.34}$$

On the other hand, the potential energy has the expression

$$V(\mathbf{q}) = \frac{k}{2} \left[q_1^2 + (q_2 - q_1)^2 + (q_3 - q_2)^2 + (q_4 - q_3)^2 + q_4^2\right],$$

Oscillations analysis

so that

$$\nabla_{\mathbf{q}} V(\mathbf{q}) = k \begin{bmatrix} 2q_1 - q_2 \\ -q_1 + 2q_2 - q_3 \\ -q_2 + 2q_3 - q_4 \\ -q_3 + 2q_4 \end{bmatrix} = k \begin{bmatrix} 2 & -1 & 0 & 0 \\ -1 & 2 & -1 & 0 \\ 0 & -1 & 2 & -1 \\ 0 & 0 & -1 & 2 \end{bmatrix} \mathbf{q},$$

implying

$$\mathbf{q}^* = \mathbf{0},$$

which is stable, since $\mathbf{q}^*$ is a point that minimizes $V(\mathbf{q})$ with

$$C := \nabla_{\mathbf{q}}^2 V(\mathbf{0}) = k \begin{bmatrix} 2 & -1 & 0 & 0 \\ -1 & 2 & -1 & 0 \\ 0 & -1 & 2 & -1 \\ 0 & 0 & -1 & 2 \end{bmatrix} > 0.$$

Since not all the masses in the system of Fig. 8.3 have the same freedom of movement, it is not possible in advance to know the amplitude vectors by inspection. With the obtained matrices A and C we find the equation

$$\det\left(\lambda^2 A + C\right) = 0,$$

which results in

$$\det \begin{bmatrix} 2k + m\lambda^2 & -k & 0 & 0 \\ -k & 2k + m\lambda^2 & -k & 0 \\ 0 & -k & 2k + m\lambda^2 & -k \\ 0 & 0 & -k & 2k + m\lambda^2 \end{bmatrix} = 0,$$

or equivalently,

$$5\left(\frac{k}{m}\right)^4 + 20\left(\frac{k}{m}\right)^3 \lambda + 21\left(\frac{k}{m}\right)^2 \lambda^2 + 8\frac{k}{m}\lambda^3 + \lambda^4 = 0,$$

whose solutions are

$$\lambda_1^2 = -\frac{k}{2m}\left(5 - \sqrt{5}\right), \quad \lambda_2^2 = -\frac{k}{2m}\left(5 + \sqrt{5}\right),$$
$$\lambda_3^2 = -\frac{k}{2m}\left(3 - \sqrt{5}\right), \quad \lambda_4^2 = -\frac{k}{2m}\left(3 + \sqrt{5}\right).$$

The solutions obtained for the corresponding amplitude vectors are as follows:

$$\mathbf{u}_1 = \begin{bmatrix} -2 \\ 1-\sqrt{5} \\ \dfrac{-1+\sqrt{5}}{2} \end{bmatrix}, \quad \mathbf{u}_2 = \begin{bmatrix} -2 \\ 1+\sqrt{5} \\ \dfrac{-1-\sqrt{5}}{2} \end{bmatrix},$$

$$\mathbf{u}_3 = \begin{bmatrix} 2 \\ 1+\sqrt{5} \\ \dfrac{1+\sqrt{5}}{2} \end{bmatrix}, \quad \mathbf{u}_4 = \begin{bmatrix} 2 \\ 1-\sqrt{5} \\ \dfrac{1-\sqrt{5}}{2} \end{bmatrix}.$$

Therefore, the sought expression is

$$\mathbf{q}(t) = \gamma_1 \sin\left(\sqrt{\dfrac{k}{2m}(5-\sqrt{5})}\, t + \varphi_1\right) \begin{bmatrix} -2 \\ 1-\sqrt{5} \\ \dfrac{-1+\sqrt{5}}{2} \end{bmatrix} +$$

$$\gamma_2 \sin\left(\sqrt{\dfrac{k}{2m}(5+\sqrt{5})}\, t + \varphi_2\right) \begin{bmatrix} -2 \\ 1+\sqrt{5} \\ \dfrac{-1-\sqrt{5}}{2} \end{bmatrix} +$$

$$\gamma_3 \sin\left(\sqrt{\dfrac{k}{2m}(3-\sqrt{5})}\, t + \varphi_3\right) \begin{bmatrix} 2 \\ 1+\sqrt{5} \\ \dfrac{1+\sqrt{5}}{2} \end{bmatrix} +$$

$$\gamma_4 \sin\left(\sqrt{\dfrac{k}{2m}(3+\sqrt{5})}\, t + \varphi_4\right) \begin{bmatrix} 2 \\ 1-\sqrt{5} \\ \dfrac{1-\sqrt{5}}{2} \end{bmatrix}.$$

8.3.4 Three identical pendula held by springs

Let us obtain the expressions of mass movements for the system of three identical pendula shown in Fig. 8.4. The pendula have mass m and are held by stiffness springs k that do not undergo deformation when all pendula are vertical. Consider as generalized coordinates the angular displacements of the arms of the pendula with respect to the vertical, that is,

$$\mathbf{q} := (\varphi_1 \quad \varphi_2 \quad \varphi_3)^\mathsf{T}.$$

Oscillations analysis

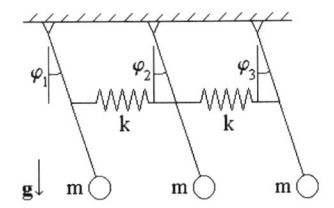

Figure 8.4 Three identical pendula held by springs.

The kinetic energy is given by

$$T(\dot{\mathbf{q}}) = \frac{1}{2}ml^2\left(\dot{q}_1^2 + \dot{q}_2^2 + \dot{q}_3^2\right) = \frac{1}{2}ml^2\dot{\mathbf{q}}^T I_{n\times n}\dot{\mathbf{q}}.$$

That is why

$$A := ml^2 I_{n\times n}.$$

The potential energy is

$$V(\mathbf{q}) = \frac{kl^2}{8}\left[(\sin q_2 - \sin q_1)^2 + (\sin q_3 - \sin q_2)^2\right] - mgl\sum_{i=1}^{3}\cos q_i,$$

so that

$$\nabla_{\mathbf{q}}V = \frac{kl^2}{4}\begin{bmatrix}(\sin q_1 - \sin q_2)\cos q_1 \\ (-\sin q_1 + 2\sin q_2 - \sin q_3)\cos q_2 \\ (-\sin q_2 + \sin q_3)\cos q_3\end{bmatrix} + mgl\begin{bmatrix}\sin q_1 \\ \sin q_2 \\ \sin q_3\end{bmatrix},$$

implying that

$$\mathbf{q}^* = \mathbf{0},$$

which is stable, since $\mathbf{q}^*$ is a point that minimizes V with

$$C := \nabla_{\mathbf{q}}^2 V(\mathbf{0}) = \begin{bmatrix}\frac{kl^2}{4} + mgl & -\frac{kl^2}{4} & 0 \\ -\frac{kl^2}{4} & \frac{kl^2}{2} + mgl & -\frac{kl^2}{4} \\ 0 & -\frac{kl^2}{4} & \frac{kl^2}{4} + mgl\end{bmatrix} > 0.$$

From the inspection analysis of possible movements we have as amplitude vectors

$$\mathbf{u}_1 = \begin{bmatrix}1 \\ 1 \\ 1\end{bmatrix}, \quad \mathbf{u}_2 = \begin{bmatrix}1 \\ 0 \\ -1\end{bmatrix},$$

which, as easily verified, are orthogonal with respect to A and must be orthogonal with respect to this same matrix, that is, if

$$\mathbf{u}_3 = \begin{pmatrix} \alpha & \beta & \gamma \end{pmatrix}^\mathsf{T},$$

then the following equations must be satisfied:

$$(\mathbf{u}_1, \mathbf{u}_3) = \alpha + \beta + \gamma = 0,$$
$$(\mathbf{u}_2, \mathbf{u}_3) = \alpha - \gamma = 0.$$

Their solution, if γ is chosen as the free parameter, is

$$\alpha = \gamma, \quad \beta = -2\gamma.$$

Particularly, if $\gamma = -1$, then

$$\mathbf{u}_3 = \begin{pmatrix} -1 & 2 & -1 \end{pmatrix}^\mathsf{T}.$$

Given that

$$\lambda_j^2 = -\frac{(\mathbf{u}_j, C\mathbf{u}_j)}{(\mathbf{u}_j, A\mathbf{u}_j)}, \quad i = 1, ..., 3,$$

we get

$$\lambda_1^2 = -\frac{g}{l}, \quad \lambda_2^2 = -\frac{kl + 4mg}{4ml}, \quad \lambda_3^2 = -\frac{3kl + 4mg}{4ml}.$$

Therefore, it is concluded that

$$\mathbf{q}(t) = \gamma_1 \sin\left(\sqrt{\frac{g}{l}}t + \varphi_1\right)\begin{bmatrix}1\\1\\1\end{bmatrix} + \gamma_2 \sin\left(\sqrt{\frac{kl + 4mg}{4ml}}t + \varphi_2\right)\begin{bmatrix}1\\0\\-1\end{bmatrix} +$$

$$\gamma_3 \sin\left(\sqrt{\frac{3kl + 4mg}{4ml}}t + \varphi_3\right)\begin{bmatrix}-1\\2\\-1\end{bmatrix}\Bigg\}.$$

8.3.5 Four-loop LC circuits

Here we will determine the law followed by load movements in the circuit shown in Fig. 8.5. Consider as generalized coordinates

$$q_j(t) := q_j(0) + \int_{\tau=0}^{t} i_j(\tau) d\tau, \quad j = 1, ..., 4,$$

where i_j denotes the current flowing through the j-th loop. From the table of analogies in Chapter 6 we have, on the one hand,

$$T(\dot{\mathbf{q}}) = \frac{1}{2}L\left(\dot{q}_1^2 + \dot{q}_2^2 + \dot{q}_3^2 + \dot{q}_4^2\right) = \frac{1}{2}L\dot{\mathbf{q}}^\mathsf{T} I_{n \times n}\dot{\mathbf{q}}, \tag{8.35}$$

Oscillations analysis

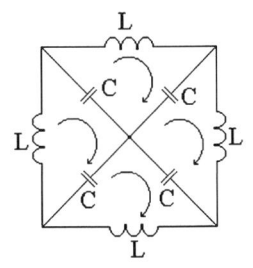

Figure 8.5 Four-loop LC circuit.

so that

$$A = \nabla^2 T(\dot{\mathbf{q}}) = LI_{n \times n},$$

and, on the other hand,

$$V(\mathbf{q}) = \frac{1}{2C}\left[(q_1 - q_2)^2 + (q_2 - q_3)^2 + (q_3 - q_4)^2 + (q_4 - q_1)^2\right]. \qquad (8.36)$$

Hence,

$$\nabla_{\mathbf{q}} V(\mathbf{q}) = \frac{1}{C}\begin{bmatrix} 2q_1 - q_2 - q_4 \\ -q_1 + 2q_2 - q_3 \\ -q_2 + 2q_3 - q_4 \\ -q_1 - q_3 + 2q_4 \end{bmatrix} = \frac{1}{C}\begin{bmatrix} 2 & -1 & 0 & -1 \\ -1 & 2 & -1 & 0 \\ 0 & -1 & 2 & -1 \\ -1 & 0 & -1 & 2 \end{bmatrix}\mathbf{q},$$

implying

$$\mathbf{q}^* = \mathbf{0},$$

which is not stable, since if the four initial currents are non-zero, equal, and in the same direction, according to Fig. 8.5, they remain so indefinitely. Therefore,

$$C := \nabla_{\mathbf{q}}^2 V(\mathbf{0}) = \frac{1}{C}\begin{bmatrix} 2 & -1 & 0 & -1 \\ -1 & 2 & -1 & 0 \\ 0 & -1 & 2 & -1 \\ -1 & 0 & -1 & 2 \end{bmatrix}.$$

By inspection, it is seen that the first three amplitude vectors can be chosen as

$$\mathbf{u}_1 = \begin{bmatrix} 1 \\ 1 \\ 1 \\ 1 \end{bmatrix}, \quad \mathbf{u}_2 = \begin{bmatrix} -1 \\ 1 \\ -1 \\ 1 \end{bmatrix}, \quad \mathbf{u}_3 = \begin{bmatrix} 1 \\ 1 \\ -1 \\ -1 \end{bmatrix},$$

which, as easily verified, are orthogonal with respect to A and must be orthogonal with respect to the same matrix. If

$$\mathbf{u}_4 = \begin{pmatrix} \alpha & \beta & \gamma & \delta \end{pmatrix}^\mathsf{T},$$

then the following simultaneous equations must be satisfied:

$$\left.\begin{aligned} (\mathbf{u}_1, \mathbf{u}_4) &= \alpha + \beta + \gamma + \delta = 0, \\ (\mathbf{u}_2, \mathbf{u}_4) &= -\alpha + \beta - \gamma + \delta = 0, \\ (\mathbf{u}_3, \mathbf{u}_4) &= \alpha + \beta - \gamma - \delta = 0, \end{aligned}\right\}$$

whose solution is

$$\alpha = \delta, \quad \beta = -\delta, \quad \gamma = -\delta,$$

and if $\delta = 1$ is chosen, then

$$\mathbf{u}_4 = \begin{pmatrix} 1 & -1 & -1 & 1 \end{pmatrix}^\mathsf{T}.$$

With the vectors obtained and since

$$\lambda_j^2 = -\frac{(\mathbf{u}_j, C\mathbf{u}_j)}{(\mathbf{u}_j, A\mathbf{u}_j)}, \quad i = 1, \ldots, 4,$$

it follows that

$$\lambda_1^2 = 0, \quad \lambda_2^2 = -\frac{4}{LC}, \quad \lambda_3^2 = -\frac{2}{LC}, \quad \lambda_4^2 = -\frac{2}{LC}.$$

Accordingly, we may conclude that

$$\mathbf{q}(t) = \left(\gamma_1^{(1)} + \gamma_1^{(2)} t\right) \begin{bmatrix} 1 \\ 1 \\ 1 \\ 1 \end{bmatrix} + \gamma_2 \sin\left(2\sqrt{\frac{1}{LC}}t + \varphi_2\right) \begin{bmatrix} -1 \\ 1 \\ -1 \\ 1 \end{bmatrix} +$$

$$\gamma_3 \sin\left(\sqrt{\frac{2}{LC}}t + \varphi_3\right) \begin{bmatrix} 1 \\ 1 \\ -1 \\ -1 \end{bmatrix} + \gamma_4 \sin\left(\sqrt{\frac{2}{LC}}t + \varphi_4\right) \begin{bmatrix} 1 \\ -1 \\ -1 \\ 1 \end{bmatrix}.$$

8.3.6 Finding one polynomial root using other known roots

The following example may be useful for obtaining a root of the characteristic polynomial in (8.30).

Suppose the values $\det A$ and $\det C$ are known and that

$$0 < \mu_1^2, \mu_2^2, \ldots, \mu_{n-1}^2 \tag{8.37}$$

are solutions of the equation

$$\det\left(-\mu^2 A + C\right) = 0 \qquad (8.38)$$

in (8.30). The unknown root is given by

$$\mu_n^2 = \frac{1}{\mu_1^2 \mu_2^2 \cdots \mu_{n-1}^2} \frac{\det C}{\det A}. \qquad (8.39)$$

Let us prove this. First, three necessary results from the matrix theory are required:
1. If $M_1, M_2 \in R^{n \times n}$, then

$$\det(M_1 M_2) = \det(M_1) \det(M_2). \qquad (8.40)$$

2. Let $M \in R^{n \times n}$ be such that $\det(M) \neq 0$. Then

$$\det\left(M^{-1}\right) = \frac{1}{\det(M)}.$$

3. The set of $\{\beta_i\}_{i=1}^n$ of the eigenvalues of a matrix $M \in R^{n \times n}$ are given by the solutions of the equation

$$\det(M - \beta I_{n \times n}) = 0$$

and they comply with

$$\det(M) = \prod_{i=1}^n \beta_i.$$

Since $A > 0$ (and hence it is invertible), by (8.40) we have

$$\det\left(-\mu^2 A + C\right) = \det\left[A\left(-\mu^2 I + A^{-1}C\right)\right] = \det(A) \det\left(-\mu^2 I + A^{-1}C\right).$$

So (8.38) is true if and only if

$$\det\left(-\mu^2 I + A^{-1}C\right) = 0,$$

from which it follows that the solutions of (8.37) are the eigenvalues of the $A^{-1}C$ matrix. But in view of the three results just stated,

$$\prod_{i=1}^n \mu_i^2 = \det\left(A^{-1}C\right) = \det\left(A^{-1}\right) \det(C) = \frac{\det(C)}{\det(A)},$$

which implies (8.39).

8.3.7 Hint: how to resolve analytically cubic equations

In the following development a technique that is useful for the resolution of characteristic equations is proposed to find the roots (which may be complex) of polynomials of the third degree. Suppose we want to find the solutions of the equation

$$ax^3 + bx^2 + cx + d = 0 \qquad (8.41)$$

with $a, b, c, d \in R$ and $a \neq 0$. With the representation

$$x := k + ly, \quad k, l \in R,$$

(8.41) may be rewritten with respect to y as

$$py^3 + qy^2 + ry + s = 0, \qquad (8.42)$$

where

$$p = al^3, \quad q = 3akl^2 + bl^2,$$
$$r = 3ak^2l + 2bkl + cl, \quad s = ak^3 + bk^2 + ck + d.$$

Selecting

$$l = \frac{1}{a^{1/3}}, \quad k = -\frac{b}{3a},$$

we obtain

$$p = 1, \quad q = 0.$$

Then (8.42) becomes

$$y^3 + ry + s = 0, \qquad (8.43)$$

with

$$r = \frac{1}{a^{1/3}} \left(c - \frac{b^2}{3a} \right)$$

and

$$s = \frac{2b^3}{27a^2} - c\frac{b}{3a} + d.$$

Eq. (8.43) may be rewritten as

$$y^3 = -ry - s. \qquad (8.44)$$

Represent now y as

$$y := u + v, \quad u, v \in R.$$

Oscillations analysis

Then we have

$$y^3 = u^3 + 3uv(u+v) + v^3 = u^3 + 3uvy + v^3. \tag{8.45}$$

Comparing (8.44) and (8.45) we conclude that

$$\left.\begin{array}{l} 3uv = -r, \\ u^3 + v^3 = -s. \end{array}\right\}$$

From the first equation above, assuming that $v \neq 0$, we have

$$u = -\frac{r}{3v},$$

which substituted in the second equation gives

$$-\left(\frac{r}{3v}\right)^3 + v^3 + s = 0.$$

Defining

$$z := v^3,$$

it follows that

$$z^2 + sz - \left(\frac{r}{3}\right)^3 = 0.$$

The solutions of this last equation lead to the following solutions of (8.41):

$$\left.\begin{array}{l} z_{1,2} = -\dfrac{s}{2} \pm \sqrt{\dfrac{s^2}{4} + \left(\dfrac{r}{3}\right)^3}, \\[2ex] v_{1,2} = \sqrt[3]{z_{1,2}} = \sqrt[3]{-\dfrac{s}{2} \pm \sqrt{\dfrac{s^2}{4} + \left(\dfrac{r}{3}\right)^3}}, \\[2ex] u_{1,2} = -\dfrac{r}{3v_{1,2}}, \\[2ex] y_{1,2} := u_{1,2} + v_{1,2} = \\[2ex] \qquad -\dfrac{r}{3\sqrt[3]{-\dfrac{s}{2} \pm \sqrt{\dfrac{s^2}{4} + \left(\dfrac{r}{3}\right)^3}}} + \sqrt[3]{-\dfrac{s}{2} \pm \sqrt{\dfrac{s^2}{4} + \left(\dfrac{r}{3}\right)^3}}, \end{array}\right\}$$

and

$$x = k + ly = \left\{ -\frac{b}{3a} + \frac{1}{a^{1/3}} \left(\sqrt[3]{-\frac{s}{2} \pm \sqrt{\frac{s^2}{4} + \left(\frac{r}{3}\right)^3}} - \frac{r}{3\sqrt[3]{-\frac{s}{2} \pm \sqrt{\frac{s^2}{4} + \left(\frac{r}{3}\right)^3}}} \right) \right\}.$$

In general, in the last expression *one real* solution and *two complex* solutions, which in particular may also be real, should be selected.

8.4 Exercises

Exercise 8.1. Show that if in the expressions of the kinetic and potential energies (8.13) of the conservative system the coefficients are constant and related as

$$c_{ik} = \lambda a_{ik}, \; i, k = 1, \ldots, n, \; \lambda > 0,$$

then oscillations with only one frequency $\omega = \sqrt{\lambda}$ are possible in this system.

Exercise 8.2. An inhomogeneous disk of radius R and mass M, the center of mass of which is located at a distance a from its geometric center O, can roll without slipping along the horizontal guide x (see Fig. 8.6). The moment of inertia of the disk relative to the axis perpendicular to its plane and passing through the center of mass is equal to J. Show that small oscillations of the system near a stable equilibrium are given by the formula

$$\theta(t) = \theta_0 \cos \omega t + \frac{\dot{\theta}_0}{\omega} \sin \omega t,$$

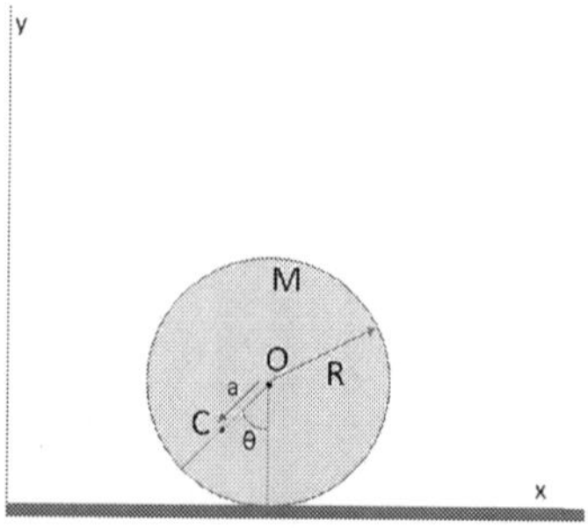

Figure 8.6 An inhomogeneous disk rolling without slipping along the horizontal guide.

with

$$\omega = \sqrt{\frac{Mga}{M(R-a)^2 + J}}.$$

Exercise 8.3. A double mathematical pendulum is suspended from a bar of mass M that can move along a smooth horizontal guide (see Fig. 8.7), such that

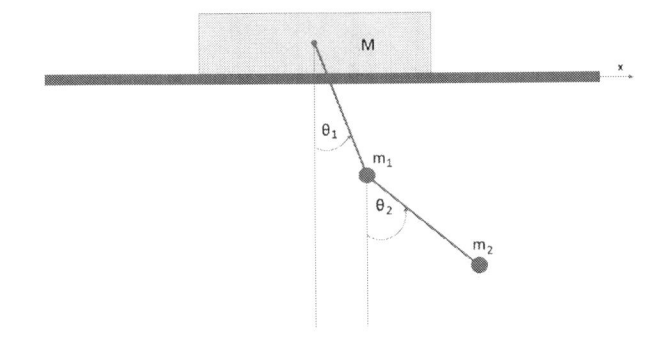

$$m_1 = m_2 = M/2, \quad l_1 = l_2 = l.$$

Figure 8.7 A double mathematical pendulum suspended from a bar.

Show that small oscillations of the system are described by the expression

$$\begin{pmatrix} x \\ l\theta_1 \\ l\theta_2 \end{pmatrix} = \left(\gamma_1^{(1)} + \gamma_2^{(2)} t\right) \begin{pmatrix} 1 \\ 0 \\ 0 \end{pmatrix} + \gamma_1^{(2)} \begin{pmatrix} 1 \\ -3 \\ -2 \end{pmatrix} \sin\left(\sqrt{\frac{g}{l}} t + \gamma_2^{(2)}\right) + \\ \gamma_1^{(3)} \begin{pmatrix} 1 \\ -4 \\ 4 \end{pmatrix} \sin\left(2\sqrt{\frac{g}{l}} t + \gamma_2^{(3)}\right).$$

Exercise 8.4. A homogeneous elliptical cylinder can roll without slipping on a horizontal plane (see Fig. 8.8). The major and minor semi-axes of the ellipse in the section of the cylinder are equal to a and b, respectively. Show that the period T of small oscillations of the cylinder near a stable equilibrium is

$$T = \pi \sqrt{\frac{b}{g} \frac{a^2 + b^2}{a^2 - b^2}}.$$

Exercise 8.5. Show that the law of change in time of charges in the electric circuit, shown in Fig. 8.9, is described by the expression

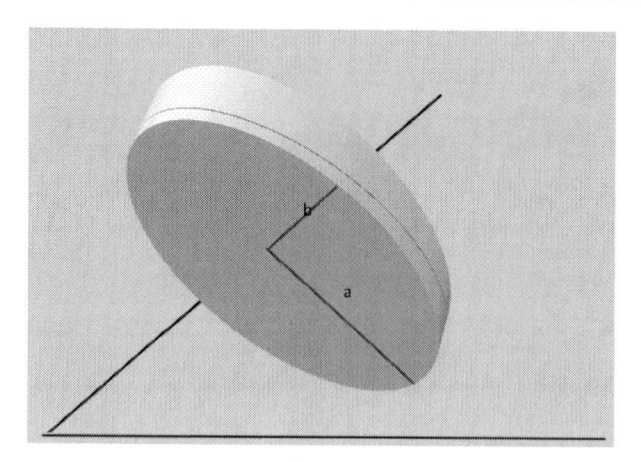

Figure 8.8 A homogeneous elliptical cylinder rolling without slipping on a horizontal plane.

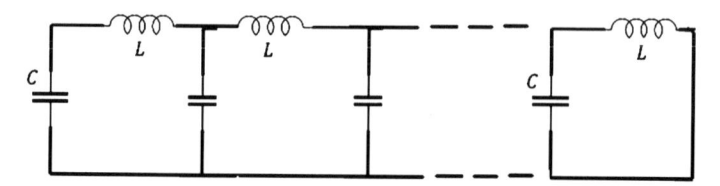

Figure 8.9 Electric circuit.

$$q_i(t) = \sum_{j=1}^{n} A_j \sin\left(\frac{i(2j-1)}{2n+1}\pi\right) \sin\left(\frac{2t}{\sqrt{LC}}\left(\frac{i(2j-1)}{2n+1}\pi\right) + \alpha_j\right),$$

where A_j and α_j, some constants, depend on the initial conditions.

Linear systems of second order

Contents

9.1 Models governed by second order differential equations   273
9.2 Frequency response   274
9.3 Examples   277
     9.3.1 Three-variable systems   277
     9.3.2 Electrical circuit   279
     9.3.3 Linear system with input delay   281
     9.3.4 Mechanical system with friction   282
     9.3.5 Electric circuit with variable elements   284
9.4 Asymptotic stability   286
     9.4.1 Algebraic criteria   286
     9.4.2 Geometric criteria of asymptotic stability   294
9.5 Polynomial robust stability   304
     9.5.1 Parametric uncertainty and robust stability   304
     9.5.2 The Kharitonov theorem   305
9.6 Exercises   308

This chapter continues the study of the linear systems obtained from the linearization process of the Lagrange equations, which was introduced in the previous chapter. This continuation covers two aspects: first, the consideration of non-potential time-dependent forces allows the use of the important Fourier transformation tool, which leads to the consideration of the system's frequency response; second, dissipative systems are considered, which generalize those of the conservative type and allow the introduction of the concept of asymptotically stable equilibrium as an extension of the previously discussed idea of equilibrium. The algebraic and geometric criteria of asymptotic stability are considered in detail. The polynomial robust stability analysis is also presented.

9.1 Models governed by second order differential equations

The main objective of this chapter is the study of the solutions of the following type of equations:

$$A\ddot{\mathbf{q}} + B\dot{\mathbf{q}} + C\mathbf{q} = \mathbf{f}(t), \tag{9.1}$$

where $\mathbf{q} \in R^n$ and $A, B, C \in R^{n \times n}$ are constant matrices and the function $\mathbf{f} : [0, \infty) \to R^n$ characterizes an external input (or perturbation) of the system under the consideration.

Remark 9.1. Through a development similar to that followed in Chapter 8 if

$$\|\Delta \mathbf{q}(t)\| \ll 1, \quad \|\Delta \dot{\mathbf{q}}(t)\| \ll 1$$

and the system under study is stationary and the potential energy is too, then the Lagrange equation can be approximated around the equilibrium $\mathbf{q}^* = \mathbf{0}$ by

$$A\ddot{\mathbf{q}} + B\dot{\mathbf{q}} + C\mathbf{q} = \mathbf{Q}_{non\text{-}pot}(t, \mathbf{q}, \dot{\mathbf{q}}) - \nabla_{\Delta \dot{\mathbf{q}}} \mathbf{Q}_{non\text{-}pot}(\mathbf{0}, \mathbf{0}, \mathbf{0}), \tag{9.2}$$

where

$$\left.\begin{aligned}
A &:= \nabla_{\mathbf{q}}^2 T(\mathbf{0}, \mathbf{0}), \\
B &:= -\nabla_{\Delta \dot{\mathbf{q}}} \mathbf{Q}_{non\text{-}pot}(\mathbf{0}, \mathbf{0}, \mathbf{0}), \\
C &:= \nabla_{\mathbf{q}}^2 V(\mathbf{0}),
\end{aligned}\right\}$$

with $A = A^T > 0$ and $C = C^T$. Therefore, if $\mathbf{Q}_{non\text{-}pot}(t, \mathbf{q}, \dot{\mathbf{q}})$ does not depend on the positions $\mathbf{q}$ and the speeds $\dot{\mathbf{q}}$, then (9.2) has the form (9.1).

9.2 Frequency response

One of the most powerful techniques for studying the response of forced linear systems with constant coefficients (of which (9.1) is only one subclass) is the Fourier transformation.

Definition 9.1. Given the vector function $\mathbf{g} : [0, \infty) \to R^n$, if

$$\mathbf{G}(j\omega) = \mathfrak{F}\{\mathbf{g}\} := \int_0^\infty e^{-j\omega t} \mathbf{g}(t)\, dt, \quad j := \sqrt{-1},$$

exists, i.e., it is finite, the complex vector function $\mathbf{G} : \mathfrak{R} \to \mathbb{C}^n$ is referred to as the **Fourier transformation** of $\mathbf{g}$. From $\mathbf{G}(j\omega)$, the function $\mathbf{g}(t)$ is recovered by

$$\mathbf{g}(t) = \mathfrak{F}^{-1}\{\mathbf{G}\} := \frac{1}{2\pi} \int_{-\infty}^\infty e^{j\omega t} \mathbf{G}(j\omega)\, d\omega.$$

Reasonably, $\mathbf{g}(t)$ is called the **inverse Fourier transformation** of $\mathbf{G}(j\omega)$.

Remark 9.2. The following **Dirichlet's conditions** guarantee the existence of $\mathfrak{F}\{\mathbf{g}\}$:
1. integrability of $\mathbf{g}$, that is, $\mathbf{g}$ has a countable number of discontinuities and extremes;

2. absolute convergence of $e^{-j\omega t}\mathbf{g}(t)$, namely,

$$\int_0^\infty \left\| e^{-j\omega t}\mathbf{g}(t) \right\| dt < \infty,$$

which, given that

$$\left\| e^{-j\omega t}\mathbf{g}(t) \right\| = \left| e^{-j\omega t} \right| \left\| \mathbf{g}(t) \right\|$$

and since

$$\left| e^{-j\omega t} \right| = 1,$$

is satisfied if

$$\int_0^\infty \left\| \mathbf{g}(t) \right\| dt < \infty,$$

that is, $\mathbf{g}$ is absolutely convergent, which implies that

$$\mathbf{g}(t) \to 0 \text{ if } t \to \infty, \tag{9.3}$$

except some points $\{t_k\}_{k=1,2,\ldots}$, where $\mathbf{g}(t_k) \underset{k\to\infty}{\not\to} 0$ and the set of such moments is construed as the set of zero-measure on the time axis.

Two properties of the Fourier transform that will be necessary are listed below. The test of the first is elementary and is omitted.

Property 1: Let $\mathbf{g}: [0, \infty) \to R^n$ and $\mathbf{h}: [0, \infty) \to R^n$ be functions with the Fourier transformations $\mathfrak{F}\{\mathbf{g}\}$ and $\mathfrak{F}\{\mathbf{h}\}$, respectively. Then the following property holds:

$$\mathfrak{F}\{A\mathbf{g} + B\mathbf{h}\} = A\mathfrak{F}\{\mathbf{g}\} + B\mathfrak{F}\{\mathbf{h}\},$$

where $A \in R^{n\times n}$ and $B \in R^{n\times n}$ are any constant matrices.

Property 2: If $\mathbf{g}: [0, \infty) \to R^n$ is differentiable and meets the conditions of Dirichlet, then $\mathfrak{F}\{\dot{\mathbf{g}}\}$ exists and

$$\mathfrak{F}\{\dot{\mathbf{g}}\} = j\omega \mathfrak{F}\{\mathbf{g}\} - \mathbf{g}(0). \tag{9.4}$$

Proof. By the definition

$$\mathfrak{F}\{\dot{\mathbf{g}}\} = \int_0^\infty e^{-j\omega t}\dot{\mathbf{g}}(t)\,dt = \int_0^\infty e^{-j\omega t}\,d\mathbf{g}(t),$$

integration by parts leads to the following relation:

$$\mathfrak{F}\{\dot{\mathbf{g}}\} = \left. e^{-j\omega t}\mathbf{g}(t) \right|_0^\infty + j\omega \int_0^\infty e^{-j\omega t}\mathbf{g}(t)\,dt.$$

In view of (9.3) the property (9.4) follows. □

Remark 9.3. By an induction process it is concluded that if $\mathbf{g}$ is k times differentiable and $\mathfrak{F}(\mathbf{g})$ exists, then $\mathfrak{F}\left\{\mathbf{g}^{(k)}\right\}$ exists too and it is equal to

$$\mathfrak{F}\left\{\mathbf{g}^{(k)}\right\} = (j\omega)^k \mathfrak{F}\{\mathbf{g}\} - (j\omega)^{k-1} \mathbf{g}(0) - \cdots - j\omega \mathbf{g}^{(k-1)}(0) - \mathbf{g}^{(k-1)}(0).$$

Definition 9.2. Consider the input-output system shown in Fig. 9.1, where

$$\mathbf{f} : [0, \infty) \to R^k, \quad \mathbf{q} : [0, \infty) \to R^n,$$

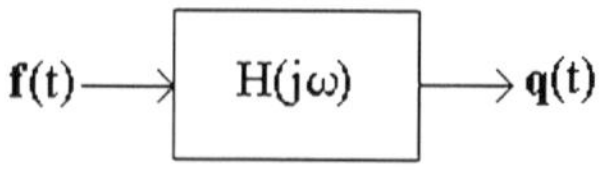

$$\mathbf{f}(t) \longrightarrow \boxed{H(j\omega)} \longrightarrow \mathbf{q}(t)$$

Figure 9.1 Input-output system.

and

$$H(j\omega) : \mathbb{C} \to \mathbb{C}^k.$$

The matrix $H(j\omega)$ is called the **frequency characteristic matrix** if it connects the Fourier transformations $\mathfrak{F}\{\mathbf{q}(t)\}$ and $\mathfrak{F}\{\mathbf{f}(t)\}$ of $\mathbf{q}(t)$ and $\mathbf{f}(t)$ as

$$\mathfrak{F}\{\mathbf{q}(t)\} = H(j\omega) \mathfrak{F}\{\mathbf{f}(t)\}$$

with zero initial conditions.

Remark 9.4. The component

$$H_{rs}(j\omega), \quad r = 1, \ldots, n, \quad s = 1, \ldots, k,$$

of the matrix $H(j\omega)$ matches the entrance $f_s(t)$ and the exit $q_r(t)$. In particular, if

$$f_s(t) = \sin(\omega t) \text{ and } f_i(t) = 0 \; \forall i \neq s,$$

then

$$q_r(t) = |H_{rs}(j\omega)| \sin(\omega t + \arg H_{rs}(j\omega)),$$

that is, $|H_{rs}(j\omega)|$ is the amplification and $\arg H_{rs}(j\omega)$ is the **phase shift** of the signal $q_r(t)$ with respect to $f_s(t)$, both relative to the frequency ω.

Lemma 9.1. *For systems of the type (9.1), we have*

$$H(j\omega) = \left[-\omega^2 A + j\omega B + C \right]^{-1}. \tag{9.5}$$

Proof. The Fourier transform of (9.1) results in

$$\mathfrak{F}\{A\ddot{\mathbf{q}} + B\dot{\mathbf{q}} + C\mathbf{q}\} = \mathfrak{F}\{\mathbf{f}(t)\},$$

ated above,

$$\begin{aligned}&\mathfrak{F}\{A\ddot{\mathbf{q}}+B\dot{\mathbf{q}}+C\mathbf{q}\}=\\&(j\omega)^2\,A\mathfrak{F}\{\mathbf{q}\}+(j\omega)\,B\mathfrak{F}\{\mathbf{q}\}+C\mathfrak{F}\{\mathbf{q}\}=\\&\left[-\omega^2 A+j\omega B+C\right]\mathfrak{F}\{\mathbf{q},\}\end{aligned}$$

from which (9.5) follows. □

9.3 Examples

Below are some examples that make use of the results obtained.

9.3.1 Three-variable systems

Example 9.1. Consider the system described by the following set of equations:

$$\left.\begin{aligned}\dot{x}&=-2x+y-z+f_1,\\\dot{y}&=x-y+f_2,\\\dot{z}&=x+y-z+f_3.\end{aligned}\right\} \quad (9.6)$$

Let us find the frequency characteristic matrix and determine the effect of the input f_3 on the output y. Let

$$\mathbf{q}:=\begin{bmatrix}x & y & z\end{bmatrix}^T.$$

Then (9.6) may be rewritten as

$$\dot{\mathbf{q}} = D\mathbf{q} + \mathbf{f}, \quad (9.7)$$

with

$$D:=\begin{bmatrix}-2 & 1 & -1\\ 1 & -1 & 0\\ 1 & 1 & -1\end{bmatrix},\quad \mathbf{f}:=\begin{bmatrix}f_1\\ f_2\\ f_3\end{bmatrix}.$$

Applying the Fourier transformation to (9.7), we get

$$(j\omega I_{3\times 3} - D)\,\mathfrak{F}\{\mathbf{q}\} = \mathfrak{F}\{\mathbf{f}\}.$$

Therefore

$$H(j\omega) = (j\omega I_{3\times 3} - D)^{-1},$$

or, in the open format,

$$H(j\omega) = \frac{\begin{bmatrix} (1+j\omega)^2 & j\omega & -(1+j\omega) \\ 1+j\omega & 3-\omega^2+j3\omega & -1 \\ 2+j\omega & 3+j\omega & 1-\omega^2+j3\omega \end{bmatrix}}{3-4\omega^2+j(5\omega-\omega^3)}. \qquad (9.8)$$

So, the influence of the input f_3 on y is described by

$$H_{23}(j\omega) = \frac{1}{-3+4\omega^2 - j(5\omega-\omega^3)} = \frac{-3+4\omega^2 + j(5\omega-\omega^3)}{(-3+4\omega^2)^2 + (5\omega-\omega^3)^2}.$$

The graphical representation of $H_{23}(j\omega)$, with $\mathrm{Re}(H_{23}(j\omega))$ on the horizontal axis, $\mathrm{Im}(H_{23}(j\omega))$ on the vertical axis, and ω as a parameter, is called **amplitude-phase characteristic** (or **Nyquist hodograph**),[1] and it is shown in Fig. 9.2. The point corresponding to $\omega = 0$ is the intersection with the horizontal axis at $(-1/3)$ and hence the positive branch of the hodograph moves (with increasing ω) to the positive part of the vertical axis, while the negative one does so in the other direction symmetrically. The magnitude of $H_{23}(j\omega)$ is given by

$$|H_{23}(j\omega)| = \frac{1}{\sqrt{(-3+4\omega^2)^2 + (5\omega-\omega^3)^2}}.$$

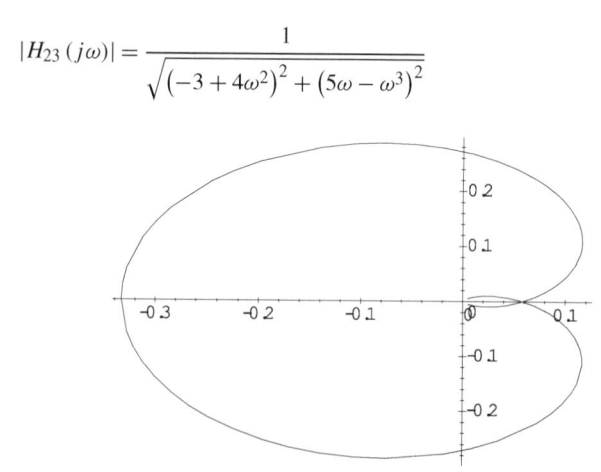

Figure 9.2 Hodograph of H_{23}.

[1] Amplitude-phase characteristic (Nyquist hodograph) – graphic display for all frequencies of the spectrum of the relations of the output signal of a stationary linear system to the input, presented in a complex form. The value of the segment from the origin to each point of the hodograph shows how many times at a given frequency the output signal is greater than the input, and the phase shift between the signals is determined by the angle to the said segment.

Linear systems of second order

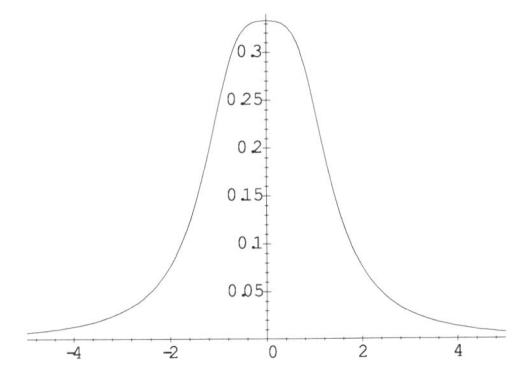

Figure 9.3 Amplitude characteristic diagram of H_{23}.

and its graphical representation as a function of ω, called **amplitude characteristic diagram**, is shown in Fig. 9.3. Finally, in Fig. 9.4 the **phase characteristic diagram** of $H_{23}(j\omega)$ is depicted, which is given by

$$\arg H_{23}(j\omega) = \arctan \frac{\operatorname{Im} H_{23}(j\omega)}{\operatorname{Re} H_{23}(j\omega)} = \arctan \frac{5\omega - \omega^3}{-3 + 4\omega^2}.$$

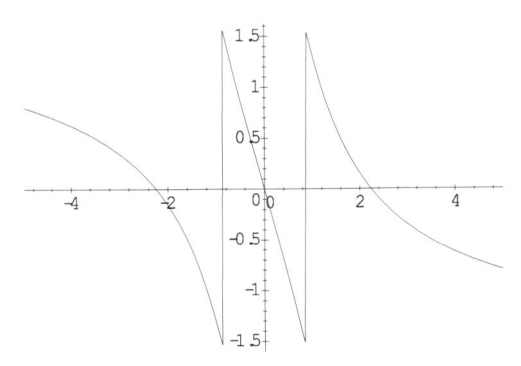

Figure 9.4 Phase characteristic diagram of H_{23}.

9.3.2 Electrical circuit

Example 9.2. Consider the electrical circuit shown in Fig. 9.5. Obtain the amplitude-phase characteristic matrix and determine the effect that the input e has on the output u_{C_2} shown. The first step is to obtain the dynamic equations of the system. To do that they are considered as generalized coordinates,

$$q_i(t) = q_i(0) + \int_{\tau=0}^{t} i_i(\tau)\,d\tau, \quad i = 1, 2,$$

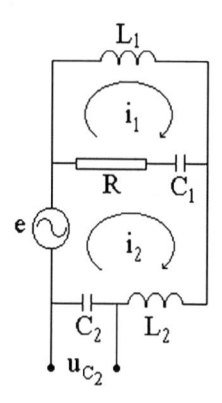

Figure 9.5 Electrical circuit.

where i_i is the current flowing through the loop i. From the table of electromechanical analogies of Chapter 6, for the kinetic and potential energies we have

$$T(\dot{\mathbf{q}}) = \frac{1}{2}L_1\dot{q}_1^2 + \frac{1}{2}L_2\dot{q}_2^2,$$

$$V(\mathbf{q}) = \frac{1}{2C_1}(q_1 - q_2)^2 + \frac{1}{2C_2}q_2^2.$$

Therefore

$$L(\mathbf{q}, \dot{\mathbf{q}}) = T(\dot{\mathbf{q}}) - V(\mathbf{q}) = \frac{1}{2}L_1\dot{q}_1^2 + \frac{1}{2}L_2\dot{q}_2^2 - \frac{1}{2C_1}(q_1 - q_2)^2 - \frac{1}{2C_2}q_2^2,$$

so the following Lagrange equations result:

$$L_1\ddot{q}_1 + \frac{1}{C_1}(q_1 - q_2) = -R(\dot{q}_1 - \dot{q}_2),$$

$$L_2\ddot{q}_2 + \frac{1}{C_1}(q_2 - q_1) + \frac{1}{C_2}q_2 = e - R(\dot{q}_2 - \dot{q}_1),$$

or equivalently,

$$\begin{bmatrix} L_1 & 0 \\ 0 & L_2 \end{bmatrix}\ddot{\mathbf{q}} + R\begin{bmatrix} 1 & -1 \\ -1 & 1 \end{bmatrix}\dot{\mathbf{q}} + \begin{bmatrix} \frac{1}{C_1} & -\frac{1}{C_1} \\ -\frac{1}{C_1} & \frac{1}{C_1} + \frac{1}{C_2} \end{bmatrix}\mathbf{q} = \begin{bmatrix} 0 \\ e \end{bmatrix}.$$

Then, in view of (9.5), it follows that

$$H(j\omega) = \begin{bmatrix} -\omega^2 L_1 + j\omega R + \frac{1}{C_1} & -j\omega R - \frac{1}{C_1} \\ -j\omega R - \frac{1}{C_1} & -\omega^2 L_2 + j\omega R + \frac{1}{C_1} + \frac{1}{C_2} \end{bmatrix}^{-1} =$$

$$\frac{1}{a(\omega)} \begin{bmatrix} -\omega^2 L_2 C_1 C_2 + C_2 + C_1 + j\omega R C_1 C_2 & (1+j\omega R C_1) C_2 \\ (1+j\omega R C_1) C_2 & (-\omega^2 L_1 C_1 + 1 + j\omega R C_1) C_2 \end{bmatrix},$$

where

$$a(\omega) := \omega^4 L_1 C_1 L_2 C_2 - \omega^2 (L_1 C_1 + (L_1 + L_2) C_2) + 1 +$$
$$j\omega R C_1 \left(-\omega^2 (L_1 + L_2) C_2 + 1\right).$$

The output u_{C_2} is obtained from q_2 by the relation

$$u_{C_2} = \frac{q_2}{C_2},$$

whose Fourier transform is

$$\mathfrak{F}\{u_{C_2}\} = \frac{1}{C_2} \mathfrak{F}\{q_2\}.$$

But

$$\mathfrak{F}\{q_2\} = H_{22}(j\omega) \mathfrak{F}\{e\},$$

which gives

$$\mathfrak{F}\{u_{C_2}\} = \frac{H_{22}(j\omega)}{C_2} \mathfrak{F}\{e\} = \frac{-\omega^2 L_1 C_1 + 1 + j\omega R C_1}{a(\omega)} \mathfrak{F}\{e\},$$

from which it follows that $\dfrac{H_{22}(j\omega)}{C_2}$ determines the effect that e has on u_{C_2}, which can be seen from the hodograph and the amplitude and phase characteristic graphs once the values of the components present in the circuit are known.

9.3.3 Linear system with input delay

Example 9.3. Fig. 9.6 shows a linear system whose output is the input signal delayed at the time τ. Let us obtain $H(j\omega)$ of the system as well as the hodograph and the amplitude and phase characteristic graphs. By Definition 9.2 we have

$$H(j\omega) = \frac{\mathfrak{F}\{q(t)\}}{\mathfrak{F}\{f(t)\}},$$

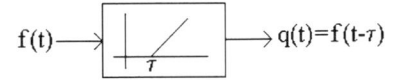

Figure 9.6 Linear system with input delay.

where

$$\mathfrak{F}\{q(t)\} = \mathfrak{F}\{f(t-\tau)\}.$$

Recall that

$$\mathfrak{F}\{f(t-\tau)\} = \int_0^\infty e^{-j\omega t} f(t-\tau) dt.$$

Using the new variable

$$s := t - \tau,$$

we have

$$\mathfrak{F}\{f(t-\tau)\} = \int_{-\tau}^\infty e^{-j\omega(s+\tau)} f(s) ds =$$

$$e^{-j\omega\tau} \left(\int_{-\tau}^0 e^{-j\omega s} f(s) ds + \int_0^\infty e^{-j\omega s} f(s) ds \right).$$

Considering that $f(t) = 0$ for any $t < 0$, we get

$$\mathfrak{F}\{f(t-\tau)\} = e^{-j\omega\tau} \mathfrak{F}\{f(t)\},$$

which implies

$$H(j\omega) = e^{-j\omega\tau}.$$

Fig. 9.7 shows the corresponding hodograph. The point for $\omega = 0$ is at (1.0) and the positive branch of ω describes a circle of radius 1 clockwise; the negative branch is symmetric with respect to the horizontal axis. On the other hand, since

$$|H(j\omega)| \equiv 1$$

the amplitude characteristic diagram is constant and equals 1. Finally, as

$$\arg H(j\omega) = -\omega\tau,$$

the characteristic phase diagram is a line of slope $-\tau$ that passes through the origin.

9.3.4 Mechanical system with friction

Example 9.4. Fig. 9.8 shows a system of two masses m joined by stiffness springs k. One of the masses is submerged in water, and both may be subject to the action of external forces. Considering that the frictional force acting on the submerged mass is proportional to its velocity, let us determine the frequency characteristic matrix. In order to obtain the Lagrange equations, consider the generalized coordinates of the

Linear systems of second order 283

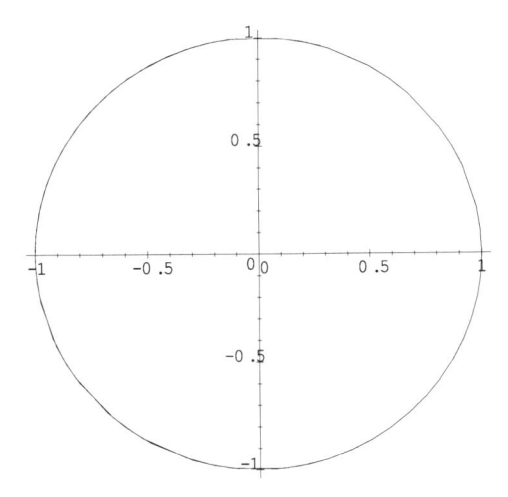

Figure 9.7 Hodograph of $e^{-j\omega\tau}$.

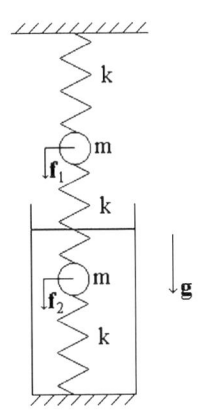

Figure 9.8 Mechanical system with friction.

displacements q_1 and q_2 of the upper and lower masses, respectively, in relation to the positions in which the springs have their natural lengths. Then the following kinetic and potential energies result:

$$T(\dot{\mathbf{q}}) = \frac{1}{2}m\left(\dot{q}_1^2 + \dot{q}_2^2\right),$$
$$V(\mathbf{q}) = \frac{1}{2}k\left[q_1^2 + (q_2 - q_1)^2 + q_2^2\right] - mg\left(2q_1 + 2q_{1,0} + q_2 + q_{2,0}\right),$$

where $q_{1,0}$ and $q_{2,0}$ represent the natural lengths of the upper and intermediate springs, respectively. It follows that the Lagrangian L has as an expression

$$L(\mathbf{q}, \dot{\mathbf{q}}) = \frac{1}{2}m\left(\dot{q}_1^2 + \dot{q}_2^2\right) - \frac{1}{2}k\left[q_1^2 + (q_2 - q_1)^2 + q_2^2\right] + mg(2q_1 + 2q_{1,0} + q_2 + q_{2,0}),$$

and therefore the Lagrange equations are given by

$$\begin{aligned} m\ddot{q}_1 + k(2q_1 - q_2) - 2mg &= f_1, \\ m\ddot{q}_2 + k(2q_2 - q_1) - mg &= f_2 - \beta\dot{q}_2, \end{aligned} \quad (9.9)$$

where f_1 and f_2 are the external actions on the upper and lower masses, respectively, and β is the coefficient of friction between the submerged mass and the water. In vector notation, (9.9) may be rewritten as

$$m\begin{bmatrix}1 & 0 \\ 0 & 1\end{bmatrix}\ddot{\mathbf{q}} + \beta\begin{bmatrix}0 & 0 \\ 0 & 1\end{bmatrix}\dot{\mathbf{q}} + k\begin{bmatrix}2 & -1 \\ -1 & 2\end{bmatrix}\mathbf{q} = \begin{bmatrix}f_1 + 2mg \\ f_2 + mg\end{bmatrix},$$

and therefore expression (9.5) becomes

$$H(j\omega) = \frac{1}{a(\omega)}\begin{bmatrix}-\omega^2 m + 2k + i\omega\beta & k \\ k & -\omega^2 m + 2k\end{bmatrix},$$

with

$$a(\omega) := \omega^4 m^2 - 4\omega^2 mk + 3k^2 + j\beta\omega\left(-\omega^2 m + 2k\right).$$

9.3.5 Electric circuit with variable elements

Example 9.5. A circuit with some variable-value elements is shown in Fig. 9.9. The system output is u_{C_1}, which is the voltage at capacitor C_1. We need to determine the values of C_{12} and L_2 such that $u_{C_1} = 0$ when

$$e = e_0 \sin(\omega_0 t + \varphi).$$

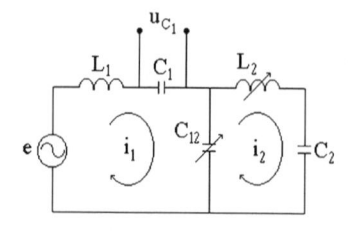

Figure 9.9 Electric circuit with variable elements.

Linear systems of second order

Defining the generalized coordinates as

$$q_i(t) = \int_{\tau=0}^{t} i_i(\tau)\,d\tau, \quad i=1,2,$$

where i_i is the current in the i-th loop, from the table of electromagnetic analogies in Chapter 6, the following kinetic and potential energies result:

$$T(\dot{\mathbf{q}}) = \frac{1}{2}L_1\dot{q}_1^2 + \frac{1}{2}L_2\dot{q}_2^2,$$

$$V(\mathbf{q}) = \frac{1}{2C_1}q_1^2 + \frac{1}{2C_{12}}(q_1-q_2)^2 + \frac{1}{2C_2}q_2^2,$$

so

$$L(\mathbf{q},\dot{\mathbf{q}}) = T - V = \frac{1}{2}L_1\dot{q}_1^2 + \frac{1}{2}L_2\dot{q}_2^2 -$$

$$\frac{1}{2C_1}q_1^2 - \frac{1}{2C_{12}}(q_1-q_2)^2 - \frac{1}{2C_2}q_2^2.$$

Therefore the Lagrange equations are given by

$$L_1\ddot{q}_1 + \frac{1}{C_1}q_1 + \frac{1}{C_{12}}(q_1-q_2) = e,$$

$$L_2\ddot{q}_2 + \frac{1}{C_{12}}(q_2-q_1) + \frac{1}{C_2}q_2 = 0,$$

or in vector format,

$$\begin{bmatrix} L_1 & 0 \\ 0 & L_2 \end{bmatrix}\ddot{\mathbf{q}} + \begin{bmatrix} \dfrac{1}{C_1}+\dfrac{1}{C_{12}} & -\dfrac{1}{C_{12}} \\ -\dfrac{1}{C_{12}} & \dfrac{1}{C_{12}}+\dfrac{1}{C_2} \end{bmatrix}\mathbf{q} = \begin{bmatrix} e \\ 0 \end{bmatrix}.$$

The representation of (9.5) is

$$H(j\omega) = \begin{bmatrix} -\omega^2 L_1 + \dfrac{1}{C_1} + \dfrac{1}{C_{12}} & -\dfrac{1}{C_{12}} \\ -\dfrac{1}{C_{12}} & -\omega^2 L_2 + \dfrac{1}{C_{12}} + \dfrac{1}{C_2} \end{bmatrix}^{-1}$$

$$= \frac{1}{a(\omega)}\begin{bmatrix} -\omega^2 L_2 + \frac{1}{C_{12}} + \frac{1}{C_2} & \frac{1}{C_{12}} \\ \frac{1}{C_{12}} & -\omega^2 L_1 + \frac{1}{C_1} + \frac{1}{C_{12}} \end{bmatrix},$$

where

$$a(\omega) := \left(-\omega^2 L_1 + \frac{1}{C_1} + \frac{1}{C_{12}}\right)\left(-\omega^2 L_2 + \frac{1}{C_{12}} + \frac{1}{C_2}\right) - \frac{1}{C_{12}^2}.$$

Since

$$u_{C_1} = \frac{q_1}{C_1},$$

the Fourier transformation leads to

$$\mathfrak{F}\{u_{C_1}\} = \frac{1}{C_1}\mathfrak{F}\{q_1\}.$$

But since

$$\mathfrak{F}\{q_1\} = H_{11}(j\omega)\mathfrak{F}\{e\},$$

we get

$$\mathfrak{F}\{u_{C_1}\} = \frac{H_{11}(j\omega)}{C_1}\mathfrak{F}\{e\},$$

from which it follows that $u_{C_1} = 0$ if $|H_{11}(j\omega_0)| = 0$, which gives

$$\frac{1}{L_2}\left(\frac{1}{C_{12}} + \frac{1}{C_2}\right) = \omega_0^2,$$

which is equivalent to the natural frequency of the second loop being equal to the excitation frequency.

9.4 Asymptotic stability

9.4.1 Algebraic criteria

This subsection studies the stability of the equilibrium positions of (9.1). Of what has been said in Chapter 7, it follows that $\mathbf{q}^*$ is an equilibrium of (9.1) if and only if at that point $\mathbf{f} \equiv \mathbf{0}$, and therefore the system to study is reduced to

$$A\ddot{\mathbf{q}} + B\dot{\mathbf{q}} + C\mathbf{q} = 0. \tag{9.10}$$

In particular, we will focus on *dissipative systems*, that is, those in which

$$B \neq 0.$$

From the definition of equilibrium, given in Chapter 7, (9.10) has only one equilibrium point and it is

$$\mathbf{q}^* = \mathbf{0}. \tag{9.11}$$

Definition 9.3. The equilibrium (9.11) is said to be asymptotically stable if it is stable and

$$\lim_{t \to \infty} \|\mathbf{q}(t)\| = 0, \quad \lim_{t \to \infty} \|\dot{\mathbf{q}}(t)\| = 0.$$

Remark 9.5. Fig. 9.10 illustrates in a phase diagram the condition imposed at $\mathbf{q}^* = \mathbf{0}$ to be asymptotically stable.

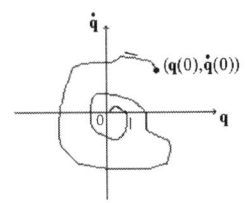

Figure 9.10 Illustration of the asymptotic stability concept.

Remember from Chapter 8 that if $\{\lambda_k\}_{k=1}^{K}$ are the different K roots of the characteristic polynomial

$$p(\lambda) := \det\left(\lambda^2 A + \lambda B + C\right), \tag{9.12}$$

the general solution of (9.10) is given by

$$\mathbf{q}(t) = \sum_{k=1}^{K} \mathbf{q}_k(t),$$

with

$$\mathbf{q}_k(t) = e^{\lambda_k t} \sum_{i=1}^{n_k} \sum_{j=1}^{n_{k,i}} \gamma_{k,i,j} t^{j-1} p_i(t) \mathbf{u}_{k,i}, \tag{9.13}$$

where n_k is the geometric multiplicity of the root λ_k and $\{\mathbf{u}_{k,i}\}_{i=1}^{n_k}$ are the amplitude vectors corresponding to this root and given by the expression

$$(\lambda_k A + \lambda_k B + C)\mathbf{u}_k = \mathbf{0},$$

while $p_i(t)$ is a polynomial in t of degree less than $n_{k,i}$, which is the number of repetitions of the root λ_k corresponding to the vector of amplitudes $\mathbf{u}_{k,i}$.

General criterion of asymptotic stability

Lemma 9.2 (Asymptotic stability criterion). *Let $\{\lambda_k\}_{k=1}^{K}$ be the set of different roots of $p(\lambda)$. The equilibrium $\mathbf{q}^* = \mathbf{0}$ (9.10) is **asymptotically stable** if and only if*

$$\text{Re}(\lambda_k) < 0 \quad \forall k = 1, ..., K.$$

Proof. In view of the form (9.13), $e^{\lambda_k t}$ appears multiplying a polynomial and since

$$\exp(\lambda_k t) = \exp\left[(\text{Re}(\lambda_k) + j\,\text{Im}(\lambda_k))t\right] =$$

$$\exp(j \operatorname{Im}(\lambda_k)t) \cdot \exp(\operatorname{Re}(\lambda_k)t) =$$
$$[\cos(\operatorname{Im}(\lambda_k)t) + j\sin(\operatorname{Im}(\lambda_k)t)]\exp(\operatorname{Re}(\lambda_k)t).$$

The result follows trivially. □

The previous theorem allows to translate the problem of the determination of the stability of the system (9.10) to the qualification of the sign of the roots of the corresponding characteristic polynomial. The following theorem allows us to go further.

Necessity condition of Stodola

Theorem 9.1 (Necessity condition: Stodola). *If the equilibrium* $\mathbf{q}^* = \mathbf{0}$ *of (9.10) is asymptotically stable, then all the coefficients of the characteristic polynomial* $p(\lambda)$ *are strictly positive.*[2]

Proof. More generally, let $\{\lambda_i\}_{i=1}^m$ be the root set of the polynomial of degree m

$$p(\lambda) = \lambda^m \rho_0 + \lambda^{m-1}\rho_1 + \ldots + \rho_m, \quad \rho_i \in R,$$

or in the equivalent form

$$p(\lambda) = \rho_0(\lambda - \lambda_1)(\lambda - \lambda_2)\cdots(\lambda - \lambda_m). \tag{9.14}$$

Suppose that $p(\lambda)$ has n_r real and n_c complex roots, of which the latter appear in conjugate pairs because $p(\lambda)$ has real coefficients. Without loss of generality, suppose that in (9.14) the first n_r factors correspond to real roots. Then (9.14) can be rewritten in the following form:

$$p(\lambda) = \rho_0 \prod_{i=1}^{n_r}(\lambda - \lambda_i) \prod_{k=1}^{n_c/2}(\lambda - \lambda_k)(\lambda - \bar{\lambda}_k), \tag{9.15}$$

with

$$\lambda_i = -u_i, \quad i = 1, \ldots, n_r,$$
$$\lambda_k = -u_k + jv_k, \quad k = 1, \ldots, n_c/2,$$
$$\bar{\lambda}_k = -u_k - jv_k, \quad k = 1, \ldots, n_c/2,$$

where, by Lemma 9.2,

$$u_i > 0, \ i = 1, \ldots, n_r, \text{ and } u_k > 0, \ k = 1, \ldots, n_c/2.$$

---

[2] Aurel Boleslav Stodola (May 11, 1859–December 25, 1942) was a Slovak engineer, physicist, and inventor. He was a pioneer in the area of technical thermodynamics and its applications and published his book Die Dampfturbine (the steam turbine) in 1903. In addition to the thermodynamic issues involved in turbine design the book discussed aspects of fluid flow, vibration, stress analysis of plates, shells, and rotating discs, and stress concentrations at holes and fillets. Stodola was a professor of mechanical engineering at the Swiss Polytechnical Institute (now ETH) in Zurich. He maintained friendly contact with Albert Einstein. In 1892, Stodola founded the Laboratory for Energy Conversion.

Then
$$p(\lambda) = \rho_0 \prod_{i=1}^{n_r} (\lambda + u_i) \prod_{k=1}^{n_c/2} \left(\lambda^2 + 2u_k\lambda + u_k^2 + v_k^2\right),$$

which contains only factors with positive terms and in whose development only positive (non-zero) coefficients arise. □

An immediate consequence is obtained.

Conclusion 9.1. *If $p(\lambda)$ has different sign coefficients or some of them are zero, then the equilibrium $\mathbf{q}^* = \mathbf{0}$ of (9.10) is not asymptotically stable.*

Based on the two previous results, some techniques are now presented to know the sign of the real parts of a polynomial

$$p(\lambda) = \lambda^m \rho_0 + \lambda^{m-1} \rho_1 + \ldots + \rho_m, \quad 0 < \rho_i \in R, \tag{9.16}$$

without their analytical finding.

The Routh–Hurwitz criterion

Definition 9.4. The matrix

$$G = \begin{bmatrix} \rho_1 & \rho_3 & \rho_5 & \rho_7 & \rho_9 & \cdots & 0 \\ \rho_0 & \rho_2 & \rho_4 & \rho_6 & \rho_8 & \cdots & 0 \\ 0 & \rho_1 & \rho_3 & \rho_5 & \rho_7 & \cdots & 0 \\ 0 & \rho_0 & \rho_2 & \rho_4 & \rho_6 & \cdots & 0 \\ 0 & 0 & \rho_1 & \rho_3 & \rho_5 & \cdots & 0 \\ 0 & 0 & \rho_0 & \rho_2 & \rho_4 & \ddots & \vdots \\ \vdots & \vdots & \vdots & \vdots & \vdots & \vdots & \rho_m \end{bmatrix} \in R^{m \times m}$$

is called the **Hurwitz matrix**, associated with $p(\lambda)$.

Definition 9.5. The polynomial $p(\lambda)$ is called **Hurwitz polynomial** if

$$\mathrm{Re}(\lambda_i) < 0, \quad i = 1, \ldots, m.$$

The criterion that appears immediately without demonstration (the proof can be found in (Poznyak, 2008)) allows to determine if a polynomial is Hurwitz.[3]

---

[3] In control system theory, the Routh–Hurwitz stability criterion is a mathematical test that is a necessary and sufficient condition for the stability of a linear time-invariant (LTI) control system. The Routh test is an efficient recursive algorithm that English mathematician Edward John Routh proposed in 1876 to determine whether all the roots of the characteristic polynomial of a linear system have negative real parts. German mathematician Adolf Hurwitz independently proposed in 1895 to arrange the coefficients of the

Theorem 9.2 (The Routh–Hurwitz criterion). *The polynomial $p(\lambda)$ given in (9.16) is Hurwitzian if and only if each principle minor of* $\det(G)$ *is strictly positive, that is,*

$$m_i > 0, \quad i = 1, \ldots, n,$$

where

$$m_i := \det \begin{bmatrix} p_1 & p_3 & p_5 & \cdots & 0 \\ p_0 & p_2 & p_4 & \cdots & 0 \\ 0 & p_1 & p_3 & \cdots & 0 \\ 0 & p_0 & p_2 & \ddots & \vdots \\ \vdots & \vdots & \vdots & \vdots & p_i \end{bmatrix}. \qquad (9.17)$$

An example illustrates the above criterion.

Example 9.6. Let us determine if the polynomial

$$p(\lambda) = \lambda^4 + 8\lambda^3 + 18\lambda^2 + 16\lambda + 5 = 0$$

is Hurwitz.

The corresponding Hurwitz matrix is

$$G = \begin{bmatrix} 8 & 16 & 0 & 0 \\ 1 & 18 & 5 & 0 \\ 0 & 8 & 16 & 0 \\ 0 & 1 & 18 & 5 \end{bmatrix},$$

from where the principle minors are

$$m_1 := \det[8] = 8 > 0,$$

$$m_2 := \det \begin{bmatrix} 8 & 16 \\ 1 & 18 \end{bmatrix} = 128 > 0,$$

$$m_3 := \det \begin{bmatrix} 8 & 16 & 0 \\ 1 & 18 & 5 \\ 0 & 8 & 16 \end{bmatrix} = 1728 > 0,$$

$$m_4 := \det \begin{bmatrix} 8 & 16 & 0 & 0 \\ 1 & 18 & 5 & 0 \\ 0 & 8 & 16 & 0 \\ 0 & 1 & 18 & 5 \end{bmatrix} = 8640 > 0.$$

Hence $p(\lambda)$ is Hurwitz.

The Routh–Hurwitz criterion has a simpler formulation.

polynomial into a square matrix, called the Hurwitz matrix, and showed that the polynomial is stable if and only if the sequence of determinants of its principal submatrices are all positive. The two procedures are equivalent, with the Routh test providing a more efficient way to compute the Hurwitz determinants than computing them directly.

The Liénard–Chipart criterion

Theorem 9.3 (Liénard–Chipart criterion). *The polynomial $p(\lambda)$ given in (9.16) is Hurwitz if and only if the following conditions are satisfied:*

1. Condition on the coefficients:

$$\rho_i > 0 \text{ for all } i = 1, \ldots, m.$$

2. Condition on the main minors of the determinant of the corresponding Hurwitz matrix:

$$m_i > 0 \text{ for all } i = m - 1, m - 3, m - 5, \ldots,$$

with m_i given in (9.17).

This criterion is illustrated by the following example.

Example 9.7. Given

$$p(\lambda) = \lambda^5 + 3\lambda^4 + \alpha\lambda^3 + \lambda^2 + \lambda + 1, \qquad (9.18)$$

we need to find the values of α that guarantee that $p(\lambda)$ is Hurwitzian.

Given that Hurwitz's matrix is

$$G = \begin{bmatrix} 3 & 1 & 1 & 0 & 0 \\ 1 & \alpha & 1 & 0 & 0 \\ 0 & 3 & 1 & 1 & 0 \\ 0 & 1 & \alpha & 1 & 0 \\ 0 & 0 & 3 & 1 & 1 \end{bmatrix},$$

two conditions on the main minors of the previous criterion are met:

$$m_4 = \det \begin{bmatrix} 3 & 1 & 1 & 0 \\ 1 & \alpha & 1 & 0 \\ 0 & 3 & 1 & 1 \\ 0 & 1 & \alpha & 1 \end{bmatrix} = 4\alpha - 3\alpha^2 - 5 > 0,$$

$$m_2 = \det \begin{bmatrix} 3 & 1 \\ 1 & \alpha \end{bmatrix} = 3\alpha - 1 > 0.$$

However, since the equation

$$4\alpha - 3\alpha^2 - 5 = 0$$

has as solutions

$$\alpha = \frac{2}{3} \pm \frac{1}{3}i\sqrt{11},$$

it follows that there are no real values of α that provide the condition $m_4 < 0$ and make the polynomial $p(\lambda)$ given in (9.18) Hurwitzian.

Example 9.8. Consider two dynamic systems

$$\left.\begin{array}{l}\ddot{x} + \dot{x} + x - \alpha y = 0, \\ \ddot{y} + \dot{y} - \beta x + y = 0\end{array}\right\} \quad (9.19)$$

and

$$\left.\begin{array}{l}\ddot{x} + \dot{x} + x - \alpha y = 0, \\ \dot{y} - \beta x + y = 0.\end{array}\right\} \quad (9.20)$$

Let us calculate the values of the parameters α and β such that the corresponding characteristic polynomials are Hurwitz.

a) Define the vector

$$\mathbf{q} := \begin{bmatrix} x & y \end{bmatrix}^\mathsf{T}. \quad (9.21)$$

With (9.21), the system (9.19) may be rewritten as

$$\begin{bmatrix} 1 & 0 \\ 0 & 1 \end{bmatrix} \ddot{\mathbf{q}} + \begin{bmatrix} 1 & 0 \\ 0 & 1 \end{bmatrix} \dot{\mathbf{q}} + \begin{bmatrix} 1 & -\alpha \\ -\beta & 1 \end{bmatrix} \mathbf{q} = \mathbf{0},$$

so that

$$\lambda^2 A + \lambda B + C = \begin{bmatrix} \lambda^2 + \lambda + 1 & -\alpha \\ -\beta & \lambda^2 + \lambda + 1 \end{bmatrix},$$

and its characteristic polynomial is

$$p(\lambda) = \lambda^4 + 2\lambda^3 + 3\lambda^2 + 2\lambda + 1 - \alpha\beta. \quad (9.22)$$

The Hurwitz matrix, corresponding to (9.22), is

$$G = \begin{bmatrix} 2 & 2 & 0 & 0 \\ 1 & 3 & 1-\alpha\beta & 0 \\ 0 & 2 & 2 & 0 \\ 0 & 1 & 3 & 1-\alpha\beta \end{bmatrix},$$

so that the condition of the Liénard–Chipart criterion gives on the one hand

$$1 - \alpha\beta > 0,$$

and on the other

$$m_3 = 4 + 4\alpha\beta > 0,$$

Linear systems of second order

that is,

$$|\alpha\beta| < 1. \tag{9.23}$$

In Fig. 9.11 the graphic representation of (9.23) corresponds to the area enclosed by the hyperbolas shown.

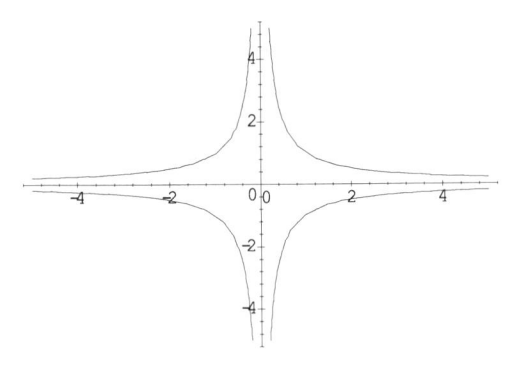

Figure 9.11 Graphics illustrating the zone (9.23).

b) With the representation (9.21), the system (9.20) is rewritten as

$$\begin{bmatrix} 1 & 0 \\ 0 & 0 \end{bmatrix} \ddot{\mathbf{q}} + \begin{bmatrix} 1 & 0 \\ 0 & 1 \end{bmatrix} \dot{\mathbf{q}} + \begin{bmatrix} 1 & -\alpha \\ -\beta & 1 \end{bmatrix} \mathbf{q} = \mathbf{0},$$

so that

$$\lambda^2 A + \lambda B + C = \begin{bmatrix} \lambda^2 + \lambda + 1 & -\alpha \\ -\beta & \lambda + 1 \end{bmatrix},$$

with the characteristic polynomial

$$p(\lambda) = \lambda^3 + 2\lambda^2 + 2\lambda + 1 - \alpha\beta. \tag{9.24}$$

The Hurwitz matrix corresponding to (9.24) is

$$G = \begin{bmatrix} 2 & 1-\alpha\beta & 0 \\ 1 & 2 & 0 \\ 0 & 2 & 1-\alpha\beta \end{bmatrix},$$

and the Liénard–Chipart criterion is on the one hand

$$1 - \alpha\beta > 0,$$

and on the other

$$3 + \alpha\beta > 0,$$

which implies

$$-3 < \alpha\beta < 1. \tag{9.25}$$

The hyperbolas shown in Fig. 9.12 enclose the area that corresponds to (9.25).

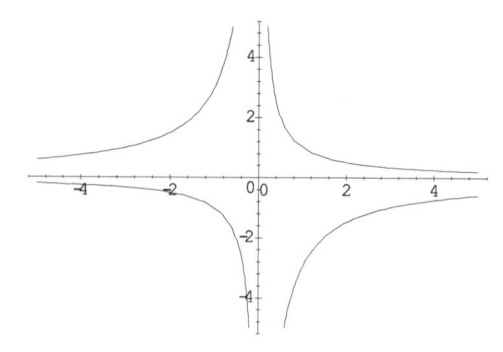

Figure 9.12 Function graphics of the zone (9.25).

9.4.2 Geometric criteria of asymptotic stability

In addition to the analytical criteria to determine the quality of a polynomial of being Hurwitzian, ones of geometric type have been developed. For the introduction to this topic some observations are required.

Argument principle

Consider the representation

$$p(\lambda) = p_0 \prod_{i=1}^{m} (\lambda - \lambda_i) \tag{9.26}$$

of the polynomial (9.16). Suppose the roots of (9.26) comply with

$$\operatorname{Re} \lambda_i \neq 0, \quad i = 1, \ldots, m.$$

Then it can be rewritten as

$$p(\lambda) = p_0 \prod_{i=1}^{l} (\lambda - \lambda_i) \prod_{k=1}^{r} (\lambda - \lambda_k), \tag{9.27}$$

where

$l :=$ number of roots of the **negative** real part,
$r :=$ number of roots of the **positive** real part.

Linear systems of second order

From the theory of complex numbers, remember that $z \in \mathbb{C}$ can be represented in terms of its magnitude and its argument in the form

$$z = |z| e^{j \arg z},$$

where $\arg z$ is the angle that the radius vector of z forms with the real axis, measured counter-clockwise with the clockwise (positive direction). Moreover, for $|z| < \infty$ and $\operatorname{Re} z \neq 0$, the complex function

$$f(j\omega) := j\omega - z = |j\omega - z| e^{j \arg(j\omega - z)}$$

is subject to the following change when the argument ω varies from $-\infty$ up to ∞ (see Fig. 9.13):

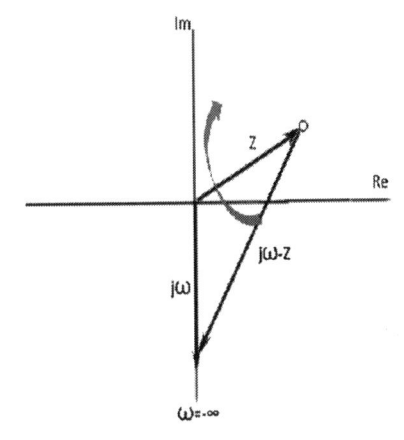

$$\underset{\omega = -\infty}{\overset{\infty}{\Delta}} \arg f(j\omega) = \begin{cases} \pi, & \text{if } \operatorname{Re} z < 0, \\ -\pi, & \text{if } \operatorname{Re} z > 0. \end{cases} \tag{9.28}$$

Figure 9.13 $f(j\omega)$ changes when the argument ω varies from $-\infty$ up to ∞.

Lemma 9.3. *The polynomial (9.27) complies with*

$$\underset{\omega \to \infty}{\overset{\infty}{\Delta}} \arg p(j\omega) = (l - r) \pi. \tag{9.29}$$

Proof. The evaluation of (9.27) implies

$$p(j\omega) = \rho_0 \prod_{i=1}^{l} (j\omega - \lambda_i) \prod_{k=1}^{r} (j\omega - \lambda_k) =$$

$$\rho_0 \prod_{i=1}^{m} (|j\omega - \lambda_i|) \exp j \left(\sum_{i=1}^{l} \arg(j\omega - \lambda_i) + \sum_{k=1}^{r} \arg(j\omega - \lambda_k) \right)$$

and by (9.28) the statement follows. □

The previous Lemma 9.3 leads directly to the result that follows.

Mikhailov criterion

Theorem 9.4 (Mikhailov criterion, 1938). *The polynomial $p(\lambda)$ of degree m, given in (9.16), is Hurwitz if and only if the hodograph of $p(\lambda)$ has counter-clockwise rotation and passes exactly m quadrants without crossing the origin when ω goes from 0 to ∞.*

Proof. By the definition $p(\lambda)$ is Hurwitzian if and only if it has a representation (9.27) with $l = m$. Then, by the previous lemma

$$\underset{\omega=\infty}{\overset{\infty}{\Delta}} \arg p(j\omega) = m\pi,$$

or, in view of the symmetry property,

$$\underset{\omega=0}{\overset{\infty}{\Delta}} \arg p(j\omega) = \frac{m}{2}\pi.$$

$\square$

A series of examples related to the Mikhailov criterion are included.

Example 9.9. Consider the polynomial

$$p(\lambda) = \lambda^5 + 5\lambda^4 + 10\lambda^3 + 11\lambda^2 + 7\lambda + 2. \tag{9.30}$$

Let us determine, by geometric criteria, if it is Hurwitz. Defining

$$\lambda = j\omega$$

for (9.30) we have

$$P(j\omega) = j\omega^5 + 5\omega^4 - j10\omega^3 - 11\omega^2 + j7\omega + 2$$
$$= 5\omega^4 - 11\omega^2 + 2 + j\omega\left(\omega^4 - 10\omega^2 + 7\right),$$

whose hodograph is shown in Fig. 9.14, of which Fig. 9.15 is a magnification. Due to the shape of the hodograph obtained, it is concluded that the polynomial (9.30) passes exactly five quadrants and, hence, it is Hurwitz.

Example 9.10. Suppose that a certain polynomial $p(\lambda)$ is of degree $m = 5$ and that its hodograph has the form given in Fig. 9.16. We need to determine l and r, the amounts of roots with negative and positive real parts, respectively. Since the hodograph does not cross the point $(0,0)$, the polynomial $p(\lambda)$ has no roots with null real parts, and considering (9.29), it follows that the following two simultaneous equations must be satisfied:

$$\left.\begin{array}{l} l + r = m = 5, \\ l - r = 3, \end{array}\right\}$$

Linear systems of second order

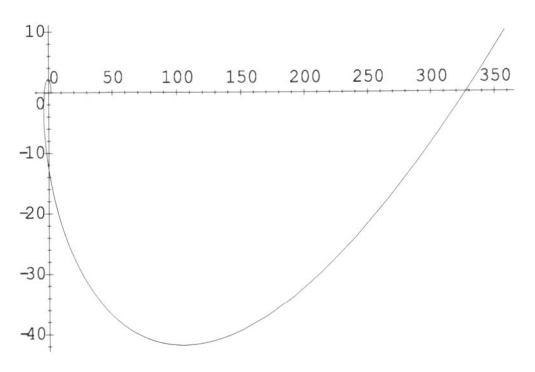

Figure 9.14 Hodograph of $p(j\omega)$.

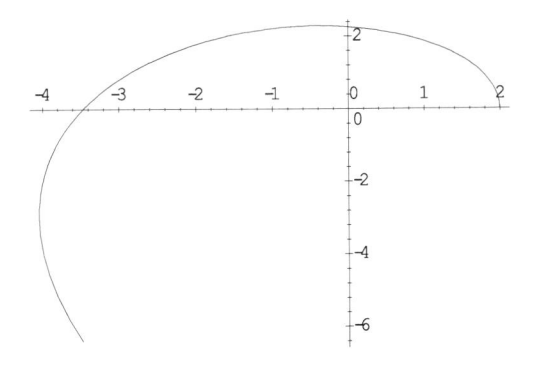

Figure 9.15 Zoom of the hodograph of $p(j\omega)$.

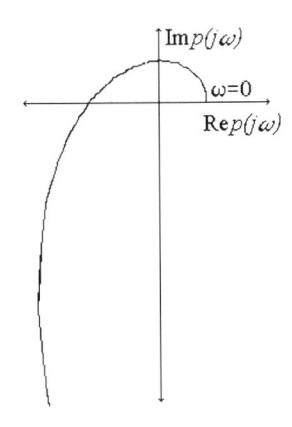

Figure 9.16 Hodograph of $p(j\omega)$.

which gives

$$l = 4, \ r = 1.$$

Example 9.11. Suppose that the hodograph given in Fig. 9.17 corresponds to a polynomial $p(\lambda)$ of degree $m = 6$. Let us calculate l, r, and n, the amounts of roots with negative, positive, and null real parts, respectively.

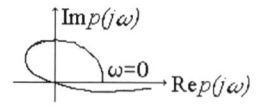

Figure 9.17 Hodograph of $p(j\omega)$.

Since the hodograph crosses the point $(0, 0)$, the polynomial $p(\lambda)$ has roots with null real parts, and since there is only one cross, which also does not occur for $\omega = 0$, there exists $\omega_0 \neq 0$ such that

$$\lambda_i = j\omega_0, \quad \bar{\lambda}_i = -j\omega_0$$

are the roots with a null real part of $p(\lambda)$, that is, $n = 2$. So we have

$$\left. \begin{array}{l} l + r + n = m = 6, \\ l - r = 2, \\ n = 2, \end{array} \right\}$$

which gives

$$l = 3, \quad r = 1, \quad n = 2.$$

Example 9.12. Using the Mikhailov criterion, obtain l, r, and n, the amounts of roots with negative, positive, and null real parts, respectively, for the following polynomials:

a)

$$p(\lambda) = \lambda^5 + 2\lambda^4 + 2\lambda^3 - 7\lambda^2 - 44\lambda - 4, \tag{9.31}$$

b)

$$p(\lambda) = \lambda^4 + \lambda^3 - 2\lambda^2 + 4\lambda + 2. \tag{9.32}$$

a) Taking $\lambda = j\omega$ in (9.31) results in

$$p(j\omega) = 2\omega^4 + 7\omega^2 - 4 + j\omega\left(\omega^4 - 2\omega^2 - 44\right),$$

whose hodograph is given in Fig. 9.18. Since the hodograph does not cross the point $(0, 0)$, $p(\lambda)$ has no roots with null real parts. So, by (9.29) the following equations are

Linear systems of second order

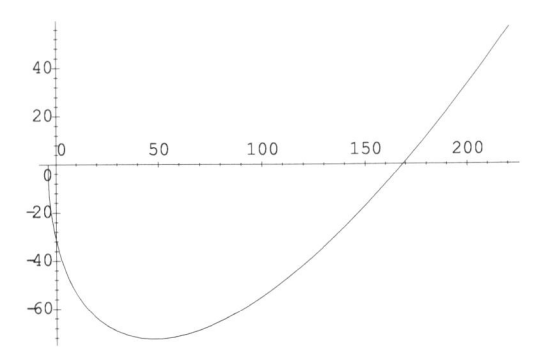

[Figure: graph with horizontal axis from 0 to 200 and vertical axis from -60 to 40, showing a curve]

Figure 9.18 Hodograph of $p(j\omega)$.

reached:

$$\left.\begin{array}{l}l+r=m=5,\\ l-r=3.\end{array}\right\}$$

That is why

$$l=4,\ r=1,\ n=0.$$

b) Taking $\lambda = j\omega$, the polynomial (9.32) may be written in the form

$$p(j\omega) = \omega^4 + 2\omega^2 + 2 + j\omega\left(-\omega^2 + 4\right),$$

which hodograph is depicted in Fig. 9.19. Since the hodograph does not pass through the point $(0,0)$, $p(\lambda)$ has no roots with null real parts. Then, by (9.29), the following equations are reached:

$$\left.\begin{array}{l}l+r=m=4,\\ l-r=0,\end{array}\right\}$$

implying

$$l=2,\ r=2.$$

Example 9.13. Given the polynomial

$$p(\lambda) = \lambda^4 + \lambda^3 + 4\lambda^2 + 2\lambda + 3 + k, \qquad (9.33)$$

using the Mikhailov criterion, let us try to obtain l, r, and n, the amounts of roots with negative, positive, and null real parts, respectively, based on the values of k.

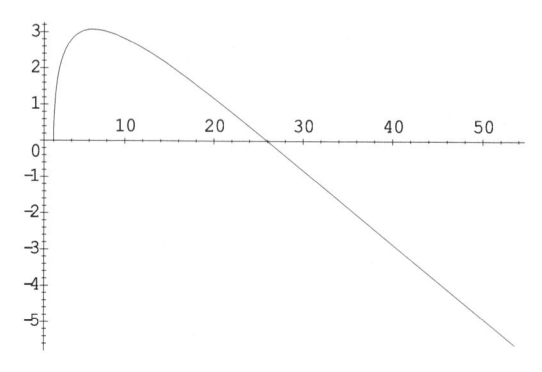

Figure 9.19 Hodograph of $p(j\omega)$.

Taking $\lambda = j\omega$, (9.33) can be represented as

$$p(j\omega) = \omega^4 - 4\omega^2 + 3 + k + j\omega\left(-\omega^2 + 2\right) =$$
$$\left(\omega^2 - \left(2 + \sqrt{1-k}\right)\right)\left(\omega^2 - \left(2 - \sqrt{1-k}\right)\right) + j\omega\left(-\omega^2 + 2\right).$$

Note that $\operatorname{Im} p(j\omega) = 0$ when

$$\omega = 0, \quad \omega = \pm\sqrt{2}.$$

This means that for $\omega \geq 0$ the hodograph crosses the real axis twice. On the other hand, crosses with the imaginary axis depend on the value of k and several cases occur.

i) If $k < -3$, then the crosses with the imaginary axis occur when

$$\omega_{1,2} = \pm\sqrt{2 + \sqrt{1-k}}.$$

In particular, for $\omega \geq 0$ the hodograph crosses the imaginary axis once. In short, the hodograph contains the crosses as in Table 9.1:

Table 9.1 Hodograph crosses for $k < -3$.

ω	Cross	Semi-axis
0	$(3+k, 0)$	real negative
$\sqrt{2}$	$(-1+k, 0)$	real negative
$\sqrt{2+\sqrt{1-k}}$	$\left(0, -\sqrt{(1-k)(2+\sqrt{1-k})}\right)$	imaginary negative

and has the form as in Fig. 9.20; that is why $l = 3, r = 1, n = 0$.

Linear systems of second order

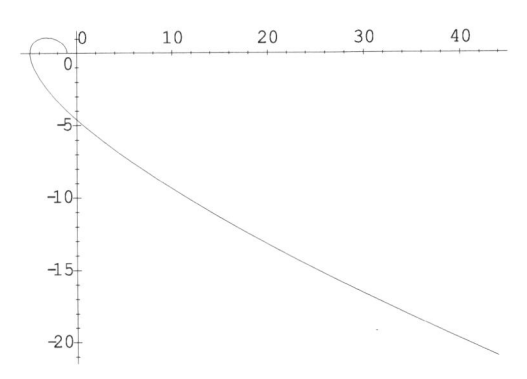

Figure 9.20 Hodograph for the case $k < -3$.

ii) If $k = -3$, the crosses with the imaginary axis correspond to

$$\omega_1 = 0, \quad \omega_1 = \pm 2.$$

In particular, for $\omega \geq 0$ the hodograph crosses the imaginary axis twice. Therefore, we have the situation as in Table 9.2:

Table 9.2 Hodograph crosses for $k = -3$.

ω	Cross	Semi-axis
0	(0, 0)	origin
$\sqrt{2}$	(−4, 0)	real negative
2	(0, −4)	imaginary negative

and the hodograph has the form as given in Fig. 9.21. Hence,

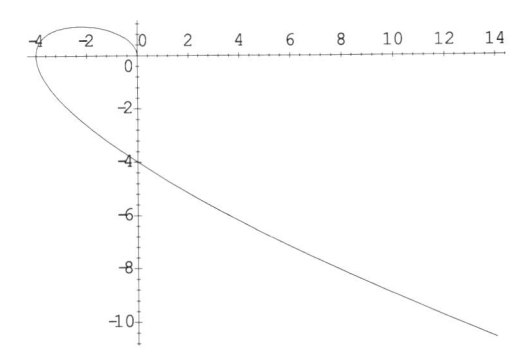

$$l = 3, \quad r = 0, \quad n = 1.$$

Figure 9.21 Hodograph of $p(j\omega)$ for $k = -3$.

iii) If $-3 < k < 1$, we have to cross the imaginary axis when

$$\omega = \omega_{1,2} = \pm\sqrt{2 - \sqrt{1-k}}, \ \omega = \omega_{3,4} = \pm\sqrt{2 + \sqrt{1-k}},$$

and for $\omega \geq 0$, the hodograph crosses the imaginary axis twice. Table 9.3 summarizes the crossroads of the hodograph with the axes,

Table 9.3 Hodograph crosses for $-3 < k < 1$.

ω	Cross	Semi-axis
0	$(3+k, 0)$	real positive
$\sqrt{2-\sqrt{1-k}}$	$\left(0, \sqrt{(1-k)(2-\sqrt{1-k})}\right)$	imaginary positive
$\sqrt{2}$	$(-1+k, 0)$	real negative
$\sqrt{2+\sqrt{1-k}}$	$-\sqrt{(1-k)(2+\sqrt{1-k})}$	imaginary negative

whereby the curve has the shape given in Fig. 9.22, and we deduce that

$$l = 4, \ r = 0, \ n = 0.$$

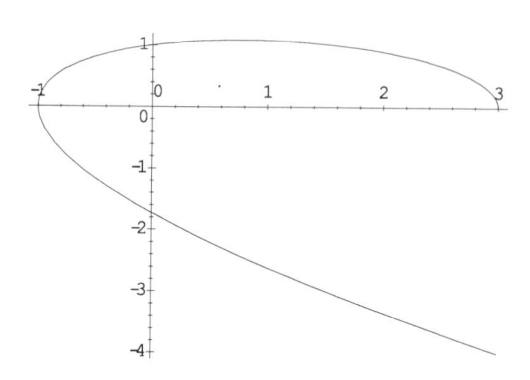

Figure 9.22 Hodograph of $p(j\omega)$ for $-3 < k < 1$.

iv) For $k = 1$ the following crosses with the imaginary axis are presented:

$$\omega_{1,2} = \pm\sqrt{2},$$

which is why for $\omega \geq 0$ the hodograph crosses the imaginary axis once. The hodograph now presents the crosses in Table 9.4,

Table 9.4 Hodograph crosses for $k = 1$.

ω	Cross	Semi-axis
0	$(4, 0)$	real positive
$\sqrt{2}$	$(0, 0)$	origin

with the form depicted in Fig. 9.23, which gives

$$l = 2,\ r = 0,\ n = 2.$$

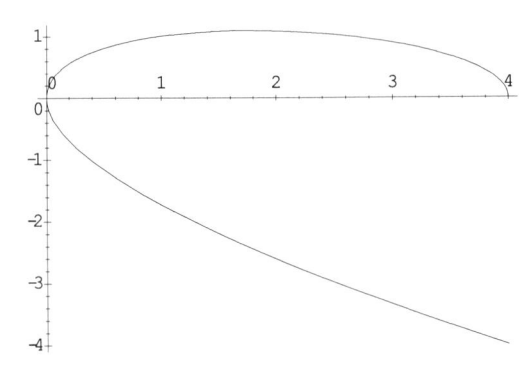

Figure 9.23 Hodograph of $p(j\omega)$ for $k = 1$.

v) If $k > 1$ the hodograph has no crosses with the imaginary axis, so we have Table 9.5,

Table 9.5 Hodograph crosses for $k > 1$.

ω	Cross	Semi-axis
0	$(3+k, 0)$	real positive
$\sqrt{2}$	$(-1+k, 0)$	real positive

with the image given in Fig. 9.24, so that

$$l = 2,\ r = 2,\ n = 0.$$

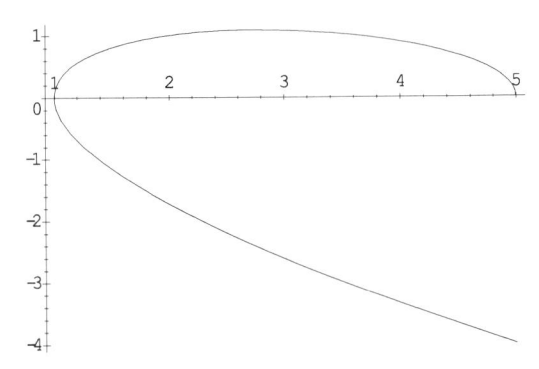

Figure 9.24 Hodograph of $p(j\omega)$ for $k > 1$.

Table 9.6 summarizes the results obtained, based on the value of k.

Table 9.6 Roots distribution for different values of k.

k	l	r	n
$k < -3$	3	1	0
$k = -3$	3	0	1
$-3 < k < 1$	4	0	0
$k = 1$	2	0	2
$k > 1$	2	2	0

9.5 Polynomial robust stability

9.5.1 Parametric uncertainty and robust stability

As shown before, the stability property of the second order system (9.1) is characterized by the root locations of the corresponding characteristic polynomial $p(\lambda)$ (see (9.16)). Evidently, any variations ΔA, ΔB, and ΔC of the matrices A, B, and C, namely,

$$A = A_0 + \Delta A, \ B = B_0 + \Delta B, \ C = C_0 + \Delta C,$$

are transformed into the variations of the coefficients ρ_j ($j = 0, ..., m$) of the corresponding characteristic polynomial,

$$p_\rho(\lambda) := \lambda^m \rho_0 + \lambda^{m-1} \rho_1 + ... + \rho_m, \quad \rho_i \in R. \tag{9.34}$$

Denote the collection of its coefficients by

$$\rho := (\rho_0, ..., \rho_m)^\mathsf{T} \in R^{m+1} \tag{9.35}$$

and suppose that this vector of coefficients belongs to a *connected set* $\mathcal{R} \in \mathbb{R}^n$ that corresponds to possible variations ΔA, ΔB, and ΔC, that is,

$$\rho \in \mathcal{R}. \tag{9.36}$$

Definition 9.6. A characteristic polynomial $p_\rho(\lambda)$ (9.34) is said to be **robust stable** if for any $\rho \in \mathcal{R}$ the roots of the corresponding polynomial belongs to the left-hand side of the complex plane $\mathbb{C}$, i.e.,

$$\operatorname{Re} \lambda_j(\rho) < 0 \ (j = 1, ..., m), \tag{9.37}$$

for all $\rho \in \mathcal{R}$.

Definition 9.7. Denote by $\mathcal{Q}\rho(\omega)$ the set of all values of the vector

$$p_\rho(i\omega) = U_\rho(\omega) + i V_\rho(\omega)$$

given in $\mathbb{C}$ under a fixed $\omega \in [0, \infty)$ when the parameters ρ take all possible values in $\mathcal{R}$, that is,

$$\mathcal{Q}\rho(\omega) := \{z : z = p_\rho(i\omega) \mid \rho \in \mathcal{R}\}. \tag{9.38}$$

The next result represents the criterion of *polynomial robust stability*.

Theorem 9.5 (The criterion of polynomial robust stability). *The characteristic polynomial $p_\rho(\lambda)$ (9.34) is* **robust stable** *if and only if:*
1. *The class $\mathcal{R}$ of polynomials $p_\rho(\lambda)$ contains at least one Hurwitz polynomial $p_\rho^*(\lambda)$, named a* **basic one**.
2. *The following* **principle of "zero-excluding"** *holds: the set $\mathcal{Q}\rho(\omega)$ does not contain the origin ("zero-point"), i.e.,*

$$0 \notin \mathcal{Q}\rho(\omega). \tag{9.39}$$

Proof. Since the vector $z = p_\rho(j\omega) \in \mathbb{C}$ is continually dependent on the vector parameter ρ, a *"transition"* from stable polynomial to unstable polynomial (when we are varying the coefficients ρ) may occur (this is always possible since the set $\mathcal{A}$ of parameters is a connected set) only when one of its roots crosses the imaginary axis, or, in other words, when there exists $\omega_0 \in [0, \infty)$ such that

$$p_\rho(i\omega_0) = U(\omega_0) + iV(\omega_0) = 0.$$

But this is equivalent to the following identity:

$$U(\omega_0) = V(\omega_0) = 0,$$

which means exactly that

$$0 \in \mathcal{Q}\rho(\omega).$$

Evidently, to avoid this effect it is necessary and sufficient to satisfy conditions 1 and 2 of this theorem. The theorem is proven. □

9.5.2 The Kharitonov theorem

Theorem 9.6 ((Kharitonov, 1978)). *Let the set $\mathcal{R}$, characterizing a parametric uncertainty, be defined as*

$$\mathcal{R} := \{a \in \mathbb{R}^n : a_i^- \leq a_i \leq a_i^+ \ (i = 1, ...n)\}. \tag{9.40}$$

Then the polynomial $p_\rho(\lambda)$ (9.34) is robust stable if and only if:

1) *the central polynomial $p_{\mathring{\rho}}(\lambda)$ with the coefficients $\mathring{a}_i = \dfrac{1}{2}\left(a_i^- + a_i^+\right)$ is Hurwitz;*

2) *the following **four polynomials** are stable (Hurwitz):*

$$\left.\begin{array}{l} p_\rho^{(1)}(\lambda) := 1 + a_1^- \lambda + a_2^+ \lambda^2 + a_3^+ \lambda^3 + a_4^- \lambda^4 + a_5^- \lambda^5 + \cdots, \\ p_\rho^{(2)}(\lambda) := 1 + a_1^+ \lambda + a_2^+ \lambda^2 + a_3^- \lambda^3 + a_4^- \lambda^4 + a_5^+ \lambda^5 + \cdots, \\ p_\rho^{(3)}(\lambda) := 1 + a_1^+ \lambda + a_2^- \lambda^2 + a_3^- \lambda^3 + a_4^+ \lambda^4 + a_5^+ \lambda^5 + \cdots, \\ p_\rho^{(4)}(\lambda) := 1 + a_1^- \lambda + a_2^- \lambda^2 + a_3^+ \lambda^3 + a_4^+ \lambda^4 + a_5^- \lambda^5 + \cdots. \end{array}\right\} \quad (9.41)$$

Proof. For any $a \in \mathcal{R}$,

$$U(\omega) = 1 - a_2 \omega^2 + a_4 \omega^4 - \cdots,$$
$$V(\omega) = a_1 \omega - a_3 \omega^3 + a_5 \omega^5 \cdots,$$

and hence for any $\omega \in [0, \infty)$,

$$U^-(\omega) \le U(\omega) \le U^+(\omega) \text{ and } V^-(\omega) \le V(\omega) \le V^+(\omega),$$

where

$$U^-(\omega) = 1 - a_2^+ \omega^2 + a_4^- \omega^4 - \cdots,$$
$$U^+(\omega) = 1 - a_2^- \omega^2 + a_4^+ \omega^4 - \cdots,$$

and

$$V^-(\omega) = a_1^- \omega - a_3^+ \omega^3 + a_5^- \omega^5 \cdots,$$
$$V^+(\omega) = a_1^+ \omega - a_3^- \omega^3 + a_5^+ \omega^5 \cdots.$$

That is why for any $\omega \in [0, \infty)$ the set $\mathcal{Q}_\rho(\omega)$ (9.38) is the rectangular (see Fig. 9.25) with width

$$\left[U^+(\omega) - U^-(\omega) \right]$$

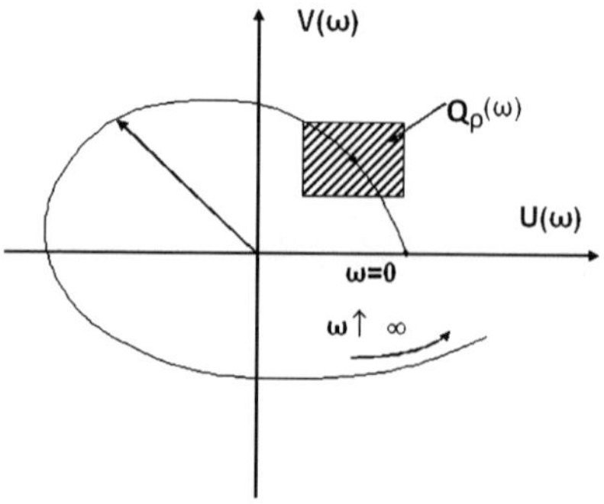

Figure 9.25 The illustration of the Kharitonov's criterion.

and height
$$\left[V^+(\omega) - V^-(\omega)\right]$$
and with the center in the point $p_{\mathring{a}}(j\omega)$, corresponding to the stable polynomial with the parameters
$$\mathring{a}_i = \frac{1}{2}\left(a_i^- + a_i^+\right).$$
Note that the vertices of the set $\mathcal{Q}\rho(\omega)$ exactly correspond to the polynomials (9.41). Suppose now that this rectangular touches the origin by one of its sides. By the monotonically increasing property the vertices of this touching side will rotate in the clockwise direction, and, hence, will become non-vertical, contradicting our previous concept. So, direct application of Theorem 9.5 leads to the formulated result. The theorem is proven. □

Example 9.14. Let us find the parameter β for which the polynomial
$$p_\rho(\lambda) = 1 + a_1\lambda + a_2\lambda^2 + a_3\lambda^3,$$
$$1 - \beta \leq a_1 \leq 1 + \beta,$$
$$1.5 \leq a_2 \leq 2, \ a_3 = 1,$$
is robust stable. To do that construct four polynomials (see (9.41)):
$$p_\rho^{(1)}(\lambda) := 1 + (1-\beta)\lambda + 2\lambda^2 + \lambda^3,$$
$$p_\rho^{(2)}(\lambda) := 1 + (1+\beta)\lambda + 2\lambda^2 + \lambda^3,$$
$$p_\rho^{(3)}(\lambda) := 1 + (1+\beta)\lambda + 1.5\lambda^2 + \lambda^3,$$
$$p_\rho^{(4)}(\lambda) := 1 + (1-\beta)\lambda + 1.5\lambda^2 + \lambda^3.$$

The corresponding *Hurwitz* matrices are as follows:
$$\begin{bmatrix} 1-\beta & 1 & 0 \\ 1 & 2 & 0 \\ 0 & (1-\beta) & 1 \end{bmatrix}, \begin{bmatrix} 1+\beta & 1 & 0 \\ 1 & 2 & 0 \\ 0 & (1+\beta) & 1 \end{bmatrix},$$
$$\begin{bmatrix} 1+\beta & 1 & 0 \\ 1 & 1.5 & 0 \\ 0 & 1+\beta & 1 \end{bmatrix}, \begin{bmatrix} 1-\beta & 1 & 0 \\ 1 & 1.5 & 0 \\ 0 & 1-\beta & 1 \end{bmatrix},$$

By the Liénard–Chipart criterion we find that the conditions of the robust stability are
$$1 - \beta > 0, \ 1 + \beta > 0, \ \text{or, equivalently, } |\beta| < 1$$
and
$$2(1-\beta) - 1 > 0, \ 2(1+\beta) - 1 > 0,$$

$$1.5(1+\beta) - 1 > 0, \quad 1.5(1-\beta) - 1 > 0,$$

which leads to the following:

$$\beta < 0.5, \quad \beta > -0.5, \quad \beta > \frac{2}{3} - 1 = -\frac{1}{3}, \quad \beta < 1 - \frac{2}{3} = \frac{1}{3},$$

or, equivalently,

$$|\beta| < 0.5, \quad |\beta| < \frac{1}{3}.$$

Finally, all constraints, taken together, give

$$|\beta| < \frac{1}{3}.$$

9.6 Exercises

Exercise 9.1. Show that for parameter values satisfying the relation

$$\frac{c_1 + c_2}{m_1} = \frac{c_3 + c_4}{m_3},$$

the equilibrium position of the dissipative system of the system, shown in Fig. 9.26, will not be asymptotically stable if there is no friction between the masses and the guide.

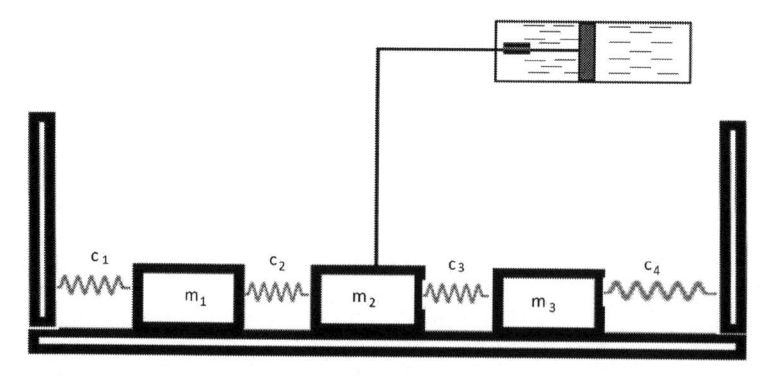

Figure 9.26 Dissipative system of three masses.

Exercise 9.2. Show that with the relations

$$L_3 \frac{(C_1 + C_2)}{C_1 C_2} = L_1 \frac{(C_3 + C_4)}{C_3 C_4},$$

Linear systems of second order

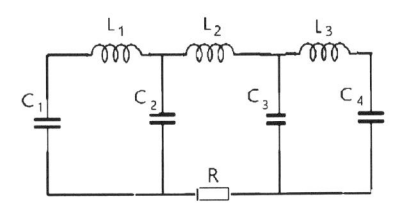

Figure 9.27 Electric circuit with 4 capacities, 3 inductions and 1 ohmic resistance.

between the parameters of the electric circuit shown in Fig. 9.27, undamped oscillations are possible in the system, despite the presence of an ohmic resistance R.

Exercise 9.3. Prove that the number l of roots with the negative real part of the polynomial $f(\lambda)$ of degree n, whose Mikhailov hodograph does not pass through the zero point and satisfies the condition

$$\Delta_{\omega=0}^{\omega=\infty} \arg f(j\omega) = k\frac{\pi}{2}, \quad |k| \leq n,$$

is

$$l = \frac{n+k}{2}.$$

Exercise 9.4. Confirm that for all values of the parameters α and β, satisfying the condition

$$\alpha < 1, \quad \alpha + \beta > 0,$$

the equilibrium position of the system

$$\left.\begin{array}{l}\ddot{x} + 2\dot{x} + x - \alpha y = 0, \\ \ddot{y} + \beta \dot{y} - x + y = 0\end{array}\right\}$$

is asymptotically stable.

Exercise 9.5. The natural frequency of a linear oscillator is equal to ω_0. Show that the frequency of damped oscillations of the same oscillator in a medium with a resistance proportional to velocity is equal to

$$\omega = \omega_0 \frac{2\pi n}{\ln k} \left[1 + \left(\frac{2\pi n}{\ln k}\right)^2\right]^{-1/2},$$

if after n oscillations its amplitude decreases by k times.

Hamiltonian formalism

Contents

10.1 Hamiltonian function 311
10.2 Hamiltonian canonical form 316
10.3 First integrals 321
10.4 Some properties of first integrals 323
    10.4.1 Cyclic coordinates 323
    10.4.2 Some properties of the Poisson brackets 324
    10.4.3 First integrals by inspection 330
10.5 Exercises 335

In this chapter, conservative systems are considered and the generalized impulses are introduced. Hamilton's variables are also considered. Here it is demonstrated that they can completely describe the dynamics of a system in the Hamiltonian canonical form. Some properties of the canonical equations are studied. Cyclic coordinates and the first integrals of Hamiltonian systems are introduced and analyzed. Some useful properties (such as Poisson brackets), helping to test if some function is a first integral, are discussed.

10.1 Hamiltonian function

In this chapter, *conservative systems* will be considered, that is, systems where the generalized forces present are represented by the relationship

$$\mathbf{Q} = -\nabla_{\mathbf{q}} V(t, \mathbf{q}),$$

with $V(t, \mathbf{q})$ as the potential energy function, and whose dynamics is described by the Lagrange equations

$$\frac{d}{dt} \nabla_{\dot{\mathbf{q}}} L(t, \mathbf{q}, \dot{\mathbf{q}}) - \nabla_{\mathbf{q}} L(t, \mathbf{q}, \dot{\mathbf{q}}) = \mathbf{0}, \tag{10.1}$$

where the Lagrangian is defined as

$$L(t, \mathbf{q}, \dot{\mathbf{q}}) := T(t, \mathbf{q}, \dot{\mathbf{q}}) - V(t, \mathbf{q}), \tag{10.2}$$

with $T(t, \mathbf{q}, \dot{\mathbf{q}})$ as the function of kinetic energy.

The following definition is fundamental in this chapter.

Classical and Analytical Mechanics. https://doi.org/10.1016/B978-0-32-389816-4.00021-1
Copyright © 2021 Elsevier Inc. All rights reserved.

Definition 10.1. The vector

$$\mathbf{p} := \nabla_{\dot{\mathbf{q}}} L(t, \mathbf{q}, \dot{\mathbf{q}}) \tag{10.3}$$

is referred to as the **generalized impulse**.

Remark 10.1. Let

$$\dot{\mathbf{q}} = \dot{\mathbf{q}}(t, \mathbf{q}, \mathbf{p}) \tag{10.4}$$

be the inverse function of the transformation (10.3), that is, $\dot{\mathbf{q}}$ is obtained from (10.3). With the definition (10.3) of $\mathbf{p}$ and in view of (10.4), Eqs. (10.1) can be expressed in the form

$$\dot{\mathbf{p}} - \left[\nabla_{\mathbf{q}} L(t, \mathbf{q}, \dot{\mathbf{q}})\right]_{\dot{\mathbf{q}} = \dot{\mathbf{q}}(t, \mathbf{q}, \mathbf{p})} = \mathbf{0}. \tag{10.5}$$

Clearly (10.1) and (10.5) are two equivalent ways of describing the dynamics of a mechanical system: the difference lies in the set of variables chosen to make the description. The set $\{t, \mathbf{q}, \dot{\mathbf{q}}\}$ is called the **Lagrange variables** and the set $\{t, \mathbf{q}, \mathbf{p}\}$ is called the **Hamiltonian variables**.

The concept of the generalized impulse (10.3) and the Lagrange function (10.2) allow defining the very useful function in the following definition.

Definition 10.2. The function

$$H(t, \mathbf{q}, \mathbf{p}) := \left[(\mathbf{p}, \dot{\mathbf{q}}) - L(t, \mathbf{q}, \dot{\mathbf{q}})\right]_{\dot{\mathbf{q}} = \dot{\mathbf{q}}(t, \mathbf{q}, \mathbf{p})} \tag{10.6}$$

is referred to as the **Hamiltonian function**. It is also known as the **Hamiltonian** or the **energy function**.

The following examples illustrate the construction of H.

Example 10.1. Consider a particle of mass m in Euclidean three-dimensional coordinate space (x, y, z). If the system's Lagrange function is

$$L = \frac{m}{2}\left(\dot{x}^2 + \dot{y}^2 + \dot{z}^2\right) - V(x, y, z),$$

let us obtain the corresponding Hamilton function. Defining

$$\mathbf{q} := \begin{bmatrix} x & y & z \end{bmatrix}^T,$$

we have

$$L(\mathbf{q}, \dot{\mathbf{q}}) = \frac{m}{2} \|\dot{\mathbf{q}}\|^2 - V(\mathbf{q}).$$

Using (10.3) we get

$$\mathbf{p} = m\dot{\mathbf{q}},$$

which leads to
$$\dot{\mathbf{q}} = \frac{1}{m}\mathbf{p},$$
and hence
$$[L(\mathbf{q},\dot{\mathbf{q}})]_{\dot{\mathbf{q}}=\frac{1}{m}\mathbf{p}} = \frac{1}{2m}\|\mathbf{p}\|^2 - V(\mathbf{q}).$$
By the definition (10.6) we get
$$H(\mathbf{q},\mathbf{p}) = \left(\mathbf{p}, \frac{1}{m}\mathbf{p}\right) - [L(\mathbf{q},\dot{\mathbf{q}})]_{\dot{\mathbf{q}}=\frac{1}{m}\mathbf{p}},$$
resulting in
$$H(\mathbf{q},\mathbf{p}) = \frac{1}{2m}\|\mathbf{p}\|^2 + V(\mathbf{q}). \tag{10.7}$$

Example 10.2. Suppose a mechanical system has the function of Lagrange
$$L(\mathbf{q},\dot{\mathbf{q}}) = \frac{3}{2}\dot{q}_1^2 + \frac{1}{2}\dot{q}_2^2 - q_1^2 - \frac{1}{2}q_2^2 - q_1 q_2.$$
Let us calculate H. The generalized impulses are
$$p_1 = \frac{\partial L}{\partial \dot{q}_1} = 3\dot{q}_1,$$
$$p_2 = \frac{\partial L}{\partial \dot{q}_2} = \dot{q}_2,$$
implying
$$\dot{q}_1 = \frac{1}{3}p_1,$$
$$\dot{q}_2 = p_2.$$
So,
$$[L(\mathbf{q},\dot{\mathbf{q}})]_{\dot{\mathbf{q}}=\dot{\mathbf{q}}(\mathbf{p})} = \frac{1}{6}p_1^2 + \frac{1}{2}p_2^2 - q_1^2 - \frac{1}{2}q_2^2 - q_1 q_2$$
and by (10.6) we get
$$H(\mathbf{q},\mathbf{p}) = \left(\begin{bmatrix}p_1\\p_2\end{bmatrix}, \begin{bmatrix}\frac{1}{3}p_1\\p_2\end{bmatrix}\right) - [L(\mathbf{q},\dot{\mathbf{q}})]_{\dot{\mathbf{q}}=\dot{\mathbf{q}}(\mathbf{p})},$$
which leads to
$$H(\mathbf{q},\mathbf{p}) = \frac{1}{6}p_1^2 + \frac{1}{2}p_2^2 + q_1^2 + \frac{1}{2}q_2^2 + q_1 q_2.$$

Example 10.3. We need to design H for the following Lagrange functions:

a)
$$L(\mathbf{q}, \dot{\mathbf{q}}) = \frac{5}{2}\dot{q}_1^2 + \frac{1}{2}\dot{q}_2^2 + \dot{q}_1\dot{q}_2 \cos(q_1 - q_2) + 3\cos q_1 + \cos q_2,$$

b)
$$L(\mathbf{q}, \dot{\mathbf{q}}) = a\dot{q}_1^2 + \left(c^2 + b^2 \cos^2 q_1\right)\dot{q}_2^2.$$

Let us present here the sequential steps for both Lagrange functions.

a) From the definition (10.3) for generalized impulses it follows that
$$p_1 = 5\dot{q}_1 + \dot{q}_2 \cos(q_1 - q_2),$$
$$p_2 = \dot{q}_2 + \dot{q}_1 \cos(q_1 - q_2),$$

from which we have
$$\dot{q}_1 = \frac{-p_1 + p_2 \cos(q_1 - q_2)}{-5 + \cos^2(q_1 - q_2)},$$
$$\dot{q}_2 = \frac{-5p_2 + p_1 \cos(q_1 - q_2)}{-5 + \cos^2(q_1 - q_2)},$$

and
$$[L(\mathbf{q}, \dot{\mathbf{q}})]_{\dot{\mathbf{q}} = \dot{\mathbf{q}}(\mathbf{p})} = -\frac{1}{2}\frac{-2p_1p_2 \cos(q_1 - q_2) + p_1^2 + 5p_2^2}{-5 + \cos^2(q_1 - q_2)} +$$
$$3\cos q_1 + \cos q_2.$$

Using (10.6) we obtain
$$H(\mathbf{q}, \mathbf{p}) = \left(\begin{bmatrix} p_1 \\ p_2 \end{bmatrix}, \begin{bmatrix} \frac{-p_1 + p_2 \cos(q_1 - q_2)}{-5 + \cos^2(q_1 - q_2)} \\ \frac{-5p_2 + p_1 \cos(q_1 - q_2)}{-5 + \cos^2(q_1 - q_2)} \end{bmatrix} \right) - [L(\mathbf{q}, \dot{\mathbf{q}})]_{\dot{\mathbf{q}} = \dot{\mathbf{q}}(\mathbf{p})} =$$
$$-\frac{1}{2}\frac{-2p_1p_2 \cos(q_1 - q_2) + p_1^2 + 5p_2^2}{-5 + \cos^2(q_1 - q_2)} - 3\cos q_1 - \cos q_2.$$

b) The vector $\mathbf{p}$ has the components
$$p_1 = 2a\dot{q}_1,$$
$$p_2 = 2\left(c^2 + b^2 \cos^2 q_1\right)\dot{q}_2,$$

which is why
$$\dot{q}_1 = \frac{1}{2a}p_1,$$

Hamiltonian formalism

$$\dot{q}_2 = \frac{1}{2\left(c^2 + b^2 \cos^2 q_1\right)} p_2,$$

and

$$[L(\mathbf{q}, \dot{\mathbf{q}})]_{\dot{\mathbf{q}}=\dot{\mathbf{q}}(\mathbf{p})} = \frac{1}{4a} p_1^2 + \frac{1}{4\left(c^2 + b^2 \cos^2 q_1\right)} p_2^2,$$

implying

$$H(\mathbf{q}, \mathbf{p}) = \left(\begin{bmatrix} p_1 \\ p_2 \end{bmatrix}, \begin{bmatrix} \frac{1}{2a} p_1 \\ \frac{1}{2\left(c^2 + b^2 \cos^2 q_1\right)} p_2 \end{bmatrix}\right) - [L(\mathbf{q}, \dot{\mathbf{q}})]_{\dot{\mathbf{q}}=\dot{\mathbf{q}}(\mathbf{p})} =$$

$$\frac{1}{4a} p_1^2 + \frac{1}{4\left(c^2 + b^2 \cos^2 q_1\right)} p_2^2.$$

Example 10.4. Fig. 10.1 shows a simple pendulum at whose distal end is a ball of mass M and radius R. The pendulum arm is a solid cylinder of mass m and with radius r and length l. The masses are uniformly distributed. Let us determine the Hamiltonian function of this system. Let

$$q := \varphi.$$

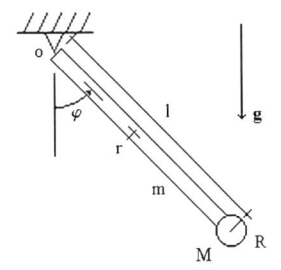

Figure 10.1 Pendulum with non-negligible-mass arm.

From Chapter 3 (Section 3.6) it follows that for this system the kinetic energy has the expression

$$T(\dot{q}) = \frac{1}{2} I_O \dot{q}^2,$$

where I_O is the moment of inertia of the arm and the ball with respect to the axis perpendicular to the plane of the movement and passing through the fixed point O and whose value is

$$I_O = m \left(\frac{r^2}{4} + \frac{l^2}{3}\right) + M \left(\frac{2R^2}{5} + l^2\right).$$

If the horizontal plane passing through O is considered as the reference level for potential energy, then

$$V(q) = -gl\left(M + \frac{m}{2}\right)\cos q.$$

So, the Lagrange function turns out to be equal to

$$L(t, q, \dot{q}) := T - V = \frac{1}{2}I_O\dot{q}^2 + gl\left(M + \frac{m}{2}\right)\cos q,$$

from where the generalized moment is calculated, using (10.3), as

$$p = \frac{\partial L}{\partial \dot{q}} = I_O\dot{q},$$

giving

$$\dot{q} = \frac{p}{I_O}.$$

By the definition (10.6),

$$H(q, p) = [p\dot{q} - L(t, q, \dot{q})]_{\dot{q}=\frac{p}{I_O}},$$

implying

$$H(q, p) = \frac{p^2}{2I_O} - gl\left(M + \frac{m}{2}\right)\cos q =$$

$$\frac{p^2}{2\left[m\left(\frac{r^2}{4} + \frac{l^2}{3}\right) + M\left(\frac{2R^2}{5} + l^2\right)\right]} - gl\left(M + \frac{m}{2}\right)\cos q.$$

10.2 Hamiltonian canonical form

Hamilton variables can fully describe the dynamics of a system. The following theorem addresses this point.

Theorem 10.1 (Hamilton, around 1835). *The Hamilton variables $\{t, \mathbf{q}, \mathbf{p}\}$ satisfy the following system of equations:*

$$\frac{\partial H(t, \mathbf{q}, \mathbf{p})}{\partial t} = \left[-\frac{\partial L(t, \mathbf{q}, \dot{\mathbf{q}})}{\partial t}\right]_{\dot{\mathbf{q}}=\dot{\mathbf{q}}(\mathbf{p})} \quad (10.8)$$

and

$$\left.\begin{array}{l} \dot{\mathbf{q}} = \nabla_\mathbf{p} H(t, \mathbf{q}, \mathbf{p}), \\ \dot{\mathbf{p}} = -\nabla_\mathbf{q} H(t, \mathbf{q}, \mathbf{p}). \end{array}\right\} \quad (10.9)$$

Hamiltonian formalism

Proof. By the definition (10.6) the Hamiltonian $H(t, \mathbf{q}, \mathbf{p})$ is defined as

$$H(t, \mathbf{q}, \mathbf{p}) := [(\mathbf{p}, \dot{\mathbf{q}}) - L(t, \mathbf{q}, \dot{\mathbf{q}})]_{\dot{\mathbf{q}} = \dot{\mathbf{q}}(t, \mathbf{q}, \mathbf{p})}, \qquad (10.10)$$

whose partial derivatives with respect to the Hamilton variables are

$$\left.\begin{aligned}\frac{\partial H(t, \mathbf{q}, \mathbf{p})}{\partial t} &= \left[-\frac{\partial L(t, \mathbf{q}, \dot{\mathbf{q}})}{\partial t}\right]_{\dot{\mathbf{q}} = \dot{\mathbf{q}}(t, \mathbf{q}, \mathbf{p})}, \\ \nabla_{\mathbf{q}} H(t, \mathbf{q}, \mathbf{p}) &= \left[-\nabla_{\mathbf{q}} L(t, \mathbf{q}, \dot{\mathbf{q}})\right]_{\dot{\mathbf{q}} = \dot{\mathbf{q}}(t, \mathbf{q}, \mathbf{p})}, \\ \nabla_{\mathbf{p}} H(t, \mathbf{q}, \mathbf{p}) &= \dot{\mathbf{q}} + \left(\frac{\partial \dot{\mathbf{q}}}{\partial \mathbf{p}}\right)^T \left(\mathbf{p} - \nabla_{\dot{\mathbf{q}}} L(t, \mathbf{q}, \dot{\mathbf{q}})\right), \end{aligned}\right\}$$

from which (10.8) and (10.9) follow if we take into account (10.5) and (10.3). □

Remark 10.2. When a dynamic system is expressed in the form (10.9) with respect to the Hamiltonian $H(t, \mathbf{q}, \mathbf{p})$, it is said to be given in the **Hamiltonian canonical form**. For brevity, a system that is in the form (10.9) will be called a **Hamiltonian system**.

The first relation in (10.9) allows to obtain $L(t, \mathbf{q}, \dot{\mathbf{q}})$ from $H(t, \mathbf{q}, \mathbf{p})$.

Example 10.5. Suppose a system is Hamiltonian with

$$H(t, \mathbf{q}, \mathbf{p}) = p_1 p_2 + q_2 q_1.$$

Show how we can recuperate the corresponding Lagrange function.

From (10.6) we have

$$L(t, \mathbf{q}, \dot{\mathbf{q}}) = [(\mathbf{p}, \dot{\mathbf{q}}) - H(t, \mathbf{q}, \mathbf{p})]_{\mathbf{p} = \mathbf{p}(\mathbf{q}, \dot{\mathbf{q}})},$$

where $\mathbf{p} = \mathbf{p}(\mathbf{q}, \dot{\mathbf{q}})$ denotes the inverse function to the transformation $\dot{\mathbf{q}} = \nabla_{\mathbf{p}} H(\mathbf{q}, \mathbf{p})$. Direct calculation gives

$$\dot{\mathbf{q}} = \nabla_{\mathbf{p}} H(t, \mathbf{q}, \mathbf{p}) = \begin{bmatrix} p_2 & p_1 \end{bmatrix}^T,$$

that is,

$$\mathbf{p} = \begin{bmatrix} \dot{q}_2 & \dot{q}_1 \end{bmatrix}^T,$$

which leads to the representation

$$[H(t, \mathbf{q}, \mathbf{p})]_{\mathbf{p} = \mathbf{p}(\mathbf{q}, \dot{\mathbf{q}})} = \dot{q}_1 \dot{q}_2 + q_2 q_1,$$

and hence

$$L(t, \mathbf{q}, \dot{\mathbf{q}}) = \left[\left(\begin{bmatrix} \dot{q}_2 \\ \dot{q}_1 \end{bmatrix}, \begin{bmatrix} \dot{q}_1 \\ \dot{q}_2 \end{bmatrix}\right) - H(t, \mathbf{q}, \mathbf{p})\right]_{\mathbf{p} = \mathbf{p}(\mathbf{q}, \dot{\mathbf{q}})}.$$

Finally we get

$$L(t, \mathbf{q}, \dot{\mathbf{q}}) = \dot{q}_1 \dot{q}_2 - q_2 q_1.$$

Example 10.6. Let us build the Lagrange functions for the following Hamiltonians:
a)
$$H(t, \mathbf{q}, \mathbf{p}) = q_1 p_2 - q_2 p_1 + a \left(p_1^2 + p_2^2 \right),$$

b)
$$H(t, \mathbf{q}, \mathbf{p}) = \frac{1}{2} \left(p_1^2 + \frac{p_2^2}{\sin^2 q_1} \right) - a \cos q_1.$$

Let us present the required construction for each function of Hamilton.
a) In view of the relation
$$\dot{\mathbf{q}} = \nabla_\mathbf{p} H(t, \mathbf{q}, \mathbf{p}) = \begin{bmatrix} -q_2 + 2ap_1 \\ q_1 + 2ap_2 \end{bmatrix},$$

we have
$$\mathbf{p} = \frac{1}{2a} \begin{bmatrix} \dot{q}_1 + q_2 \\ \dot{q}_2 - q_1 \end{bmatrix},$$

which gives
$$[H(t, \mathbf{q}, \mathbf{p})]_{\mathbf{p}=\mathbf{p}(\mathbf{q},\dot{\mathbf{q}})} = \frac{1}{2a} \left[q_1 (\dot{q}_2 - q_1) - q_2 (\dot{q}_1 + q_2) \right] +$$
$$\frac{1}{4a} \left[(\dot{q}_1 + q_2)^2 + (\dot{q}_2 - q_1)^2 \right].$$

That is why (in view of (10.6))
$$L(t, \mathbf{q}, \dot{\mathbf{q}}) = \left[\frac{1}{2a} \left(\begin{bmatrix} \dot{q}_1 + q_2 \\ \dot{q}_2 - q_1 \end{bmatrix}, \begin{bmatrix} \dot{q}_1 \\ \dot{q}_2 \end{bmatrix} \right) - H(t, \mathbf{q}, \mathbf{p}) \right]_{\mathbf{p}=\mathbf{p}(\mathbf{q},\dot{\mathbf{q}})},$$

implying
$$L(t, \mathbf{q}, \dot{\mathbf{q}}) = \frac{1}{4a} \left[(\dot{q}_1 + q_2)^2 + (\dot{q}_2 - q_1)^2 \right].$$

b) We have
$$\dot{\mathbf{q}} = \nabla_\mathbf{p} H(t, \mathbf{q}, \mathbf{p}) = \begin{bmatrix} p_1 \\ \dfrac{p_2}{\sin^2 q_1} \end{bmatrix},$$

which gives
$$\mathbf{p} = \begin{bmatrix} \dot{q}_1 \\ \dot{q}_2 \sin^2 q_1 \end{bmatrix}.$$

Hence

$$[H(t, \mathbf{q}, \mathbf{p})]_{\mathbf{p}=\mathbf{p}(\mathbf{q}, \dot{\mathbf{q}})} = \frac{1}{2}\left(\dot{q}_1^2 + (\dot{q}_2 \sin q_1)^2\right) - a \cos q_1,$$

and therefore

$$L(t, \mathbf{q}, \dot{\mathbf{q}}) = \left[\left(\begin{bmatrix}\dot{q}_1 \\ \dot{q}_2 \sin^2 q_1\end{bmatrix}, \begin{bmatrix}\dot{q}_1 \\ \dot{q}_2\end{bmatrix}\right) - H(t, \mathbf{q}, \mathbf{p})\right]_{\mathbf{p}=\mathbf{p}(\mathbf{q}, \dot{\mathbf{q}})} =$$
$$\frac{1}{2}\left[\dot{q}_1^2 + (\dot{q}_2 \sin q_1)^2\right] + a \cos q_1.$$

The canonical descriptions (10.9) must meet certain conditions; these are given in the result that follows.

Theorem 10.2. *If a system of equations*

$$\left.\begin{aligned}\dot{\mathbf{q}} &= \mathbf{f}(t, \mathbf{q}, \mathbf{p}), \\ \dot{\mathbf{p}} &= \mathbf{g}(t, \mathbf{q}, \mathbf{p}),\end{aligned}\right\} \tag{10.11}$$

with vector functions $\mathbf{f}, \mathbf{g} : [0, \infty) \times R^n \times R^n \to R^n$, *which are some continuously differentiable given functions, is the canonical description of some dynamic system, then the following two relationships must be satisfied:*

1.

$$\frac{\partial \mathbf{f}}{\partial \mathbf{p}} = \left(\frac{\partial \mathbf{f}}{\partial \mathbf{p}}\right)^T, \quad \frac{\partial \mathbf{g}}{\partial \mathbf{q}} = \left(\frac{\partial \mathbf{g}}{\partial \mathbf{q}}\right)^T, \tag{10.12}$$

2.

$$\frac{\partial \mathbf{f}}{\partial \mathbf{q}} = -\left(\frac{\partial \mathbf{g}}{\partial \mathbf{p}}\right)^T. \tag{10.13}$$

Proof. If (10.11) is a canonical form of Hamilton, then, in view of (10.9), there is a Hamilton function $H(t, \mathbf{q}, \mathbf{p})$ such that

$$\left.\begin{aligned}\mathbf{f} &= \nabla_\mathbf{p} H(t, \mathbf{q}, \mathbf{p}), \\ \mathbf{g} &= -\nabla_\mathbf{q} H(t, \mathbf{q}, \mathbf{p}).\end{aligned}\right\} \tag{10.14}$$

Then

$$\left.\begin{aligned}\frac{\partial \mathbf{f}}{\partial \mathbf{p}} &= \nabla_\mathbf{p}^2 H(t, \mathbf{q}, \mathbf{p}), \\ \frac{\partial \mathbf{g}}{\partial \mathbf{q}} &= -\nabla_\mathbf{q}^2 H(t, \mathbf{q}, \mathbf{p}),\end{aligned}\right\}$$

where the symmetry condition (10.12) is followed. Now cross-deriving (10.14) we get

$$\left.\begin{array}{l}\dfrac{\partial \mathbf{f}}{\partial \mathbf{q}} = \dfrac{\partial}{\partial \mathbf{q}} \left(\nabla_\mathbf{p} H\left(t, \mathbf{q}, \mathbf{p}\right)\right), \\ \dfrac{\partial \mathbf{g}}{\partial \mathbf{p}} = -\dfrac{\partial}{\partial \mathbf{p}} \left(\nabla_\mathbf{q} H\left(t, \mathbf{q}, \mathbf{p}\right)\right),\end{array}\right\}$$

and the property

$$\dfrac{\partial}{\partial \mathbf{q}} \left(\nabla_\mathbf{p} H\left(t, \mathbf{q}, \mathbf{p}\right)\right) = \dfrac{\partial}{\partial \mathbf{p}} \left(\nabla_\mathbf{q} H\left(t, \mathbf{q}, \mathbf{p}\right)\right)$$

leads to the condition (10.13). □

The previous result allows to solve the following problem.

Example 10.7. Determine the conditions that the system of equations

$$\left.\begin{array}{l}\dot{\mathbf{q}} = A\mathbf{q} + B\mathbf{p}, \\ \dot{\mathbf{p}} = C\mathbf{q} + D\mathbf{p}\end{array}\right\} \tag{10.15}$$

must meet in order to be a Hamilton's canonical description of a dynamic system. Apply the conditions obtained for the system

$$\left.\begin{array}{l}\dot{x}_1 = x_1 + x_2, \\ \dot{x}_2 = 3x_1 + \alpha x_2,\end{array}\right\} \tag{10.16}$$

and calculate the value of α that makes it in the canonical form of Hamilton.

In the notation of the previous theorem, for (10.15) we have

$$\left.\begin{array}{l}\mathbf{f} = A\mathbf{q} + B\mathbf{p}, \\ \mathbf{g} = C\mathbf{q} + D\mathbf{p},\end{array}\right\}$$

which is why

$$\left.\begin{array}{ll}\dfrac{\partial \mathbf{f}}{\partial \mathbf{q}} = A, & \dfrac{\partial \mathbf{f}}{\partial \mathbf{p}} = B, \\ \dfrac{\partial \mathbf{g}}{\partial \mathbf{q}} = C, & \dfrac{\partial \mathbf{g}}{\partial \mathbf{p}} = D.\end{array}\right\}$$

So, we must have, by condition (10.12),

$$B = B^T, \quad C = C^T, \tag{10.17}$$

and by condition (10.13),

$$A = -D^T. \tag{10.18}$$

The conclusions (10.17) and (10.18), applied to the system (10.16), under the identification

$$q = x_1, \quad p = x_2,$$

lead to the condition

$$\alpha = -1.$$

10.3 First integrals

The expression of a dynamic system in Hamilton's canonical form has consequences that go beyond being a mere alternative form to Lagrange's equations. Remember from Chapter 6 that (10.1) can be expressed in the normal format, namely,

$$\ddot{\mathbf{q}} = \mathcal{F}(t, \mathbf{q}, \dot{\mathbf{q}}), \quad \mathbf{q} \in R^n.$$

Therefore, the dynamics of the system represented by (10.1) is described by a system of n nonlinear differential equations of the second order. However, in the alternative form of Hamilton (10.9), the same system is governed by $2n$ nonlinear differential equations of the first order. This observation is the great contribution of the technique addressed in this chapter, since it is known that, comparatively, there are many more results applicable to this last type of equations than for the first. In fact, this section presents several such results.

It has already been mentioned that in mechanics the functions are smooth, so in the following, this fact is used without explicitly mentioning it.

Definition 10.3. A function $f(t, \mathbf{q}, \mathbf{p})$ that is constant in the trajectories of a Hamiltonian system is called the **first integral** of that system.

The previous definition says that if the total temporal derivative of f, given by

$$\frac{df}{dt} = \frac{\partial f}{\partial t} + \sum_{i=1}^{n} \left(\frac{\partial f}{\partial q_i} \dot{q}_i + \frac{\partial f}{\partial p_i} \dot{p}_i \right),$$

along the paths of a system in the form (10.9)

$$\dot{q}_i = \frac{\partial H}{\partial p_i}, \quad \dot{p}_i = -\frac{\partial H}{\partial q_i}, \quad i = 1, \dots, n, \tag{10.19}$$

for some function $H(t, \mathbf{q}, \mathbf{p})$, that is,

$$\frac{df}{dt} = \frac{\partial f}{\partial t} + \sum_{i=1}^{n} \left(\frac{\partial f}{\partial q_i} \frac{\partial H}{\partial p_i} - \frac{\partial f}{\partial p_i} \frac{\partial H}{\partial q_i} \right),$$

is identically zero, then f is the first integral of the system (10.19).

The previous discussion allows to formulate in a trivial way the following result.

Theorem 10.3. *A function $f(t, \mathbf{q}, \mathbf{p})$ is the first integral of a Hamiltonian system if and only if*

$$\frac{\partial f}{\partial t} + (f, H) \equiv 0,$$

where

$$(f, H) := \sum_{i=1}^{n} \left(\frac{\partial f}{\partial q_i} \frac{\partial H}{\partial p_i} - \frac{\partial f}{\partial p_i} \frac{\partial H}{\partial q_i} \right) \tag{10.20}$$

receives the name of the **Lee–Poisson (or simply Poisson) bracket.**

The importance of the first integrals is that they allow reducing the number of differential equations to solve in a Hamiltonian system. In other words, suppose the Hamiltonian system

$$\left. \begin{array}{l} \dot{\mathbf{q}} = \nabla_{\mathbf{p}} H(t, \mathbf{q}, \mathbf{p}), \\ \dot{\mathbf{p}} = -\nabla_{\mathbf{q}} H(t, \mathbf{q}, \mathbf{p}) \end{array} \right\} \tag{10.21}$$

has as first integrals the following linearly independent relationships:[1]

$$\left. \begin{array}{l} f_1(t, \mathbf{q}, \mathbf{p}) \equiv C_1, \\ \vdots \\ f_l(t, \mathbf{q}, \mathbf{p}) \equiv C_l, \end{array} \right\} \tag{10.22}$$

where C_i, $i = 1, ..., l$, denote constants. Then a Hamilton $\mathbf{q}$ or $\mathbf{p}$ variable can be found from each of the relationships, which eliminates the need to solve its corresponding differential equation (10.21).

Conclusion 10.1. *Based on the above, it can be said that the first l integrals (10.22) reduce the order of the system (10.21) to $2n - l$. So, if $l = 2n$, then it is not necessary to solve any differential equation at all.*

---
[1] With the notation

$$\mathbf{z} := \begin{bmatrix} \mathbf{q} \\ \mathbf{p} \end{bmatrix}$$

the set of expressions (10.22) is linearly independent if and only if all the possible determinants of order $l \times l$ of the matrix

$$\begin{bmatrix} \frac{\partial f_1}{\partial z_1} & \frac{\partial f_1}{\partial z_2} & \cdots & \frac{\partial f_1}{\partial z_{2n}} \\ \frac{\partial f_2}{\partial z_1} & \frac{\partial f_2}{\partial z_2} & \cdots & \frac{\partial f_2}{\partial z_{2n}} \\ \vdots & \vdots & \ddots & \vdots \\ \frac{\partial f_l}{\partial z_1} & \frac{\partial f_l}{\partial z_2} & \cdots & \frac{\partial f_l}{\partial z_{2n}} \end{bmatrix}$$

are non-zero for any $(t, \mathbf{z})$.

10.4 Some properties of first integrals

Given the importance of the first integrals, several of its properties are now investigated.

10.4.1 Cyclic coordinates

Lemma 10.1. *If among the arguments of the Lagrange function $L(t, \mathbf{q}, \dot{\mathbf{q}})$ one of the coordinates q_α does not appear, that is,*

$$L = L(t, q_1, \cdots, q_{\alpha-1}, q_{\alpha+1}, \cdots, q_n, \dot{\mathbf{q}}),$$

which is true if and only if

$$\frac{\partial L}{\partial q_\alpha} = 0 \ \forall (t, \mathbf{q}, \dot{\mathbf{q}}),$$

then there is the first integral

$$f(t, \mathbf{q}, \mathbf{p}) \equiv C_\alpha,$$

where, in addition,

$$f(t, \mathbf{q}, \mathbf{p}) = p_\alpha.$$

Proof. The Lagrange equation in (10.1), corresponding to α, is given by

$$\frac{d}{dt}\frac{\partial L}{\partial \dot{q}_\alpha} - \frac{\partial L}{\partial q_\alpha} = 0,$$

but in view of

$$\frac{\partial L}{\partial \dot{q}_\alpha} = p_\alpha, \quad \frac{\partial L}{\partial q_\alpha} = 0,$$

it is reduced to

$$\frac{dp_\alpha}{dt} = 0,$$

which gives

$$p_\alpha = \operatorname*{const}_{t} = C_\alpha.$$

□

Definition 10.4. If for some $\alpha \in \{1, 2, ..., n\}$ we have

$$\frac{\partial L}{\partial q_\alpha} \equiv 0,$$

then the coordinate q_α is called **cyclic**.

By the previous lemma, if q_α is cyclic, then

$$p_\alpha \equiv C_\alpha.$$

Therefore, in the Hamilton function corresponding to the system in question, both q_α and p_α do not appear, that is,

$$H = H(t, q_1, \cdots, q_{\alpha-1}, q_{\alpha+1}, \cdots, q_n; p_1, \cdots, p_{\alpha-1}, C_\alpha, p_{\alpha+1}, \cdots, p_n), \tag{10.23}$$

and since

$$\dot{q}_\alpha = \frac{\partial H}{\partial p_\alpha} = \frac{\partial H}{\partial C_\alpha},$$

by direct integration we get

$$q_\alpha(t) = q_\alpha(0) + \int_{\tau=0}^{t} \frac{\partial}{\partial C_\alpha} H_\alpha\left(\tau, q_{i \neq \alpha}(\tau), p_{i \neq \alpha}(\tau), C_\alpha\right) d\tau,$$

where $H_\alpha\left(\tau, q_{i \neq \alpha}(\tau), p_{i \neq \alpha}(\tau), C_\alpha\right)$ denotes (10.23).

Conclusion 10.2. *It may be concluded that each cyclic coordinate reduces the number of differential equations of the description (10.9) by 2.*

10.4.2 Some properties of the Poisson brackets

From the last theorem we know that the function $f(t, \mathbf{q}, \mathbf{p})$ is the first integral of the dynamic system with canonical description (10.9) with respect to some function of Hamilton $H(t, \mathbf{q}, \mathbf{p})$ if and only if

$$\frac{\partial f}{\partial t} + (f, H) \equiv 0, \tag{10.24}$$

and if in particular $\dfrac{\partial f}{\partial t} \equiv 0$, then the condition (10.24) results in

$$(f, H) \equiv 0, \tag{10.25}$$

which shows the importance of the Lee–Poisson brackets in the characterization of the first time-independent integrals.

Lemma 10.2. *Consider the functions $\varphi(t, \mathbf{q}, \mathbf{p})$, $\psi(t, \mathbf{q}, \mathbf{p})$, and $\chi(t, \mathbf{q}, \mathbf{p})$. Then the following properties hold:*
a)

$$(\varphi, \psi) = -(\psi, \varphi),$$

b)
$$(c\varphi, \psi) = c(\varphi, \psi), \quad c = \text{const},$$

c)
$$(\varphi + \psi, \chi) = (\varphi, \chi) + (\psi, \chi),$$

d) *if s is a Hamilton variable, that is,*
$$s \in \{t, p_1, q_1, \cdots, p_n, q_n\},$$
then
$$\frac{\partial}{\partial s}(\varphi, \psi) = \left(\frac{\partial \varphi}{\partial s}, \psi\right) + \left(\varphi, \frac{\partial \psi}{\partial s}\right), \qquad (10.26)$$

e) *the Poisson identity*
$$((\varphi, \psi), \chi) + ((\psi, \chi), \varphi) + ((\chi, \varphi), \psi) = 0.$$

Proof. Sections (a)–(d) follow directly from the definition:

a)
$$(\varphi, \psi) = \sum_{i=1}^{n} \left(\frac{\partial \varphi}{\partial q_i} \frac{\partial \psi}{\partial p_i} - \frac{\partial \varphi}{\partial p_i} \frac{\partial \psi}{\partial q_i}\right) =$$
$$-\sum_{i=1}^{n} \left(\frac{\partial \psi}{\partial q_i} \frac{\partial \varphi}{\partial p_i} - \frac{\partial \psi}{\partial p_i} \frac{\partial \varphi}{\partial q_i}\right) = -(\psi, \varphi),$$

b)
$$(c\varphi, \psi) = \sum_{i=1}^{n} \left(\frac{\partial c\varphi}{\partial q_i} \frac{\partial \psi}{\partial p_i} - \frac{\partial c\varphi}{\partial p_i} \frac{\partial \psi}{\partial q_i}\right) =$$
$$c \sum_{i=1}^{n} \left(\frac{\partial \varphi}{\partial q_i} \frac{\partial \psi}{\partial p_i} - \frac{\partial \varphi}{\partial p_i} \frac{\partial \psi}{\partial q_i}\right) = c(\varphi, \psi),$$

c)
$$(\varphi + \psi, \chi) = \sum_{i=1}^{n} \left(\frac{\partial(\varphi + \psi)}{\partial q_i} \frac{\partial \chi}{\partial p_i} - \frac{\partial(\varphi + \psi)}{\partial p_i} \frac{\partial \chi}{\partial q_i}\right) =$$
$$\sum_{i=1}^{n} \left(\frac{\partial \varphi}{\partial q_i} \frac{\partial \chi}{\partial p_i} - \frac{\partial \varphi}{\partial p_i} \frac{\partial \chi}{\partial q_i}\right) + \sum_{i=1}^{n} \left(\frac{\partial \psi}{\partial q_i} \frac{\partial \chi}{\partial p_i} - \frac{\partial \psi}{\partial p_i} \frac{\partial \chi}{\partial q_i}\right) =$$
$$(\varphi, \chi) + (\psi, \chi),$$

d)

$$\frac{\partial}{\partial s}(\varphi, \psi) = \frac{\partial}{\partial s} \sum_{i=1}^{n} \left(\frac{\partial \varphi}{\partial q_i} \frac{\partial \psi}{\partial p_i} - \frac{\partial \varphi}{\partial p_i} \frac{\partial \psi}{\partial q_i} \right) =$$

$$\sum_{i=1}^{n} \left[\left(\frac{\partial}{\partial s} \frac{\partial \varphi}{\partial q_i} \right) \frac{\partial \psi}{\partial p_i} - \left(\frac{\partial}{\partial s} \frac{\partial \varphi}{\partial p_i} \right) \frac{\partial \psi}{\partial q_i} \right] +$$

$$\sum_{i=1}^{n} \left[\frac{\partial \varphi}{\partial q_i} \left(\frac{\partial}{\partial s} \frac{\partial \psi}{\partial p_i} \right) - \frac{\partial \varphi}{\partial p_i} \left(\frac{\partial}{\partial s} \frac{\partial \psi}{\partial q_i} \right) \right].$$

This result (10.26) follows trivially, because due to the smoothness of φ and ψ,

$$\frac{\partial}{\partial s} \frac{\partial \theta}{\partial r_i} = \frac{\partial}{\partial r_i} \frac{\partial \theta}{\partial s}, \quad \theta \in \{\varphi, \psi\}, \quad r \in \{p, q\}, \quad i = 1, \ldots, n.$$

e) We have

$$((\varphi, \psi), \chi) =$$

$$\sum_{i=1}^{n} \left(\frac{\partial \sum_{j=1}^{n} \left(\frac{\partial \varphi}{\partial q_j} \frac{\partial \psi}{\partial p_j} - \frac{\partial \varphi}{\partial p_j} \frac{\partial \psi}{\partial q_j} \right)}{\partial q_i} \frac{\partial \chi}{\partial p_i} - \frac{\partial \sum_{j=1}^{n} \left(\frac{\partial \varphi}{\partial q_j} \frac{\partial \psi}{\partial p_j} - \frac{\partial \varphi}{\partial p_j} \frac{\partial \psi}{\partial q_j} \right)}{\partial p_i} \frac{\partial \chi}{\partial q_i} \right) =$$

$$\sum_{i=1}^{n} \left(\sum_{j=1}^{n} \frac{\partial \left(\frac{\partial \varphi}{\partial q_j} \frac{\partial \psi}{\partial p_j} - \frac{\partial \varphi}{\partial p_j} \frac{\partial \psi}{\partial q_j} \right)}{\partial q_i} \frac{\partial \chi}{\partial p_i} - \sum_{j=1}^{n} \frac{\partial \left(\frac{\partial \varphi}{\partial q_j} \frac{\partial \psi}{\partial p_j} - \frac{\partial \varphi}{\partial p_j} \frac{\partial \psi}{\partial q_j} \right)}{\partial p_i} \frac{\partial \chi}{\partial q_i} \right)$$

$$= \sum_{i=1}^{n} \sum_{j=1}^{n} \left(\frac{\partial^2 \varphi}{\partial q_j \partial q_i} \frac{\partial \psi}{\partial p_j} + \frac{\partial \varphi}{\partial q_j} \frac{\partial^2 \psi}{\partial p_j \partial q_i} - \frac{\partial^2 \varphi}{\partial p_j \partial q_i} \frac{\partial \psi}{\partial q_j} - \frac{\partial \varphi}{\partial p_j} \frac{\partial^2 \psi}{\partial q_j \partial q_i} \right) \frac{\partial \chi}{\partial p_i}$$

$$- \sum_{i=1}^{n} \sum_{j=1}^{n} \left(\frac{\partial^2 \varphi}{\partial q_j \partial p_i} \frac{\partial \psi}{\partial p_j} + \frac{\partial \varphi}{\partial q_j} \frac{\partial^2 \psi}{\partial p_j \partial p_i} - \frac{\partial^2 \varphi}{\partial p_j \partial p_i} \frac{\partial \psi}{\partial q_j} - \frac{\partial \varphi}{\partial p_j} \frac{\partial^2 \psi}{\partial q_j \partial p_i} \right) \frac{\partial \chi}{\partial q_i},$$

and analogously,

$$((\varphi, \psi), \chi) =$$

$$\left(\frac{\partial^2 \varphi}{\partial q^2} \frac{\partial \psi}{\partial p} + \frac{\partial \varphi}{\partial q} \frac{\partial^2 \psi}{\partial p \partial q} - \frac{\partial^2 \varphi}{\partial p \partial q} \frac{\partial \psi}{\partial q} - \frac{\partial \varphi}{\partial p} \frac{\partial^2 \psi}{\partial q \partial q} \right) \frac{\partial \chi}{\partial p} -$$

$$\left(\frac{\partial^2 \varphi}{\partial q \partial p} \frac{\partial \psi}{\partial p} + \frac{\partial \varphi}{\partial q} \frac{\partial^2 \psi}{\partial p \partial p} - \frac{\partial^2 \varphi}{\partial p \partial p} \frac{\partial \psi}{\partial q} - \frac{\partial \varphi}{\partial p} \frac{\partial^2 \psi}{\partial q \partial p} \right) \frac{\partial \chi}{\partial q} =$$

$$\frac{\partial^2 \varphi}{\partial q^2} \frac{\partial \psi}{\partial p} \frac{\partial \chi}{\partial p} - \frac{\partial^2 \varphi}{\partial p \partial q} \frac{\partial \psi}{\partial q} \frac{\partial \chi}{\partial p} - \frac{\partial^2 \varphi}{\partial q \partial p} \frac{\partial \psi}{\partial p} \frac{\partial \chi}{\partial q} + \frac{\partial^2 \varphi}{\partial p \partial p} \frac{\partial \psi}{\partial q} \frac{\partial \chi}{\partial q} +$$

$$\frac{\partial \chi}{\partial q} \frac{\partial^2 \varphi}{\partial p \partial q} \frac{\partial \psi}{\partial p} - \frac{\partial \chi}{\partial p} \frac{\partial^2 \varphi}{\partial q \partial q} \frac{\partial \psi}{\partial p} - \frac{\partial \chi}{\partial q} \frac{\partial^2 \varphi}{\partial p \partial p} \frac{\partial \psi}{\partial q} + \frac{\partial \chi}{\partial p} \frac{\partial^2 \varphi}{\partial q \partial p} \frac{\partial \psi}{\partial q}$$

and

$$((\psi,\chi),\varphi) = \left(\frac{\partial^2\psi}{\partial q^2}\frac{\partial\chi}{\partial p} + \frac{\partial\psi}{\partial q}\frac{\partial^2\chi}{\partial p\partial q} - \frac{\partial^2\psi}{\partial p\partial q}\frac{\partial\chi}{\partial q} - \frac{\partial\psi}{\partial p}\frac{\partial^2\chi}{\partial q\partial q}\right)\frac{\partial\varphi}{\partial p}$$
$$- \left(\frac{\partial^2\psi}{\partial q\partial p}\frac{\partial\chi}{\partial p} + \frac{\partial\psi}{\partial q}\frac{\partial^2\chi}{\partial p\partial p} - \frac{\partial^2\psi}{\partial p\partial p}\frac{\partial\chi}{\partial q} - \frac{\partial\psi}{\partial p}\frac{\partial^2\chi}{\partial q\partial p}\right)\frac{\partial\varphi}{\partial q},$$
$$((\chi,\varphi),\psi) = \left(\frac{\partial^2\chi}{\partial q^2}\frac{\partial\varphi}{\partial p} + \frac{\partial\chi}{\partial q}\frac{\partial^2\varphi}{\partial p\partial q} - \frac{\partial^2\chi}{\partial p\partial q}\frac{\partial\varphi}{\partial q} - \frac{\partial\chi}{\partial p}\frac{\partial^2\varphi}{\partial q\partial q}\right)\frac{\partial\psi}{\partial p}$$
$$- \left(\frac{\partial^2\chi}{\partial q\partial p}\frac{\partial\varphi}{\partial p} + \frac{\partial\chi}{\partial q}\frac{\partial^2\varphi}{\partial p\partial p} - \frac{\partial^2\chi}{\partial p\partial p}\frac{\partial\varphi}{\partial q} - \frac{\partial\chi}{\partial p}\frac{\partial^2\varphi}{\partial q\partial p}\right)\frac{\partial\psi}{\partial q},$$

implying

$$((\varphi,\psi),\chi) + ((\psi,\chi),\varphi) + ((\chi,\varphi),\psi) = 0.$$

It is also easy to check that

$$(\varphi+\psi,\chi) + (\psi+\chi,\varphi) + (\chi+\varphi,\psi) =$$
$$(\varphi,\chi) + (\psi,\chi) + (\psi,\varphi) + (\chi,\varphi) + (\chi,\psi) + (\varphi,\psi).$$

$\square$

The newly characterized bracket allows to obtain first integrals from others, as shown in the following two results.

Theorem 10.4 (Jacobi–Poisson). *If f and g are first integrals of a system with Hamiltonian H, then the function*

$$\psi := (f,g) \tag{10.27}$$

is also the first integral of the same system.

Proof. Since f and g are first integrals, by (10.24) we have

$$\left.\begin{array}{l}\dfrac{\partial f}{\partial t} + (f,H) = 0,\\[2mm] \dfrac{\partial g}{\partial t} + (g,H) = 0,\end{array}\right\} \tag{10.28}$$

and for ψ, if it is also a first integral, we must have

$$\frac{\partial \psi}{\partial t} + (\psi, H) = 0. \tag{10.29}$$

Now, by the definition of ψ, given in (10.27), and by the properties of the Lee–Poisson brackets we have

$$\frac{\partial \psi}{\partial t} = \left(\frac{\partial f}{\partial t}, g\right) + \left(f, \frac{\partial g}{\partial t}\right)$$

and

$$(\psi, H) = ((f, g), H) = -((g, H), f) - ((H, f), g).$$

Using the properties

$$-(H, f) = (f, H), \quad \left(f, \frac{\partial g}{\partial t}\right) = -\left(\frac{\partial g}{\partial t}, f\right),$$

we get

$$\frac{\partial \psi}{\partial t} + (\psi, H) = \left(\frac{\partial f}{\partial t} + (f, H), g\right) - \left(\frac{\partial g}{\partial t} + (g, H), f\right),$$

where (10.29) follows, given the premise (10.28). □

Lemma 10.3. *If $f(t, \mathbf{q}, \mathbf{p}) = \underset{t}{\mathrm{const}}$ and therefore is the first integral of a Hamiltonian system and if q_α is a cyclic coordinate of the same system, then $\dfrac{\partial^i f}{\partial q_\alpha^i}$ for any $i = 1, 2..$ are also first integrals.*

Proof. Let H be the Hamilton function of the system. Since f is the first integral, by (10.24) we have

$$\frac{\partial f}{\partial t} + (f, H) \equiv 0.$$

The derivative of this expression with respect to q_α is given by

$$\frac{\partial^2 f}{\partial t \partial q_\alpha} + \frac{\partial}{\partial q_\alpha}(f, H) \equiv 0. \tag{10.30}$$

But by the property (d) of the Lee–Poisson brackets,

$$\frac{\partial}{\partial q_\alpha}(f, H) = \left(\frac{\partial f}{\partial q_\alpha}, H\right) + \left(f, \frac{\partial H}{\partial q_\alpha}\right) = \left(\frac{\partial f}{\partial q_\alpha}, H\right),$$

where $\dfrac{\partial H}{\partial q_\alpha} = 0$ since q_α is cyclic. Using this result, (10.30) is reduced to

$$\frac{\partial}{\partial t}\frac{\partial f}{\partial q_\alpha} + \left(\frac{\partial f}{\partial q_\alpha}, H\right) \equiv 0,$$

and hence, in view of (10.24) $\dfrac{\partial f}{\partial q_\alpha}$ is the first integral, implying that there exists a constant C such that

$$\frac{\partial f}{\partial q_\alpha}(t, \mathbf{q}, \mathbf{p}) \equiv C.$$

The fact that higher order derivatives are also first integrals is established by a similar process. □

In **stationary Hamiltonian systems**, where H is not an explicit function of time, it follows that H is also a first integral.

Lemma 10.4. *If the Hamilton function H of a dynamic system is such that*

$$\frac{\partial H}{\partial t} \equiv 0,$$

then H is a first integral, namely,

$$H(\mathbf{q}(t), \mathbf{p}(t)) = \underset{t}{\mathrm{const.}}$$

Proof. By the condition (10.25), H is the first integral if and only if

$$(H, H) = 0,$$

which is trivially satisfied in view of the definition of the Lee–Poisson brackets given in (10.20). □

The previous lemma allows us to test easily a result of Chapter 3 (Section 3.3) concerning the conservation of total mechanical energy in conservative systems.

Example 10.8. Let us show that the total mechanical energy of a conservative particle system remains constant. Since the mechanical energy of a particle system is the sum of the mechanical energies of the particles, it is enough to show the statement for the terms. If it has mass m, the kinetic energy is

$$T = \frac{m}{2}\left(\dot{x}^2 + \dot{y}^2 + \dot{z}^2\right),$$

while the potential energy $V(x, y, z)$ obeys some standard expression. As obtained in the first example of this chapter, if

$$\mathbf{q} := \begin{bmatrix} x & y & z \end{bmatrix}^T,$$

the generalized impulse is given by

$$\mathbf{p} = m\dot{\mathbf{q}},$$

implying

$$T(\mathbf{p}) = \frac{1}{2m}\|\mathbf{p}\|^2,$$

and the Hamiltonian of this system is

$$H(\mathbf{q}, \mathbf{p}) = T(\mathbf{p}) + V(\mathbf{q}),$$

that is, it is the total mechanical energy of the particle, which also does not explicitly depend on time. So, in view of the previous lemma, it remains constant, that is,

$$T(\mathbf{p}) + V(\mathbf{q}) \underset{t}{=} \text{const}.$$

10.4.3 First integrals by inspection

The first integrals can be identified by the form of the Hamiltonian. The results that follow address this fact.

Lemma 10.5. *If the Hamiltonian H of a given system has the structure*

$$H = H(f(q_1, q_2, \cdots, q_m, p_1, p_2, \cdots, p_m), q_{m+1}, p_{m+1}, \cdots, q_n, p_n; t),$$

then $f(q_1, q_2, \cdots, q_m, p_1, p_2, \cdots, p_m)$ *is a first integral, that is,*

$$f(q_1(t), q_2(t), \cdots, q_m(t), p_1(t), p_2(t), \cdots, p_m(t)) = C$$

for some constant C.

Proof. Note that

$$(f, H) = \sum_{i=1}^{n} \left(\frac{\partial f}{\partial q_i} \frac{\partial H}{\partial p_i} - \frac{\partial f}{\partial p_i} \frac{\partial H}{\partial q_i} \right) = \sum_{i=1}^{n} \left(\frac{\partial f}{\partial q_i} \frac{\partial H}{\partial f} \frac{\partial f}{\partial p_i} - \frac{\partial f}{\partial p_i} \frac{\partial H}{\partial f} \frac{\partial f}{\partial q_i} \right) = 0,$$

and the result follows from the condition (10.25). □

The following simple example illustrates the usefulness of the previous result.

Example 10.9. The immediate application of the previous lemma to the Hamiltonian function

$$H = \left(\frac{\cos q_1}{p_1} + q_2^2 + p_2^2 \right) \left(\frac{1}{3} \ln q_3 + p_3 \right)$$

allows to establish that

$$\frac{\cos q_1(t)}{p_1(t)} + q_2^2(t) + p_2^{2(t)} = C_1,$$

$$\frac{1}{3} \ln q_3(t) + p_3(t) = C_2,$$

where C_1, C_2 are some constants.

Lemma 10.6. *If the Hamiltonian H of a system has the form*

$$H = H(\varphi_1(q_1, p_1), \varphi_2(q_2, p_2), \cdots, \varphi_n(q_n, p_n); t),$$

Hamiltonian formalism

then $\{\varphi_i(q_i, p_i)\}_{i=1}^{n}$ are the first integrals of the system, namely,

$$\varphi_i(q_i(t), p_i(t)) = C_i, \quad i = 1, ..., n,$$

with some constant C_i, $i = 1, ..., n$.

Proof. Since for all $l = 1, ..., n$

$$\frac{d}{dt}\varphi_l = \frac{\partial}{\partial t}\varphi_l + (\varphi_l, H) =$$

$$(\varphi_l, H) = \sum_{i=1}^{n}\left(\frac{\partial \varphi_l}{\partial q_i}\frac{\partial H}{\partial p_i} - \frac{\partial \varphi_l}{\partial p_i}\frac{\partial H}{\partial q_i}\right) =$$

$$\frac{\partial \varphi_l}{\partial q_l}\frac{\partial H}{\partial \varphi_l}\frac{\partial \varphi_l}{\partial p_l} - \frac{\partial \varphi_l}{\partial p_l}\frac{\partial H}{\partial \varphi_l}\frac{\partial \varphi_l}{\partial q_l} = 0,$$

in view of the condition (10.25) the statement follows. □

An illustrative example of the result is presented below.

Example 10.10.
a) If

$$H = \sum_{i=1}^{n} a_i(t)\sin\left(q_i^2 + p_i^2\right)$$

is a Hamiltonian function of some system, by the previous lemma, it is concluded that

$$\sin\left(q_i^2(t) + p_i^2(t)\right) = \alpha_i, \quad i = 1, ..., n,$$

for some constants α_i, $i = 1, ..., n$.

b) If the function

$$H = \sin\left(\sum_{i=1}^{n}\left(q_i^2 + p_i^2\right)\right)$$

is a Hamiltonian of some system, then

$$q_i^2(t) + p_i^2(t) = \beta_i, \quad i = 1, ..., n,$$

with some constants β_i, $i = 1, ..., n$.

Lemma 10.7 (On a telescopic structure). *Assume that the Hamiltonian function H of a system can be presented in the form*

$$H = H\left(\varphi_j\left(\cdots\varphi_2\left(\varphi_1(q_1, p_1); q_2, p_2\right); \cdots\right); q_{j+1}, p_{j+1}, \cdots, q_n, p_n; t\right),$$

for some j ($1 \leq j \leq n$). Then

$$\{\varphi_k \left(\varphi_{k-1} \left(\cdots \varphi_2 \left(\varphi_1 \left(q_1, p_1 \right); q_2, p_2 \right); \cdots ; q_{k-1}, p_{k-1} \right); q_k, p_k \right)\}_{i=1}^{j}$$

are the first integrals of this system so that there exist constants C_k, $k = 1, \ldots, j$, such that

$$\varphi_k \left(\varphi_{k-1} \left(\cdots \varphi_2 \left(\varphi_1 \left(q_1, p_1 \right); q_2, p_2 \right); \cdots ; q_{k-1}, p_{k-1} \right); q_k, p_k \right) = C_k.$$

Proof. Note that for any k ($1 \leq k \leq j$) and any i ($1 \leq i \leq n$) we have

$$\frac{\partial \varphi_k}{\partial r_i} = \frac{\partial H}{\partial \varphi_k} \left(\frac{\partial \varphi_k}{\partial \varphi_{k-1}} \cdots \frac{\partial \varphi_2}{\partial \varphi_1} \right) \frac{\partial \varphi_i}{\partial r_i}, \quad r \in \{p, q\}, \ i \leq k,$$

$$\frac{\partial H}{\partial r_i} = \frac{\partial H}{\partial \varphi_k} \left(\frac{\partial \varphi_k}{\partial \varphi_{k-1}} \cdots \frac{\partial \varphi_2}{\partial \varphi_1} \right) \frac{\partial \varphi_i}{\partial r_i},$$

and therefore

$$(\varphi_k, H) = \sum_{i=1}^{n} \left(\frac{\partial \varphi_k}{\partial q_i} \frac{\partial H}{\partial p_i} - \frac{\partial \varphi_k}{\partial p_i} \frac{\partial H}{\partial q_i} \right) = \sum_{i=1}^{j} \left(\frac{\partial \varphi_k}{\partial q_i} \frac{\partial H}{\partial p_i} - \frac{\partial \varphi_k}{\partial p_i} \frac{\partial H}{\partial q_i} \right)$$

$$\sum_{i=1}^{n} \left[\frac{\partial H}{\partial \varphi_k} \left(\frac{\partial \varphi_k}{\partial \varphi_{k-1}} \cdots \frac{\partial \varphi_2}{\partial \varphi_1} \right) \frac{\partial \varphi_i}{\partial q_i} \frac{\partial H}{\partial \varphi_k} \left(\frac{\partial \varphi_k}{\partial \varphi_{k-1}} \cdots \frac{\partial \varphi_2}{\partial \varphi_1} \right) \frac{\partial \varphi_i}{\partial p_i} - \right.$$

$$\frac{\partial H}{\partial \varphi_k} \left(\frac{\partial \varphi_k}{\partial \varphi_{k-1}} \cdots \frac{\partial \varphi_2}{\partial \varphi_1} \right) \frac{\partial \varphi_i}{\partial p_i} \frac{\partial H}{\partial \varphi_k} \left(\frac{\partial \varphi_k}{\partial \varphi_{k-1}} \cdots \frac{\partial \varphi_2}{\partial \varphi_1} \right) \frac{\partial \varphi_i}{\partial q_i} =$$

$$\left[\frac{\partial H}{\partial \varphi_k} \left(\frac{\partial \varphi_k}{\partial \varphi_{k-1}} \cdots \frac{\partial \varphi_2}{\partial \varphi_1} \right) \right]^2 \sum_{i=1}^{n} \left(\frac{\partial \varphi_i}{\partial q_i} \frac{\partial \varphi_i}{\partial p_i} - \frac{\partial \varphi_i}{\partial p_i} \frac{\partial \varphi_i}{\partial q_i} \right) = 0,$$

and again the result follows from the condition (10.25). $\square$

The newly formulated lemma serves to identify first integrals in systems as in the following example.

Example 10.11. Suppose the Hamiltonian function of a certain system is

$$H = p_1^2 + \sin q_1 + \frac{(p_2 + p_3 \cos q_2)^2}{q_2^2}.$$

Then, by the previous lemma, it follows that

$$p_3(t) = C_1,$$
$$\frac{(p_2(t) + p_3(t) \cos q_2(t))^2}{q_2^2(t)} = C_2,$$
$$p_1^2(t) + \sin q_1(t) = C_3$$

for some constants C_1, C_2, C_3.

Hamiltonian formalism

The chapter concludes with two more results on obtaining first integrals from the Hamiltonian form.

Lemma 10.8. *If a system has a Hamiltonian function of the form*

$$H = \frac{\sum_{i=1}^n f_i(q_i, p_i)}{\sum_{i=1}^n \varphi_i(q_i, p_i)}, \tag{10.31}$$

then $\{f_i(q_i, p_i) - H\varphi_i(q_i, p_i)\}_{i=1}^n$ *are the first integrals of the system, namely, there exist some constants* C_i, $i = 1, \ldots, n$, *such that*

$$f_i(q_i, p_i) - H\varphi_i(q_i, p_i) = C_i. \tag{10.32}$$

Proof. First let us show that for any functions $f(q, p)$ and $g(q, p)$, we have

$$(fg, H) = f(g, H) + g(f, H). \tag{10.33}$$

Indeed, by the definition of the Poisson brackets we have

$$(fg, H) = \sum_{i=1}^n \left[\frac{\partial(fg)}{\partial q_i} \frac{\partial H}{\partial p_i} - \frac{\partial(fg)}{\partial p_i} \frac{\partial H}{\partial q_i} \right] =$$

$$\sum_{i=1}^n \left[\left(\frac{\partial f}{\partial q_i} g + f \frac{\partial g}{\partial q_i} \right) \frac{\partial H}{\partial p_i} - \left(\frac{\partial f}{\partial p_i} g + f \frac{\partial g}{\partial p_i} \right) \frac{\partial H}{\partial q_i} \right] =$$

$$f \sum_{i=1}^n \left(\frac{\partial g}{\partial q_i} \frac{\partial H}{\partial p_i} - \frac{\partial g}{\partial p_i} \frac{\partial H}{\partial q_i} \right) + g \sum_{i=1}^n \left(\frac{\partial f}{\partial q_i} \frac{\partial H}{\partial p_i} - \frac{\partial f}{\partial p_i} \frac{\partial H}{\partial q_i} \right),$$

and the assertion (10.33) follows. In view of (10.25), formula (10.32) holds if and only if

$$(f_i(q_i, p_i) - H\varphi_i(q_i, p_i), H) = 0.$$

But, by (10.33)

$$(f_i - H\varphi_i, H) = (f_i, H) - (H\varphi_i, H) = (f_i, H) - H(\varphi_i, H) - \varphi_i(H, H),$$

and since $(H, H) = 0$ we may conclude that

$$(f_i - H\varphi_i, H) = \sum_{j=1}^n \left(\frac{\partial f_i}{\partial q_j} \frac{\partial H}{\partial p_j} - \frac{\partial f_i}{\partial p_j} \frac{\partial H}{\partial q_j} \right) - H \sum_{j=1}^n \left(\frac{\partial \varphi_i}{\partial q_j} \frac{\partial H}{\partial p_j} - \frac{\partial \varphi_i}{\partial p_j} \frac{\partial H}{\partial q_j} \right)$$

$$= \frac{\partial f_i}{\partial q_i} \frac{\partial H}{\partial p_i} - \frac{\partial f_i}{\partial p_i} \frac{\partial H}{\partial q_i} - H \left(\frac{\partial \varphi_i}{\partial q_i} \frac{\partial H}{\partial p_i} - \frac{\partial \varphi_i}{\partial p_i} \frac{\partial H}{\partial q_i} \right)$$

$$= \left(\frac{\partial f_i}{\partial q_i} - H \frac{\partial \varphi_i}{\partial q_i} \right) \frac{\partial H}{\partial p_i} - \left(\frac{\partial f_i}{\partial p_i} - H \frac{\partial \varphi_i}{\partial p_i} \right) \frac{\partial H}{\partial q_i}.$$

Now, deriving (10.31) we get

$$\frac{\partial H}{\partial r_j} = \frac{\frac{\partial f_j}{\partial r_j}\sum_{i=1}^n \varphi_i - \frac{\partial \varphi_j}{\partial r_j}\sum_{i=1}^n f_i}{\left(\sum_{i=1}^n \varphi_i\right)^2} = \frac{\frac{\partial f_j}{\partial r_j} - H\frac{\partial \varphi_j}{\partial r_j}}{\sum_{l=1}^n \varphi_l}, \qquad (10.34)$$

$r \in \{p, q\}, \quad j = 1, ..., n,$

which is why

$$(f_i - H\varphi_i, H) = \sum_{j=1}^n \left(\frac{\partial f_i}{\partial q_j} - H\frac{\partial \varphi_i}{\partial q_j}\right)\frac{\partial H}{\partial p_j} - \left(\frac{\partial f_i}{\partial p_j} - H\frac{\partial \varphi_i}{\partial p_j}\right)\frac{\partial H}{\partial q_i} =$$

$$\left(\frac{\partial f_i}{\partial q_i} - H\frac{\partial \varphi_i}{\partial q_i}\right)\frac{\partial H}{\partial p_i} - \left(\frac{\partial f_i}{\partial p_i} - H\frac{\partial \varphi_i}{\partial p_i}\right)\frac{\partial H}{\partial q_i} =$$

$$\frac{\partial f_i}{\partial q_i}\frac{\partial H}{\partial p_i} - \frac{\partial f_i}{\partial p_i}\frac{\partial H}{\partial q_i} = \frac{\partial f_i}{\partial q_i}\frac{\frac{\partial f_i}{\partial p_i} - H\frac{\partial \varphi_i}{\partial p_i}}{\sum_{l=1}^n \varphi_l} - \frac{\partial f_i}{\partial p_i}\frac{\frac{\partial f_i}{\partial q_i} - H\frac{\partial \varphi_i}{\partial q_i}}{\sum_{l=1}^n \varphi_l} = 0,$$

which concludes the proof. □

Lemma 10.9. *Consider a system with the Hamiltonian*

$$H = f(t)\frac{\sum_{i=1}^n \alpha_i\varphi_i(q_i, p_i)}{\sum_{i=1}^n \beta_i\varphi_i(q_i, p_i)},$$

where $\alpha_i, \beta_i, i = 1, ..., n$, are some constants. Then $\{\varphi_i(q_i, p_i)\}_{i=1}^n$ are the first integrals of the system, satisfying

$$\varphi_i(q_i(t), p_i(t)) = C_i, \quad i = 1, ..., n,$$

for some set of constants C_i.

Proof. Since $\varphi_i = \varphi_i(q_i, p_i)$ it follows that

$$(\varphi_i, H) = \sum_{j=1}^n \left(\frac{\partial \varphi_i}{\partial q_j}\frac{\partial H}{\partial p_j} - \frac{\partial \varphi_i}{\partial p_j}\frac{\partial H}{\partial q_j}\right) = \frac{\partial \varphi_i}{\partial q_i}\frac{\partial H}{\partial p_i} - \frac{\partial \varphi_i}{\partial p_i}\frac{\partial H}{\partial q_i}.$$

But in view of (10.34) H results in

$$\frac{\partial H}{\partial r_j} = f(t)\frac{\partial \varphi_j}{\partial r_j}\frac{\alpha_j \sum_{i=1}^n \beta_i\varphi_i - \beta_j \sum_{i=1}^n \alpha_i\varphi_i}{\left(\sum_{i=1}^n \beta_i\varphi_i\right)^2} =$$

$$\frac{\partial \varphi_j}{\partial r_j}\frac{\alpha_j f(t) - \beta_j H}{\sum_{i=1}^n \beta_i\varphi_i}, \quad r \in \{p, q\}, \quad j = 1, ..., n,$$

Hamiltonian formalism

so

$$(\varphi_i, H) = \frac{\partial \varphi_i}{\partial q_i} \frac{\partial \varphi_i}{\partial p_i} \frac{\alpha_i f(t) - \beta_i H}{\sum_{j=1}^n \beta_j \varphi_j} - \frac{\partial \varphi_i}{\partial p_i} \frac{\partial \varphi_i}{\partial q_i} \frac{\alpha_i f(t) - \beta_i H}{\sum_{j=1}^n \beta_j \varphi_j} = 0,$$

and by (10.25) the result follows. $\square$

10.5 Exercises

Exercise 10.1. The Hamiltonian H of a system is

$$H = \left(q_1^2 + p_1^2\right) F(p_2, ..., p_n, t).$$

Show that the movement of the system is given in the form

$$\left.\begin{array}{l} p_i(t) = \alpha_i, \\ q_i(t) = \alpha_1^2 \int \dfrac{\partial}{\partial \alpha_i} F(\alpha_2, ..., \alpha_n, t) \, dt + \beta_i, \end{array}\right\} \text{ for } i = 2, ..., n,$$

and

$$\left.\begin{array}{l} q_1(t) = \alpha_1 \sin\left(2 \int F(\alpha_2, ..., \alpha_n, t) \, dt + \beta_1\right), \\ q_1(t) = \alpha_1 \cos\left(2 \int F(\alpha_2, ..., \alpha_n, t) \, dt + \beta_1\right). \end{array}\right\}$$

Exercise 10.2. Show that using the Poisson brackets (10.20) the canonical Hamilton equations (10.9) can be represented as

$$\left.\begin{array}{l} \dot{q}_i = (q_i, H), \\ \dot{p}_i = (p_i, H), \end{array}\right\} i = 1, ..., n.$$

Exercise 10.3. The function $W(q, p, t)$ $(q, p \in R)$ satisfies the relation

$$\frac{\partial}{\partial t} W + (W, H) = F(t),$$

where (W, H) is the Poisson bracket. Check that the first integrals of the canonical system with the Hamiltonian H have the form

$$f_i(\mathbf{q}, \mathbf{p}, t) = W(q_i, p_i, t) - \int F(t) \, dt, \ i = 1, ..., n.$$

Exercise 10.4. A mechanical system with the Lagrangian

$$L = \frac{1}{2} \sum_{i=1}^4 \dot{q}_i^2 - \Pi(\mathbf{q})$$

has the first integral

$$f = \alpha (q_1\dot{q}_2 - \dot{q}_1 q_2) + \beta (q_3\dot{q}_4 - \dot{q}_3 q_4),$$

where α and β are constant parameters. Show that the potential energy $\Pi(q)$ of the system has the form

$$\Pi(\mathbf{q}) = F\left(q_1^2 + q_2^2, q_3^2 + q_4^2, \alpha \arctan\left(\frac{q_3}{q_4}\right) - \beta \arctan\left(\frac{q_1}{q_2}\right)\right).$$

Exercise 10.5. A dynamic system is described by a system of differential equations

$$\dot{x}_i = f_i(\mathbf{x}), \ i = 1, ..., n, \ \mathbf{x} \in R^n. \tag{10.35}$$

For each starting point $\mathbf{x}(0) = \mathbf{a}$ there is a temporary mean

$$\lim_{T \to \infty} \frac{1}{T} \int_{t=0}^{T} g(\mathbf{x}(\mathbf{a}, t)) dt = \psi(\mathbf{a}),$$

where $g: R^n \to R$ is some function and $\mathbf{x}(\mathbf{a}, t)$ is the solution of the system (10.35) with the initial value $\mathbf{a}$. Show that the function $\psi(\mathbf{x})$ will be the first integral of the system.

The Hamilton–Jacobi equation 11

Contents

11.1 Canonical transformations   337
11.2 The Hamilton–Jacobi method   339
11.3 Hamiltonian action and its variation   339
11.4 Integral invariants   343
     11.4.1 Integral invariants of Poincaré and Poincaré–Cartan   343
     11.4.2 The Lee Hwa Chung theorem   344
11.5 Canonicity criteria   347
     11.5.1 Poincaré theorem: (c, F)-criterion   347
     11.5.2 Analytical expression for the Hamiltonian after a coordinate canonical transformation   350
     11.5.3 Brackets of Lagrange   353
     11.5.4 Free canonical transformation and the S-canonicity criterion   356
11.6 The Hamilton–Jacobi equation   361
11.7 Complete integral of the Hamilton–Jacobi equation   362
     11.7.1 Complete integral   362
     11.7.2 Generalized-conservative (stationary) systems with first integrals   362
11.8 On relations with optimal control   369
     11.8.1 Problem formulation and value function   370
     11.8.2 Hamilton–Jacobi–Bellman equation   370
     11.8.3 Verification rule as a sufficient condition of optimality   371
     11.8.4 Affine dynamics with a quadratic cost   372
     11.8.5 The case when the Hamiltonian admits the existence of first integrals   375
     11.8.6 The deterministic Feynman–Kac formula: the general smooth case   376
11.9 Exercises   380

The canonical transformations of the dynamic variables, describing Hamiltonians in new variables, are considered. Several criteria of canonicity are studied. The Hamilton–Jacobi (HJ) equation that corresponds to Hamiltonian canonical equations is also considered and analyzed in detail. Their complete integrals (solutions) are found. This chapter shows also the relation between the HJ equation in mechanics of conservative systems and the dynamic programming method in optimal control theory.

11.1 Canonical transformations

The Hamiltonian H depends on the variables t, $\mathbf{q}$, and $\mathbf{p}$:

$$H = H(t, \mathbf{q}, \mathbf{p}),$$

so that

$$\dot{q}_i = \frac{\partial H}{\partial p_i}, \quad \dot{p}_i = -\frac{\partial H}{\partial q_i}, \quad i = \overline{1,n}. \qquad (11.1)$$

These variables are transformed into a set of new variables:

$$\left.\begin{array}{l} \tilde{q}_i = \varphi_i(t, \mathbf{q}, \mathbf{p}), \\ \tilde{p}_i = \psi_i(t, \mathbf{q}, \mathbf{p}). \end{array}\right\} \qquad (11.2)$$

We also will require the property of *one-to-one mappings* of (11.2) to be able to recuperate back the coordinates $(\mathbf{q}, \mathbf{p})$ from new ones $(\tilde{\mathbf{q}}, \tilde{\mathbf{p}})$. To make it possible, using the theorem on inverse functions, the following condition must be satisfied:

$$\det \begin{bmatrix} \frac{\partial}{\partial q_1}\varphi_1 & \cdots & \frac{\partial}{\partial q_1}\varphi_n & \frac{\partial}{\partial q_1}\psi_1 & \cdots & \frac{\partial}{\partial q_1}\psi_n \\ \frac{\partial}{\partial p_1}\varphi_1 & \cdots & \frac{\partial}{\partial p_1}\varphi_n & \frac{\partial}{\partial p_1}\psi_1 & \cdots & \frac{\partial}{\partial p_1}\psi_n \\ \vdots & \vdots & \vdots & \vdots & \vdots & \vdots \\ \frac{\partial}{\partial q_n}\varphi_1 & \cdots & \frac{\partial}{\partial q_n}\varphi_n & \frac{\partial}{\partial q_n}\psi_1 & \cdots & \frac{\partial}{\partial q_n}\psi_n \\ \frac{\partial}{\partial p_n}\varphi_1 & \cdots & \frac{\partial}{\partial p_n}\varphi_n & \frac{\partial}{\partial p_n}\psi_1 & \cdots & \frac{\partial}{\partial p_n}\psi_n \end{bmatrix} \neq 0 \qquad (11.3)$$

for all $(\mathbf{q}, \mathbf{p})$ and any $t \geq 0$.

Definition 11.1. The nonlinear mapping from the space $(t, \mathbf{q}, \mathbf{p})$ into the space $(t, \tilde{\mathbf{q}}, \tilde{\mathbf{p}})$ by means of the vector functions $\boldsymbol{\varphi}$ and $\boldsymbol{\psi}$ (see (11.2)), respectively, which satisfies the condition (11.3), is referred to as a **canonical transformation** if there exists a function $\tilde{H}(t, \tilde{\mathbf{q}}, \tilde{\mathbf{p}})$ such that the new variables $\tilde{\mathbf{q}}$ and $\tilde{\mathbf{p}}$ satisfy

$$\frac{d}{dt}\tilde{q}_i(t) = \frac{\partial \tilde{H}(t, \tilde{\mathbf{q}}, \tilde{\mathbf{p}})}{\partial \tilde{p}_i}, \quad \frac{d}{dt}\tilde{p}_i(t) = -\frac{\partial \tilde{H}(t, \tilde{\mathbf{q}}, \tilde{\mathbf{p}})}{\partial \tilde{q}_i}, \quad i = \overline{1,n}. \qquad (11.4)$$

The central idea of Hamilton was to look for a transformation such that the new Hamiltonian $\tilde{H}$ would be as simple as possible, for example, equal to zero or to a constant c. Therefore, following his idea, selecting

$$\tilde{H}(t, \tilde{\mathbf{q}}, \tilde{\mathbf{p}}) = 0 \quad \text{for any } (t, \tilde{\mathbf{q}}, \tilde{\mathbf{p}}),$$

we will have

$$\left.\begin{array}{l} \dfrac{d}{dt}\tilde{q}_i(t) = 0 \text{ and } \tilde{q}_i(t) = \alpha_i \underset{t}{=} \text{const}, \\ \dfrac{d}{dt}\tilde{p}_i(t) = 0 \text{ and } \tilde{q}_i(t) = \beta_i \underset{t}{=} \text{const}; \end{array}\right\} \qquad (11.5)$$

and from this point we do not need to resolve the system of differential equations (11.1), dealing now only with the system (11.5) of algebraic equations, which is significantly simpler.

11.2 The Hamilton–Jacobi method

This scheme consists in the following considerations.

Given the derivatives of the Hamilton variables

$$\dot{\mathbf{q}} = \nabla_{\bar{p}} H, \quad \dot{\mathbf{p}} = -\nabla_{\bar{q}} H,$$

as well as its canonical transformations $\tilde{\mathbf{q}} = \boldsymbol{\varphi}(t, \mathbf{q}, \mathbf{p})$ and $\tilde{\mathbf{p}} = \boldsymbol{\psi}(t, \mathbf{q}, \mathbf{p})$, respectively, which satisfy

$$\frac{d}{dt}\boldsymbol{\varphi}(t, \mathbf{q}, \mathbf{p}) = \nabla_{\bar{p}} \tilde{H}(t, \tilde{\mathbf{q}}, \tilde{\mathbf{p}}), \quad \frac{d}{dt}\boldsymbol{\psi}(t, \mathbf{q}, \mathbf{p}) = -\nabla_{\bar{q}} \tilde{H}(t, \tilde{\mathbf{q}}, \tilde{\mathbf{p}}),$$

we may realize the main idea to select these transformations in such a way that the new Hamiltonian would be as simple as possible. In the particular case where the Hamiltonian $\tilde{H}(t, \tilde{\mathbf{q}}, \tilde{\mathbf{p}})$ is equal to zero, namely,

$$\tilde{H}(t, \tilde{\mathbf{q}}, \tilde{\mathbf{p}}) = 0,$$

the transformation functions $\boldsymbol{\varphi}(t, \mathbf{q}, \mathbf{p})$ and $\tilde{\mathbf{p}} = \boldsymbol{\psi}(t, \mathbf{q}, \mathbf{p})$ are constant:

$$\tilde{\mathbf{q}} = \boldsymbol{\varphi}(t, \mathbf{q}, \mathbf{p}) = \underset{t}{\text{const}}, \quad \tilde{\mathbf{p}} = \boldsymbol{\psi}(t, \mathbf{q}, \mathbf{p})) = \underset{t}{\text{const}},$$

being able to redefine these functions by

$$\alpha_i = \varphi_i(t, \mathbf{q}, \mathbf{p}), \quad \beta_i = \psi_i(t, \mathbf{q}, \mathbf{p}).$$

So, the new algebraic system is of the type

$$q_i = q_i(t, \boldsymbol{\alpha}, \boldsymbol{\beta}), \quad p_i = p_i(t, \boldsymbol{\alpha}, \boldsymbol{\beta}).$$

Two problems arise:
1. Obtain criteria of "canonicity" for transformations $\boldsymbol{\varphi}$ and $\boldsymbol{\psi}$.
2. For which canonical transformations $\boldsymbol{\varphi}$ and $\boldsymbol{\psi}$ do we have $\tilde{H}(t, \tilde{\mathbf{q}}, \tilde{\mathbf{p}}) = 0$?

11.3 Hamiltonian action and its variation

The first problem is solved using the so-called invariant integral of Poincaré and the concept of the Hamiltonian action defined by (see Fig. 11.1)

$$I(\alpha) := \int_{t_0(\alpha)}^{t_1(\alpha)} L(\tau, \mathbf{q}(\tau, \alpha), \dot{\mathbf{q}}(\tau, \alpha)) d\tau, \quad \alpha \in [0, 1]. \tag{11.6}$$

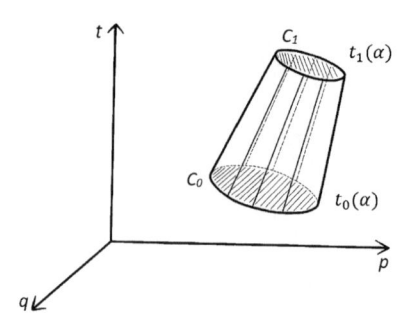

Figure 11.1 A family of trajectories of a Hamiltonian system in the extended state space with two initial and final contours.

Its derivative on α is

$$\left.\begin{aligned}I'(\alpha) &= L(t_1(\alpha), \mathbf{q}(t_1(\alpha)), \dot{\mathbf{q}}(t_1(\alpha))) t_1'(\alpha) - \\ &\quad L(t_0(\alpha), \mathbf{q}(t_0(\alpha)), \dot{\mathbf{q}}(t_0(\alpha))) t_0'(\alpha) + \\ &\quad \int_{t_0(\alpha)}^{t_1(\alpha)} \left[\nabla_{\mathbf{q}}^T L(\tau, \mathbf{q}(\tau, \alpha), \dot{\mathbf{q}}(\tau, \alpha)) \frac{\partial \mathbf{q}(\tau, \alpha)}{\partial \alpha} + \right. \\ &\quad \left. \nabla_{\dot{\mathbf{q}}}^T L(\tau, \mathbf{q}(\tau, \alpha), \dot{\mathbf{q}}(\tau, \alpha)) \frac{\partial \dot{\mathbf{q}}(\tau, \alpha)}{\partial \alpha} d\tau \right].\end{aligned}\right\} \quad (11.7)$$

Let us use the short notations:

$$L_{t=t_1(\alpha)} := L(t_1(\alpha), \mathbf{q}(t_1(\alpha)), \dot{\mathbf{q}}(t_1(\alpha))),$$
$$L_{t=t_0(\alpha)} := L(t_0(\alpha), \mathbf{q}(t_0(\alpha)), \dot{\mathbf{q}}(t_0(\alpha))),$$
$$L_{\tau,\alpha} := L(\tau, \mathbf{q}(\tau, \alpha), \dot{\mathbf{q}}(\tau, \alpha)).$$

Therefore we have

$$\left.\begin{aligned}\delta I(\alpha) &= L_{t=t_1(\alpha)} \delta t_1(\alpha) - L_{t=t_0(\alpha)} \delta t_0(\alpha) + \\ &\quad \int_{t_0(\alpha)}^{t_1(\alpha)} \sum_{i=1}^{n} \left[\frac{\partial L_{\tau,\alpha}}{\partial q_i} \delta q_i(\tau, \alpha) + \frac{\partial L_{\tau,\alpha}}{\partial \dot{q}_i} \delta \dot{q}_i(\tau, \alpha) \right] d\tau,\end{aligned}\right\} \quad (11.8)$$

where for $\tau \in (t_0(\alpha), t_1(\alpha))$

The Hamilton–Jacobi equation

$$\delta q_i(\tau, \alpha) = \frac{\partial q_i(\tau, \alpha)}{\partial \alpha} \delta \alpha := [\delta q_i]_\tau .$$

Integration by parts leads to the following expression:

$$\left.\begin{aligned}
\delta I(\alpha) &= L_{t=t_1(\alpha)} \delta t_1(\alpha) - L_{t=t_0(\alpha)} \delta t_0(\alpha) + \\
&\quad \int_{t_0(\alpha)}^{t_1(\alpha)} \sum_{i=1}^{n} \left[\frac{\partial L_{\tau,\alpha}}{\partial q_i} - \frac{d}{d\tau} \frac{\partial L_{\tau,\alpha}}{\partial \dot{q}_i} \right] \delta q_i(\tau, \alpha) d\tau + \\
&\quad \sum_{i=1}^{n} \underbrace{\frac{\partial L_{\tau,\alpha}}{\partial \dot{q}_i}}_{p_i(\tau,\alpha)} \delta q_i(\tau, \alpha) |_{t=t_1(\alpha)} - \sum_{i=1}^{n} \underbrace{\frac{\partial L_{\tau,\alpha}}{\partial \dot{q}_i}}_{p_i(\tau,\alpha)} \delta q_i(\tau, \alpha) |_{t=t_0(\alpha)} = \\
L_{t=t_1(\alpha)} &\delta t_1(\alpha) - L_{t=t_0(\alpha)} \delta t_0(\alpha) + \\
&\quad \sum_{i=1}^{n} p_i(t_1(\alpha), \alpha) \delta q_i(t_1(\alpha), \alpha) - \sum_{i=1}^{n} p_i(t_0(\alpha), \alpha) \delta q_i(t_0(\alpha), \alpha) \\
&\quad + \int_{t_0(\alpha)}^{t_1(\alpha)} \sum_{i=1}^{n} \left[\frac{\partial L_{\tau,\alpha}}{\partial q_i} - \frac{d}{d\tau} \frac{\partial L_{\tau,\alpha}}{\partial \dot{q}_i} \right] \delta q_i(\tau, \alpha) d\tau.
\end{aligned}\right\} \quad (11.9)$$

For the variations of the terminal conditions we have ($s = 0, 1$)

$$\left.\begin{aligned}
\delta q_i(t_s(\alpha), \alpha) &= \dot{q}_i(t_s(\alpha), \alpha) \delta t_s(\alpha) + \frac{\partial q_i(t_s(\alpha), \alpha)}{\partial \alpha} \delta \alpha = \\
\dot{q}_i(t_s(\alpha), \alpha) &\delta t_s(\alpha) + [\delta q_i]_{t_s(\alpha)},
\end{aligned}\right\}$$

which gives

$$\delta q_i(t_s(\alpha), \alpha) = [\delta q_i]_{t_s(\alpha)} + \dot{q}_i(t_s(\alpha), \alpha) \delta t_s(\alpha), \quad s = 0, 1. \quad (11.10)$$

Recalling that

$$H(t, \mathbf{q}, \mathbf{p}) = \left[\sum_{i=1}^{n} p_i \dot{q}_i - L(t, \mathbf{q}, \dot{\mathbf{q}}) \right]_{\dot{\mathbf{q}} = \dot{\mathbf{q}}(t, \mathbf{q}, \mathbf{p})},$$

$$[L(t, \mathbf{q}, \dot{\mathbf{q}})]_{\dot{\mathbf{q}} = \dot{\mathbf{q}}(t, \mathbf{q}, \mathbf{p})} = \sum_{i=1}^{n} p_i \dot{q}_i(t, \mathbf{q}, \mathbf{p}) - H(t, \mathbf{q}, \mathbf{p}),$$

substitution of (11.10) into (11.9) leads to the relation

$$\delta I(\alpha) = \left[\sum_{i=1}^{n} p_i \dot{q}_i(t, \mathbf{q}, \mathbf{p}) - H(t, \mathbf{q}, \mathbf{p})\right]_{t=t_1(\alpha)} \delta t_1(\alpha) +$$

$$\sum_{i=1}^{n} p_i(t_1(\alpha), \alpha) \left[[\delta q_i]_{t_1(\alpha)} + \dot{q}_i(t_1(\alpha), \alpha) \delta t_1(\alpha)\right]$$

$$- \left[\sum_{i=1}^{n} p_i \dot{q}_i(t, \mathbf{q}, \mathbf{p}) - H(t, \mathbf{q}, \mathbf{p})\right]_{t=t_0(\alpha)} \delta t_0(\alpha) -$$

$$\sum_{i=1}^{n} p_i(t_0(\alpha), \alpha) \left[[\delta q_i]_{t_0(\alpha)} + \dot{q}_i(t_0(\alpha), \alpha) \delta t_0(\alpha)\right]$$

$$\int_{t_0(\alpha)}^{t_1(\alpha)} \sum_{i=1}^{n} \left[\frac{\partial L_{\tau,\alpha}}{\partial q_i} - \frac{d}{d\tau}\frac{\partial L_{\tau,\alpha}}{\partial \dot{q}_i}\right] \delta q_i(\tau, \alpha) d\tau.$$

Simplification of the last equation gives

$$\delta I(\alpha) = \sum_{i=1}^{n} p_i(t_1(\alpha), \alpha) [\delta q_i]_{t_1(\alpha)} - [H(t, \mathbf{q}, \mathbf{p})]_{t=t_1(\alpha)} \delta t_1(\alpha)$$

$$- \left(\sum_{i=1}^{n} p_i(t_0(\alpha), \alpha) [\delta q_i]_{t_0(\alpha)} - [H(t, \mathbf{q}, \mathbf{p})]_{t=t_0(\alpha)} \delta t_0(\alpha)\right) +$$

$$\int_{t_0(\alpha)}^{t_1(\alpha)} \sum_{i=1}^{n} \left[\frac{\partial L_{\tau,\alpha}}{\partial q_i} - \frac{d}{d\tau}\frac{\partial L_{\tau,\alpha}}{\partial \dot{q}_i}\right] \delta q_i(\tau, \alpha) d\tau.$$

Defining

$$\left[\sum_{i=1}^{n} p_i [\delta q_i] - H(t, \mathbf{q}, \mathbf{p}) \delta t\right]_{t=t_0(\alpha)}^{t=t_1(\alpha)} :=$$

$$\sum_{i=1}^{n} p_i(t_1(\alpha), \alpha) [\delta q_i]_{t_1(\alpha)} - [H(t, \mathbf{q}, \mathbf{p})]_{t=t_1(\alpha)} \delta t_1(\alpha) -$$

$$\left(\sum_{i=1}^{n} p_i(t_0(\alpha), \alpha) [\delta q_i]_{t_0(\alpha)} - [H(t, \mathbf{q}, \mathbf{p})]_{t=t_0(\alpha)} \delta t_0(\alpha)\right),$$

the last relation becomes

The Hamilton–Jacobi equation

$$\delta I(\alpha) = \left[\sum_{i=1}^{n} p_i [\delta q_i] - H(t, \mathbf{q}, \mathbf{p}) \delta t\right]_{t=t_0(\alpha)}^{t=t_1(\alpha)} + \int_{t_0(\alpha)}^{t_1(\alpha)} \sum_{i=1}^{n} \left[\frac{\partial L_{\tau,\alpha}}{\partial q_i} - \frac{d}{d\tau}\frac{\partial L_{\tau,\alpha}}{\partial \dot{q}_i}\right] \delta q_i(\tau, \alpha) d\tau. \quad (11.11)$$

Remark 11.1. If the triple $(t, \mathbf{q}, \dot{\mathbf{q}})$ corresponds to the dynamic path (line) of a real mechanical system where

$$\frac{\partial L_{t,\alpha}}{\partial q_i} - \frac{d}{d\tau}\frac{\partial L_{t,\alpha}}{\partial \dot{q}_i} = 0 \ (i = 1, ..., n),$$

then

$$\delta I(\alpha) = \left[\sum_{i=1}^{n} p_i [\delta q_i] - H(t, \mathbf{q}, \mathbf{p}) \delta t\right]_{t=t_0(\alpha)}^{t=t_1(\alpha)}. \quad (11.12)$$

11.4 Integral invariants

Expression (11.12) for the variation of the Hamiltonian action in the extended space $(t, \mathbf{q}, \mathbf{p})$ allows to establish two fundamental statements of mechanical systems.

11.4.1 Integral invariants of Poincaré and Poincaré–Cartan

Theorem 11.1 (Poincaré, 1885).[1] *For any Hamiltonian system the following properties hold:*

---

[1] Jules Henri Poincaré (April 29, 1854–July 17, 1912) was a French mathematician, theoretical physicist, engineer, and philosopher of science. He is often described as a polymath, and in mathematics as "the last universalist," since he excelled in all fields of the discipline as it existed during his lifetime. As a mathematician and physicist, he made many original fundamental contributions to pure and applied mathematics, mathematical physics, and celestial mechanics. In his research on the three-body problem, Poincaré became the first person to discover a chaotic deterministic system which laid the foundations of modern chaos theory. He is also considered to be one of the founders of the field of topology.

Poincaré made the importance of paying attention to the invariance of laws of physics under different transformations clear, and was the first to present the Lorentz transformations in their modern symmetrical form. Poincaré discovered the remaining relativistic velocity transformations and recorded them in a letter to Hendrik Lorentz in 1905. Thus he obtained perfect invariance of all of Maxwell's equations, an important step in the formulation of the theory of special relativity. In 1905, Poincaré first proposed gravitational waves (*ondes gravifiques*) emanating from a body and propagating at the speed of light as being required by the Lorentz transformations.

The Poincaré group used in physics and mathematics was named after him.

1. **Integral Poincaré–Cartan invariant:**

$$I_{P-K} := \oint_{\text{any contour } C} \left[\sum_{i=1}^{n} p_i \delta q_i - H(t, \mathbf{q}, \mathbf{p}) \delta t \right] = \underset{\text{independently on contour } C}{\text{const}}. \quad (11.13)$$

2. **Integral universal Poincaré invariant:**

$$I_P := \oint_{\text{any contour } C_{t=\text{const}}} \sum_{i=1}^{n} p_i \delta q_i = \underset{\text{independently on contour } C_{t=\text{const}}}{\text{const}} \quad (11.14)$$

(it is referred to as **universal invariant** because of the independence of $H(t, \mathbf{q}, \mathbf{p})$, that is, it is valid for all Hamiltonian systems).

Proof. 1. Consider two contours, C_0 corresponding to time $t_0(\alpha)$ and C_1 corresponding to time $t_1(\alpha)$, as in Fig. 11.1. Taking into account that C is a contour $(I(\alpha = 1) = I(\alpha = 0))$, from (11.12) we have

$$0 = I(1) - I(0) = \int_{\alpha=0}^{1} dI(\alpha) =$$

$$\int_{\alpha=0}^{1} \left[\sum_{i=1}^{n} p_i [\delta q_i] - H(t, \mathbf{q}, \mathbf{p}) \delta t \right]_{t=t_1(\alpha)} d\alpha -$$

$$\int_{\alpha=0}^{1} \left[\sum_{i=1}^{n} p_i [\delta q_i] - H(t, \mathbf{q}, \mathbf{p}) \delta t \right]_{t=t_0(\alpha)} d\alpha =$$

$$\oint_{C_1} \left[\sum_{i=1}^{n} p_i [\delta q_i] - H(t, \mathbf{q}, \mathbf{p}) \delta t \right] -$$

$$\oint_{C_0} \left[\sum_{i=1}^{n} p_i [\delta q_i] - H(t, \mathbf{q}, \mathbf{p}) \delta t \right],$$

implying (11.13).

2. In this case we have $\delta t_0(\alpha) = \delta t_1(\alpha) = 0$ since both contours correspond to the constant times $t_0(\alpha) = \underset{\alpha \in [0;1]}{\text{const}}$ and $t_1(\alpha) = \underset{\alpha \in [0;1]}{\text{const}}$ (see Fig. 11.2). That is why

$$[I_{P-K}]_{C_{t=\text{const}}} = \oint_{C_{t=\text{const}}} \sum_{i=1}^{n} p_i [\delta q_i] = I_P.$$

□

11.4.2 The Lee Hwa Chung theorem

The following results (Lee, 1947) turned out to be very useful for a wide class of applications.

The Hamilton–Jacobi equation

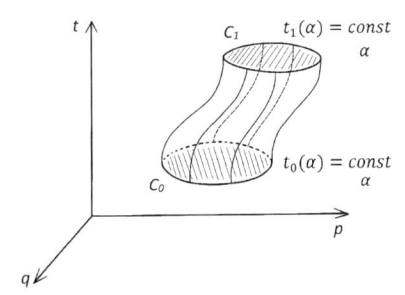

Figure 11.2 A family of trajectories of a Hamiltonian system in the extended state space with two initial and final contours: both correspond to constant times.

Theorem 11.2 (Lee Hwa Chung, 1947). *If the contour integral*

$$I_{LHC} = \oint_{C_{t=\text{const}}} \sum_{i=1}^{n} [A_i(t, \mathbf{q}, \mathbf{p}) \delta q_i + B_i(t, \mathbf{q}, \mathbf{p}) \delta p_i]$$
$$= \oint_{C_{t=\text{const}}} \left(A^\mathsf{T}(t, \mathbf{q}, \mathbf{p}) \delta \mathbf{q} + B^\mathsf{T}(t, \mathbf{q}, \mathbf{p}) \delta \mathbf{p} \right) \quad (11.15)$$

*does not depend on the contour $C_{t=\text{const}}$ for any Hamiltonian system, or in other words, it is **universal**, then there exists a constant c (independent on the considered contour) such that*

$$I_{LHC} = c I_P. \quad (11.16)$$

Proof. Since both integrals are constants, there exists a constant c_1 (may be dependent on the considered contour $C^{(1)}_{t=\text{const}}$) such that (omitting arguments)

$$\oint_{C^{(1)}_{t=\text{const}}} \left(A^\mathsf{T} \delta \mathbf{q} + B^\mathsf{T} \delta \mathbf{p} \right) = c_1 \oint_{C^{(1)}_{t=\text{const}}} \mathbf{p}^\mathsf{T} \delta \mathbf{q},$$

or equivalently,

$$\oint_{C^{(1)}_{t=\text{const}}} \left([A - c_1 \mathbf{p}]^\mathsf{T} \delta \mathbf{q} + B^\mathsf{T} \delta \mathbf{p} \right) = 0. \quad (11.17)$$

Since the integral (11.17) is equal to zero for any value of the variable $t \in C^{(1)}_{t=\text{const}}$ and for any arbitrary path $C^{(1)}_{t=\text{const}}$ of integration, the expression under the sign of the integral must be a *total differential* of some function $\Phi(\mathbf{q}, \mathbf{p})$. Therefore

$$0 = \oint_{C^{(1)}_{t=\text{const}}} \delta \Phi = \oint_{C^{(1)}_{t=\text{const}}} \left[\left(\frac{\delta \Phi}{\delta \mathbf{q}} \right)^\mathsf{T} \delta \mathbf{q} + \left(\frac{\delta \Phi}{\delta \mathbf{p}} \right)^\mathsf{T} \delta \mathbf{p} \right],$$

with

$$\frac{\delta \Phi}{\delta \mathbf{q}} = A - c_1 \mathbf{p} \text{ and } \frac{\delta \Phi}{\delta \mathbf{p}} = B. \tag{11.18}$$

Since $t = \text{const}$, here we have $\frac{\delta}{\delta q_i} = \frac{\partial}{\partial q_i}$ and $\frac{\delta}{\delta p_i} = \frac{\partial}{\partial p_i}$. Taking into account that for smooth functions

$$\frac{\partial^2 \Phi}{\partial p_i \partial q_i} = \frac{\partial^2 \Phi}{\partial q_i \partial p_i}$$

and hence

$$\frac{\partial^2 \Phi}{\partial \mathbf{p} \partial \mathbf{q}} = \left(\frac{\partial^2 \Phi}{\partial \mathbf{q} \partial \mathbf{p}} \right)^\mathsf{T},$$

from this property and in view of (11.18) it follows that

$$\frac{\partial A}{\partial \mathbf{p}} - c_1 I_{n \times n} = \left(\frac{\partial B}{\partial \mathbf{q}} \right)^\mathsf{T}. \tag{11.19}$$

But the same is true for another contour $C^{(1)}_{t=\text{const}}$ with another constant c_2, namely,

$$\frac{\partial A}{\partial \mathbf{p}} - c_2 I_{n \times n} = \left(\frac{\partial B}{\partial \mathbf{q}} \right)^\mathsf{T}. \tag{11.20}$$

Comparing (11.19) and (11.20) we conclude that $c_1 = c_2 = c$ and

$$\frac{\partial A}{\partial \mathbf{p}} - c I_{n \times n} = \left(\frac{\partial B}{\partial \mathbf{q}} \right)^\mathsf{T}, \tag{11.21}$$

which proves the theorem. □

Example 11.1. Let us consider the single-dimensional case with $n = 1$ and the integral

$$I = \oint_{C_{t=\text{const}}} [A(t, q, p) \delta q + B(t, q, p) \delta p],$$

with

$$A = \frac{p}{q} \text{ and } B = q + \ln(q).$$

We wish to know whether this integral is universal. If it is universal, then by the Lee Hwa Chung theorem, Theorem 11.2, and in view of (11.21) the following property should be satisfied:

$$\frac{\partial}{\partial p}(A - cp) = \frac{\partial}{\partial q} B,$$

or
$$\frac{\partial}{\partial p}\left(\frac{p}{q}\right) - c = \frac{\partial}{\partial q}(q + \ln q),$$

which gives
$$\frac{1}{q} - c = 1 + \frac{1}{q}$$

and
$$c = -1.$$

So, indeed there exists a constant $c = -1$, satisfying (11.21), and hence the considered integral I is universal.

11.5 Canonicity criteria

11.5.1 Poincaré theorem: (c, F)-criterion

Theorem 11.3 (Poincaré). *A pair $\boldsymbol{\varphi}$ and $\boldsymbol{\psi}$ of the coordinates transformation*

$$\tilde{q}_i = \varphi_i(t, \mathbf{q}, \mathbf{p}), \quad \tilde{p}_i = \psi_i(t, \mathbf{q}, \mathbf{p}) \quad (i = 1, ..., n) \tag{11.22}$$

*is **canonical** if and only if there exist a constant c and a function $F(t, \mathbf{q}, \mathbf{p})$ such that*

$$\sum_{i=1}^{n} \psi_i \delta \varphi_i - c \sum_{i=1}^{n} p_i \delta q_i = -\delta F, \tag{11.23}$$

where

$$\delta \varphi_i := d\varphi_i - \frac{\partial \varphi}{\partial t} dt, \quad \delta F := dF - \frac{\partial F}{\partial t} dt.$$

Proof. a) *Necessity.* Suppose that $\{\boldsymbol{\varphi}, \boldsymbol{\psi}\}$ is a canonical couple for a Hamiltonian system with the coordinates $(\mathbf{q}, \mathbf{p})$ and the Hamiltonian $H(t, \mathbf{q}, \mathbf{p})$ transforms these coordinates into new ones $(\tilde{\mathbf{q}}, \tilde{\mathbf{p}})$, satisfying the Hamiltonian equations with the Hamiltonian $\tilde{H}(t, \tilde{\mathbf{q}}, \tilde{\mathbf{p}})$. Take two arbitrary closed contours $C_{t=\text{const}}$ and $\tilde{C}_{t=\text{const}}$ in two different spaces $(t, \mathbf{q}, \mathbf{p})$ and $(t, \tilde{\mathbf{q}}, \tilde{\mathbf{p}})$ but corresponding to the same fixed time t. Then by Theorem 11.1 we have

$$\oint_{C_{t=\text{const}}} \left(\sum_{i=1}^{n} p_i \delta q_i - H(t, \mathbf{q}, \mathbf{p}) \delta t \right) = \oint_{C_{t=\text{const}}} \sum_{i=1}^{n} p_i \delta q_i,$$

$$\oint_{\tilde{C}_{t=\text{const}}} \left(\sum_{i=1}^{n} \tilde{p}_i \delta \tilde{q}_i - H(t, \tilde{\mathbf{q}}, \tilde{\mathbf{p}}) \delta t \right) = \oint_{\tilde{C}_{t=\text{const}}} \sum_{i=1}^{n} \tilde{p}_i \delta \tilde{q}_i.$$

Using (11.22) we get

$$\oint_{C_{t=\text{const}}} \sum_{i=1}^{n} \tilde{p}_i \delta \tilde{q}_i = \oint_{C_{t=\text{const}}} \sum_{i=1}^{n} \psi_i(t, \mathbf{q}, \mathbf{p}) \delta \varphi_i(t, \mathbf{q}, \mathbf{p}) =$$

$$\oint_{C_{t=\text{const}}} \sum_{i=1}^{n} \psi_i(t, \mathbf{q}, \mathbf{p}) \left[\left(\frac{\delta \varphi_i(t, \mathbf{q}, \mathbf{p})}{\delta \mathbf{q}} \right)^\mathsf{T} \delta \mathbf{q} + \left(\frac{\delta \varphi_i(t, \mathbf{q}, \mathbf{p})}{\delta \mathbf{p}} \right)^\mathsf{T} \delta \mathbf{p} \right] =$$

$$\oint_{C_{t=\text{const}}} \left[\left(\sum_{i=1}^{n} \psi_i(t, \mathbf{q}, \mathbf{p}) \frac{\delta \varphi_i(t, \mathbf{q}, \mathbf{p})}{\delta \mathbf{q}} \right)^\mathsf{T} \delta \mathbf{q} \right.$$

$$+ \left. \left(\sum_{i=1}^{n} \psi_i(t, \mathbf{q}, \mathbf{p}) \frac{\delta \varphi_i(t, \mathbf{q}, \mathbf{p})}{\delta \mathbf{p}} \right)^\mathsf{T} \delta \mathbf{p} \right]$$

$$= \oint_{C_{t=\text{const}}} \left(A^\mathsf{T}(t, \mathbf{q}, \mathbf{p}) \delta \mathbf{q} + B^\mathsf{T}(t, \mathbf{q}, \mathbf{p}) \delta \mathbf{p} \right),$$

where

$$A(t, \mathbf{q}, \mathbf{p}) = \sum_{i=1}^{n} \psi_i(t, \mathbf{q}, \mathbf{p}) \frac{\delta \varphi_i(t, \mathbf{q}, \mathbf{p})}{\delta \mathbf{q}}$$

and

$$B(t, \mathbf{q}, \mathbf{p}) = \sum_{i=1}^{n} \psi_i(t, \mathbf{q}, \mathbf{p}) \frac{\delta \varphi_i(t, \mathbf{q}, \mathbf{p})}{\delta \mathbf{p}}.$$

Then by the Lee Hwa Chung theorem, Theorem 11.15, it follows that there exists a constant c such that

$$\oint_{C_{t=\text{const}}} \sum_{i=1}^{n} \tilde{p}_i \delta \tilde{q}_i = \oint_{C_{t=\text{const}}} \left(A^\mathsf{T}(t, \mathbf{q}, \mathbf{p}) \delta \mathbf{q} + B^\mathsf{T}(t, \mathbf{q}, \mathbf{p}) \delta \mathbf{p} \right)$$

$$= c \oint_{C_{t=\text{const}}} \sum_{i=1}^{n} p_i \delta q_i,$$

which gives

$$\oint_{C_{t=\text{const}}} \left(\sum_{i=1}^{n} \psi_i \delta \varphi_i - c \sum_{i=1}^{n} p_i \delta q_i \right) = 0.$$

Since the last expression is true for any arbitrary contour $C_{t=\text{const}}$, the expression under the integral should be a total differential of some function Φ of $(\mathbf{q}, \mathbf{p})$ variables, namely,

$$\sum_{i=1}^{n} \psi_i \delta \varphi_i - c \sum_{i=1}^{n} p_i \delta q_i = \delta \Phi.$$

Defining $F := -\Phi$, we get (11.23).

The Hamilton–Jacobi equation

b) *Sufficiency.* Suppose that (11.23) holds. We need to demonstrate that there exists a Hamiltonian function $\tilde{H}(t, \tilde{\mathbf{q}}, \tilde{\mathbf{p}})$ such that

$$\frac{d}{dt}\tilde{\mathbf{q}} = \frac{\partial}{\partial \tilde{\mathbf{p}}} \tilde{H}(t, \tilde{\mathbf{q}}, \tilde{\mathbf{p}}) \text{ and } \frac{d}{dt}\tilde{\mathbf{p}} = -\frac{\partial}{\partial \tilde{\mathbf{q}}} \tilde{H}(t, \tilde{\mathbf{q}}, \tilde{\mathbf{p}}).$$

Integrating (11.23) on some contour $C_{t=\text{const}}$ we obtain

$$\oint_{C_{t=\text{const}}} \left(\sum_{i=1}^{n} \psi_i \delta \varphi_i - c \sum_{i=1}^{n} p_i \delta q_i \right) = -\oint_{C_{t=\text{const}}} \delta F = 0,$$

which implies

$$\oint_{C_{t=\text{const}}} \sum_{i=1}^{n} \tilde{p}_i \delta \tilde{q}_i = \oint_{C_{t=\text{const}}} \sum_{j=1}^{n} \psi_j \delta \varphi_j = c \oint_{C_{t=\text{const}}} \sum_{i=1}^{n} p_i \delta q_i.$$

Since the original system is Hamiltonian by the Poincaré theorem, Theorem 11.1, it follows that

$$\oint_{C_{t=\text{const}}} \sum_{i=1}^{n} \tilde{p}_i \delta \tilde{q}_i = c \oint_{C_{t=\text{const}}} \sum_{i=1}^{n} p_i \delta q_i = cI_P =$$

$$\text{const}$$

independently on contour $C_{t=\text{const}}$.

Therefore a system in new coordinates is Hamiltonian too. $\square$

Definition 11.2. The function F is referred to as the **generating function**, and the constant c as the **valence** of the canonical transformation $\{\varphi, \psi\}$.

Example 11.2. Let $n = 1$ and let the nonlinear transformation $\{\varphi, \psi\}$ be given by

$$\tilde{q} = qp, \quad \tilde{p} = \ln\left(q^{20} p^{17}\right). \tag{11.24}$$

We need to establish if this transformation is canonical, finding valence c and generating function F. We have

$$\varphi = qp, \quad \psi = \ln\left(q^{20} p^{17}\right) = 20 \ln q + 17 \ln p$$

and

$$\oint_{C_{t=\text{const}}} [\psi \delta \varphi - cp \delta q] = \oint_{C_{t=\text{const}}} [(20 \ln q + 17 \ln p) \delta (qp) - cp \delta q] =$$

$$\oint_{C_{t=\text{const}}} [(20 \ln q + 17 \ln p - c) p \delta q + (20 \ln q + 17 \ln p) q \delta p] = 0.$$

If F exists it should satisfy the equations

$$\left.\begin{array}{l}(20\ln q + 17\ln p - c)\,p = \\ \quad -\dfrac{\partial}{\partial q}F \text{ and } (20\ln q + 17\ln p)\,q = -\dfrac{\partial}{\partial p}F, \\ \dfrac{\partial^2}{\partial p \partial q}F = \dfrac{\partial^2}{\partial q \partial p}F,\end{array}\right\} \qquad (11.25)$$

which gives

$$(20\ln q + 17\ln p - c) + 17 = (20\ln q + 17\ln p) + 20,$$

which is true for

$$c = -3.$$

So, we have found the valence. Now to find the generating function F, let us integrate the second relation in (11.25):

$$-F = \int_p (20\ln q + 17\ln p)\,q\,dp = 20qp\ln q + 17q\int_p \ln p\,dp + f(q).$$

To find F we need to recuperate $f(q)$. Substituting this representation of F in the first equation of (11.25) implies

$$-\dfrac{\partial}{\partial q}F = (20\ln q + 17\ln p - c)\,p = 20p\ln q + 20p$$

$$+ 17\int_p \ln p\,dp + \dfrac{d}{dq}f(q),$$

$$\dfrac{d}{dq}f(q) = 17p\ln p - 17p - 17\int_p \ln p\,dp,$$

$$f(q) = \left(17p\ln p - 17p - 17\int_p \ln p\,dp\right)q - \text{const},$$

and, finally,

$$F = qp(17 - 37\ln q) + \text{const}.$$

So, the considered transformation (11.24) is canonical.

11.5.2 Analytical expression for the Hamiltonian after a coordinate canonical transformation

Let us prove the following useful result.

Theorem 11.4. *If some integral*

$$I_\Phi := \oint_{\text{any contour } C} \left(\sum_{i=1}^n p_i \delta q_i - \Phi(t, \mathbf{q}, \mathbf{p}) \delta t \right)$$

in the coordinate space $(t, \mathbf{q}, \mathbf{p})$ *is constant for any arbitrary contour C, that is,*

$$I_\Phi = \underset{\text{independently on contour } C}{\text{const}},$$

then this system is Hamiltonian with $H(t, \mathbf{q}, \mathbf{p})$ *such that*

$$cH(t, \mathbf{q}, \mathbf{p}) = \Phi(t, \mathbf{q}, \mathbf{p}) + \frac{\partial}{\partial t} G(t, \mathbf{q}, \mathbf{p}) \tag{11.26}$$

for some function $G(t, q, p)$ *and a constant c.*

Proof. By the Poincaré theorem, Theorem 11.1, the considered system is Hamiltonian if and only if the integral

$$I_{P-K} := \oint_{\text{any contour } C} \left[\sum_{i=1}^n p_i \delta q_i - H(t, \mathbf{q}, \mathbf{p}) \delta t \right]$$

is constant, namely,

$$I_{P-K} = \underset{\text{independently on contour } C}{\text{const}}.$$

Therefore we have

$$I_\Phi = c I_{P-K}$$

for some c, and as a result,

$$\oint_{\text{any contour } C} \left[\sum_{i=1}^n p_i \delta q_i - \Phi(t, q, p) \delta t \right] = c \oint_{\text{any contour } C} \left[\sum_{i=1}^n p_i \delta q_i - H \delta t \right],$$

which gives

$$\oint_{\text{any contour } C} (cH - \Phi) \delta t = 0.$$

From this relation it follows that it is a complete integral over this contour:

$$(cH - \Phi) \delta t = dG = \frac{\partial G}{\partial t} \delta t + \sum_{i=1}^n \left[\frac{\partial G}{\partial q_i} \delta q_i + \frac{\partial G}{\partial p_i} \delta p_i \right],$$

and since variations δt, δq_i, and δp_i are independent we get

$$cH - \Phi - \frac{\partial G}{\partial t} = \frac{\partial G}{\partial q_i} = \frac{\partial G}{\partial p_i} = 0,$$

which gives (11.26). $\square$

Now we are ready to formulate the main result of this subsection.

Theorem 11.5. *If a transformation* $\{\boldsymbol{\varphi}, \boldsymbol{\psi}\}$ *is canonical, then the Hamiltonian* $\tilde{H}(t, \tilde{\mathbf{q}}, \tilde{\mathbf{p}})$ *after a coordinate canonical transformation is as follows:*

$$c_1 \tilde{H}(t, \tilde{\mathbf{q}}, \tilde{\mathbf{p}}) = \left(cH(t, \mathbf{q}, \mathbf{p}) + \frac{\partial}{\partial t} F(t, \mathbf{q}, \mathbf{p}) \right) |_{\mathbf{q}=\mathbf{q}(\tilde{q}, \tilde{p}), \ \mathbf{p}=\mathbf{p}(\tilde{q}, \tilde{p})} \\ + \frac{\partial}{\partial t} G(t, \tilde{\mathbf{q}}, \tilde{\mathbf{p}}). \tag{11.27}$$

Proof. Since the considered pair $\{\boldsymbol{\varphi}, \boldsymbol{\psi}\}$ is canonical, the system $(t, \mathbf{q}, \mathbf{p})$ and $(t, \tilde{\mathbf{q}}, \tilde{\mathbf{p}})$ is Hamiltonian and by Theorem 11.1 we have

$$\left. \begin{aligned} & \sum_{i=1}^{n} \psi_i \delta \varphi_i - c \sum_{i=1}^{n} p_i \delta q_i = -\delta F \Leftrightarrow \\ & \sum_{i=1}^{n} \tilde{p}_i \delta \tilde{q}_i - cH \delta t = c \left(\sum_{i=1}^{n} p_i \delta q_i - H \delta t \right) - \delta F \Leftrightarrow \\ & \sum_{i=1}^{n} \tilde{p}_i \delta \tilde{q}_i - cH \delta t = c \left(\sum_{i=1}^{n} p_i \delta q_i - H \delta t \right) - \left(dF - \frac{\partial}{\partial t} F \delta t \right) \Leftrightarrow \\ & \sum_{i=1}^{n} \tilde{p}_i \delta \tilde{q}_i - \left(cH + \frac{\partial}{\partial t} F \right) \delta t = c \left(\sum_{i=1}^{n} p_i \delta q_i - H \delta t \right) - dF. \end{aligned} \right\} \tag{11.28}$$

Integrating this identity over some arbitrary contour $\tilde{C}$ in the extended space $(t, \tilde{\mathbf{q}}, \tilde{\mathbf{p}})$, which corresponds to some contour C in the space $(t, \mathbf{q}, \mathbf{p})$, we get

$$I_{P-K} := \oint_{\text{any contour } \tilde{C}} \left[\sum_{i=1}^{n} \tilde{p}_i \delta \tilde{q}_i - \left(cH + \frac{\partial}{\partial t} F \right) \delta t \right] = \underset{\text{independently on contour } \tilde{C}}{\text{const}}$$

$$= \oint_{\text{any contour } C} \left[c \left(\sum_{i=1}^{n} p_i \delta q_i - H \delta t \right) - dF \right] =$$

$$= \underbrace{c \oint_{\text{any contour } C} \left(\sum_{i=1}^{n} p_i \delta q_i - H \delta t \right)}_{I_{P-K}} - \underbrace{\oint_{\text{any contour } C} dF}_{0} =$$

$$cI_{P-K} = \underset{\text{independently of contour } C}{\text{const}}.$$

Hence, by Theorem 11.4 it follows that

$$\Phi(t, \tilde{\mathbf{q}}, \tilde{\mathbf{p}}) = \left(cH + \frac{\partial}{\partial t}F\right)\Big|_{\mathbf{q}=\mathbf{q}(\tilde{q},\tilde{p}),\ \mathbf{p}=\mathbf{p}(\tilde{q},\tilde{p})}$$

and

$$c_1\tilde{H}\left(t, \tilde{\mathbf{q}}, \tilde{\mathbf{p}}\right) = \Phi\left(t, \tilde{\mathbf{q}}, \tilde{\mathbf{p}}\right) + \frac{\partial}{\partial t}G(t, \tilde{\mathbf{q}}, \tilde{\mathbf{p}}) =$$
$$\left(cH + \frac{\partial}{\partial t}F\right)\Big|_{\mathbf{q}=\mathbf{q}(\tilde{q},\tilde{p}),\ \mathbf{p}=\mathbf{p}(\tilde{q},\tilde{p})} + \frac{\partial}{\partial t}G(t, \tilde{\mathbf{q}}, \tilde{\mathbf{p}}),$$

which proves the theorem. □

Remark 11.2. The constant c_1 is not equal to zero since $\left[\sum_{i=1}^{n}\tilde{p}_i\delta\tilde{q}_i - \tilde{H}\delta t\right]$ is not a total differential. Hence dividing (11.27) by c_1 we obtain

$$\tilde{H}\left(t, \tilde{\mathbf{q}}, \tilde{\mathbf{p}}\right) = \left(\tilde{c}H(t, \mathbf{q}, \mathbf{p}) + \frac{\partial}{\partial t}\tilde{F}(t, \mathbf{q}, \mathbf{p})\right)\Big|_{\mathbf{q}=\mathbf{q}(\tilde{q},\tilde{p}),\ \mathbf{p}=\mathbf{p}(\tilde{q},\tilde{p})}, \qquad (11.29)$$

where

$$\tilde{c} = c/c_1, \quad \tilde{F} = \frac{1}{c_1}\left(F + \tilde{G}\right).$$

11.5.3 Brackets of Lagrange

Corollary 11.1. *A functional pair* $\{\boldsymbol{\varphi}, \boldsymbol{\psi}\}$ *is canonical if and only if there exists a constant c such that:*

1.

$$\sum_{i=1}^{n}\left[\frac{\partial\varphi_i}{\partial q_j}\frac{\partial\psi_i}{\partial q_k} - \frac{\partial\varphi_i}{\partial q_k}\frac{\partial\psi_i}{\partial q_j}\right] := [q_j, q_k] = 0 \quad \forall j, k = \overline{1, n}, \qquad (11.30)$$

2.

$$\sum_{i=1}^{n}\left[\frac{\partial\varphi_i}{\partial p_j}\frac{\partial\psi_i}{\partial p_k} - \frac{\partial\varphi_i}{\partial p_k}\frac{\partial\psi_i}{\partial p_j}\right] := [p_j, q_k] = 0 \quad \forall j, k = \overline{1, n}, \qquad (11.31)$$

3.

$$\sum_{i=1}^{n}\left[\frac{\partial\varphi_i}{\partial q_j}\frac{\partial\psi_i}{\partial p_k} - \frac{\partial\varphi_i}{\partial p_k}\frac{\partial\psi_i}{\partial q_j}\right] := [q_j, p_k] = c\delta_{j,k}, \qquad (11.32)$$

with

$$c\delta_{j,k} = c \begin{cases} 1, & \text{for } j = k, \\ 0, & \text{for } j \neq k, \end{cases} \quad j, k = \overline{1, n}.$$

The expression $[q_j, p_k]$ is referred to as the **brackets of Lagrange**.

Proof. By (11.23) we have

$$\sum_{i=1}^n \psi_i \delta\varphi_i - c \sum_{i=1}^n p_i \delta q_i = -\delta F.$$

Since

$$\delta F(t, q, p) = \sum_{j=1}^n \left[\frac{\partial F}{\partial q_j} \delta q_j + \frac{\partial F}{\partial p_j} \delta p_j \right]$$

and

$$\delta\varphi_i = \sum_{j=1}^n \left[\frac{\partial \varphi_i}{\partial q_j} \delta q_j + \frac{\partial \varphi_i}{\partial p_j} \delta p_j \right],$$

we get

$$\sum_{i=1}^n \psi_i \sum_{j=1}^n \left[\frac{\partial \varphi_i}{\partial q_j} \delta q_j + \frac{\partial \varphi_i}{\partial p_j} \delta p_j \right] - c \sum_{i=1}^n p_i \delta q_i =$$

$$\sum_{j=1}^n \sum_{i=1}^n \psi_i \left[\frac{\partial \varphi_i}{\partial q_j} \delta q_j + \frac{\partial \varphi_i}{\partial p_j} \delta p_j \right] - c \sum_{i=1}^n p_i \delta q_i =$$

$$\sum_{i=1}^n \sum_{j=1}^n \psi_j \left[\frac{\partial \varphi_j}{\partial q_i} \delta q_i + \frac{\partial \varphi_j}{\partial p_i} \delta p_i \right] - c \sum_{i=1}^n p_i \delta q_i =$$

$$\sum_{i=1}^n \left[\left(\sum_{j=1}^n \psi_j \frac{\partial \varphi_j}{\partial q_i} - c p_i \right) \delta q_i + \sum_{j=1}^n \psi_j \frac{\partial \varphi_j}{\partial p_i} \delta p_i \right] =$$

$$-\sum_{i=1}^n \left[\frac{\partial F}{\partial q_i} \delta q_i + \frac{\partial F}{\partial p_i} \delta p_i \right].$$

This implies

$$-\frac{\partial F}{\partial q_i} = \sum_{j=1}^n \psi_j \frac{\partial \varphi_j}{\partial q_i} - c p_i, \quad -\frac{\partial F}{\partial p_i} = \sum_{j=1}^n \psi_j \frac{\partial \varphi_j}{\partial p_i}. \tag{11.33}$$

Since $F(t, \mathbf{q}, \mathbf{p})$ is total differential in space $(\mathbf{q}, \mathbf{p})$ under the fixed t, the following properties hold for all i, k:

$$\left.\begin{aligned}\frac{\partial}{\partial q_k}\left(\frac{\partial F}{\partial q_i}\right) &= \frac{\partial}{\partial q_i}\left(\frac{\partial F}{\partial q_k}\right),\\ \frac{\partial}{\partial p_k}\left(\frac{\partial F}{\partial p_i}\right) &= \frac{\partial}{\partial p_i}\left(\frac{\partial F}{\partial p_k}\right),\\ \frac{\partial}{\partial p_k}\left(\frac{\partial F}{\partial q_i}\right) &= \frac{\partial}{\partial q_i}\left(\frac{\partial F}{\partial p_k}\right).\end{aligned}\right\} \qquad (11.34)$$

Substituting (11.33) into (11.34) gives

$$\frac{\partial}{\partial q_k}\left(\sum_{j=1}^n \psi_j \frac{\partial \varphi_j}{\partial q_i} - c p_i\right) = \sum_{j=1}^n \left(\frac{\partial \psi_j}{\partial q_k}\frac{\partial \varphi_j}{\partial q_i} + \psi_j \frac{\partial}{\partial q_k}\left(\frac{\partial \varphi_j}{\partial q_i}\right)\right) =$$
$$\frac{\partial}{\partial q_i}\left(\sum_{j=1}^n \psi_j \frac{\partial \varphi_j}{\partial q_k} - c p_k\right) = \sum_{j=1}^n \left(\frac{\partial \psi_j}{\partial q_i}\frac{\partial \varphi_j}{\partial q_k} + \psi_j \frac{\partial}{\partial q_i}\left(\frac{\partial \varphi_j}{\partial q_k}\right)\right),$$

which leads to (11.30). Analogously we may obtain (11.31). As for (11.32) we have

$$\frac{\partial}{\partial p_k}\left(\frac{\partial F}{\partial q_i}\right) = \frac{\partial}{\partial p_k}\left(\sum_{j=1}^n \psi_j \frac{\partial \varphi_j}{\partial q_i} - c p_i\right) =$$
$$\left(\sum_{j=1}^n \left[\frac{\partial \psi_j}{\partial p_k}\frac{\partial \varphi_j}{\partial q_i} + \psi_j \frac{\partial}{\partial p_k}\left(\frac{\partial \varphi_j}{\partial q_i}\right)\right] - c\delta_{ik}\right) =$$
$$\frac{\partial}{\partial q_i}\left(\frac{\partial F}{\partial p_k}\right) = \frac{\partial}{\partial q_i}\left(\sum_{j=1}^n \psi_j \frac{\partial \varphi_j}{\partial p_k}\right) =$$
$$\sum_{j=1}^n \left(\frac{\partial \psi_j}{\partial q_i}\frac{\partial \varphi_j}{\partial p_k} + \psi_j \frac{\partial}{\partial q_i}\frac{\partial \varphi_j}{\partial p_k}\right),$$

which leads to (11.32). $\qquad \square$

Example 11.3. We have the linear transformation

$$\tilde{\mathbf{q}} = A\mathbf{q} + B\mathbf{p}, \ \tilde{\mathbf{p}} = C\mathbf{q} + D\mathbf{p}.$$

Let us find the conditions, which matrices A, B, C, and D should satisfy, guaranteeing that this transformation is canonical.

For this system

$$\varphi_i = \sum_{s=1}^n (a_{is} q_s + b_{is} p_s), \ \psi_i = \sum_{s=1}^n (c_{is} q_s + d_{is} p_s).$$

So,

$$0 = [q_j, q_k] = \sum_{i=1}^{n}\left[\frac{\partial \varphi_i}{\partial q_j}\frac{\partial \psi_i}{\partial q_k} - \frac{\partial \varphi_i}{\partial q_k}\frac{\partial \psi_i}{\partial q_j}\right] = \sum_{i=1}^{n}\left[a_{ij}c_{ik} - a_{ik}c_{ik}\right],$$

which in matrix format is

$$A^\mathsf{T} C = C^\mathsf{T} A.$$

Analogously, $0 = [p_j, p_k]$ is equivalent to the relation

$$B^\mathsf{T} D = D^\mathsf{T} B$$

and finally, $[q_j, p_k] = c\delta_{j,k}$ gives

$$A^\mathsf{T} D - C^\mathsf{T} B = c I_{n\times n}.$$

11.5.4 Free canonical transformation and the S-canonicity criterion

Let us consider stationary nonlinear transformations, which do not depend on time t, namely, let in (11.2)

$$\left.\begin{array}{l}\tilde{\mathbf{q}} = \varphi(\mathbf{q}, \mathbf{p}),\\ \tilde{\mathbf{p}} = \psi(\mathbf{q}, \mathbf{p}).\end{array}\right\} \qquad (11.35)$$

In view of the condition (11.3) there exists an inverse transformation $\tilde{\varphi}$ such that

$$\mathbf{p} = \tilde{\varphi}(\mathbf{q}, \tilde{\mathbf{q}}), \qquad (11.36)$$

which leads to the following representation of $\tilde{\mathbf{p}}$:

$$\tilde{\mathbf{p}} = \psi(\mathbf{q}, \tilde{\varphi}(\mathbf{q}, \tilde{\mathbf{q}})) = \tilde{\psi}(\mathbf{q}, \tilde{\mathbf{q}}). \qquad (11.37)$$

Definition 11.3. Nonlinear transformations $\left(\tilde{\varphi}, \tilde{\psi}\right)$, defined by the relations (11.36) and (11.37), are referred to as **free transformations**.

Theorem 11.6 (S-canonicity criterion). *The free nonlinear transformation $\left(\tilde{\varphi}, \tilde{\psi}\right)$ (see (11.35)) is* **canonical** *if and only if there exist a constant c and a function $S(t, \mathbf{q}, \tilde{\mathbf{q}})$ such that for all $i = 1, ..., n$ the following relations hold:*

$$\frac{\partial S(t, \mathbf{q}, \tilde{\mathbf{q}})}{\partial q_i} = cp_i, \quad \frac{\partial S(t, \mathbf{q}, \tilde{\mathbf{q}})}{\partial \tilde{q}_i} = -\tilde{p}_i, \qquad (11.38)$$

wherein

$$\tilde{H}(t, \mathbf{q}, \tilde{\mathbf{q}}) = cH(t, \mathbf{q}, \mathbf{p})\,|_{\mathbf{p}=\mathbf{p}(\tilde{\mathbf{q}}, \tilde{\psi}(\mathbf{q}, \tilde{\mathbf{q}}))} + \frac{\partial S(t, \mathbf{q}, \tilde{\mathbf{q}})}{\partial t}. \qquad (11.39)$$

The Hamilton–Jacobi equation

Proof. In view of the transformation (11.36) the generating function $F(t, \mathbf{q}, \mathbf{p})$ in (11.23) can be represented as

$$F(t, \mathbf{q}, \mathbf{p}) = F(t, \mathbf{q}, \tilde{\boldsymbol{\varphi}}(\mathbf{q}, \tilde{\mathbf{q}})) := S(t, \mathbf{q}, \tilde{\mathbf{q}}), \tag{11.40}$$

whose total differential has the form

$$dF = dS = \frac{\partial S}{\partial t}\delta t + \sum_{i=1}^{n} \left[\frac{\partial S}{\partial q_i} \delta q_i + \frac{\partial S}{\partial \tilde{q}_i} \delta \tilde{q}_i \right].$$

Since we also have

$$dF = \frac{\partial F}{\partial t}\delta t + \sum_{i=1}^{n} \left[\frac{\partial F}{\partial q_i} \delta q_i + \frac{\partial F}{\partial p_i} \delta p_i \right],$$

the last relation in (11.29) gives

$$\left.\begin{aligned}\sum_{i=1}^{n} \tilde{p}_i \delta \tilde{q}_i - \tilde{H}\delta t &= c\left(\sum_{i=1}^{n} p_i \delta q_i - H\delta t\right) - dF = \\ c\left(\sum_{i=1}^{n} p_i \delta q_i - H\delta t\right) - \frac{\partial S}{\partial t}\delta t - \sum_{i=1}^{n} \left[\frac{\partial S}{\partial q_i}\delta q_i + \frac{\partial S}{\partial \tilde{q}_i}\delta \tilde{q}_i\right] = \\ \sum_{i=1}^{n}\left(cp_i - \frac{\partial S}{\partial q_i}\right)\delta q_i + \sum_{i=1}^{n}\left(-\frac{\partial S}{\partial \tilde{q}_i}\right)\delta \tilde{q}_i + \left(-cH - \frac{\partial S}{\partial t}\right)\delta t.\end{aligned}\right\} \tag{11.41}$$

Equating the coefficients of δq_i, $\delta \tilde{q}_i$, and δt in (11.41) we get (11.38) and (11.39). $\square$

Example 11.4. We need to check if the transformations

$$\left.\begin{aligned}\tilde{q}_1 &= (\gamma p_1 + q_2)^{-1} - q_1, \quad \tilde{q}_2 = \frac{2}{\gamma p_2 + q_1} - 2q_2, \\ \tilde{p}_1 &= -(\gamma p_1 + q_2), \quad \tilde{p}_2 = -\frac{1}{2}(\gamma p_2 + q_1)\end{aligned}\right\} \tag{11.42}$$

with $\gamma \neq 0$ are free canonical or not. To answer the question we need to find a constant c and a function $S(t, \mathbf{q}, \tilde{\mathbf{q}})$ satisfying (11.38). Resolving (11.42) with respect to p_1 and p_2, we get

$$p_1 = \frac{1}{\gamma}\left[\frac{1}{\tilde{q}_1 + q_1} - q_2\right], \quad p_2 = \frac{1}{\gamma}\left[\frac{2}{\tilde{q}_2 + 2q_2} - q_1\right].$$

Substitution of these expressions in the formulas for $\tilde{p}_1$ and $\tilde{p}_2$ gives

$$\tilde{p}_1 = -\left[\frac{1}{\tilde{q}_1 + q_1} - q_2\right] - q_2 = -\left(\frac{1}{\tilde{q}_1 + q_1} + q_2 - q_2\right) = -\frac{1}{\tilde{q}_1 + q_1} := \frac{\partial S}{\partial \tilde{q}_1},$$

$$\tilde{p}_2 = -\frac{1}{2}(\gamma p_2 + q_1) = -\frac{1}{2}\left[\frac{2}{\tilde{q}_2 + 2q_2} - q_1\right] - \frac{1}{2}q_1$$
$$= -\frac{1}{2}\left(\frac{2}{\tilde{q}_2 + 2q_2} - q_1 + q_1\right) = -\frac{1}{\tilde{q}_2 + 2q_2} := \frac{\partial S}{\partial q_2}.$$

By Theorem 11.6 we should have

$$\frac{\partial S(t, \mathbf{q}, \tilde{\mathbf{q}})}{\partial \tilde{q}_1} = -\tilde{p}_1 = \frac{1}{\tilde{q}_1 + q_1},$$
$$\frac{\partial S(t, \mathbf{q}, \tilde{\mathbf{q}})}{\partial \tilde{q}_2} = -\tilde{p}_2 = \frac{1}{\tilde{q}_2 + 2q_2}. \tag{11.43}$$

Resolving the first equation in (11.43) with respect to the function $S(t, \mathbf{q}, \tilde{\mathbf{q}})$ we find

$$S(t, \mathbf{q}, \tilde{\mathbf{q}}) = \ln(\tilde{q}_1 + q_1) + f(\tilde{q}_2, q_1, q_2, t).$$

Substituting this formula for $S(t, \mathbf{q}, \tilde{\mathbf{q}})$ into the second formula in (11.43) gives

$$\frac{\partial f(\tilde{q}_2, q_1, q_2, t)}{\partial \tilde{q}_2} = \frac{1}{\tilde{q}_2 + 2q_2}.$$

Integrating this equation with respect to $\tilde{q}_2$ we find

$$f(\tilde{q}_2, q_1, q_2, t) = \ln(\tilde{q}_2 + 2q_2) + f_1(q_1, q_2, t),$$

which implies

$$S(t, \mathbf{q}, \tilde{\mathbf{q}}) = \ln(\tilde{q}_1 + q_1) + \ln(\tilde{q}_2 + 2q_2) + f_1(q_1, q_2, t). \tag{11.44}$$

Recalling now that

$$\frac{\partial S(t, \mathbf{q}, \tilde{\mathbf{q}})}{\partial q_1} = cp_1,$$

after the substitution of (11.44) in this relation it follows that

$$\frac{1}{\tilde{q}_1 + q_1} + \frac{\partial f_1(q_1, q_2, t)}{\partial q_1} = cp_1 = \frac{c}{\gamma}\left[\frac{1}{\tilde{q}_1 + q_1} - q_2\right].$$

Taking

$$c = \gamma,$$

we derive

$$\frac{\partial f_1(q_1, q_2, t)}{\partial q_1} = -q_2,$$

which leads to

$$f_1(q_1, q_2, t) = -q_1 q_2 + f_2(q_2, t)$$

and

$$S(t, \mathbf{q}, \tilde{\mathbf{q}}) = \ln(\tilde{q}_1 + q_1) + \ln(\tilde{q}_2 + 2q_2) - q_1 q_2 + f_2(q_2, t).$$

Analogously, we derive

$$\frac{\partial S(t, \mathbf{q}, \tilde{\mathbf{q}})}{\partial q_2} = \frac{2}{\tilde{q}_2 + 2q_2} - q_1 + \frac{\partial f_2(q_2, t)}{\partial q_2} = cp_2 =$$
$$c\frac{1}{\gamma}\left[\frac{2}{\tilde{q}_2 + 2q_2} - q_1\right].$$

Since $c = \gamma$, we get

$$\frac{\partial f_2(q_2, t)}{\partial q_2} = 0,$$

which is why

$$f_2(q_2, t) = f_3(t),$$

where $f_3(t)$ is an arbitrary function of t. So finally,

$$S(t, \mathbf{q}, \tilde{\mathbf{q}}) = S(t, \mathbf{q}, \tilde{\mathbf{q}}) = \ln(\tilde{q}_1 + q_1) + \ln(\tilde{q}_2 + 2q_2) - q_1 q_2 + f_3(t).$$

This means that we have found the constant c and the function $S(t, \mathbf{q}, \tilde{\mathbf{q}})$, satisfying (11.38), and therefore the nonlinear free transformation (11.42) is canonical.

Example 11.5. For the nonlinear transformation

$$\tilde{q}_i = \ln p_i - q_i, \quad \tilde{p}_i = -p_i \quad (i = 1, ..., n) \tag{11.45}$$

we need to check if it is a free canonical transformation finding $S(t, \mathbf{q}, \tilde{\mathbf{q}})$ and c. From (11.45) it follows that

$$p_i = \exp(\tilde{q}_i + q_i), \quad \tilde{p}_i = -\exp(\tilde{q}_i + q_i).$$

By Theorem 11.6 we should have

$$\frac{\partial S(t, \mathbf{q}, \tilde{\mathbf{q}})}{\partial \tilde{q}_i} = -\tilde{p}_i = \exp(\tilde{q}_i + q_i).$$

Integrating this equation with respect to the variables $\tilde{q}_i$ gives

$$S(t, \mathbf{q}, \tilde{\mathbf{q}}) = e^{\tilde{q}_i + q_i} + f_i(\tilde{q}_{j \neq i}, q_i, t).$$

Substituting this expression into

$$\frac{\partial S(t, \mathbf{q}, \tilde{\mathbf{q}})}{\partial \tilde{q}_{j \neq i}} = -\tilde{p}_{j \neq i}$$

leads to

$$\frac{\partial S(t, \mathbf{q}, \tilde{\mathbf{q}})}{\partial \tilde{q}_{j \neq i}} = -\tilde{p}_{j \neq i} = \frac{\partial f\left(\tilde{q}_{j \neq i}, q_i, t\right)}{\partial \tilde{q}_{j \neq i}} = \exp\left(\tilde{q}_{j \neq i} + q_{j \neq i}\right),$$

implying

$$f_i\left(\tilde{q}_{j \neq i}, q_i, t\right) = \exp\left(\tilde{q}_{j \neq i} + q_{j \neq i}\right) + f_{i,j}\left(\tilde{q}_{k \neq i, j}, q_i, q_j, t\right).$$

Iterating this process we get

$$S(t, \mathbf{q}, \tilde{\mathbf{q}}) = \sum_{i=1}^{n} e^{\tilde{q}_i + q_i} + f(\mathbf{q}, t).$$

Analogously, substitution of the last formula for $S(t, \mathbf{q}, \tilde{\mathbf{q}})$ into

$$\frac{\partial S(t, \mathbf{q}, \tilde{\mathbf{q}})}{\partial q_i} = c p_i$$

leads to

$$\frac{\partial S(t, \mathbf{q}, \tilde{\mathbf{q}})}{\partial q_i} = e^{\tilde{q}_i + q_i} + \frac{\partial f(\mathbf{q}, t)}{\partial q_i} = c p_i = c \exp\left(\tilde{q}_i + q_i\right).$$

Taking

$$c = 1$$

we obtain for all $i = 1, \ldots, n$

$$\frac{\partial f(\mathbf{q}, t)}{\partial q_i} = 0,$$

which is equivalent to

$$f(\mathbf{q}, t) = f_0(t),$$

implying

$$S(t, \mathbf{q}, \tilde{\mathbf{q}}) = \sum_{i=1}^{n} e^{\tilde{q}_i + q_i} + f_0(t),$$

which means the transformation (11.45) is free canonical.

11.6 The Hamilton–Jacobi equation

As mentioned in Section 7.2, the main idea of the Hamiltonian approach consists in finding the canonical transformation such that in new transformed variables the corresponding Hamiltonian would be equal to zero. Using free canonical transformations

$$\mathbf{p} = \tilde{\varphi}(\mathbf{q}, \tilde{\mathbf{q}}), \quad \tilde{\mathbf{p}} = \psi(\mathbf{q}, \tilde{\varphi}(\mathbf{q}, \tilde{\mathbf{q}})) = \tilde{\psi}(\mathbf{q}, \tilde{\mathbf{q}}), \tag{11.46}$$

we have the following expression for the new Hamiltonian (see (11.39)):

$$\tilde{H}(t, \mathbf{q}, \tilde{\mathbf{q}}) = cH(t, \mathbf{q}, \mathbf{p}) \mid_{\mathbf{p}=\mathbf{p}(\tilde{q}, \tilde{\psi}(\mathbf{q}, \tilde{\mathbf{q}}))} + \frac{\partial S(t, \mathbf{q}, \tilde{\mathbf{q}})}{\partial t}, \tag{11.47}$$

where the function $S(t, \mathbf{q}, \tilde{\mathbf{q}})$ satisfies the relations (11.38), namely,

$$\frac{\partial S(t, \mathbf{q}, \tilde{\mathbf{q}})}{\partial q_i} = cp_i, \quad \frac{\partial S(t, \mathbf{q}, \tilde{\mathbf{q}})}{\partial \tilde{q}_i} = -\tilde{p}_i \quad (i = 1, ..., n), \tag{11.48}$$

for some scalar $c \neq 0$.

The following result is one the main results in this chapter.

Theorem 11.7 (Hamilton–Jacobi). *If there exists a solution $S(t, \mathbf{q}, \tilde{\mathbf{q}})$ of the partial differential equation*

$$\frac{\partial}{\partial t} S(t, \mathbf{q}, \tilde{\mathbf{q}}) + cH(t, \mathbf{q}, \frac{1}{c}\frac{\partial}{\partial \mathbf{q}} S(t, \mathbf{q}, \tilde{\mathbf{q}})) = 0, \tag{11.49}$$

which satisfies the "non-singularity condition"

$$\det\left[\frac{\partial^2}{\partial \tilde{\mathbf{q}} \partial \mathbf{q}} S(t, \mathbf{q}, \tilde{\mathbf{q}}))\right] \neq 0 \quad \text{for any } \mathbf{q}, \tilde{\mathbf{q}} \in R^n \text{ and } t \geq 0$$

and satisfies the S-canonicity criterion (see Theorem 11.6), then

$$\left.\begin{array}{l} \tilde{q}_i = \alpha_i = \underset{t}{\text{const}} = \varphi_i(t, \mathbf{q}, \mathbf{p}), \\ \tilde{p}_i = \beta_i = \underset{t}{\text{const}} = \psi_i(t, \mathbf{q}, \mathbf{p}), \end{array}\right\} \tag{11.50}$$

and the dynamics in the original space $(t, \mathbf{q}, \mathbf{p})$ for all $i = 1, ..., n$ satisfies the equations

$$\left.\begin{array}{l} \dfrac{\partial S(t, \mathbf{q}, \boldsymbol{\alpha})}{\partial \alpha_i} = -\beta_i, \\ \dfrac{\partial}{\partial q_i} S(t, \mathbf{q}, \boldsymbol{\alpha}) = cp_i. \end{array}\right\} \tag{11.51}$$

Proof. Taking in (11.47) $\tilde{H}(t, \mathbf{q}, \tilde{\mathbf{q}}) = 0$, after the substitution of $\tilde{p}$ from (11.48) in the obtained equation we get the Hamilton–Jacobi equation (11.49). If there exists a non-singular solution $S(t, \mathbf{q}, \tilde{\mathbf{q}})$ of (11.49) satisfying the S-canonicity criterion (11.48), in view of the main Hamiltonian equations

$$\frac{d}{dt}\tilde{\mathbf{q}} = \frac{\partial}{\partial \tilde{\mathbf{p}}}\tilde{H} = 0, \quad \frac{d}{dt}\tilde{\mathbf{p}} = -\frac{\partial}{\partial \tilde{\mathbf{q}}}\tilde{H} = 0,$$

we find (11.50). Since our transformations are free (see Definition 11.3) and non-singular, we are able to resolve the first equation in (11.51), obtaining

$$\mathbf{q} = \mathbf{q}(t, \boldsymbol{\alpha}, \boldsymbol{\beta}) \qquad (11.52)$$

and

$$\mathbf{p} = \frac{1}{c}\frac{\partial}{\partial \mathbf{q}} S(t, \mathbf{q}(t, \boldsymbol{\alpha}, \boldsymbol{\beta}), \boldsymbol{\alpha}). \qquad (11.53)$$

□

11.7 Complete integral of the Hamilton–Jacobi equation

11.7.1 Complete integral

In this section we will present the method helping to find the solution of the Hamilton–Jacobi (HJ) equation for stationary systems.

Definition 11.4. A function $S(t, \mathbf{q}, \tilde{\mathbf{q}})$, which is a non-singular solution of (11.49), is referred to as a **complete integral** of this HJ equation.

Remark 11.3. To obtain a solution $S(t, \mathbf{q}, \tilde{\mathbf{q}})$ of the HJ equation (11.49) it is sufficient to take $c = 1$, since it follows from (11.49) that if $S(t, \mathbf{q}, \tilde{\mathbf{q}})$ is a solution, then $\tilde{S}(t, \mathbf{q}, \tilde{\mathbf{q}}) = \frac{1}{c} S(t, \mathbf{q}, \tilde{\mathbf{q}})$ is also a solution.

11.7.2 Generalized-conservative (stationary) systems with first integrals

For the class of generalized-conservative (stationary) systems

$$\frac{\partial H}{\partial t} = 0$$

we have $H = \underset{t}{\mathrm{const}}$ (see Lemma 10.4). Let us try to find the solution of (11.49) in the form

$$S(t, \mathbf{q}, \tilde{\mathbf{q}}) = S(t, \mathbf{q}, \varphi(t, \mathbf{q}, \mathbf{p})) = -ht + V(\mathbf{q}, \mathbf{p}), \qquad (11.54)$$

where h is a constant. Then the HJ equation (11.49) with $c = 1$ becomes equal to

$$h = H(\mathbf{q}, \frac{\partial}{\partial \mathbf{q}} V(\mathbf{q}, \mathbf{p})). \qquad (11.55)$$

The Hamilton–Jacobi equation

Consider now the class of conservative systems where the variables are grouped as

$$H = H(\varphi_1(q_1, p_1), \varphi_2(q_2, p_2), \cdots, \varphi_n(q_n, p_n)). \tag{11.56}$$

Then $\{\varphi_i(q_i, p_i)\}_{i=1}^n$, as follows from Lemma 10.6, are the first integrals of the system, that is,

$$\varphi_i(q_i(t), p_i(t)) = \alpha_i, \tag{11.57}$$

implying

$$p_i = \Phi_i(q_i, \alpha_i) \quad (i = 1, \ldots, n).$$

Recalling that in view of (11.48) and (11.54) for $c = 1$ we have

$$p_i = \frac{\partial S(t, \mathbf{q}, \tilde{\mathbf{q}})}{\partial q_i} = \frac{\partial}{\partial q_i} V(\mathbf{q}, \mathbf{p}) = \Phi_i(q_i, \alpha_i), \tag{11.58}$$

the following result holds.

Theorem 11.8. *In the considered case*

$$V(\mathbf{q}, \mathbf{p}) = \sum_{i=1}^n \int \Phi_i(q_i, \alpha_i) dq_i \tag{11.59}$$

and the complete integral $S(t, \mathbf{q}, \tilde{\mathbf{q}})$ (11.54) is

$$S(t, \mathbf{q}, \tilde{\mathbf{q}}) = -ht + \sum_{i=1}^n \int \Phi_i(q_i, \alpha_i) dq_i. \tag{11.60}$$

Proof. It follows directly from (11.54), which is in this case

$$h = H(\mathbf{q}, V(\mathbf{q}, \mathbf{p})), \tag{11.61}$$

and the relations (11.58). □

Remark 11.4. From (11.55) and (11.61) it follows also that

$$h = h(\alpha_1, \ldots, \alpha_n). \tag{11.62}$$

Remark 11.5. The same representation (11.60) is valid if instead of (11.57) the system

$$\varphi_i(\mathbf{q}(t), \mathbf{p}(t)) = \alpha_i$$

is a complete system of other first integrals (not obligatorily structured as in (11.56)).

We consider now several examples.

Example 11.6. Consider the Hamiltonian system with Hamiltonian

$$H = \frac{1}{2}(p_1 q_2 + 2 p_1 p_2 + q_1^2).$$

We need to realize the following steps:
- to present the dynamics of the system in the Hamiltonian canonical form;
- to find the solution of the obtained system of differential equations using the HJ equation.

1. By (10.9) the Hamiltonian canonical form is as follows:

$$\dot{q}_1 = \frac{\partial H}{\partial p_1} = \frac{1}{2}(q_2 + 2 p_2), \quad \dot{q}_2 = \frac{\partial H}{\partial p_2} = p_1,$$

$$\dot{p}_1 = -\frac{\partial H}{\partial q_1} = -q_1, \quad \dot{p}_2 = -\frac{\partial H}{\partial q_2} = \frac{1}{2} p_1.$$

2. Since $\dfrac{\partial H}{\partial t} = 0$, the corresponding HJ equation (11.49) is

$$\frac{\partial S}{\partial t} + \frac{1}{2}(\frac{\partial S}{\partial q_1} q_2 + 2 \frac{\partial S}{\partial q_1} \frac{\partial S}{\partial q_2} + q_1^2) = 0,$$

where

$$S = -ht + V(\mathbf{q}, \mathbf{p}),$$

satisfying

$$h = H(\mathbf{q}, V(\mathbf{q}, \mathbf{p})).$$

Taking into account the representation

$$H = \frac{1}{2}\left[p_1(q_2 + 2 p_2) + q_1^2\right],$$

we find that the first integrals are

$$\varphi_2 = q_2 + 2 p_2 := \alpha_2, \quad \varphi_1 = p_1 \alpha_2 + q_1^2 := \alpha_1,$$

implying

$$h = \frac{1}{2}\alpha_1.$$

Resolving the equations above, we get

$$p_1 = \frac{\alpha_1 - q_1^2}{\alpha_2} := \Phi_1, \quad p_2 = \frac{\alpha_2 - q_2}{2} := \Phi_2$$

The Hamilton–Jacobi equation

and by (11.59) and (11.60)

$$V := \int \frac{\alpha_1 - q_1^2}{\alpha_2} dq_1 + \int \frac{\alpha_2 - q_2}{2} dq_2 =$$
$$\frac{\alpha_1}{\alpha_2} q_1 - \frac{1}{3\alpha_2} q_1^3 + \frac{\alpha_2}{2} q_2 - q_2^2,$$

which gives

$$S(t, \mathbf{q}, \boldsymbol{\alpha}) = -\frac{1}{2}\alpha_1 t + \frac{\alpha_1}{\alpha_2} q_1 - \frac{1}{3\alpha_2} q_1^3 + \frac{\alpha_2}{2} q_2 - q_2^2.$$

By (11.51) it follows that

$$-\beta_1 = \frac{\partial S}{\partial \alpha_1} = -\frac{1}{2}t + \frac{1}{\alpha_2}q_1,$$
$$-\beta_2 = \frac{\partial S}{\partial \alpha_2} = \frac{1}{3\alpha_2^2}q_1^3 + \frac{1}{2}q_2,$$

which is why

$$q_1(t, \boldsymbol{\alpha}, \boldsymbol{\beta}) = \alpha_2(\frac{t}{2} - \beta_1)$$

and

$$q_2(t, \boldsymbol{\alpha}, \boldsymbol{\beta}) = -2\left[\beta_2 + \frac{1}{3\alpha_2^2}q_1^3\right] = -2\left[\beta_2 + \frac{\alpha_2}{3}(\frac{t}{2} - \beta_1)^3\right].$$

The corresponding $\mathbf{p}$ can be found using (11.51) again:

$$p_1(t, \boldsymbol{\alpha}, \boldsymbol{\beta}) = \frac{\partial S}{\partial q_1} = \frac{\alpha_1}{\alpha_2} - \frac{1}{\alpha_2}q_1^2 = \frac{\alpha_1}{\alpha_2} - \alpha_2(\frac{t}{2} - \beta_1)^2,$$
$$p_2(t, \boldsymbol{\alpha}, \boldsymbol{\beta}) = \frac{\partial S}{\partial q_2} = \frac{\alpha_2}{2} - 2q_2 = \frac{\alpha_2}{2} + 4\left[\beta_2 + \frac{\alpha_2}{3}(\frac{t}{2} - \beta_1)^3\right].$$

The constants $\boldsymbol{\alpha}, \boldsymbol{\beta}$ can be found from the following system of algebraic equations, obtained by the initial conditions:

$$\left.\begin{aligned} q_1(0, \boldsymbol{\alpha}, \boldsymbol{\beta}) &= q_{1,0} = -\alpha_2 \beta_1, \\ q_2(0, \boldsymbol{\alpha}, \boldsymbol{\beta}) &= q_{2,0} = -2\left[\beta_2 - \frac{\alpha_2}{3}\beta_1^3\right], \\ p_1(0, \boldsymbol{\alpha}, \boldsymbol{\beta}) &= p_{1,0} = \frac{\alpha_1}{\alpha_2} + \alpha_2 \beta_1^2 = \frac{\alpha_1}{\alpha_2} - q_{1,0}\beta_1, \\ p_2(0, \boldsymbol{\alpha}, \boldsymbol{\beta}) &= p_{2,0} = \frac{\alpha_2}{2} + 4\left[\beta_2 - \frac{\alpha_2}{3}\beta_1^3\right] = \frac{\alpha_2}{2} - 2q_{2,0}. \end{aligned}\right\} \quad (11.63)$$

Example 11.7. Consider the Lagrangian conservative dynamic system with the Lagrangian

$$L(t, \mathbf{q}, \dot{\mathbf{q}}) = \frac{1}{2}(\dot{q}_1^2 q_1^2 + \dot{q}_2^2 q_2^2 + \dot{q}_3^2) - \cos q_1.$$

We need to find the dynamics $\mathbf{q} = \mathbf{q}(t, \mathbf{q}(0), \dot{\mathbf{q}}(0))$ using the Hamiltonian approach. The state variable $\mathbf{q} \in R^3$ satisfies the following Lagrange equation (see (10.1)):

$$\frac{d}{dt}\frac{\partial}{\partial \dot{\mathbf{q}}} L(t, \mathbf{q}, \dot{\mathbf{q}}) - \frac{\partial}{\partial \mathbf{q}} L(t, \mathbf{q}, \dot{\mathbf{q}}) = 0,$$

or, in the open format,

$$\ddot{q}_1 \frac{q_1^2}{2} + \sin q_1 = 0, \quad \ddot{q}_2 q_2^2 + \dot{q}_2^2 q_2 = 0, \quad \ddot{q}_3 = 0.$$

The generalized impulses p_i are as follows:

$$p_1 = \frac{\partial L}{\partial \dot{q}_1} = q_1^2 \dot{q}_1, \quad p_2 = \frac{\partial L}{\partial \dot{q}_2} = q_2^2 \dot{q}_2, \quad p_3 = \frac{\partial L}{\partial \dot{q}_3} = \dot{q}_3.$$

The Hamiltonian H is equal to

$$H = \left[\sum_{i=1}^{3} p_i \dot{q}_i - L(t, \mathbf{q}, \dot{\mathbf{q}})\right]_{\dot{\mathbf{q}} = \dot{\mathbf{q}}(\mathbf{q},\mathbf{p})} =$$

$$\frac{p_1^2}{q_1^2} + \frac{p_2^2}{q_2^2} + p_3^2 - \frac{1}{2}\left(\frac{p_1^2}{q_1^2} + \frac{p_2^2}{q_2^2} + p_3^2\right) + \cos q_1 =$$

$$\frac{1}{2}\left(\frac{p_1^2}{q_1^2} + \frac{p_2^2}{q_2^2} + p_3^2\right) + \cos q_1$$

and the Hamiltonian canonical form is

$$\dot{q}_1 = \frac{\partial}{\partial p_1} H = \frac{p_1}{q_1^2}, \quad \dot{q}_2 = \frac{\partial}{\partial p_2} H = \frac{p_2}{q_2^2}, \quad \dot{q}_3 = \frac{\partial}{\partial p_3} H = p_3,$$

$$\dot{p}_1 = -\frac{\partial}{\partial q_1} H = \frac{p_1^2}{q_1^3} + \sin q_1, \quad \dot{p}_2 = -\frac{\partial}{\partial q_2} H = \frac{p_2^2}{q_2^3}, \quad \dot{p}_3 = -\frac{\partial}{\partial q_3} H = 0.$$

The HJ equation in S-format with $c = 1$ is

$$\frac{\partial S}{\partial t} + \cos q_1 + \frac{1}{2}\left[\frac{1}{q_1^2}\left(\frac{\partial S}{\partial q_1}\right)^2 + \frac{1}{q_2^2}\left(\frac{\partial S}{\partial q_2}\right)^2 + \left(\frac{\partial S}{\partial q_3}\right)^2\right] = 0.$$

For this system the first integrals are

$$\frac{1}{2}\frac{p_1^2}{q_1^2} + \cos q_1 = \alpha_1, \quad \frac{1}{2}\frac{p_2^2}{q_2^2} = \alpha_2, \quad \frac{1}{2}p_3^2 = \alpha_3,$$

which allows to express the generalized impulses as

$$p_1 = \pm\sqrt{2q_1^2(\alpha_1 - \cos q_1)} := \Phi_1, \ \alpha_1 \geq \cos q_1,$$
$$p_2 = \pm\sqrt{2\alpha_2}|q_2| := \Phi_2, \ p_3 = \pm\sqrt{2\alpha_3} := \Phi_3.$$

By (11.60) we have

$$S(t, \mathbf{q}, \boldsymbol{\alpha}) = -ht + \sum_{i=1}^{n} \int \Phi_i(q_i, \alpha_i) dq_i = -ht$$
$$\pm \int \sqrt{2q_1^2(\alpha_1 - \cos q_1)} dq_1 \pm \int \sqrt{2\alpha_2}|q_2| dq_2 \pm \int \sqrt{2\alpha_3} dq_3 =$$
$$-ht \pm \int \sqrt{2q_1^2(\alpha_1 - \cos q_1)} dq_1 \pm \sqrt{\frac{\alpha_2}{2}} q_2^2 \pm \sqrt{2\alpha_3} q_3.$$

Note that in view of the relation $h = H$ it follows that

$$h = \alpha_1 + \alpha_2 + \alpha_3.$$

Then, by (11.51), we get

$$\frac{\partial S}{\partial \alpha_1} = -\beta_1 = -t \pm \int \frac{q_1^2}{\sqrt{2q_1^2(\alpha_1 - \cos q_1)}} dq_1,$$

$$\frac{\partial S}{\partial \alpha_2} = -\beta_2 = -t \pm \frac{1}{2}\sqrt{\frac{1}{2\alpha_2}} q_2^2,$$

$$\frac{\partial S}{\partial \alpha_3} = -\beta_3 = -t \pm \sqrt{\frac{1}{2\alpha_3}} q_3.$$

From these relations we find

$$q_2 = \mp\sqrt{2\sqrt{2\alpha_2}(t - \beta_2)}, \ t \geq \beta_2,$$
$$q_3 = \mp\sqrt{2\alpha_3}(t - \beta_3),$$

and

$$p_1(t, \boldsymbol{\alpha}, \boldsymbol{\beta}) = \frac{\partial S}{\partial q_1} = \pm\sqrt{2q_1^2(\alpha_1 - \cos q_1)},$$
$$p_2(t, \boldsymbol{\alpha}, \boldsymbol{\beta}) = \frac{\partial S}{\partial q_2} = \pm\sqrt{2\alpha_2} q_2 = (2\alpha_2)^{3/4}\sqrt{2(t - \beta_2)},$$
$$p_3(t, \boldsymbol{\alpha}, \boldsymbol{\beta}) = \frac{\partial S}{\partial q_3} = \pm\sqrt{2\alpha_3}.$$

The constants α_1, α_2, and α_3 can be found from these last relations putting $t = 0$ analogously as has been done in (11.63).

Example 11.8. Let for a Hamiltonian system

$$S(t, \mathbf{q}, \boldsymbol{\alpha}) = -\sum_{i=1}^{n} q_i f_i(t) - \sum_{i=1}^{n} \alpha_i q_i - \sum_{i=1}^{n} \int [f_i(t) + \alpha_i]^2 \, dt.$$

We need:
- to find the Hamiltonian $H(t, \mathbf{q}, \mathbf{p})$;
- to obtain its first integrals;
- to define the differential equation and dynamics for $\mathbf{q}(t, \boldsymbol{\alpha}, \boldsymbol{\beta})$ in the Lagrange form.

1. Since the considered system is Hamiltonian we have

$$-\beta_i = \frac{\partial S}{\partial \alpha_i} = -q_i - 2 \sum_{i=1}^{n} \int [f_i(t) + \alpha_i] \, dt,$$

from which it follows that

$$q_i = \beta_i - 2 \sum_{i=1}^{n} \int [f_i(t) + \alpha_i] \, dt = \beta_i - 2 \sum_{i=1}^{n} \int f_i(t) \, dt - 2t \sum_{i=1}^{n} \alpha_i.$$

We also have

$$p_i = \frac{\partial S}{\partial q_i} = -f_i(t) - \alpha_i,$$

which gives

$$\alpha_i = -f_i(t) - p_i.$$

Recall that

$$\frac{\partial S}{\partial t} + H = 0,$$

which is why

$$H = -\left[\frac{\partial S(t, \mathbf{q}, \boldsymbol{\alpha})}{\partial t}\right]_{\alpha=\alpha(t,\mathbf{p})} =$$

$$\left[\sum_{i=1}^{n} q_i \frac{d}{dt} f_i(t) + \sum_{i=1}^{n} [f_i(t) + \alpha_i]^2\right]_{\alpha=\alpha(t,\mathbf{p})},$$

and, finally,

$$H(t, \mathbf{q}, \mathbf{p}) = \sum_{i=1}^{n} \left(q_i \frac{d}{dt} f_i(t) + p_i^2\right).$$

2. By (10.6) we have

$$L(t, \mathbf{q}, \dot{\mathbf{q}}) = \left[H(t, \mathbf{q}, \mathbf{p}) - \sum_{i=1}^{n} p_i \dot{q}_i \right]_{\mathbf{p}=\mathbf{p}(t,\mathbf{q},\dot{\mathbf{q}})} =$$

$$\left[\sum_{i=1}^{n} \left(q_i \frac{d}{dt} f_i(t) + p_i^2 \right) - \sum_{i=1}^{n} p_i \dot{q}_i \right]_{\mathbf{p}=\mathbf{p}(t,\mathbf{q},\dot{\mathbf{q}})}.$$

But

$$\dot{q}_i = \frac{\partial H}{\partial p_i} = 2 p_i,$$

and therefore

$$L(t, \mathbf{q}, \dot{\mathbf{q}}) = \sum_{i=1}^{n} \left(q_i \frac{d}{dt} f_i(t) - \frac{\dot{q}_i^2}{4} \right).$$

3. Recalling that

$$\frac{d}{dt} \frac{\partial L}{\partial \dot{q}_i} - \frac{\partial L}{\partial q_i} = 0,$$

we get the following dynamic equations:

$$\frac{1}{2} \ddot{q}_i + \frac{d}{dt} f_i(t) = 0.$$

After double integration we finally obtain

$$\dot{q}_i(t) = -2 f_i(t) + c_{i,1}$$

and

$$q_i(t) = -2 \int_{\tau=0}^{t} f_i(\tau) d\tau + c_{i,1} t + c_{i,2} =$$

$$q_i(0) - 2 \int_{\tau=0}^{t} f_i(\tau) d\tau + [\dot{q}_i(0) + 2 f_i(0)] t$$

for all $i = 1, ..., n$.

11.8 On relations with optimal control

This section shows the relation between the HJ equation in mechanics of conservative systems and the dynamic programming method (DPM) in optimal control theory. Here we follow Chapter 3 of (Boltyanski and Poznyak, 2012). See also (Kwatny and Blankenship, 2000) and (Levi, 2014).

11.8.1 Problem formulation and value function

Let $(s, y) \in [0, T) \times R^n$ be "*an initial time and state pair*" to the following control system over $[s, T]$:

$$\left.\begin{aligned} \dot{x}(t) &= f(x(t), u(t), t), \ t \in [s, T], \\ x(s) &= y, \end{aligned}\right\} \quad (11.64)$$

where $x \in R^n$ is its state vector and $u \in \mathbb{R}^r$ is the control that may run over a given control region $U \subset R^r$ with the cost functional

$$J(s, y; u(\cdot)) = h_0(x(T)) + \int_{t=s}^{T} h(x(t), u(t), t) \, dt \quad (11.65)$$

containing the integral term. Suppose that $u(t)$ is partially continuous and functions f, h, and h_0 are sufficiently smooth (for the details see (Boltyanski and Poznyak, 2012)). Under these assumptions, for any initial $(s, y) \in [0, T) \times R^n$ and any admissible $u(\cdot)$ the optimization problem

$$J(s, y; u(\cdot)) \to \min_{u(\cdot) \in U} \quad (11.66)$$

formulated for the plant (11.64) and for the cost functional $J(s, y; u(\cdot))$ (see (11.65)) admits a unique solution

$$x(\cdot) := x(\cdot, s, y, u(\cdot)),$$

and the functional (11.65) is well defined.

Definition 11.5 (Value function). The function $V(s, y)$ defined for any $(s, y) \in [0, T) \times \mathbb{R}^n$ as

$$\left.\begin{aligned} V(s, y) &:= \inf_{u(\cdot) \in U} J(s, y; u(\cdot)), \\ V(T, y) &= h_0(y) \end{aligned}\right\} \quad (11.67)$$

is called the **value function** of the optimization problem (11.66).

We will be interested in the solution of the optimal control problem (11.66) when $s = 0$ and $y = x(0)$.

11.8.2 Hamilton–Jacobi–Bellman equation

The following theorem presents the conditions for the admissible control u which makes the value function $V(0, x(0))$ minimal.

Suppose that under the accepted assumptions the value function $V(s, y)$ (11.67) is continuously differentiable, that is, $V \in C^1([0, T) \times \mathbb{R}^n)$. Then $V(s, y)$ is a solution to the following terminal value problem of a first order partial differential equation, named below the **Hamilton–Jacobi–Bellman (HJB) equation**, associated with the original optimization problem (11.66):

$$\left. \begin{array}{l} -\dfrac{\partial}{\partial t} V(t,x) + \sup\limits_{u \in U} H(-\dfrac{\partial}{\partial x} V(t,x), x(t), u(t), t) = 0, \\ (t,x) \in [0,T) \times \mathbb{R}^n, \\ V(T,x) = h_0(x), \ x \in \mathbb{R}^n, \end{array} \right\} \quad (11.68)$$

where

$$\begin{array}{l} H(\psi, x, u, t) := \psi^\top f(x, u, t) - h(x(t), u(t), t) \\ \left(t, x, u, \psi \in [0, T] \times \mathbb{R}^n \times \mathbb{R}^r \times \mathbb{R}^n\right) \end{array} \quad (11.69)$$

is the **Hamiltonian** of the system (11.64) containing the adjoint vector $\psi(t) \in \mathbb{R}^n$, which satisfies the following system of ordinary differential equations:

$$\left. \begin{array}{l} \dot{x}(t) = \dfrac{\partial}{\partial \psi} H(\psi, x, u^*(\cdot), t) = f\left(x, u^*(\cdot), t\right), \ x(0) = x_0, \\ \dot{\psi}(t) = -\dfrac{\partial}{\partial x} H(\psi, x, u^*(\cdot), t), \ \psi(T) = -\dfrac{\partial}{\partial x} h_0(x(T)). \end{array} \right\} \quad (11.70)$$

Here

$$u^*(\cdot) := u^*\left(t, x, \dfrac{\partial}{\partial x} V(t,x)\right) \quad (11.71)$$

is a solution to the optimization problem

$$H(-\dfrac{\partial}{\partial x} V(t,x), x, u, t) \rightarrow \sup_{u \in U} \quad (11.72)$$

with fixed values x, t, and $\dfrac{\partial}{\partial x} V(t,x)$.

The proof can be found in (Boltyanski and Poznyak, 2012, Section 3.3) (see the proof of Theorem 3.3).

Remark 11.6. As follows from this theorem:
- any dynamic system from the considered class, controlled by the optimal control $u^*(\cdot)$ (see (11.71)), is Hamiltonian;
- the state vector x corresponds to the generalized coordinate **q**;
- the adjoint variable ψ plays the same role as the generalized impulses **p** in (10.9);
- the function h in (11.65) corresponds to the Lagrange function L.

11.8.3 Verification rule as a sufficient condition of optimality

The theorem below, representing the *sufficient conditions* of *optimality*, is referred to as the *verification rule*.

Theorem 11.9. *We make the following assumptions:*

1. *Suppose that we can obtain the solution $V(t,x)$ to the HJB equation*

$$\left.\begin{aligned}-\frac{\partial}{\partial t}V(t,x)+H(-\frac{\partial}{\partial x}V(t,x),x,u^*(\cdot),t)=0,\\ V(T,x)=h_0(x),\ (t,x)\in[0,T)\times\mathbb{R}^n,\end{aligned}\right\} \quad (11.73)$$

which for any $(t,x)\in[0,T)\times\mathbb{R}^n$ is unique and smooth, that is,

$$V\in C^1\left([0,T)\times\mathbb{R}^n\right);$$

2. *Suppose that for any pair $(s,x)\in[0,T)\times\mathbb{R}^n$ there exists a solution $x(s,x)$ to the following ordinary differential equation:*

$$\left.\begin{aligned}\dot{x}(t)=f\left(x(t),u^*\left(t,x(t),\frac{\partial}{\partial x}V(t,x(t))\right),t\right),\\ x^*(s)=x,\end{aligned}\right\} \quad (11.74)$$

satisfied for $t\in[s,T]$.
Then with $(s,x)=(0,x)$ the pair $(x(t),u^*(\cdot))$ is optimal, that is,

$$u^*(\cdot)=u^*\left(t,x(t),\frac{\partial}{\partial x}V(t,x)\mid_{x=x^*(t)}\right) \quad (11.75)$$

*is an **optimal control**.*

To apply the optimal control $u^*(\cdot)$ (see (11.75)) we need to have at our disposition the function $V(t,x)$, or in other words, we should be able to resolve the HJB equation (11.73).

11.8.4 Affine dynamics with a quadratic cost

Definition 11.6. The plant (11.64) is called **stationary** and **affine in control** if the right-hand side does not depend on t and is linear on $u\in U=\mathbb{R}^r$, that is,

$$f(x(t),u(t))=f_0(x(t))+f_1(x(t))u(t). \quad (11.76)$$

Consider the loss functional (11.65) with $h_0(x)=0$ and the quadratic cost function

$$\begin{aligned}h(x,u):=\|x\|_Q^2+\|u\|_R^2,\\ 0\leq Q\in\mathbb{R}^n,\ 0<R\in\mathbb{R}^r\end{aligned} \quad (11.77)$$

The Hamilton–Jacobi equation

on the infinite horizon ($T = \infty$). Then (11.71) gives

$$u^*(x) = \arg\sup_{u \in \mathbb{R}^r} H\left(-\frac{\partial}{\partial x}V(x), x, u\right) =$$
$$\max_{u \in \mathbb{R}^r}\left(-\frac{\partial}{\partial x}V(x)^\mathsf{T}[f_0(x) + f_1(x)u] - \|x\|_Q^2 - \|u\|_R^2\right) \quad (11.78)$$
$$= -\frac{1}{2}R^{-1}f_1(x)^\mathsf{T}\frac{\partial}{\partial x}V(x)$$

and the corresponding HJ equation becomes as follows:

$$-\frac{\partial}{\partial x}V(x)^\mathsf{T} f_0(x) - \|x\|_Q^2 + \frac{1}{4}\left\|R^{-1}f_1(x)^\mathsf{T}\frac{\partial}{\partial x}V(x)\right\|_R^2 = 0. \quad (11.79)$$

Suppose also, for simplicity, that we deal with the special subclass of the affine systems (11.76) for which the matrix $\left[f_1(x)R^{-1}f_1(x)^\mathsf{T}\right]$ is invertible for any $x \in \mathbb{R}^n$, that is,

$$\operatorname{rank}\left[f_1(x)R^{-1}f_1(x)^\mathsf{T}\right] = n. \quad (11.80)$$

Denote

$$R_f(x) := \left[f_1(x)R^{-1}f_1(x)^\mathsf{T}\right]^{1/2} > 0, \quad (11.81)$$

which, by (11.80), is strictly positive, and, hence, $R_f^{-1}(x)$ exists. Then Eq. (11.79) may be rewritten as

$$r^2(x) := \|x\|_Q^2 + \left\|R_f^{-1/2}(x)f_0(x)\right\|^2 =$$
$$\frac{1}{4}\left\|R_f^{1/2}(x)\frac{\partial}{\partial x}V(x) - 2R_f^{-1/2}(x)f_0(x)\right\|^2,$$

where

$$r(x) := \sqrt{\|x\|_Q^2 + \left\|R_f^{-1/2}(x)f_0(x)\right\|^2}.$$

This implies the following representation:

$$\frac{1}{2}R_f^{1/2}(x)\frac{\partial}{\partial x}V(x) - R_f^{-1/2}(x)f_0(x) = \bar{e}(x)r(x),$$
$\bar{e}(x)$ being a unitary vector ($\|\bar{e}(x)\| = 1$),

or, equivalently,

$$\begin{aligned}\frac{1}{2}\frac{\partial}{\partial x}V(x) &= R_f^{-1}(x)f_0(x) + R_f^{-1/2}(x)\bar{e}(x)r(x) \\ &= \left[f_1(x)R^{-1}f_1(x)^\mathsf{T}\right]^{-1}f_0(x) + R_f^{-1/2}(x)\bar{e}(x)r(x).\end{aligned} \qquad (11.82)$$

So, substitution of (11.82) into the optimal control (11.78) gives

$$\begin{aligned}u^*(x) = -R^{-1}f_1(x)^\mathsf{T}\left[\frac{1}{2}\frac{\partial}{\partial x}V(x)\right] = \\ -R^{-1}f_1(x)^\mathsf{T}\left(\left[f_1(x)R^{-1}f_1(x)^\mathsf{T}\right]^{-1}f_0(x) + \right. \\ \left. r(x)\left[f_1(x)R^{-1}f_1(x)^\mathsf{T}\right]^{-1}\bar{e}(x)\right).\end{aligned} \qquad (11.83)$$

There exist many ways to select $\bar{e}(x)$, but all of them have to guarantee the property

$$J(0, x(0); u(\cdot)) = \int_{t=0}^{\infty}\left(\|x(t)\|_Q^2 + \|u(t)\|_R^2\right)dt < \infty. \qquad (11.84)$$

Substituting (11.83) into (11.64) with $s = 0$ and $y = x(0)$ leads to the final expression for the optimal trajectory:

$$\dot{x}(t) = f(x(t), u(t)) = f_0(x(t)) + f_1(x(t))u^*(t) = -r(x)\bar{e}(x).$$

To guarantee (11.84), we need at least to satisfy the asymptotic stability property $\|x(T)\| \underset{T\to\infty}{\to} 0$. To select $\bar{e}(x)$ satisfying this requirement, let us consider the function $V(x) = \frac{1}{2}\|x\|^2$, for which we have

$$\dot{V}(x(t)) = x^\mathsf{T}(t)\dot{x}(t) = -r(x)x^\mathsf{T}(t)\bar{e}(x).$$

Taking, for example,

$$\bar{e}(x) := \frac{1}{\sqrt{n}}\mathrm{SIGN}(x),$$
$$\mathrm{SIGN}(x) := (\mathrm{sign}(x_1), \ldots, \mathrm{sign}(x_n)),$$

we get

$$\dot{V}(x(t)) = -\frac{r(x)}{\sqrt{n}}x^\mathsf{T}(t)\mathrm{SIGN}(x(t)) = \\ -\frac{r(x)}{\sqrt{n}}\sum_{i=1}^{n}|x_i(t)| \leq -\frac{r(x)}{\sqrt{n}}\sqrt{2V(x(t))} < 0$$

for $x(t) \neq 0$. If, additionally, $r(x) \geq c > 0$, then this inequality implies

$$\dot V(x(t)) \leq -\sqrt{\frac{2}{n}c}\sqrt{V(x(t))}$$

and, as a result, $V(x(t)) \to 0$ in the finite time

$$t_{reach} = \sqrt{nV(x(0))}/c,$$

so that $V(x(t)) = 0$ and hence $x(t) = 0$ for any $t \geq t_{reach}$, satisfying (11.84) with the optimal control

$$u^*(x) = -R^{-1}f_1(x)^\mathsf{T}\left[f_1(x)R^{-1}f_1(x)^\mathsf{T}\right]^{-1}[f_0(x) + r(x)\,\mathrm{SIGN}(x)].$$
(11.85)

11.8.5 The case when the Hamiltonian admits the existence of first integrals

The next theorem represent the main idea of finding the HJB solution using the notion of the first integrals. Let the system of n first integrals of a **stationary** $\left(\dfrac{\partial}{\partial t}f(x, u^*, t) = 0\right)$ H-Hamiltonian system

$$\varphi_i(x(t), \psi(t)) = \alpha_i \quad (i = 1, ..., n)$$
(11.86)

be solvable with respect to the vectors $\psi_1(t), ..., \psi_n(t)$, that is, for any $x^* \in \mathbb{R}^n$

$$\det \begin{Vmatrix} \dfrac{\partial}{\partial \psi_1}\varphi_1(x^*, \psi) & \cdots & \dfrac{\partial}{\partial \psi_n}\varphi_1(x^*, \psi) \\ \vdots & \cdots & \vdots \\ \dfrac{\partial}{\partial \psi_1}\varphi_n(x^*, \psi) & \cdots & \dfrac{\partial}{\partial \psi_n}\varphi_n(x^*, \psi) \end{Vmatrix} \neq 0.$$
(11.87)

Denote this solution by

$$\psi_i := S_i(x, \boldsymbol{\alpha}) \quad (i = 1, ..., n).$$
(11.88)

Then, as in (11.60), the solution to the HJB equation (11.73) is given by

$$V(s, y) = -\tilde{h}s - \sum_{i=1}^{n}\int S_i(y, \boldsymbol{\alpha})\,dy_i,$$
(11.89)

where the constants α_i $(i = 1, ..., n)$ and $\tilde{h}$ are related (for a given initial state x_0) by the equation

$$\sum_{i=1}^{n}\int S_i(y, \boldsymbol{\alpha})\,dy_i\,\big|_{y=x^*(T\mid T, x_0)=x_0} = -\tilde{h}T - h_0(x_0).$$
(11.90)

The optimal control $u^*(\cdot)$ (see (11.75)) has the following form:

$$\left.\begin{array}{l} u^*(\cdot) = u^*\left(t, x(t), -S(x, \alpha)\mid_{x=x^*(t)}\right), \\ S(x, \alpha) = (S_1(x, \alpha), \ldots, S_n(x, \alpha))^\mathsf{T}. \end{array}\right\} \quad (11.91)$$

Remark 11.7. It follows from (11.90) that the constant $\tilde{h}$ is a function of constants α and initial conditions x_0, namely,

$$\tilde{h} = \tilde{h}(\alpha, x_0).$$

11.8.6 The deterministic Feynman–Kac formula: the general smooth case

In the general case, when the system of n first integrals is not available, the solution to the HJB equation (11.68), after substituting in it $u^*(\cdot)$ (see (11.75)), is given by the deterministic *Feynman–Kac formula*.

Suppose that the solution

$$x = x(t \mid 0, y)$$

of the Hamiltonian canonical equations (11.70) is solvable with respect to an initial condition y for any $t \in [0, T]$, that is, there exists a function $Y(t, x)$ such that for any $t \in [0, T]$ and any x

$$x(t \mid 0, Y(t, x)) = x,$$
$$y = Y(t, x(t \mid 0, y)),$$

and, in particularly,

$$x(t \mid 0, Y(t, y)) = y,$$

which implies

$$x(t \mid 0, Y(t, y)) = y = Y(t, x(t \mid 0, y))$$

and

$$\frac{\partial}{\partial y} x(t \mid 0, Y(t, y)) = I.$$

Define the function $v(t, y)$ as

$$\begin{aligned} v(t, y) = {} & h_0(y) + \\ & \int_{\tau=t}^{T} \left[\frac{\partial}{\partial \psi} H(\tau, x(\tau \mid 0, y), \psi(\tau \mid 0, y))^\mathsf{T} \psi(\tau \mid 0, y) \right. \\ & \left. - H(\tau, x(\tau \mid 0, y), \psi(\tau \mid 0, y)) \right] d\tau. \end{aligned} \quad (11.92)$$

The Hamilton–Jacobi equation

Then the function $V(t, x)$ given by the formula

$$V(t, x) := v(t, Y(t, x)) \tag{11.93}$$

is a solution (local) to the HJB equation (11.73).

Proof. a) Since

$$\frac{\partial}{\partial y} x(t \mid 0, y)\mid_{t=0} = I,$$

the equation $x = x(t \mid 0, y)$ is solvable with respect to y for any small enough $t \in (0, \varepsilon)$, that is, there exists a function $Y(t, x)$ such that $y = Y(t, x)$. Define the function

$$s(t, y) := -H\left(t, y, \frac{\partial}{\partial y} h_0(y)\right) - \int_{\tau=0}^{t} \frac{\partial}{\partial \tau} H(\tau, x(\tau \mid 0, y), \psi(\tau \mid 0, y)) \, d\tau. \tag{11.94}$$

In view of (11.70), it follows that

$$\frac{d}{dt}[s(t, y) + H(t, x(t \mid 0, y), \psi(t \mid 0, y))] =$$
$$-\frac{\partial}{\partial t} H(t, x(t \mid 0, y), \psi(t \mid 0, y)) + \frac{\partial}{\partial t} H(t, x(t \mid 0, y), \psi(t \mid 0, y)) +$$
$$\frac{\partial}{\partial x} H(t, x(t \mid 0, y), \psi(t \mid 0, y))^\mathsf{T} \underbrace{\dot{x}(t \mid 0, y)}_{\frac{\partial}{\partial \psi} H} +$$
$$\frac{\partial}{\partial \psi} H(t, x(t \mid 0, y), \psi(t \mid 0, y))^\mathsf{T} \underbrace{\dot{\psi}(t \mid 0, y)}_{-\frac{\partial}{\partial x} H} = 0.$$

$$\tag{11.95}$$

From (11.94) for $t = 0$ we get

$$s(0, y) + H\left(0, y, \frac{\partial}{\partial y} h_0(y)\right) = 0,$$

which together with (11.95) implies

$$s(t, y) + H(t, x(t \mid 0, y), \psi(t \mid 0, y)) \equiv 0 \tag{11.96}$$

for any $t \in [0, \varepsilon]$. In view of the property

$$H(-\frac{\partial}{\partial x} V(t, x), x, u, t) = -H(\frac{\partial}{\partial x} V(t, x), x, u, t),$$

to prove the theorem it is sufficient to show that

$$\psi(t \mid 0, y) = -\frac{\partial}{\partial y} V(t, y) \tag{11.97}$$

and

$$s(t, y) = -\frac{\partial}{\partial t} V(t, y). \tag{11.98}$$

b) For the vector function

$$U(t, y) := \frac{\partial}{\partial y} V(t, y) + \psi(t \mid 0, y), \tag{11.99}$$

which in view of (11.70) verifies

$$U(T, y) = \frac{\partial}{\partial y} V(T, y) + \psi(T \mid 0, y) = \frac{\partial}{\partial y} h_0(y) + \psi(T \mid 0, y) = 0, \tag{11.100}$$

we have

$$\frac{d}{dt} U(t, y) = \frac{\partial}{\partial y} \frac{d}{dt} V(t, y) + \dot{\psi}(\cdot) \mid_{(\cdot)=(t\mid 0, y)} =$$
$$-\frac{\partial}{\partial y} \left[\frac{\partial}{\partial \psi} H(t, x(\cdot), \psi(\cdot))^\mathsf{T} \psi(\cdot) - H(t, x(\cdot), \psi(\cdot)) \right] \mid_{(\cdot)=(t\mid 0, y)}$$
$$-\frac{\partial}{\partial x} H(t, x(\cdot), \psi(\cdot)) \mid_{(\cdot)=(t\mid 0, y)} =$$
$$-\frac{\partial}{\partial y} \left[\frac{\partial}{\partial \psi} H(t, x(\cdot), \psi(\cdot))^\mathsf{T} \psi(\cdot) \right] \mid_{(\cdot)=(t\mid 0, y)} +$$
$$\frac{\partial}{\partial x} H(t, x(\cdot), \psi(\cdot)) \mid_{(\cdot)=(t\mid 0, y)} \underbrace{\frac{\partial}{\partial y} x(t \mid 0, Y(t, y))}_{I}$$
$$-\frac{\partial}{\partial x} H(t, x(\cdot), \psi(\cdot)) \mid_{(\cdot)=(t\mid 0, y)} =$$
$$-\frac{\partial}{\partial y} \left[\underbrace{\frac{\partial}{\partial \psi} H(t, x(\cdot), \psi(\cdot))^\mathsf{T}}_{f(x, u^*(\cdot), t)} \psi(\cdot) \right] \mid_{(\cdot)=(t\mid 0, y)} = 0$$

$$\tag{11.101}$$

since the term $f(x, u^*(\cdot), t) \psi(\cdot) \mid_{(\cdot)=(t\mid 0, y)}$ does not depend on y as well as $\psi(\cdot) \mid_{(\cdot)=(t\mid 0, y)}$ (it depends on the terminal, but not the initial condition y). Both properties (11.100) and (11.101) give

$$U(t, y) \equiv 0,$$

which proves (11.97).

The Hamilton–Jacobi equation

c) By (11.73) it follows that

$$H(-\frac{\partial}{\partial y}V(t,y),y,u^*(\cdot),t) = \frac{\partial}{\partial t}V(t,y),$$

but in view of (11.96) we have

$$s(t,y) = -H\left(t, x(t\mid 0,y), \underbrace{\psi(t\mid 0,y)}_{-\frac{\partial}{\partial y}V(t,y)}\right) = -\frac{\partial}{\partial t}V(t,y),$$

which gives (11.98). □

Corollary 11.2. *It follows from (11.75) that*

$$u^*(\cdot) = u^*\left(t, x(t), \frac{\partial}{\partial x}V(t,x)\mid_{x=x^*(t)}\right) = u^*\left(t, x, -\psi(t\mid 0,x)\mid_{x=x^*(t)}\right).$$

Remark 11.8. Note that the integral term

$$\mathcal{I}_t := \int_{\tau=t}^{T}\left[\frac{\partial}{\partial\psi}H(\tau, x(\cdot), \psi(\cdot))^\mathsf{T}\psi(\tau\mid 0,y) - H(\tau, x(\cdot), \psi(\cdot))\right]_{(\cdot)=(\tau\mid 0,y)} d\tau \qquad (11.102)$$

in (11.92) is exactly equal to the Hamiltonian action (see (11.6))

$$I(\alpha) := \int_{t_0(\alpha)}^{t_1(\alpha)} L(\tau, \mathbf{q}(\tau,\alpha), \dot{\mathbf{q}}(\tau,\alpha)) d\tau, \quad \alpha \in [0,1]$$

with $t_0(\alpha) = t$ and $t_1(\alpha) = T$. Indeed, recalling that

$$\dot{x} = \frac{\partial}{\partial\psi}H(\tau, x(\tau\mid 0,y), \psi(\tau\mid 0,y))$$

and associating x with $\mathbf{q}$ and ψ with $\mathbf{p}$, from (11.102) we get

$$\mathcal{I}_t := \int_{\tau=t}^{T}\left[\dot{x}^\mathsf{T}(\tau)\psi(\cdot) - H(\tau, x(\cdot), \psi(\cdot))\right]_{(\cdot)=(\tau\mid 0,y)} d\tau =$$

$$\int_{\tau=t}^{T}\left[\dot{\mathbf{q}}^\mathsf{T}(\tau)\mathbf{p}(\tau) - H(\tau, \mathbf{q}(\tau), \mathbf{p}(\tau))\right]_{(\cdot)=(\tau\mid 0,y),\ \mathbf{p}=\mathbf{p}(t,\mathbf{q},\dot{\mathbf{q}})} =$$

$$\int_{t=t_0(\alpha)}^{T=t_1(\alpha)} L(\tau, \mathbf{q}(\tau,\alpha), \dot{\mathbf{q}}(\tau,\alpha)) d\tau = I(\alpha).$$

Remark 11.9. Of course, the direct use of formula (11.92) in most real cases is not possible, since analytical expressions for the integral term cannot be written out completely. However, the expression may be of interest in terms of its digital computer implementation.

11.9 Exercises

Exercise 11.1. Show that the curvilinear integral

$$I = \oint \sum_{i=1}^{n} ([\alpha_i p_i + \varphi_i(q_i)]\delta q_i + [\beta_i q_i + \psi_i(p_i)]\delta p_i)$$

is a universal Poincaré invariant (see (11.14)) if

$$\alpha_1 - \beta_1 = \alpha_2 - \beta_2 = \ldots = \alpha_n - \beta_n.$$

Exercise 11.2. The function $f(\mathbf{q}, \mathbf{p}, t)$ is the first integral of the canonical system with the Hamiltonian $H(\mathbf{q}, \mathbf{p}, t)$. Prove that the integral

$$I = \int \cdots \int f(\mathbf{q}, \mathbf{p}, t) \, dq_1 dq_2 \cdots dq_n dp_1 dp_2 \cdots dp_n$$

is an integral invariant (see (11.13)).

Exercise 11.3. Show that the necessary and sufficient condition for maintaining the phase volume

$$v(t) = \iint_{C \subset R^2} dx_1(t) \, dx_2(t)$$

of the stationary linear dynamical system

$$\left.\begin{array}{l} \dot{x}_1 = a_{11}x_1 + a_{12}x_2, \\ \dot{x}_2 = a_{21}x_1 + a_{22}x_2, \\ x_1(0) = x_{10}, \; x_2(0) = x_{20} \end{array}\right\}$$

is that this system is Hamiltonian with the Hamiltonian function

$$H = \frac{1}{2}a_{12}x_2^2 + a_{11}x_1x_2 - \frac{1}{2}a_{21}x_1^2,$$

satisfying the condition

$$a_{11} + a_{22} = 0.$$

Hint. Use the following steps:
1.
$$\dot{x} = Ax \Longrightarrow x(t) = e^{At}x(0),$$

2.
$$v(t) = \iint_{C_t \subset R^2} dx(t) = \iint_{C_0 \subset R^2} \left|\det \frac{\partial x(t)}{\partial x(0)}\right| dx(0) =$$
$$\iint_{C_0 \subset R^2} \left|\det e^{At}\right| dx(0) = \left|\det e^{At}\right| \iint_{C_0 \subset R^2} dx(0) = \left|\det e^{At}\right| v(0)$$
(C_t is the contour corresponding C_0),

3. apply the Liouville formula

$$\det e^{At} = \exp\{(\operatorname{tr} A) t\} = 1 \text{ for all } t \text{ if an only if } \operatorname{tr} A = 0,$$

which by Example 10.7 shows that the system is Hamiltonian.

Exercise 11.4. Find the full integral of the system with the Hamiltonian

$$H = pq f(t) + p\psi(t)$$

and show that the law of the system's motion is described by the relations

$$q(t) = \exp\left(\int f(t) dt\right)\left[c + \int \psi(t) \exp\left(-\int f(t) dt\right) dt\right],$$
$$p(t) = c_1 \exp\left(-\int f(t) dt\right).$$

Exercise 11.5. A body of mass m, connected to a fixed wall by a stiff spring c, can slide along a smooth horizontal guide Ox. The force $F(t)$ ($|F(t)| \le F^+$) acts on the body. At the initial moment of time, the body is motionless and is at a distance $6\dfrac{F^+}{c}$ from the equilibrium position. Demonstrate that the change in the force $F(t)$ at which the body returns to the equilibrium position at zero speed in a minimum time is given by the formula

$$F(t) = \begin{cases} F^+, & \text{for } 0 \le t \le \pi\sqrt{\dfrac{m}{c}}, \\ -F^+, & \text{for } \pi\sqrt{\dfrac{m}{c}} < t \le 2\pi\sqrt{\dfrac{m}{c}}, \\ F^+, & \text{for } 2\pi\sqrt{\dfrac{m}{c}} < t \le 3\pi\sqrt{\dfrac{m}{c}}. \end{cases}$$

Hint. In (11.65) take

$$h_0(x(T)) = 0, \quad h(x(t), u(t), t) = 1,$$

and apply Theorem 11.8.2.

Collection of electromechanical models 12

Contents

12.1 Cylindrical manipulator (2-PJ and 1-R) 384
12.2 Rectangular (Cartesian) robot manipulator 387
12.3 Scaffolding type robot manipulator 389
12.4 Spherical (polar) robot manipulator 392
12.5 Articulated robot manipulator 1 396
12.6 Universal programmable manipulator 399
12.7 Cincinnati Milacron T^3 manipulator 404
12.8 CD motor, gear, and load train 412
12.9 Stanford/JPL robot manipulator 414
12.10 Unimate 2000 manipulator 418
12.11 Robot manipulator with swivel base 424
12.12 Cylindrical robot with spring 427
12.13 Non-ordinary manipulator with shock absorber 429
12.14 Planar manipulator with two joints 434
12.15 Double "crank-turn" swivel manipulator 437
12.16 Robot manipulator of multicylinder type 442
12.17 Arm manipulator with springs 445
12.18 Articulated robot manipulator 2 460
12.19 Maker 110 465
12.20 Manipulator on a horizontal platform 468
12.21 Two-arm planar manipulator 471
12.22 Manipulator with three degrees of freedom 476
12.23 CD motor with load 478
12.24 Models of power converters with switching-mode power supply 479
    12.24.1 Buck type DC-DC converter 480
    12.24.2 Boost type DC-DC converter 482
12.25 Induction motor 483

This chapter is dedicated to the implementation of the methods given previously for the construction of mathematical models of different mechanical systems (mostly robots) and also some electrical systems such as power converters and DC and AC machines. All models analyzed here are presented in the uniform Lagrangian format

$$D(q)\ddot{q} + C(q,\dot{q})\dot{q} + g(q) = \tau.$$

with a detailed description and deriving of the corresponding Lagrange functions $L = T - V$.

12.1 Cylindrical manipulator (2-PJ and 1-R)

Consider the cylindrical manipulator with two prismatic joints (PJs) and a rotating joint (R) represented in Fig. 12.1.

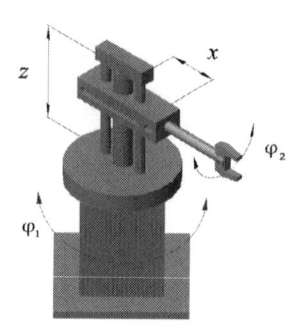

Figure 12.1 Manipulator with two prismatic joints (PJ) and a rotating joint (R).

Generalized coordinates

The generalized coordinates for this mechanical system are as follows:

$$q_1 := \varphi_1, \quad q_2 := \varphi_2, \quad q_3 := z, \quad q_4 := x.$$

Kinetic energy

The kinetic energy T of this system is given by the following expression:

$$T = \sum_{i=1}^{4} T_{m_i}, \qquad (12.1)$$

where T_{m_i} can be calculated using the König formula (see Section 3.6 in Chapter 3)

$$T_{m_i} = T_{m_i,0} + T_{m_i,rot-0} + 2m_i \left(\mathbf{v}_{m_i-c.i.-0}, \mathbf{v}_0 \right),$$

$$T_{m_i,0} = \frac{1}{2} m_i \|\mathbf{v}_0\|^2, \quad T_{m_i,rot-0} = \frac{1}{2} \left(\boldsymbol{\omega}, I_{i,0} \boldsymbol{\omega} \right).$$

Here $I_{i,0}$ is the inertia tensor with respect to a coordinate system with the origin at point O, the vector $\mathbf{v}_{m_i-c.i.-0}$ is the speed of the center of inertia with respect to the coordinate system with the origin at point O, and the vector $\mathbf{v}_0$ is the speed of the origin of the coordinate system. In our case we have

$$T_{m_1} = T_{m_1,0} + T_{m_1,rot-0} + m_1 \left(\mathbf{v}_{m_2-c.i.-0}, \mathbf{v}_0 \right) =$$

Collection of electromechanical models

$$T_{m_1,rot-0} = \frac{1}{2}\dot{\varphi}_1^2 \left(\frac{m_1 r_1^2}{2}\right) = \frac{m_1 r_1^2}{4}\dot{q}_1^2$$

and

$$T_{m_2} = T_{m_2,0} + T_{m_2,rot-0} + m_2\left(\mathbf{v}_{m_2-c.i.-0}, \mathbf{v}_0\right) =$$
$$T_{m_1,rot-0} = \frac{1}{2}\dot{\varphi}_1^2 \left(\frac{m_2 r_2^2}{2}\right) = \frac{m_2 r_2^2}{4}\dot{q}_1^2.$$

Also the following relations hold:

$$T_{m_3} = T_{m_3,0} + T_{m_3,rot-0} + m_3\left(\mathbf{v}_{m_3-c.i.-0}, \mathbf{v}_0\right) = T_{m_3,0} + T_{m_3,rot-0} =$$
$$\frac{1}{2}m_3\dot{z}^2 + \frac{1}{2}\dot{\varphi}_1^2 \frac{m_3 l_3^2}{12} = \frac{1}{2}m_3\dot{q}_3^2 + \frac{1}{2}\dot{q}_1^2 \frac{m_3 l_3^2}{24}$$

and

$$T_{m_4} = T_{m_4,0} + T_{m_4,rot-0} + m_4\left(\mathbf{v}_{m_4-c.i.-0}, \mathbf{v}_0\right) =$$
$$\frac{1}{2}m_4\left[(x\dot{\varphi}_1)^2 + \dot{x}^2 + \dot{z}^2\right] +$$

$$\frac{1}{2}\begin{pmatrix}\dot{\varphi}_2 \\ 0 \\ \dot{\varphi}_1\end{pmatrix}^T \begin{Vmatrix} \frac{1}{2}m_4 r_4^2 & 0 & 0 \\ 0 & m_4\left(\frac{r_4^2}{4}+\frac{l_4^2}{12}\right) & 0 \\ 0 & 0 & m_4\left(\frac{r_4^2}{4}+\frac{l_4^2}{12}\right) \end{Vmatrix} \begin{pmatrix}\dot{\varphi}_2 \\ 0 \\ \dot{\varphi}_1\end{pmatrix}$$

$$= \frac{1}{2}m_4\left(q_4^2\dot{q}_1^2 + \dot{q}_4^2 + \dot{q}_3^2 + \frac{1}{2}r_4^2\dot{q}_2^2 + \left[\frac{r_4^2}{4}+\frac{l_4^2}{12}\right]\dot{q}_1^2\right).$$

Substituting all derived terms in (12.1) gives

$$T = \frac{1}{4}\left(m_1 r_1^2 + m_2 r_2^2 + \frac{m_3 l_3^2}{12} + \frac{1}{2}m_4\left[r_4^2 + \frac{l_4^2}{3}\right]\right)\dot{q}_1^2 +$$
$$\frac{1}{4}m_4 r_4^2 \dot{q}_2^2 + \left(\frac{1}{2}m_3 + \frac{1}{2}m_4\right)\dot{q}_3^2 + \frac{1}{2}m_4\left(q_4^2 \dot{q}_1^2 + \dot{q}_4^2\right).$$
(12.2)

Potential energy

The potential energy V is calculated as

$$V = \sum_{i=1}^{4} V_{m_i},$$

$V_{m_1} = \text{const}, \ V_{m_2} = \text{const},$
$V_{m_3} = m_3 g z = m_3 g q_3, \ V_{m_4} = m_4 g z = m_4 g q_3,$

which finally gives

$$V = g(m_3 q_3 + m_4 q_3) + \text{const.} \tag{12.3}$$

Non-potential generalized forces

The generalized forces are given by the following formulas:

$$\left.\begin{aligned}
& Q_{non\text{-}pot,1} = \tau_1 - f_{fric-1}\dot{\varphi}_1 = \tau_1 - f_{fric-1}\dot{q}_1, \\
& \tau_1 \text{ is a torsion force,} \\
& Q_{non\text{-}pot,2} = \tau_2 - f_{fric-2}\dot{\varphi}_2 = \tau_2 - f_{fric-2}\dot{q}_2, \\
& \tau_2 \text{ is a torsion force,} \\
& Q_{non\text{-}pot,3} = F_3 - f_{fric-3}\dot{z} = F_3 - f_{fric-3}\dot{q}_3, \\
& F_3 \text{ is a force of vertical movement,} \\
& Q_{non\text{-}pot,4} = F_4 - f_{fric-4}\dot{x} = F_4 - f_{fric-4}\dot{q}_4, \\
& F_4 \text{ is a force of horizontal movement.}
\end{aligned}\right\} \tag{12.4}$$

Lagrange equations

Based on the expressions for T (12.2) and V (12.3), we can derive Lagrange's equations for this system:

$$\frac{d}{dt}\frac{\partial}{\partial \dot{q}_i}L - \frac{\partial}{\partial q_i}L = Q_{non\text{-}pot,i} \ (i = 1, ..., 4), \ L = T - V,$$

which can be represented in the format

$$D(q)\ddot{q} + C(q, \dot{q})\dot{q} + g(q) = \tau, \tag{12.5}$$

with

$$D(q) = \begin{Vmatrix} \dfrac{m_1 r_1^2}{2} + \dfrac{m_2 r_2^2}{2} + \\ \dfrac{m_3 l_3^2}{12} + m_4 \left(q_4^2 + \dfrac{r_4^2}{4} + \dfrac{l_4^2}{12} \right) & 0 & 0 & 0 \\ 0 & \dfrac{1}{2}m_4 r_4^2 & 0 & 0 \\ 0 & 0 & m_3 + m_4 & 0 \\ 0 & 0 & 0 & m_4 \end{Vmatrix},$$

$$C(q, \dot{q}) = \begin{Vmatrix} 2m_4 q_4 \dot{q}_4 + f_{fric-1} & 0 & 0 & 0 \\ 0 & f_{fric-2} & 0 & 0 \\ 0 & 0 & f_{fric-3} & 0 \\ m_4 q_4 \dot{q}_1 & 0 & 0 & f_{fric-4} \end{Vmatrix},$$

$$g(q) = \begin{Vmatrix} 0 \\ 0 \\ 0 \\ 0 \end{Vmatrix}, \quad \tau = \begin{Vmatrix} \tau_1 \\ \tau_2 \\ -2gm_3 + F_3 \\ F_4 \end{Vmatrix}.$$

12.2 Rectangular (Cartesian) robot manipulator

Consider the rectangular (Cartesian) manipulator with two prismatic joints and a rotating joint, represented in Fig. 12.2.

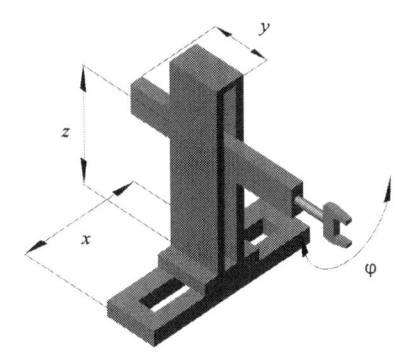

Figure 12.2 Rectangular (Cartesian) robot manipulator.

Generalized coordinates

The generalized coordinates of this system are

$$q_1 := \varphi, \quad q_2 := x, \quad q_3 := y, \quad q_4 := z.$$

Kinetic energy

The kinetic energy T of this system is $T = \sum_{i=1}^{4} T_{m_i}$, where T_{m_i} can be calculated using the König formula:

$$T_{m_i} = T_{m_i,0} + T_{m_i,rot-0} + 2m_i \left(\mathbf{v}_{m_i-c.i.-0}, \mathbf{v}_0 \right),$$
$$T_{m_i,0} = \frac{1}{2} m_i \|\mathbf{v}_0\|^2, \quad T_{m_i,rot-0} = \frac{1}{2} \left(\boldsymbol{\omega}, I_{i,0} \boldsymbol{\omega} \right).$$

Here $I_{i,0}$ is the inertia tensor with respect to a coordinate system with the origin at point O, $\mathbf{v}_{m_i-c.i.-0}$ is the speed of the center of inertia with respect to the system of coordinates with the origin at point O, and $\mathbf{v}_0$ is the velocity of the origin of the coordinate system.

In our case we have

$$T_{m_1} = T_{m_1,0} + T_{m_1,rot-0} + m_1\left(\mathbf{v}_{m_2-c.i.-0}, \mathbf{v}_0\right)$$
$$= T_{m_1,0} = \frac{1}{2}m_1\dot{x}^2 = \frac{1}{2}m_1\dot{q}_2^2,$$
$$T_{m_2} = T_{m_2,0} + T_{m_2,rot-0} + m_2\left(\mathbf{v}_{m_2-c.i.-0}, \mathbf{v}_0\right)$$
$$= T_{m_2,0} = \frac{1}{2}m_2\dot{x}^2 = \frac{1}{2}m_2\dot{q}_2^2,$$
$$T_{m_3} = T_{m_3,0} + T_{m_3,rot-0} + m_3\left(\mathbf{v}_{m_3-c.i.-0}, \mathbf{v}_0\right)$$
$$= T_{m_3,0} = \frac{1}{2}m_3\left[\dot{x}^2 + \dot{z}^2\right] = \frac{1}{2}m_3\left(\dot{q}_2^2 + \dot{q}_4^2\right),$$
$$T_{m_4} = T_{m_4,0} + T_{m_4,rot-0} + m_4\left(\mathbf{v}_{m_4-c.i.-0}, \mathbf{v}_0\right) =$$
$$T_{m_4,0} + T_{m_4,rot-0} = \frac{1}{2}m_4\left[\dot{x}^2 + \dot{y}^2 + \dot{z}^2\right] + \frac{1}{2}\dot{\varphi}^2\left(\frac{1}{2}m_4 r_4^2\right)$$
$$= \frac{1}{2}m_4\left(\dot{q}_2^2 + \dot{q}_3^2 + \dot{q}_4^2\right) + \frac{1}{4}m_4 r_4^2 \dot{q}_1^2.$$

So finally,

$$T = \frac{1}{4}m_4 r_4^2 \dot{q}_1^2 + \frac{1}{2}\left(m_1 + m_2 + m_3 + m_4\right)\dot{q}_2^2 + \frac{1}{2}m_4 \dot{q}_3^2 + \frac{1}{2}\left(m_3 + m_4\right)\dot{q}_4^2. \tag{12.6}$$

Potential energy

The potential energy V is as follows:

$$V = \sum_{i=1}^{4} V_{m_i},$$
$$V_{m_1} = \text{const}, \quad V_{m_2} = \text{const},$$
$$V_{m_3} = m_3 g z = m_3 g q_4, \quad V_{m_4} = m_4 g z = m_4 g q_4,$$

which gives

$$V = g\left(m_3 + m_4\right)q_4 + \text{const}. \tag{12.7}$$

Non-potential forces

The generalized forces are given by the following formulas:

$$Q_{non\text{-}pot,1} = \tau_1 - f_{fric-1}\dot{\varphi} = \tau_1 - f_{fric-1}\dot{q}_1,$$

τ_1 is a torsion force,

$$Q_{non\text{-}pot,2} = F_2 - f_{fric-2}\dot{x} = F_2 - f_{fric-2}\dot{q}_2,$$

F_2 is a force of horizontal movement,

$Q_{non-pot,3} = F_3 - f_{fric-3}\dot{y} = F_3 - f_{fric-3}\dot{q}_3$,

F_3 is a force of transverse movement,

$Q_{non-pot,4} = F_4 - f_{fric-4}\dot{z} = F_4 - f_{fric-4}\dot{q}_4$,

F_4 is a force of vertical movement.

Lagrange equations

Based on the expressions for T (12.6) and V (12.7) we can derive Lagrange's equations:

$$\frac{d}{dt}\frac{\partial}{\partial \dot{q}_i}L - \frac{\partial}{\partial q_i}L = Q_{non-pot,i} \quad (i = 1, ..., 4), \quad L = T - V.$$

In the standard format (12.5)

$$D(q)\ddot{q} + C(q,\dot{q})\dot{q} + g(q) = \tau,$$

with

$$D(q) = \begin{Vmatrix} \frac{1}{2}m_4 r_4^2 & 0 & 0 & 0 \\ 0 & \begin{bmatrix} m_1 + m_2 + \\ m_3 + m_4 \end{bmatrix} & 0 & 0 \\ 0 & 0 & m_4 & 0 \\ 0 & 0 & 0 & (m_3 + m_4) \end{Vmatrix},$$

$$C(q,\dot{q}) = \begin{Vmatrix} f_{fric-1} & 0 & 0 & 0 \\ 0 & f_{fric-2} & 0 & 0 \\ 0 & 0 & f_{fric-3} & 0 \\ 0 & 0 & 0 & f_{fric-4} \end{Vmatrix},$$

$$g(q) = \begin{Vmatrix} 0 \\ 0 \\ 0 \\ 0 \end{Vmatrix}, \quad \tau = \begin{Vmatrix} \tau_1 \\ F_2 \\ F_3 \\ F_4 - g(m_3 + m_4) \end{Vmatrix}.$$

12.3 Scaffolding type robot manipulator

Consider the scaffolding robot manipulator represented in Fig. 12.3.

Generalized coordinates

The generalized coordinates for this mechanical system are as follows:

$$q_1 := \varphi_1, \quad q_2 := \varphi_2, \quad q_3 := z, \quad q_4 := x.$$

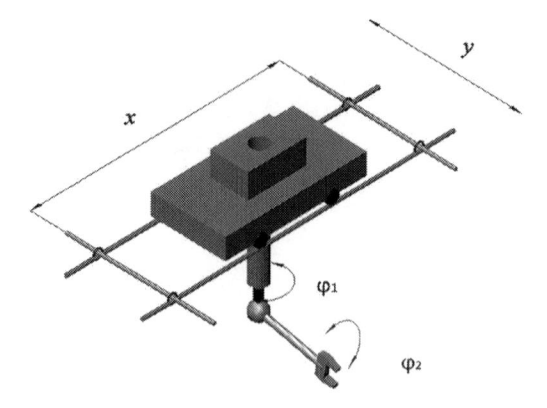

Figure 12.3 Scaffolding robot manipulator.

Kinetic energy

The kinetic energy $T = \sum_{i=1}^{4} T_{m_i}$ of this system is given by the following components:

$$T_{m_i} = T_{m_i,0} + T_{m_i,rot-0} + 2m_i \left(\mathbf{v}_{m_i-c.i.-0}, \mathbf{v}_0\right),$$
$$T_{m_i,0} = \frac{1}{2} m_i \|\mathbf{v}_0\|^2, \quad T_{m_i,rot-0} = \frac{1}{2} \left(\boldsymbol{\omega}, I_{i,0}\boldsymbol{\omega}\right).$$

In our case

$$\left.\begin{aligned}
T_{m_1} &= T_{m_1,0} + T_{m_1,rot-0} + m_1 \left(\mathbf{v}_{m_2-c.i.-0}, \mathbf{v}_0\right) \\
&= T_{m_1,0} = \frac{1}{2} m_1 \left(\dot{x}^2 + \dot{y}^2\right) = \frac{1}{2} m_1 \left(\dot{q}_1^2 + \dot{q}_2^2\right),
\end{aligned}\right\}$$

$$\left.\begin{aligned}
T_{m_2} &= T_{m_2,0} + T_{m_2,rot-0} + m_2 \left(\mathbf{v}_{m_2-c.i.-0}, \mathbf{v}_0\right) = \\
T_{m_2,0} + T_{m_2,rot-0} &= \frac{1}{2} m_2 \left(\dot{x}^2 + \dot{y}^2\right) + \frac{1}{2} m_2 \left(r_2 + r_3\right) \frac{1}{2} \dot{\varphi}_1^2 \\
&= \frac{1}{2} m_2 \left(\dot{q}_1^2 + \dot{q}_2^2\right) + \frac{1}{4} m_2 \left(r_2 + r_3\right)^2 \dot{q}_3^2,
\end{aligned}\right\}$$

$$\left.\begin{aligned}
T_{m_3} &= T_{m_3,0} + T_{m_3,rot-0} + m_3 \left(\mathbf{v}_{m_3-c.i.-0}, \mathbf{v}_0\right) = \\
T_{m_3,0} + T_{m_3,rot-0} &= \frac{1}{2} m_3 \left(\dot{x}^2 + \dot{y}^2\right) + \frac{1}{2} m_3 r_3 \frac{1}{2} \dot{\varphi}_1^2 \\
\frac{1}{2} m_3 \left(\dot{q}_1^2 + \dot{q}_2^2\right) &+ \frac{1}{4} m_3 r_3^2 \dot{q}_3^2,
\end{aligned}\right\}$$

and

$$T_{m4} = T_{m4,0} + T_{m4,rot-0} + m_4 \left(\mathbf{v}_{m4-c.i.-0}, \mathbf{v}_0\right) = \frac{1}{2} m_4 \left(\dot{x}^2 + \dot{y}^2\right) +$$

$$\frac{1}{2} \begin{pmatrix} \dot{\varphi}_1 \\ 0 \\ \dot{\varphi}_2 \end{pmatrix}^\mathsf{T} \left\| \begin{array}{ccc} m_4 \left(\frac{r_4^2}{4} + \frac{l_4^2}{3}\right) & 0 & 0 \\ 0 & m_4 \left(\frac{r_4^2}{4} + \frac{l_4^2}{3}\right) & 0 \\ 0 & 0 & m_4 \frac{r_4^2}{2} \end{array} \right\| \begin{pmatrix} \dot{\varphi}_1 \\ 0 \\ \dot{\varphi}_2 \end{pmatrix}$$

$$+ m_4 \begin{pmatrix} \dot{x} & \dot{y} & 0 \end{pmatrix} \begin{pmatrix} \dot{\varphi}_1 \frac{l_4}{2} \sin\varphi \\ \dot{\varphi}_1 \frac{l_4}{2} \cos\varphi \\ 0 \end{pmatrix}$$

$$= \frac{1}{2} m_4 \left[\left(\dot{q}_1^2 + \dot{q}_2^2\right) + \left(\frac{r_4^2}{4} + \frac{l_4^2}{3}\right) \dot{q}_3^2 + \right.$$

$$\left. \frac{r_4^2}{2} \dot{q}_4^2 + l_4 \left(\dot{q}_1 \sin q_3 + \dot{q}_2 \cos q_3\right) \dot{q}_3 \right].$$

Potential energy
The potential energy $V = \sum_{i=1}^{4} V_{m_i}$ with $V_{m_i} = \text{const}$ gives $V = \text{const}$.

Non-potential forces
The generalized non-potential forces are as follows:

$Q_{non-pot,1} = F_1 - f_{fric-1}\dot{x} = F_1 - f_{fric-1}\dot{q}_1$,
F_1 is a force of horizontal movement,

$Q_{non-pot,2} = F_2 - f_{fric-2}\dot{y} = F_2 - f_{fric-2}\dot{q}_2$,
F_2 is a force of transverse movement,

$Q_{non-pot,3} = \tau_3 - f_{fric-3}\dot{\varphi}_1 = \tau_3 - f_{fric-3}\dot{q}_3$,
τ_3 is a torsion force,

$Q_{non-pot,4} = \tau_4 - f_{fric-4}\dot{\varphi}_2 = \tau_4 - f_{fric-4}\dot{q}_4$,
τ_4 is a torsion force.

Lagrange equations
The Lagrange equations

$$\frac{d}{dt}\frac{\partial}{\partial \dot{q}_i} L - \frac{\partial}{\partial q_i} L = Q_{non-pot,i} \quad (i = 1, 4), \quad L = T - V$$

may be represented in the format (12.5)

$$D(q)\ddot{q} + C(q, \dot{q})\dot{q} + g(q) = \tau,$$

with

$$D(q) = \begin{Vmatrix} \begin{pmatrix} m_1+m_2 \\ +m_3+m_4 \end{pmatrix} & 0 & \dfrac{1}{2}m_4l_4\sin q_3 & 0 \\ 0 & \begin{pmatrix} m_1+m_2 \\ +m_3+m_4 \end{pmatrix} & \dfrac{1}{2}m_4l_4\cos q_3 & 0 \\ \dfrac{m_4l_4}{2}\sin q_3 & \dfrac{m_4l_4}{2}\cos q_3 & \begin{bmatrix} m_4\left(\dfrac{r_4^2}{4}+\dfrac{l_4^2}{3}\right)+ \\ \dfrac{1}{2}m_2(r_2+r_3)^2 \\ +\dfrac{1}{2}m_3r_3^2 \end{bmatrix} & 0 \\ 0 & 0 & 0 & m_4\dfrac{r_4^2}{2} \end{Vmatrix},$$

$$C(q,\dot{q}) = \begin{Vmatrix} f_{fric-1} & 0 & \dfrac{1}{2}m_4l_4(\cos q_3)\dot{q}_3 & 0 \\ 0 & f_{fric-2} & \dfrac{1}{2}m_4l_4(\sin q_3)\dot{q}_3 & 0 \\ 0 & 0 & f_{fric-3} & 0 \\ 0 & 0 & 0 & f_{fric-4} \end{Vmatrix},$$

and

$$g(q) = \begin{Vmatrix} 0 \\ 0 \\ 0 \\ 0 \end{Vmatrix}, \quad \tau = \begin{Vmatrix} F_1 \\ F_2 \\ \tau_3 \\ \tau_4 \end{Vmatrix}.$$

12.4 Spherical (polar) robot manipulator

Consider the spherical (polar) manipulator with three rotating joints represented in Fig. 12.4.

Generalized coordinates

The generalized coordinates for this mechanical system are as follows:

$$q_1 := \varphi_1, \quad q_2 := \varphi_2, \quad q_3 := \varphi_3.$$

Kinetic energy

The kinetic energy T of this system is given by the following expression, $T = \sum_{i=1}^{4} T_{m_i}$, where T_{m_i} can be calculated based on the König formula:

$$T_{m_i} = T_{m_i,0} + T_{m_i,rot-0} + 2m_i\left(\mathbf{v}_{m_i-c.i.-0}, \mathbf{v}_0\right),$$

Collection of electromechanical models

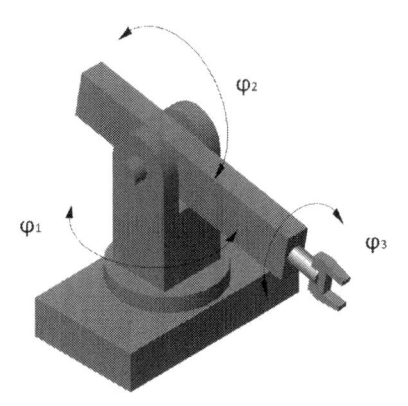

Figure 12.4 Spherical (polar) manipulator with three rotating joints.

$$T_{m_i,0} = \frac{1}{2} m_i \|\mathbf{v}_0\|^2.$$

Here $\mathbf{v}_{m_i-c.i.-0}$ is the velocity of the center of inertia with respect to the coordinate system with the origin in the point O, v_0 is the velocity of the origin of the coordinate systems, and $T_{m_i,rot-0} = \frac{1}{2}(\boldsymbol{\omega}, I_{i,0}\boldsymbol{\omega})$, where $I_{i,0}$ is the tensor of inertia with respect to the coordinate system with the origin in the point O.

In our case we have

$$T_{m_1} = T_{m_1,0} + T_{m_1,rot-0} + m_1\left(\mathbf{v}_{m_2-c.i.-0}, \mathbf{v}_0\right)$$

$$= T_{m_1,rot-0} = \frac{1}{2}\dot{\varphi}_1^2 \left(\frac{m_1 r_1^2}{2}\right) = \frac{m_1 r_1^2}{4}\dot{q}_1^2,$$

$$T_{m_2} = T_{m_2,0} + T_{m_2,rot-0} + m_2\left(\mathbf{v}_{m_2-c.i.-0}, \mathbf{v}_0\right) = T_{m_1,rot-0}$$

$$= \frac{m_2 \dot{\varphi}_1^2}{2V}\left[v_{21}\frac{(a_1^2 + a_2^2)}{12} + v_{22}\left(\frac{(a_1 - a_3)^2}{2} + \frac{a_2^2}{6}\right)\right.$$

$$\left. + v_{23}\left(\frac{(a_1 - a_3)^2}{2} + \frac{r_2^2}{4}\right)\right],$$

$$v_{21} = a_1 a_2 h_1, \quad v_{22} = 2a_2 a_3 h_2, \quad v_{23} = a_1 a_2 h_1,$$

$$T_{m_3} = T_{m_3,0} + T_{m_3,rot-0} + m_3\left(\mathbf{v}_{m_3-c.i.-0}, \mathbf{v}_0\right) = T_{m_3,rot-0} =$$

$$\frac{1}{2}\begin{pmatrix}\dot{\varphi}_1\sin\varphi_1 \\ \dot{\varphi}_1\cos\varphi_1 \\ \dot{\varphi}_2\end{pmatrix}^T \begin{Vmatrix}\frac{m_3 b^2}{12} & 0 & 0 \\ 0 & \frac{m_3 l_3^2}{12} & 0 \\ 0 & 0 & \frac{m_3(b_{m3}^2 + l_3^2)}{12}\end{Vmatrix}\begin{pmatrix}\dot{\varphi}_1\sin\varphi_1 \\ \dot{\varphi}_1\cos\varphi_1 \\ \dot{\varphi}_2\end{pmatrix}$$

$$= \frac{1}{24}m_3\left[\left(b^2\sin^2 q_1 + l_3^2\cos^2 q_1\right)\dot{q}_1^2 + \left(b_{m3}^2 + l_3^2\right)\dot{q}_2^2\right],$$

and

$$T_{m4} = T_{m4,0} + T_{m4,rot-0} + m_4\left(\mathbf{v}_{m4-c.i.-0}, \mathbf{v}_0\right)$$

$$= T_{m4,0} + T_{m4,rot-0} = \frac{1}{2}m_4\left[\frac{(l_3+l_4)^2\dot{\varphi}_1^2}{4} + \frac{(l_3+l_4)^2\dot{\varphi}_2^2}{4}\right] +$$

$$\frac{1}{2}\begin{pmatrix}\dot{\varphi}_1\sin\varphi_1 + \dot{\varphi}_3 \\ \dot{\varphi}_1\cos\varphi_1 \\ \dot{\varphi}_2\end{pmatrix}^T \begin{Vmatrix} \frac{m_4 r_4^2}{2} & 0 & 0 \\ 0 & m_4\left(\frac{r_4^2}{4} + \frac{l_4^2}{12}\right) & 0 \\ 0 & 0 & m_4\left(\frac{r_4^2}{4} + \frac{l_4^2}{12}\right)\end{Vmatrix}$$

$$\times \begin{pmatrix}\dot{\varphi}_1\sin\varphi_1 + \dot{\varphi}_3 \\ \dot{\varphi}_1\cos\varphi_1 \\ \dot{\varphi}_2\end{pmatrix}$$

$$= \frac{1}{8}m_4\left[(l_3+l_4)^2\left(\dot{q}_1^2 + \dot{q}_2^2\right) + \left(r_4^2 + \frac{l_4^2}{3}\right)\left(\dot{q}_1^2\cos^2 q_1 + \dot{q}_2^2\right)\right.$$

$$\left. + 2r_4^2\left(\dot{q}_1^2\sin^2 q_1 + \dot{q}_3^2 + 2\dot{q}_3\dot{q}_1\sin q_1\right)\right].$$

Potential energy

The potential energy $V = \sum_{i=1}^{4} V_{m_i}$ contains $V_{m_1} = \text{const}$, $V_{m_2} = \text{const}$, $V_{m_3} = \text{const}$, and

$$V_{m4} = m_4 g\left(h_1 + h_2 + r\sin\varphi_2\right) = m_4 g\left(h_1 + h_2 + r\sin q_2\right),$$

which leads to

$$V = m_4 g\left(h_1 + h_2 + r\sin q_2\right) + \text{const.}$$

Non-potential forces

The generalized non-potential forces are

$$Q_{non\text{-}pot,1} = \tau_1 - f_{fric-1}\dot{\varphi}_1 = \tau_1 - f_{fric-1}\dot{q}_1,$$
τ_1 is a torsion force,
$$Q_{non\text{-}pot,2} = \tau_2 - f_{fric-2}\dot{\varphi}_2 = \tau_2 - f_{fric-2}\dot{q}_2,$$
τ_2 is a torsion force,
$$Q_{non\text{-}pot,3} = \tau_3 - f_{fric-3}\dot{\varphi}_3 = \tau_3 - f_{fric-3}\dot{q}_3,$$
τ_3 is a torsion force.

Lagrange equations

Based on the expressions for T and V, we can derive Lagrange's equations

$$\frac{d}{dt}\frac{\partial}{\partial \dot{q}_i}L - \frac{\partial}{\partial q_i}L = Q_{non-pot,i}, \; i = 1, 2, 3, \; L = T - V$$

for the considered system in the format (12.5)

$$D(q)\ddot{q} + C(q, \dot{q})\dot{q} + g(q) = \tau,$$

where

$$D(q) = \begin{Vmatrix} d_{11} & 0 & d_{13} \\ 0 & d_{22} & 0 \\ d_{31} & 0 & d_{33} \end{Vmatrix},$$

with

$$V = v_{21} + v_{22} + v_{23},$$

$$d_{11} = \frac{m_2}{V}\left[v_{21}\frac{a_1^2 + a_2^2}{12} + \right.$$

$$v_{22}\left(\frac{(a_1 - a_3)^2}{2} + \frac{a_2^2}{6}\right) + v_{23}\left(\frac{(a_1 - a_3)^2}{2} + \frac{r_2^2}{4}\right)\Bigg]$$

$$+ \frac{1}{2}m_1 r_1^2 + m_3 \left(\frac{b^2}{12}\sin^2 q_1 + \frac{l_3^2}{12}\cos^2 q_1\right) +$$

$$m_4\left(\left(\frac{r_4^2}{4} + \frac{l_4^2}{12}\right)\cos^2 q_1 + \frac{1}{2}r_4^2 \sin^2 q_1 + \frac{(l_3 + l_4)^2}{4}\right),$$

$$v_{21} = a_1 a_2 h_1, \; v_{22} = 2a_2 a_3 h_2, \; v_{23} = a_1 a_2 h_1,$$

$$d_{22} = \left[m_3\frac{(b_{m3}^2 + l^2)}{12} + m_4\left(\frac{r_4^2}{4} + \frac{l_4^2}{12} + \frac{(l_3 + l_4)^2}{4}\right)\right],$$

$$d_{33} = \frac{1}{2}m_4 r_4^2, \; d_{13} = d_{31} = \frac{1}{2}m_4 r_4^2 \sin(q_1),$$

and

$$C(q, \dot{q}) = \begin{Vmatrix} f_{fric-1} & 0 & 0 \\ 0 & f_{fric-2} & 0 \\ \frac{1}{2}m_4 r_4^2 \cos(q_1) & 0 & f_{fric-3} \end{Vmatrix},$$

$$g(q) = \begin{Vmatrix} 0 \\ m_4 gr \cos q_2 \\ 0 \end{Vmatrix}, \; \tau = \begin{Vmatrix} \tau_1 \\ \tau_2 \\ \tau_3 \end{Vmatrix}.$$

12.5 Articulated robot manipulator 1

Consider now the articulated robot manipulator represented in Fig. 12.5.

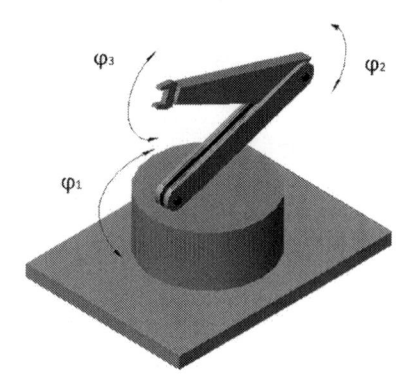

Figure 12.5 Articulated robot manipulator.

Generalized coordinates

The generalized coordinates for this mechanical system are as follows:

$$q_1 := \varphi_1, \; q_2 := \varphi_2, \; q_3 := \varphi_3.$$

Kinetic energy

The kinetic energy $T = \sum_{i=1}^{3} T_{m_i}$ of this system consists of three components T_{m_i}, which are calculated as

$$T_{m_i} = T_{m_i,0} + T_{m_i,rot-0} + 2m_i \left(\mathbf{v}_{m_i-c.i.-0}, \mathbf{v}_0\right),$$

$$T_{m_i,0} = \frac{1}{2} m_i \|\mathbf{v}_0\|^2, \; T_{m_i,rot-0} = \frac{1}{2} \left(\boldsymbol{\omega}, I_{i,0}\boldsymbol{\omega}\right),$$

where $I_{i,0}$ is the inertia tensor with respect to the coordinate system with the origin in the point O, $\mathbf{v}_{m_i-c.i.-0}$ is the velocity of the inertia tensor with respect to the coordinate system with the origin in the point O, and $\mathbf{v}_0$ is the velocity of the origin of the coordinate system. In our case we have

$$T_{m_1} = T_{m_1,0} + T_{m_1,rot-0} + m_1 \left(\mathbf{v}_{m_2-c.i.-0}, \mathbf{v}_0\right)$$

$$= T_{m_1,rot-0} = \frac{1}{2}\dot{\varphi}_1^2 \left(\frac{m_1 r_1^2}{2}\right) = \frac{m_1 r_1^2}{4}\dot{q}_1^2,$$

$$T_{m_2} = T_{m_2,0} + T_{m_2,rot-0} + m_2 \left(\mathbf{v}_{m_2-c.i.-0}, \mathbf{v}_0\right) = T_{m_2,rot-0} =$$

$$\frac{1}{2}\begin{pmatrix}\dot\varphi_1\sin\varphi_2\\ \dot\varphi_1\cos\varphi_2\\ \dot\varphi_2\end{pmatrix}^T\begin{Vmatrix}I_{2xx} & 0 & 0\\ 0 & \frac{1}{4}m_2l_2^2+I_{2yy} & 0\\ 0 & 0 & \frac{1}{4}m_2l_2^2+I_{2zz}\end{Vmatrix}\begin{pmatrix}\dot\varphi_1\sin\varphi_2\\ \dot\varphi_1\cos\varphi_2\\ \dot\varphi_2\end{pmatrix}$$

$$=\frac{1}{2}\left[I_{2xx}\dot q_1^2\sin^2 q_2+\left(\frac{m_2l_2^2}{4}+I_{2yy}\right)\dot q_1^2\cos^2 q_2+\left(\frac{m_2l_2^2}{4}+I_{2zz}\right)\dot q_2^2\right],$$

$$T_{m_3}=T_{m_3,0}+T_{m_3,rot-0}+m_3\left(\mathbf{v}_{m_3-c.i.-0},\mathbf{v}_0\right)=\frac{1}{2}m_3\left(\dot\varphi_2^2l_2^2+\dot\varphi_1^2l_2^2\cos^2\varphi_2\right)+$$

$$\frac{1}{2}\begin{pmatrix}\dot\varphi_1\sin(\varphi_3-\varphi_2)\\ \dot\varphi_1\cos(\varphi_3-\varphi_2)\\ \dot\varphi_2+\dot\varphi_3\end{pmatrix}^T\begin{Vmatrix}I_{3xx} & 0 & 0\\ 0 & m_3d_3^2+I_{3yy} & 0\\ 0 & 0 & md_3^2+I_{3zz}\end{Vmatrix}\times$$

$$\begin{pmatrix}\dot\varphi_1\sin(\varphi_3-\varphi_2)\\ \dot\varphi_1\cos(\varphi_3-\varphi_2)\\ \dot\varphi_2+\dot\varphi_3\end{pmatrix}+m_3\begin{pmatrix}[-(\dot\varphi_2+\dot\varphi_3)d_3\cos\varphi_1\sin(\varphi_3-\varphi_2)+)\\ \dot\varphi_1d_3\sin\varphi_1\cos(\varphi_3-\varphi_2)-\\ (\dot\varphi_2+\dot\varphi_3)d_3\cos(\varphi_3-\varphi_2)]\\ [\dot\varphi_1d_3\cos\varphi_1\cos(\varphi_3-\varphi_2)+\\ (\dot\varphi_2+\dot\varphi_3)d_3\sin\varphi_1\sin(\varphi_3-\varphi_2)]\end{pmatrix}^T\times$$

$$\begin{pmatrix}-\dot\varphi_2l_2\cos\varphi_1\sin\varphi_2-\dot\varphi_1l_2\sin\varphi_1\cos\varphi_2\\ \dot\varphi_2l_2\cos\varphi_2\\ -\dot\varphi_1l_2\cos\varphi_1\cos\varphi_2+\dot\varphi_2l_2\sin\varphi_1\sin\varphi_2\end{pmatrix}=$$

$$=\frac{1}{2}m_3l_2^2\left(\dot q_1^2\cos^2 q_2+\dot q_2^2\right)+\frac{1}{2}I_{3xx}\dot q_1^2\sin^2(q_3-q_2)+$$

$$\frac{1}{2}\left(m_3d_3^2+I_{3yy}\right)\dot q_1^2\cos^2(q_3-q_2)+\frac{1}{2}\left(m_3d_3^2+I_{3zz}\right)\left(\dot q_2^2+\dot q_3^2+2\dot q_2\dot q_3\right)$$

$$-m_3\left[(\dot q_2+\dot q_3)\dot q_2l_2d_3\cos q_3\right]-m_3\left[\dot q_1^2l_2d_3\cos q_2\cos(q_3-q_2)\right].$$

Here d_3 is the distance of the union of links 2 and 3 to the center of inertia of link 3.

Potential energy

The potential energy $V=\sum_{i=1}^3 V_{m_i}$ has components

$$V_{m_1}=\text{const},\quad V_{m_2}=m_2g\left(\frac{l_2}{2}\sin\varphi_2\right)=m_2g\frac{l_2}{2}\sin q_2,$$

$$V_{m_3}=m_3g\left(d_3\sin(\varphi_3-\varphi_2)+l_2\sin\varphi_2\right)$$
$$=m_3g\left(d_3\sin(q_3-q_2)+l_2\sin q_2\right),$$

which gives

$$V=\frac{1}{2}m_2gl_2\sin q_2+m_3g\left(d_3\sin q_3+l_2\sin q_2\right).$$

Non-potential forces

The non-potential forces are given by

$$Q_{non\text{-}pot,i} = \tau_i - f_{fric-i}\dot{\varphi}_i = \tau_i - f_{fric-i}\dot{q}_i,$$

where τ_i ($i = 1, 2, 3$) are torsion forces.

Lagrange equations

Based on the obtained expressions for T, V, and $L = T - V$ we can derive Lagrange's equations for the considered system:

$$\frac{d}{dt}\frac{\partial}{\partial \dot{q}_i}L - \frac{\partial}{\partial q_i}L = Q_{non\text{-}pot,i}, \quad i = 1, 2, 3,$$

which leads to the following dynamic model in the format (12.5):

$$D(q)\ddot{q} + C(q,\dot{q})\dot{q} + g(q) = \tau,$$

where

$$D(q) = \begin{Vmatrix} D_{11}(q) & 0 & 0 \\ 0 & D_{22}(q) & D_{23}(q) \\ 0 & D_{32}(q) & D_{33}(q) \end{Vmatrix},$$

with

$$D_{11}(q) = \left[\frac{1}{2}m_1 r_1^2 + I_{2xx}\sin^2 q_2 + \left(\frac{m_2 l_2^2}{4} + I_{2yy} + m_3 l_2^2\right)\cos^2 q_2\right]$$
$$+ \left[I_{3xx}\sin^2(q_3 - q_2) + \left(m_3 d_3^2 + I_{3yy}\right)\cos^2(q_3 - q_2)\right]$$
$$- 2m_3 \left[l_2 d_3 \cos q_2 \cos(q_3 - q_2)\right],$$

$$D_{22}(q) = \frac{m_2 l_2^2}{4} + I_{2zz} + m_3(l_2 + r_2)^2 + m_3 d_3^2 + I_{3zz}$$
$$- 2m_3 \left[(l_2 + r_2) d_3 \cos q_3\right],$$

$$D_{23}(q) = D_{32}(q) = \left[m_3 d_3^2 + I_{3zz}\right] - m_3 \left[(l_2 + r_2) d_3 \cos q_3\right],$$

$$D_{33}(q) = m_3 d_3^2 + I_{3zz},$$

and

$$C(q,\dot{q}) = \begin{Vmatrix} C_{11}(q,\dot{q}) & 0 & 0 \\ C_{21}(q,\dot{q}) & C_{22}(q,\dot{q}) & C_{23}(q,\dot{q}) \\ C_{31}(q,\dot{q}) & C_{32}(q,\dot{q}) & C_{33}(q,\dot{q}) \end{Vmatrix},$$

where

$$C_{11}(q,\dot{q}) = f_{fric-1} + 2\left[\left(I_{2xx} - \frac{m_2 l_2^2}{4} - I_{2yy} - m_3 l_2^2\right)\sin q_2 \cos q_2\right]\dot{q}_2$$
$$+ 2\left[\left(I_{3xx} - m_3 d_3^2 - I_{3yy}\right)\sin q_3 \cos q_3\right](\dot{q}_3 - \dot{q}_2) +$$
$$2m_3\left[l_2 d_3 \cos q_2 \sin(q_3 - q_2)\right]\dot{q}_3,$$
$$C_{21}(q,\dot{q}) = -\left[\left(I_{2xx} - \frac{m_2 l_2^2}{4} - I_{2yy} - m_3 l_2^2\right)\sin q_2 \cos q_2\right]\dot{q}_1 -$$
$$\left[\left(m_3 d_3^2 + I_{3yy} - I_{3xx}\right)\sin(q_3 - q_2)\cos(q_3 - q_2) - m_3 l_2 d_3 \sin q_3\right]\dot{q}_1,$$
$$C_{22}(q,\dot{q}) = f_{fric-2} + 2m_3[l_2 d_3 \sin q_3]\dot{q}_3,$$
$$C_{31}(q,\dot{q}) = -\left[\left(I_{3xx} - m_3 d_3^2 - I_{3yy}\right)\sin q_3 \cos q_3\right]\dot{q}_1 -$$
$$m_3 l_2 d_3 \cos q_2 \sin(q_3 - q_2)\dot{q}_1,$$
$$C_{23}(q,\dot{q}) = m_3[l_2 d_3 \sin q_3]\dot{q}_3,$$
$$C_{32}(q,\dot{q}) = 2m_3[l_2 d_3 \sin q_3]\dot{q}_2 - m_3 l_2 d_3 \sin q_3 \dot{q}_2,$$
$$C_{33}(q,\dot{q}) = f_{fric-3} + m_3[l_2 d_3 \sin q_3]\dot{q}_3 - m_3 l_2 d_3 \sin q_3 \dot{q}_2,$$
$$g(q) = \left\|\begin{array}{c} 0 \\ \left(\frac{1}{2}m_2 + m_3\right)gl_2 \cos q_2 - m_3 g[d_3 \cos(q_3 - q_2)] \\ m_3 g d_3 \cos q_3 \end{array}\right\|, \quad \tau = \left\|\begin{array}{c} \tau_1 \\ \tau_2 \\ \tau_3 \end{array}\right\|.$$

12.6 Universal programmable manipulator

Consider the *universal programmable manipulator* (PUMA) for assembly shown in Fig. 12.6.

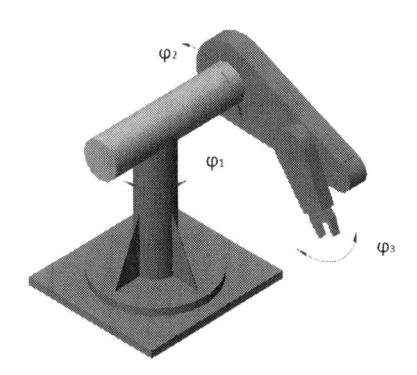

Figure 12.6 Universal programmable manipulator (PUMA).

Generalized coordinates

The generalized coordinates for this mechanical system are as follows:

$$q_1 := \varphi_1, \ q_2 := \varphi_2, \ q_3 := \varphi_3.$$

Kinetic energy

The components of the kinetic energy $T = \sum_{i=1}^{5} T_{m_i}$ of this system are calculated as

$$T_{m_i} = T_{m_i,0} + T_{m_i,rot-0} + 2m_i \left(\mathbf{v}_{m_i-c.i.-0}, \mathbf{v}_0\right),$$

$$T_{m_i,0} = \frac{1}{2} m_i \|\mathbf{v}_0\|^2, \ T_{m_i,rot-0} = \frac{1}{2}\left(\boldsymbol{\omega}, I_{i,0}\boldsymbol{\omega}\right),$$

which in our case gives

$$T_{m_1} = T_{m_1,0} + T_{m_1,rot-0} + m_1 \left(\mathbf{v}_{m_1-c.i.-0}, \mathbf{v}_0\right) =$$

$$T_{m_1,rot-0} = \frac{1}{2}\dot{\varphi}_1^2 \left(\frac{m_1 r_1^2}{2}\right) = \frac{m_1 r_1^2}{4} \dot{q}_1^2,$$

$$T_{m_2} = T_{m_2,0} + T_{m_2,rot-0} + m_2 \left(\mathbf{v}_{m_2-c.i.-0}, \mathbf{v}_0\right) =$$

$$T_{m_2,rot-0} = \frac{1}{2}\dot{\varphi}_1^2 \left(\frac{m_2 r_2^2}{2}\right) = \frac{m_2 r_2^2}{4} \dot{q}_1^2,$$

$$T_{m_3} = T_{m_3,0} + T_{m_3,rot-0} + m_2 \left(\mathbf{v}_{m_3-c.i.-0}, \mathbf{v}_0\right) = T_{m_3,rot-0} =$$

$$\frac{1}{2}\begin{pmatrix}\dot{\varphi}_2\\0\\\dot{\varphi}_1\end{pmatrix}^T \begin{Vmatrix} \frac{1}{2}m_3 r_3^2 & 0 & 0 \\ 0 & \frac{1}{4}m_3\left(r_3^2+\frac{l_3^2}{3}\right) & 0 \\ 0 & 0 & \frac{1}{4}m_3\left(r_3^2+\frac{l_3^2}{3}\right) \end{Vmatrix} \begin{pmatrix}\dot{\varphi}_2\\0\\\dot{\varphi}_1\end{pmatrix}$$

$$= \frac{1}{2}m_3 \left[\frac{1}{4}\left(r_3^2+\frac{l_3^2}{3}\right)\dot{q}_1^2 + \frac{1}{2}r_3^2 \dot{q}_2^2\right],$$

$$T_{m_4} = T_{m_4,0} + T_{m_4,rot-0} + m_4 \left(\mathbf{v}_{m_4-c.i.-0}, \mathbf{v}_0\right) = \frac{1}{2}m_4 \frac{l_3^2 \dot{\varphi}_1^2}{4} +$$

$$\frac{1}{2}\begin{pmatrix}\dot{\varphi}_1 \sin\varphi_2\\\dot{\varphi}_1 \cos\varphi_2\\\dot{\varphi}_2\end{pmatrix}^T \begin{Vmatrix} I_{4xx} & 0 & 0 \\ 0 & m_4 d_4^2 + I_{4yy} & 0 \\ 0 & 0 & m_4 d_4^2 + I_{4zz} \end{Vmatrix} \begin{pmatrix}\dot{\varphi}_1 \sin\varphi_2\\\dot{\varphi}_1 \cos\varphi_2\\\dot{\varphi}_2\end{pmatrix}$$

$$+ \begin{pmatrix}-\frac{1}{2}\dot{\varphi}_1 l_3 \cos\varphi_1 \\ 0 \\ \frac{1}{2}\dot{\varphi}_1 l_3 \sin\varphi_1\end{pmatrix}^T \begin{pmatrix}-\dot{\varphi}_2 d_4 \cos\varphi_1 \sin\varphi_2 - \dot{\varphi}_1 d_4 \sin\varphi_1 \cos\varphi_2 \\ \dot{\varphi}_2 d_4 \cos\varphi_2 \\ -\dot{\varphi}_1 d_4 \cos\varphi_1 \cos\varphi_2 + \dot{\varphi}_2 d_4 \sin\varphi_1 \sin\varphi_2\end{pmatrix} =$$

$$\frac{1}{2}\left[m_4\frac{l_3^2\dot{q}_1^2}{4}+I_{4xx}\dot{q}_1^2\sin^2 q_2+\left(m_4d_4^2+I_{4yy}\right)\dot{q}_1^2\cos^2 q_2+\right.$$
$$\left.\left(m_4d_4^2+I_{4zz}\right)\dot{q}_2^2\right]+\frac{1}{2}m_4l_3d_4\dot{q}_1\dot{q}_2\cos q_2,$$

and

$$T_{m_5}=T_{m_5,0}+T_{m_5,rot-0}+m_5\left(\mathbf{v}_{m_5-c.i.-0},\mathbf{v}_0\right)=$$
$$\frac{1}{2}m_5\left[\left(\frac{l_3^2}{4}+b_5^2\left(\cos^2\varphi_2\right)\right)\dot{\varphi}_1^2+b_5^2\dot{\varphi}_2^2+\dot{\varphi}_1\dot{\varphi}_2l_3b_5\sin\varphi_2\right]$$
$$\frac{1}{2}\begin{pmatrix}\dot{\varphi}_1\sin(\varphi_2+\varphi_3)\\\dot{\varphi}_1\cos(\varphi_2+\varphi_3)\\\dot{\varphi}_2+\dot{\varphi}_3\end{pmatrix}^T\begin{Vmatrix}I_{5xx}&0&0\\0&m_5d_5^2+I_{5yy}&0\\0&0&m_5d_5^2+I_{5zz}\end{Vmatrix}\cdot$$
$$\cdot\begin{pmatrix}\dot{\varphi}_1\sin(\varphi_2+\varphi_3)\\\dot{\varphi}_1\cos(\varphi_2+\varphi_3)\\\dot{\varphi}_2+\dot{\varphi}_3\end{pmatrix}+$$
$$m_5\begin{pmatrix}\dot{\varphi}_1d_5\sin\varphi_1\cos(\varphi_2+\varphi_3)+(\dot{\varphi}_2+\dot{\varphi}_3)d_5\cos\varphi_1\sin(\varphi_2+\varphi_3)\\-(\dot{\varphi}_2+\dot{\varphi}_3)d_5\cos(\varphi_2+\varphi_3)\\\dot{\varphi}_1d_5\cos\varphi_1\cos(\varphi_2+\varphi_3)-(\dot{\varphi}_2+\dot{\varphi}_3)d_5\sin\varphi_1\sin(\varphi_2+\varphi_3)\end{pmatrix}^T$$
$$\begin{pmatrix}-\frac{1}{2}\dot{\varphi}_1l_3\cos\varphi_1-\dot{\varphi}_2b_5\cos\varphi_1\sin\varphi_2-\dot{\varphi}_1b_5\sin\varphi_1\cos\varphi_2\\\dot{\varphi}_2b_5\cos\varphi_2\\\frac{1}{2}\dot{\varphi}_1l_3\sin\varphi_1-\dot{\varphi}_1b_5\cos\varphi_1\cos\varphi_2+\dot{\varphi}_2b_5\sin\varphi_1\sin\varphi_2\end{pmatrix}$$
$$=\frac{1}{2}m_5\left[\left(\frac{l_3^2}{4}+b_5^2\left(\cos^2 q_2\right)\right)\dot{q}_1^2+b_5^2\dot{q}_2^2+\dot{q}_1\dot{q}_2l_3b_5\sin q_2\right]$$
$$+\frac{1}{2}\left[\left[I_{5xx}\sin^2(q_2+q_3)+\left(m_5d_5^2+I_{5yy}\right)\cos^2(q_2+q_3)\right]\dot{q}_1^2+\right.$$
$$\left.\left(m_5d_5^2+I_{5zz}\right)(\dot{q}_2+\dot{q}_3)^2\right]-$$
$$\frac{m_5}{2}\dot{q}_1(\dot{q}_2+\dot{q}_3)d_5l_3\sin(q_2+q_3)-m_5\dot{q}_1^2d_5b_5\cos q_2\cos(q_2+q_3)$$
$$-m_5(\dot{q}_2+\dot{q}_3)\dot{q}_2d_5b_5\cos q_3,$$

where b_5 is the distance from the union of m_3 and m_4 to the union of m_4 and m_5.

Potential energy

The potential energy $V=\sum_{i=1}^5 V_{m_i}$ has components

$$V_{m_1}=\text{const},\quad V_{m_2}=\text{const}t,\quad V_{m_3}=\text{const},$$

$$V_{m_4} = m_4g\,(d_4\sin\varphi_2 + d_4) = m_4gd_4\,(\sin q_2 + \text{const}),$$
$$V_{m_5} = m_5g\,[-d_5\sin(\varphi_2 + \varphi_3) + l_4\sin\varphi_2 + \text{const}]$$
$$= m_5g\,[-d_5\sin(q_2 + q_3) + l_4\sin q_2 + \text{const}],$$

so that

$$V = m_4gd_4\sin q_2 + m_5g\,[-d_5\sin(q_2 + q_3) + l_4\sin q_2] + \text{const}.$$

Non-potential forces
The non-potential generalized forces are

$$Q_{non\text{-}pot,i} = \tau_i - f_{fric-i}\dot\varphi_i = \tau_i - f_{fric-i}\dot q_i,$$
$$\tau_i \text{ is a torsion force } (i = 1, 2, 3).$$

Lagrange equations
Based on the expressions for T, V, and $L = T - V$ we can derive the Lagrange equations for the system:

$$\frac{d}{dt}\frac{\partial}{\partial \dot q_i}L - \frac{\partial}{\partial q_i}L = Q_{non\text{-}pot,i},\quad i = 1, 2, 3,$$

which can be represented in the standard format (12.5)

$$D(q)\ddot q + C(q,\dot q)\dot q + g(q) = \tau,$$

where

$$D(q) = \begin{Vmatrix} D_{11}(q) & D_{12}(q) & D_{13}(q) \\ D_{21}(q) & D_{22}(q) & D_{23}(q) \\ D_{31}(q) & D_{32}(q) & m_5d_5^2 + I_{5zz} \end{Vmatrix},$$

with

$$D_{11}(q) = \frac{1}{2}\left[m_1r_1^2 + m_2r_2^2 + \frac{m_3}{2}\left(r_3^2 + \frac{l_3^2}{3}\right) + \frac{m_4l_3^2}{2}\right] +$$
$$I_{4xx}\sin^2 q_2 + \left(m_4d_4^2 + I_{4yy}\right)\cos^2 q_2 + m_5\left(\frac{l_3^2}{4} + b_5^2\left(\cos^2 q_2\right)\right) +$$
$$I_{5xx}\sin^2(q_2 + q_3) + \left(m_5d_5^2 + I_{5yy}\right)\cos^2(q_2 + q_3)$$
$$-2m_5d_5b_5\cos q_2\cos(q_2 + q_3),$$

$$D_{12}(q) = D_{21}(q) = \frac{1}{2}[m_4l_3d_4\cos\varphi_2 + m_5l_3b_5\sin q_2]$$
$$-\frac{1}{2}m_5d_5l_3\sin(q_2 + q_3),$$

$$D_{22}(q) = \frac{1}{2}m_3r_3^2 + m_4d_4^2 + I_{4zz} + m_5b_5^2 + m_5d_5^2 + I_{5zz} - m_5d_5b_5\cos q_3,$$
$$D_{13}(q) = D_{31}(q) = -\frac{1}{2}m_5d_5l_3\sin(q_2+q_3),$$
$$D_{23}(q) = D_{32}(q) = m_5\left(d_5^2 - d_5b_5\cos q_3 + I_{5zz}\right),$$

and

$$C(q,\dot{q}) = \begin{Vmatrix} C_{11}(q,\dot{q}) & C_{12}(q,\dot{q}) & C_{31}(q,\dot{q}) \\ C_{21}(q,\dot{q}) & C_{22}(q,\dot{q}) & m_5d_5b_5\sin q_3 \dot{q}_3 \\ C_{31}(q,\dot{q}) & C_{32}(q,\dot{q}) & f_{fric-3} \end{Vmatrix},$$

with

$$C_{11}(q,\dot{q}) = f_{fric-1} + 2\left[I_{4xx}\sin q_2\cos q_2 + m_5b_5^2\sin q_2\cos q_2\right]\dot{q}_2 -$$
$$2\left(m_4d_4^2 + I_{4yy}\right)\sin q_2\cos q_2\dot{q}_2 +$$
$$2[I_{5xx}\sin(q_2+q_3)\cos(q_2+q_3)](\dot{q}_2+\dot{q}_3) -$$
$$2\left(m_5d_5^2 + I_{5yy}\right)\sin(q_2+q_3)\cos(q_2+q_3)(\dot{q}_2+\dot{q}_3) +$$
$$2m_5d_5b_5[\sin(2q_2+q_3)\dot{q}_2 + \cos q_2\sin(q_2+q_3)\dot{q}_3],$$

$$C_{21}(q,\dot{q}) = \frac{1}{2}[m_5l_3b_5\cos q_2 - m_4l_3d_4\sin q_2]\dot{q}_2$$
$$-\left[I_{4xx}\sin q_2\cos q_2 + \left(m_4d_4^2 + I_{4yy}\right)\cos q_2\right]\dot{q}_1 -$$
$$\left[\left(\frac{1}{2}I_{5xx} - m_5d_5^2 - I_{5yy}\right)\sin(q_2+q_3)\cos(q_2+q_3)\right]\dot{q}_1$$
$$-\left[m_5b_5^2\sin q_2\cos q_2\right]\dot{q}_1 - [m_5d_5b_5\sin(2q_2+q_3)]\dot{q}_1,$$

$$C_{12}(q,\dot{q}) = \frac{1}{2}[m_5l_3b_5\cos q_2 - m_4l_3d_4\sin q_2]\dot{q}_2$$
$$-\frac{1}{2}m_5d_5l_3\cos(q_2+q_3)(\dot{q}_2+\dot{q}_3),$$

$$C_{22}(q,\dot{q}) = f_{fric-2} + [m_5d_5b_5\sin q_3]\dot{q}_3 -$$
$$\frac{1}{2}[m_5l_3b_5\cos q_2 - m_4d_4l_3\sin q_2]\dot{q}_1,$$

$$C_{31}(q,\dot{q}) = -[I_{5xx}\sin(q_2+q_3)\cos(q_2+q_3)]\dot{q}_1 +$$
$$\left[\left(m_5d_5^2 + I_{5yy}\right)\sin(q_2+q_3)\cos(q_2+q_3)\right]\dot{q}_1 -$$
$$m_5d_5b_5\cos q_2\sin(q_2+q_3)\dot{q}_1,$$

$$C_{32}(q,\dot{q}) = m_5d_5b_5\sin q_3\dot{q}_3 + m_5d_5b_5\cos q_3(\dot{q}_2+\dot{q}_3),$$

$$C_{33}(q,\dot{q}) = -\frac{1}{2}m_5d_5l_3\cos(q_2+q_3)(\dot{q}_2+\dot{q}_3),$$

$$g(q) = \left\| \begin{array}{c} 0 \\ m_4 g d_4 \cos q_2 + m_5 g \left(d_4 \cos q_2 - d_5 \cos(q_2 + q_3) \right) \\ -m_5 g d_5 \cos q_3 \end{array} \right\|, \quad \tau = \left\| \begin{array}{c} \tau_1 \\ \tau_2 \\ \tau_3 \end{array} \right\|.$$

12.7 Cincinnati Milacron T³ manipulator

Consider the manipulator *"Cincinnati Milacron T³"* represented in Fig. 12.7.

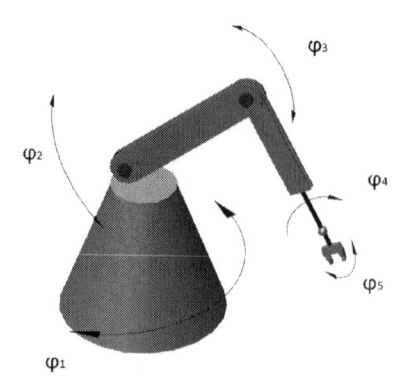

Figure 12.7 Manipulator "Cincinnati Milacron T³".

Generalized coordinates

The following variables are selected as the generalized coordinates for this system:

$$q_1 := \varphi_1, \quad q_2 := \varphi_2, \quad q_3 := \varphi_3, \quad q_4 := \varphi_4, \quad q_5 := \varphi_5.$$

Kinetic energy

The components of the kinetic energy $T = \sum_{i=1}^{5} T_{m_i}$ of this system may be calculated by the König formula

$$T_{m_i} = T_{m_i,0} + T_{m_i,rot-0} + 2 m_i \left(\mathbf{v}_{m_i - c.i. - 0}, \mathbf{v}_0 \right),$$
$$T_{m_i,0} = \frac{1}{2} m_i \| \mathbf{v}_0 \|^2, \quad T_{m_i,rot-0} = \frac{1}{2} \left(\boldsymbol{\omega}, I_{i,0} \boldsymbol{\omega} \right),$$

which in our case leads to

$$T_{m_1} = T_{m_1,0} + T_{m_1,rot-0} + m_1 \left(\mathbf{v}_{m_1 - c.i. - 0}, \mathbf{v}_0 \right)$$
$$= T_{m_1,rot-0} = \frac{1}{2} \dot{\varphi}_1^2 \left(\frac{m_1 r_1^2}{2} \right) = \frac{m_1 r_1^2}{4} \dot{q}_1^2,$$

$$T_{m_2} = T_{m_2,0} + T_{m_2,rot-0} + m_2 \left(\mathbf{v}_{m_2 - c.i. - 0}, \mathbf{v}_0 \right) = T_{m_2,rot-0} =$$

$$\frac{1}{2}\begin{pmatrix}\dot{\varphi}_1\sin\varphi_2\\\dot{\varphi}_1\cos\varphi_2\\\dot{\varphi}_2\end{pmatrix}^{\top}\begin{Vmatrix}I_{2xx}&0&0\\0&m_2d_2^2+I_{2yy}&0\\0&0&m_2d_2^2+I_{2zz}\end{Vmatrix}\begin{pmatrix}\dot{\varphi}_1\sin\varphi_2\\\dot{\varphi}_1\cos\varphi_2\\\dot{\varphi}_2\end{pmatrix}$$

$$=\frac{1}{2}\left[I_{2xx}\dot{q}_1^2\sin^2 q_2+\left(m_2d_2^2+I_{2yy}\right)\dot{q}_1^2\cos^2 q_2+\left(m_2d_2^2+I_{2zz}\right)\dot{q}_2^2\right],$$

$$T_{m_3}=T_{m_3,0}+T_{m_3,rot-0}+m_2\left(\mathbf{v}_{m_3-c.i.-0},\mathbf{v}_0\right)=$$

$$\frac{1}{2}m_3\left[l_2^2\left(\cos^2\varphi_2\right)\dot{\varphi}_1^2+l_2^2\dot{\varphi}_2^2\right]+$$

$$\frac{1}{2}\boldsymbol{\omega}_3^{\top}\begin{Vmatrix}I_{3xx}&0&0\\0&m_3d_3^2+I_{3yy}&0\\0&0&m_3d_3^2+I_{3zz}\end{Vmatrix}\boldsymbol{\omega}_3+m_3\mathbf{a}_3^{\top}\mathbf{b}_3=$$

$$\frac{1}{2}m_3\left[l_2^2\cos^2 q_2\dot{q}_1^2+l_2^2\dot{q}_2^2\right]+$$

$$\frac{1}{2}\left[I_{3xx}\sin^2(q_2+q_3)+\left(m_3d_3^2+I_{3yy}\right)\cos^2(q_2+q_3)\right]\dot{q}_1^2+$$

$$\frac{1}{2}\left(m_3d_3^2+I_{3zz}\right)(\dot{q}_2+\dot{q}_3)^2-$$

$$m_3\left[d_3l_2\cos q_2\cos(q_2+q_3)\,\dot{q}_1^2+d_3l_2\cos q_3\,(\dot{q}_2+\dot{q}_3)\,\dot{q}_2\right].$$

where

$$\boldsymbol{\omega}_3=\begin{pmatrix}\dot{\varphi}_1\sin(\varphi_2+\varphi_3)\\\dot{\varphi}_1\cos(\varphi_2+\varphi_3)\\\dot{\varphi}_2+\dot{\varphi}_3\end{pmatrix},$$

$$\mathbf{a}_3=\begin{pmatrix}\dot{\varphi}_1d_3\sin\varphi_1\cos(\varphi_2+\varphi_3)+(\dot{\varphi}_2+\dot{\varphi}_3)d_3\cos\varphi_1\sin(\varphi_2+\varphi_3)\\-(\dot{\varphi}_2+\dot{\varphi}_3)d_3\cos(\varphi_2+\varphi_3)\\\dot{\varphi}_1d_3\cos\varphi_1\cos(\varphi_2+\varphi_3)-(\dot{\varphi}_2+\dot{\varphi}_3)d_3\sin\varphi_1\sin(\varphi_2+\varphi_3)\end{pmatrix},$$

$$\mathbf{b}_3=\begin{pmatrix}-\dot{\varphi}_2l_2\cos\varphi_1\sin\varphi_2-\dot{\varphi}_1l_2\sin\varphi_1\cos\varphi_2\\\dot{\varphi}_2l_2\cos\varphi_2\\-\dot{\varphi}_1l_2\cos\varphi_1\cos\varphi_2+\dot{\varphi}_2l_2\sin\varphi_1\sin\varphi_2\end{pmatrix},$$

$$T_{m_4}=T_{m_4,0}+T_{m_4,rot-0}+m_4\left(\mathbf{v}_{m_4-c.i.-0},\mathbf{v}_0\right)$$

$$=\frac{1}{2}m_4\left[l_2^2\left(\cos^2\varphi_2\right)\dot{\varphi}_1^2+l_2^2\dot{\varphi}_2^2\right]+$$

$$\frac{1}{2}\boldsymbol{\omega}_4^{\top}\begin{Vmatrix}I_{4xx}&0&0\\0&\left[m_4(l_3+d_4)^2+I_{4yy}\right]&0\\0&0&\left[m_4(l_3+d_4)^2+I_{4zz}\right]\end{Vmatrix}\boldsymbol{\omega}_4+m_4\mathbf{a}_4^{\top}\mathbf{b}_4$$

$$=\frac{1}{2}m_4l_2^2\dot{q}_2^2+\frac{1}{2}\left[I_{4xx}\sin^2(q_2+q_3)+\right.$$

$$\left(m_4\left(l_3+d_4\right)^2+I_{4yy}\right)\cos^2(q_2+q_3)\right]\dot{q}_1^2+$$
$$m_4\left[\frac{1}{2}l_2^2\cos^2 q_2-(l_3+d_4)\,l_2\cos q_2\cos(q_2+q_3)\right]\dot{q}_1^2+$$
$$\frac{1}{2}I_{4xx}\left[2\sin(q_2+q_3)\,\dot{q}_1\dot{q}_4+\dot{q}_4^2\right]+$$
$$\frac{1}{2}\left[\left(m_4\left(l_3+d_4\right)^2+I_{4zz}\right)\right](\dot{q}_2+\dot{q}_3)^2-$$
$$[m_4\,(l_3+d_4)\,l_2\cos q_3]\,(\dot{q}_2+\dot{q}_3)\,\dot{q}_2,$$

where

$$\boldsymbol{\omega}_4=\begin{pmatrix}\dot{\varphi}_1\sin(\varphi_2+\varphi_3)+\dot{\varphi}_4\\ \dot{\varphi}_1\cos(\varphi_2+\varphi_3)\\ \dot{\varphi}_2+\dot{\varphi}_3\end{pmatrix},$$

$$\mathbf{a}_4=\begin{pmatrix}\left[\begin{array}{l}\dot{\varphi}_1\,(l_3+d_4)\sin\varphi_1\cos(\varphi_2+\varphi_3)\\ +(\dot{\varphi}_2+\dot{\varphi}_3)\,(l_3+d_4)\cos\varphi_1\sin(\varphi_2+\varphi_3)\end{array}\right]\\ \left[-(\dot{\varphi}_2+\dot{\varphi}_3)\,(l_3+d_4)\cos(\varphi_2+\varphi_3)\right]\\ \left[\begin{array}{l}\dot{\varphi}_1\,(l_3+d_4)\cos\varphi_1\cos(\varphi_2+\varphi_3)\\ -(\dot{\varphi}_2+\dot{\varphi}_3)\,(l_3+d_4)\sin\varphi_1\sin(\varphi_2+\varphi_3)\end{array}\right]\end{pmatrix},$$

$$\mathbf{b}_4=\begin{pmatrix}-\dot{\varphi}_2 l_2\cos\varphi_1\sin\varphi_2-\dot{\varphi}_1 l_2\sin\varphi_1\cos\varphi_2\\ \dot{\varphi}_2 l_2\cos\varphi_2\\ -\dot{\varphi}_1 l_2\cos\varphi_1\cos\varphi_2+\dot{\varphi}_2 l_2\sin\varphi_1\sin\varphi_2\end{pmatrix},$$

and

$$T_{m_5}=T_{m_5,0}+T_{m_5,rot-0}+m_5\left(\mathbf{v}_{m_5-c.i.-0},\mathbf{v}_0\right)=$$
$$\frac{1}{2}m_5\left[\dot{\varphi}_1^2\left[(l_3+l_4)\cos(\varphi_2+\varphi_3)-l_2\cos\varphi_2\right]^2+\right.$$
$$\left.\dot{\varphi}_2^2 l_2^2+\left(\dot{\varphi}_2^2+\dot{\varphi}_3^2+2\dot{\varphi}_2\dot{\varphi}_3\right)(l_3+l_4)^2\sin^2(\varphi_2+\varphi_3)\right]-$$
$$\frac{1}{2}m_5\left[2\dot{\varphi}_2\,(\dot{\varphi}_2+\dot{\varphi}_3)\,l_2\,(l_3+l_4)\sin\varphi_2\sin(\varphi_2+\varphi_3)\right]+$$
$$\frac{1}{2}\boldsymbol{\omega}_5^\top\left\|\begin{array}{ccc}I_{5xx}&0&0\\ 0&m_5 d_5^2+I_{5yy}&0\\ 0&0&m_5 d_5^2+I_{5zz}\end{array}\right\|\boldsymbol{\omega}_5+m_5\mathbf{a}_5^\top\mathbf{b}_5=$$
$$\frac{1}{2}m_5\left[\left[(l_3+l_4)\cos(q_2+q_3)-l_2\cos q_2\right]^2\dot{q}_1^2\right]+\frac{1}{2}m_5 l_2^2\dot{q}_2^2+$$
$$\frac{1}{2}\left[I_{5xx}\sin^2(q_2+q_3+q_5)+\left(m_5 d_5^2+I_{5yy}\right)\cos^2(q_2+q_3+q_5)\right]\dot{q}_1^2-$$
$$m_5\left[d_5\cos(q_2+q_3+q_5)\right]\left[(l_3+l_4)\cos(q_2+q_3)-l_2\cos q_2\right]\dot{q}_1^2+$$

$$\frac{1}{2}m_5\left[(l_3+l_4)^2\sin^2(q_2+q_3)\right](\dot{q}_2+\dot{q}_3)^2-$$
$$m_5\left[l_2(l_3+l_4)\sin q_2\sin(q_2+q_3)\right](\dot{q}_2+\dot{q}_3)\dot{q}_2+$$
$$\frac{[m_5d_5^2+I_{5zz}]}{2}(\dot{q}_2+\dot{q}_3+\dot{q}_5)^2+m_5[d_5l_2\cos(q_3+q_5)](\dot{q}_2+\dot{q}_3+\dot{q}_5)\dot{q}_2-$$
$$m_5[d_5(l_3+l_4)\cos(q_5)](\dot{q}_2+\dot{q}_3+\dot{q}_5)(\dot{q}_2+\dot{q}_3),$$

with

$$\boldsymbol{\omega}_5=\begin{pmatrix}\dot{\varphi}_1\sin(\varphi_2+\varphi_3+\varphi_5)\\ \dot{\varphi}_1\cos(\varphi_2+\varphi_3+\varphi_5)\\ \dot{\varphi}_2+\dot{\varphi}_3+\dot{\varphi}_5\end{pmatrix},$$

$$\mathbf{a}_5=\begin{pmatrix}-\dot{\varphi}_1 d_5\sin\varphi_1\cos(\varphi_2+\varphi_3+\varphi_5)-\\(\dot{\varphi}_2+\dot{\varphi}_3+\dot{\varphi}_5)d_5\cos\varphi_1\sin(\varphi_2+\varphi_3+\varphi_5)\\(\dot{\varphi}_2+\dot{\varphi}_3+\dot{\varphi}_5)d_5\cos(\varphi_2+\varphi_3+\varphi_5)\\-\dot{\varphi}_1 d_5\cos\varphi_1\cos(\varphi_2+\varphi_3+\varphi_5)+\\(\dot{\varphi}_2+\dot{\varphi}_3+\dot{\varphi}_5)d_5\sin\varphi_1\sin(\varphi_2+\varphi_3+\varphi_5)\end{pmatrix},$$

$$\mathbf{b}_5=\begin{pmatrix}-\dot{\varphi}_2 l_2\cos\varphi_1\sin\varphi_2+(\dot{\varphi}_2+\dot{\varphi}_3)(l_3+l_4)\cos\varphi_1\sin(\varphi_2+\varphi_3)+\\\dot{\varphi}_1\sin\varphi_1((l_3+l_4)\cos(\varphi_2+\varphi_3)-l_2\cos\varphi_2)\\\dot{\varphi}_2 l_2\cos\varphi_2-(\dot{\varphi}_2+\dot{\varphi}_3)(l_3+l_4)\cos(\varphi_2+\varphi_3)\\\dot{\varphi}_2 l_2\sin\varphi_1\sin\varphi_2-(\dot{\varphi}_2+\dot{\varphi}_3)(l_3+l_4)\sin\varphi_1\sin(\varphi_2+\varphi_3)+\\\dot{\varphi}_1\cos\varphi_1((l_3+l_4)\cos(\varphi_2+\varphi_3)-l_2\cos\varphi_2)\end{pmatrix}.$$

Potential energy

The potential energy $V=\sum_{i=1}^{5}V_{m_i}$ consists of

$$V_{m_1}=\text{const},$$
$$V_{m_2}=m_2g(d_2\sin\varphi_2+\text{const})=m_2gd_2(\sin q_2+\text{const}),$$
$$V_{m_3}=m_3g(-d_3\sin(\varphi_2+\varphi_3)+l_2\sin\varphi_2+\text{const})=$$
$$m_3g(-d_3\sin(q_2+q_3)+\sin q_2+\text{const}),$$
$$V_{m_4}=m_4g(-(l_3+d_4)\sin(\varphi_2+\varphi_3)+l_2\sin\varphi_2+\text{const})$$
$$=m_4g(-(l_3+d_4)\sin(q_2+q_3)+l_2\sin q_2+\text{const}),$$
$$V_{m_5}=m_5g[-(l_3+l_4)\sin(\varphi_2+\varphi_3)+$$
$$l_2\sin\varphi_2+d_5\sin(\varphi_2+\varphi_3+\varphi_5)+\text{const}]=$$
$$m_5g[-(l_3+l_4)\sin(q_2+q_3)+l_2\sin q_2+$$
$$d_5\sin(q_2+q_3+q_5)+\text{const}],$$

which gives

$$V=m_2gd_2\sin q_2+m_3g(-d_3\sin(q_2+q_3)+l_2\sin q_2)+$$

$$m_4 g \left(- (l_3 + d_4) \sin(q_2 + q_3) + l_2 \sin q_2 \right) +$$
$$m_5 g \left[- (l_3 + l_4) \sin(q_2 + q_3) + \right.$$
$$\left. l_2 \sin q_2 + d_5 \sin(q_2 + q_3 + q_5) \right] + \text{const}.$$

Non-potential forces

The non-potential generalized forces are as follows:

$$Q_{non\text{-}pot,i} = \tau_i - f_{fric-i} \dot{\varphi}_i = \tau_i - f_{fric-i} \dot{q}_i,$$
τ_i is a torsion force $(i = 1, ..., 5)$.

Lagrange equations

The dynamic equations of the considered manipulator are represented by the corresponding system of Lagrange equations:

$$\frac{d}{dt} \frac{\partial}{\partial \dot{q}_i} L - \frac{\partial}{\partial q_i} L = Q_{non\text{-}pot,i}, \quad i = 1, ..., 5,$$
$$L = T - V,$$

which leads to the following standard matrix format (12.5) of the dynamic model:

$$D(q)\ddot{q} + C(q, \dot{q})\dot{q} + g(q) = \tau,$$

where

$$D(q) = \begin{Vmatrix} D_{11}(q) & 0 & 0 & D_{14}(q) & 0 \\ 0 & D_{22}(q) & D_{23}(q) & 0 & D_{25}(q) \\ 0 & D_{32}(q) & D_{33}(q) & 0 & D_{35}(q) \\ D_{41}(q) & 0 & 0 & I_{4xx} & 0 \\ 0 & D_{52}(q) & D_{53}(q) & 0 & m_5 d_5^2 + I_{5zz} \end{Vmatrix},$$

with

$$D_{11}(q) = \frac{1}{2} m_1 r_1^2 + I_{2xx} \sin^2 q_2 + \left(m_2 d_2^2 + I_{2yy} \right) \cos^2 q_2 +$$
$$m_3 l_2^2 \cos^2 q_2 + I_{3xx} \sin^2 (q_2 + q_3) + \left(m_3 d_3^2 + I_{3yy} \right) \cos^2 (q_2 + q_3) -$$
$$2 m_3 d_3 l_2 \cos q_2 \cos(q_2 + q_3) + I_{4xx} \sin^2 (q_2 + q_3) + m_4 (l_3 + d_4)^2 +$$
$$I_{4yy} \cos^2 (q_2 + q_3) + m_4 l_2^2 \cos^2 q_2 - 2 m_4 [(l_3 + d_4) l_2 \cos q_2 \cos(q_2 + q_3)] +$$
$$m_5 [(l_3 + l_4) \cos(q_2 + q_3) - l_2 \cos q_2]^2 -$$
$$2 m_5 d_5 (l_3 + l_4) \cos(q_2 + q_3) \cos(q_2 + q_3 + q_5) +$$
$$2 m_5 [d_5 l_2 \cos q_2 \cos(q_2 + q_3 + q_5)] + I_{5xx} \sin^2 (q_2 + q_3 + q_5) +$$
$$\left(m_5 d_5^2 + I_{5yy} \right) \cos^2 (q_2 + q_3 + q_5),$$

$$D_{41}(q) = I_{4xx} \sin(q_2 + q_3),$$
$$D_{22}(q) = m_2 d_2^2 + I_{4zz} + [m_3 + m_4 + m_5] l_2^2 + m_3 d_3^2 + I_{3zz}$$
$$+ m_4 (l_3 + d_4)^2 + I_{4zz} + m_5 (l_3 + l_4)^2 \sin^2(q_2 + q_3) -$$
$$2 l_2 [m_3 d_3 \cos q_3 + m_4 (l_3 + d_4) \cos q_3] -$$
$$2 m_5 l_2 (l_3 + l_4) \sin q_2 \sin(q_2 + q_3) + m_5 d_5^2 + I_{5zz}$$
$$+ 2 m_5 d_5 l_2 \cos(q_3 + q_5) - 2 m_5 d_5 (l_3 + l_4) \cos(q_5),$$
$$D_{32}(q) = m_3 d_3^2 + I_{3zz} + m_4 (l_3 + d_4)^2 + I_{4zz} +$$
$$m_5 (l_3 + l_4)^2 \sin^2(q_2 + q_3) -$$
$$l_2 [m_3 d_3 \cos q_3 + m_4 (l_3 + d_4) \cos q_3] -$$
$$m_5 l_2 (l_3 + l_4) \sin q_2 \sin(q_2 + q_3) - 2 m_5 d_5 (l_3 + l_4) \cos(q_5) +$$
$$\left[m_5 d_5^2 + I_{5zz} \right] + m_5 [d_5 l_2 \cos(q_3 + q_5)],$$
$$D_{52}(q) = m_5 d_5^2 + I_{5zz} + m_5 d_5 l_2 \cos(q_3 + q_5) - m_5 d_5 (l_3 + l_4) \cos(q_5),$$
$$D_{23}(q) = m_3 d_3^2 + I_{3zz} + m_4 (l_3 + d_4)^2 + I_{4zz} +$$
$$m_5 (l_3 + l_4)^2 \sin^2(q_2 + q_3) - m_3 d_3 l_2 \cos q_3 -$$
$$m_4 (l_3 + d_4) l_2 \cos q_3 - m_5 l_2 (l_3 + l_4) \sin q_2 \sin(q_2 + q_3) +$$
$$m_5 l_2 d_5 \cos(q_3 + q_5) + m_5 d_5^2 + I_{5zz} - 2 m_5 d_5 (l_3 + l_4) \cos q_5,$$
$$D_{33}(q) = m_3 d_3^2 + I_{3zz} + m_4 (l_3 + d_4)^2 + I_{4zz} + \left[m_5 d_5^2 + I_{5zz} \right] +$$
$$m_5 (l_3 + l_4)^2 \sin^2(q_2 + q_3) - 2 m_5 [d_5 (l_3 + l_4) \cos q_5],$$
$$D_{53}(q) = m_5 d_5^2 + I_{5zz} - m_5 d_5 (l_3 + l_4) \cos q_5,$$
$$D_{14}(q) = I_{4xx} \sin(q_2 + q_3),$$
$$D_{25}(q) = m_5 d_5^2 + I_{5zz} + m_5 d_5 l_2 \cos(q_3 + q_5) - m_5 d_5 (l_3 + l_4) \cos q_5,$$
$$D_{35}(q) = m_5 d_5^2 + I_{5zz} - m_5 d_5 (l_3 + l_4) \cos q_5,$$

and

$$C(q, \dot{q}) = \begin{Vmatrix} C_{11}(q, \dot{q}) & 0 & 0 & C_{14}(q, \dot{q}) & 0 \\ C_{21}(q, \dot{q}) & C_{22}(q, \dot{q}) & C_{23}(q, \dot{q}) & C_{24}(q, \dot{q}) & C_{25}(q, \dot{q}) \\ C_{31}(q, \dot{q}) & C_{32}(q, \dot{q}) & C_{33}(q, \dot{q}) & C_{34}(q, \dot{q}) & C_{35}(q, \dot{q}) \\ C_{41}(q, \dot{q}) & 0 & 0 & f_{fric-4} & 0 \\ C_{51}(q, \dot{q}) & C_{52}(q, \dot{q}) & C_{53}(q, \dot{q}) & 0 & C_{55}(q, \dot{q}) \end{Vmatrix},$$

containing

$$C_{11}(q, \dot{q}) = f_{fric-1} + 2 \left[\left(I_{2xx} - m_2 d_2^2 - I_{2yy} - m_3 l_2^2 \right) \sin q_2 \cos q_2 \right] \dot{q}_2 +$$
$$2 \left[\left(I_{3xx} - m_3 d_3^2 - I_{3yy} \right) \sin(q_2 + q_3) \cos(q_2 + q_3) \right] (\dot{q}_2 + \dot{q}_3) +$$

$$2\left[-m_4\left(l_3+d_4\right)^2\sin\left(q_2+q_3\right)\cos\left(q_2+q_3\right)\right]\left(\dot{q}_2+\dot{q}_3\right)+$$
$$2\left[\left(I_{4xx}-I_{4yy}\right)\sin\left(q_2+q_3\right)\cos\left(q_2+q_3\right)\right]\left(\dot{q}_2+\dot{q}_3\right)+$$
$$2m_3d_3l_2\cos q_2\sin\left(q_2+q_3\right)\dot{q}_3+\left[2m_3d_3l_2\sin\left(2q_2+q_3\right)\right]\dot{q}_2+$$
$$2m_4\left[\left(l_3+d_4\right)l_2\sin\left(2q_2+q_3\right)-l_2^2\sin q_2\cos q_2\right]\dot{q}_2+$$
$$2m_5\left[\left(l_3+l_4\right)\cos\left(q_2+q_3\right)-l_2\cos q_2\right]\times$$
$$\left[-\left(l_3+l_4\right)\sin\left(q_2+q_3\right)\right]\left(\dot{q}_2+\dot{q}_3\right)+l_2\sin q_2\dot{q}_2$$
$$2m_4\left[\left(l_3+d_4\right)l_2\cos q_2\sin\left(q_2+q_3\right)\right]\dot{q}_3+\left[2\left(I_{5xx}-m_5d_5^2-I_{5yy}\right)\right]\times$$
$$\sin\left(q_2+q_3+q_5\right)\cos\left(q_2+q_3+q_5\right)\left(\dot{q}_2+\dot{q}_3+\dot{q}_5\right)+$$
$$2m_5d_5\sin\left(2q_2+2q_3+q_5\right)\left[l_3+l_4-l_2\right]\dot{q}_2+$$
$$2m_5d_5\left[\left(l_3+l_4\right)\sin\left(2q_2+2q_3+q_5\right)\right]\dot{q}_3-$$
$$2m_5d_5\left[l_2\sin\left(q_2+q_3+q_5\right)\cos q_2\right]\left(\dot{q}_3+\dot{q}_5\right)$$
$$+2m_5d_5\left[\left(l_3+l_4\right)\sin\left(q_2+q_3+q_5\right)\cos\left(q_2+q_3\right)\right]\dot{q}_5,$$

$$C_{21}\left(q,\dot{q}\right)=\left[\left(-I_{2xx}+m_2d_2^2+I_{2yy}\right)\sin q_2\cos q_2\right]\dot{q}_1$$
$$\left[\left(m_3l_2^2+m_4l_2^2\right)\sin q_2\cos q_2\right]\dot{q}_1$$
$$\left[\left(I_{3yy}-I_{3xx}+m_3d_3^2+m_4\left(l_3+d_4\right)^2+I_{4yy}-I_{4xx}\right)\right]\times$$
$$\left[\sin\left(q_2+q_3\right)\cos\left(q_2+q_3\right)\right]\dot{q}_1-$$
$$\left[m_4\left(l_3+d_4\right)l_2\sin\left(2q_2+q_3\right)\right]\dot{q}_1-$$
$$\left[m_3d_3l_2\sin\left(2q_2+q_3\right)-m_5l_2^2\cos q_2\sin q_2\right]\dot{q}_1-$$
$$m_5\left[\left(l_3+l_4\right)l_2\sin\left(2q_2+q_3\right)\right]\dot{q}_1+$$
$$m_5\left[\left(l_3+l_4\right)^2\cos\left(q_2+q_3\right)\sin\left(q_2+q_3\right)\right]\dot{q}_1-$$
$$\left[\sin\left(q_2+q_3+q_5\right)\cos\left(q_2+q_3+q_5\right)\right]\times$$
$$\left[\left(I_{5xx}-m_5d_5^2-I_{5yy}\right)\right]\dot{q}_1-$$
$$m_5d_5\left[\left(l_3+l_4\right)\sin\left(2q_2+2q_3+q_5\right)\right]\dot{q}_1+$$
$$m_5d_5\left[l_2\sin\left(2q_2+q_3+q_5\right)\right]\dot{q}_1,$$

$$C_{31}\left(q,\dot{q}\right)=-\sin\left(q_2+q_3\right)\cos\left(q_2+q_3\right)\left[I_{3xx}-m_3d_3^2-I_{3yy}\right]\dot{q}_1-$$
$$\left[I_{4xx}-m_4\left(l_3+d_4\right)^2-I_{4yy}\right]\sin\left(q_2+q_3\right)\cos\left(q_2+q_3\right)\dot{q}_1-$$
$$\left[m_3d_3l_2+m_4\left(l_3+d_4\right)l_2\right]\cos q_2\sin\left(q_2+q_3\right)\dot{q}_1+$$
$$m_5\left[\left(l_3+l_4\right)\sin\left(q_2+q_3\right)\right]\left[\left(l_3+l_4\right)\cos\left(q_2+q_3\right)-l_2\cos q_2\right]\dot{q}_1-$$
$$\left[I_{5xx}-m_5d_5^2-I_{5yy}\right]\sin\left(q_2+q_3+q_5\right)\cos\left(q_2+q_3+q_5\right)\dot{q}_1-$$
$$m_5\left[d_5\sin\left(q_2+q_3+q_5\right)\right]\left[\left(l_3+l_4\right)\cos\left(q_2+q_3\right)-l_2\cos q_2\right]\dot{q}_1-$$

$$m_5 \left[d_5 \cos(q_2 + q_3 + q_5) \right] (l_3 + l_4) \sin(q_2 + q_3) \dot{q}_1,$$

$$C_{41}(q, \dot{q}) = I_{4xx} \cos(q_2 + q_3)(\dot{q}_2 + \dot{q}_3),$$

$$C_{51}(q, \dot{q}) = - \left[\left(I_{5xx} - m_5 d_5^2 - I_{5yy} \right) \right] \times$$
$$\left[\sin(q_2 + q_3 + q_5) \cos(q_2 + q_3 + q_5) \right] \dot{q}_1 -$$
$$m_5 \left[d_5 \sin(q_2 + q_3 + q_5) \right] \times \left[(l_3 + l_4) \cos(q_2 + q_3) - l_2 \cos q_2 \right] \dot{q}_1,$$

$$C_{22}(q, \dot{q}) = f_{fric-2} + \left[m_5 (l_3 + l_4)^2 \right] \times$$
$$\left[\sin(q_2 + q_3) \cos(q_2 + q_3) \right] (\dot{q}_2 + 2\dot{q}_3) - m_5 l_2 (l_3 + l_4) \sin(2q_2 + q_3)(\dot{q}_2),$$

$$C_{23}(q, \dot{q}) = - \left[m_3 d_3 l_2 \sin q_3 + m_4 (l_3 + d_4) l_2 \sin q_3 \right] (\dot{q}_2 + \dot{q}_3) +$$
$$\left[\sin(q_2 + q_3) \cos(q_2 + q_3) \right] \times \left[m_5 (l_3 + l_4)^2 \right] \dot{q}_2 -$$
$$\left[m_5 l_2 (l_3 + l_4) \sin(2q_2 + q_3) \right] \dot{q}_2 + \left[m_5 d_5 l_2 \sin(q_3 + q_5) \right] \dot{q}_2 +$$
$$m_5 \left[l_2 (l_3 + l_4) \sin q_2 \cos(q_2 + q_3) \right] (\dot{q}_2 + \dot{q}_3),$$

$$C_{52}(q, \dot{q}) = m_5 \left[d_5 l_2 \sin(q_3 + q_5) \right] \dot{q}_2$$
$$- m_5 \left[d_5 (l_3 + l_4) \sin q_5 \right] (\dot{q}_2 + \dot{q}_3 + \dot{q}_5),$$

$$C_{23}(q, \dot{q}) = l_2 \left[m_3 d_3 \sin q_3 + (l_3 + d_4) m_4 \sin q_3 \right] (2\dot{q}_2 + \dot{q}_3) -$$
$$m_5 l_2 (l_3 + d_4) \sin q_2 \cos(q_2 + q_3)(2\dot{q}_2 + \dot{q}_3) -$$
$$m_5 \left[d_5 l_2 \sin(q_3 + q_5) \right] (2\dot{q}_2 + \dot{q}_3 + \dot{q}_5) +$$
$$\left[m_5 (l_3 + l_4)^2 \sin(q_2 + q_3) \cos(q_2 + q_3) \right] \dot{q}_3,$$

$$C_{33}(q, \dot{q}) = f_{fric-3} + \left[\sin(q_2 + q_3) \cos(q_2 + q_3) \right] \times$$
$$\left[m_5 (l_3 + l_4)^2 \right] (\dot{q}_3 + 2\dot{q}_2) + \left[(m_3 d_3 l_2 + m_4 (l_3 + d_4) l_2) \sin q_3 \right] \dot{q}_2$$
$$- \left[m_5 l_2 (l_3 + l_4) \sin q_2 \cos(q_2 + q_3) \right] \dot{q}_2,$$

$$C_{53}(q, \dot{q}) = -m_5 \left[d_5 (l_3 + l_4) \sin q_5 \right] (\dot{q}_2 + \dot{q}_3 + \dot{q}_5),$$

$$C_{14}(q, \dot{q}) = I_{4xx} \cos(q_2 + q_3)(\dot{q}_2 + \dot{q}_3),$$

$$C_{24}(q, \dot{q}) = -I_{4xx} \cos(q_2 + q_3) \dot{q}_1,$$

$$C_{34}(q, \dot{q}) = -I_{4xx} \cos(q_2 + q_3) \dot{q}_1,$$

$$C_{25}(q, \dot{q}) = m_5 d_5 (l_3 + l_4) \sin q_5 (2\dot{q}_2 + 2\dot{q}_3 + \dot{q}_5) -$$
$$m_5 \left[d_5 l_2 \sin(q_3 + q_5) \right] (2\dot{q}_2 + \dot{q}_3 + \dot{q}_5),$$

$$C_{35}(q, \dot{q}) = m_5 \left[d_5 (l_3 + l_4) \sin q_5 \right] (2\dot{q}_2 + 2\dot{q}_3 + \dot{q}_5),$$

$$C_{55}(q, \dot{q}) = f_{fric-5} + m_5 \left[d_5 (l_3 + l_4) \sin q_5 \right] (\dot{q}_2 + \dot{q}_3),$$

and

$$g(q) = \begin{Vmatrix} 0 \\ g_2(q) \\ g_3(q) \\ 0 \\ -m_5 g d_5 \cos(q_2 + q_3 + q_5) \end{Vmatrix}, \quad \tau = \begin{Vmatrix} \tau_1 \\ \tau_2 \\ \tau_3 \\ \tau_4 \\ \tau_5 \end{Vmatrix},$$

where

$$g_2(q) = -m_2 g d_2 \cos q_2 + m_3 g (d_3 \cos(q_2 + q_3) - l_2 \cos q_2) +$$
$$m_4 g ((l_3 + d_4) \cos(q_2 + q_3) - l_2 \cos q_2) +$$
$$m_5 g [(l_3 + l_4) \cos(q_2 + q_3) - l_2 \cos q_2] -$$
$$m_5 g [d_5 \cos(q_2 + q_3 + q_5)],$$
$$g_3(q) = m_3 g d_3 \cos(q_2 + q_3) + m_4 g (l_3 + d_4) \cos(q_2 + q_3)$$
$$+ m_5 g [(l_3 + l_4) \cos(q_2 + q_3) - d_5 \cos(q_2 + q_3 + q_5)].$$

12.8 CD motor, gear, and load train

Consider the following electromechanical system containing a *CD* motor, gear train, and load, which is represented in Fig. 12.8.

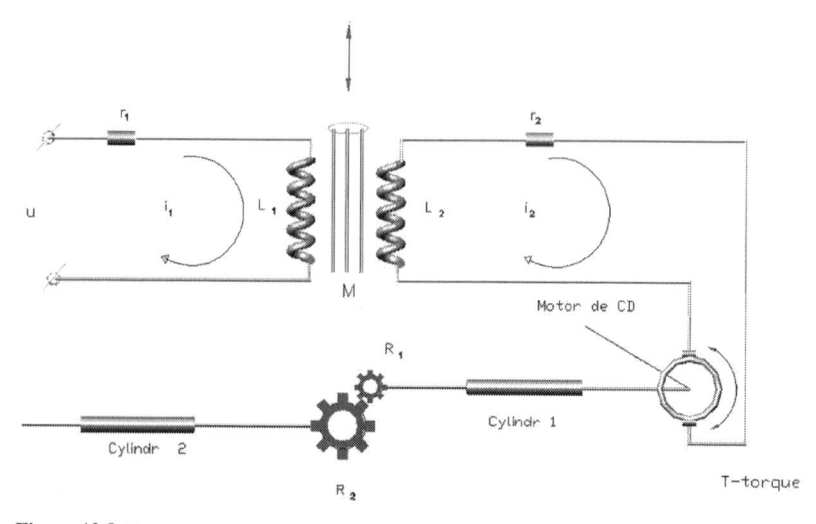

Figure 12.8 CD motor, gear train, and load.

Generalized coordinates

In this system the generalized coordinates are

$$q_1 := \int_{\tau=0}^{t} i_1(\tau)\,d\tau, \quad q_2 := \int_{\tau=0}^{t} i_2(\tau)\,d\tau, \quad q_3 := \varphi.$$

Kinetic energy

The kinetic energy T is given here by the expression

$$T = T_e + \sum_{i=1}^{4} T_{m_i},$$

where T_{m_i} can be calculated based on the König formula:

$$T_{m_i} = T_{m_i,0} + T_{m_i,rot-0} + 2m_i\left(\mathbf{v}_{m_i-c.i.-0}, \mathbf{v}_0\right),$$

$$T_{m_i,0} = \frac{1}{2} m_i \|\mathbf{v}_0\|^2, \quad T_{m_i,rot-0} = \frac{1}{2}\left(\boldsymbol{\omega}, I_{i,0}\boldsymbol{\omega}\right).$$

In our case (see also Section 6.5) we have

$$T_e = \frac{L_1}{2} i_1^2 + \frac{L_2}{2} i_2^2 - \mu i_1 i_2 = \frac{1}{2} L_1 \dot{q}_1^2 + \frac{1}{2} L_2 \dot{q}_2^2 - \mu \dot{q}_1 \dot{q}_2,$$

μ is a constant,

$$T_{m_1} = T_{m_1,0} + T_{m_1,rot-0} + m_1\left(\mathbf{v}_{m_1-c.i.-0}, \mathbf{v}_0\right)$$
$$= T_{m_1,rot-0} = \frac{1}{2}\frac{m_1 r_1^2}{2}\dot{\varphi}^2 = \frac{1}{4} m_1 r_1^2 \dot{q}_3^2,$$

$$T_{m_2} = T_{m_2,0} + T_{m_2,rot-0} + m_2\left(\mathbf{v}_{m_1-c.i.-0}, \mathbf{v}_0\right)$$
$$= T_{m_2,rot-0} = \frac{1}{2}\frac{m_2 r_2^2}{2}\dot{\varphi}^2 = \frac{1}{4} m_2 r_2^2 \dot{q}_3^2,$$

$$T_{m_3} = T_{m_3,0} + T_{m_3,rot-0} + m_3\left(\mathbf{v}_{m_3-c.i.-0}, \mathbf{v}_0\right)$$
$$= T_{m_3,rot-0} = \frac{1}{2}\frac{m_3 r_3^2}{2}\frac{r_2^2 \dot{\varphi}^2}{r_3^2} = \frac{1}{4} m_3 r_2^2 \dot{q}_3^2,$$

$$T_{m_4} = T_{m_4,0} + T_{m_4,rot-0} + m_4\left(\mathbf{v}_{m_4-c.i.-0}, \mathbf{v}_0\right)$$
$$= T_{m_4,rot-0} = \frac{1}{2}\frac{m_4 r_4^2}{2}\frac{r_2^2 \dot{\varphi}^2}{r_3^2} = \frac{1}{4} m_4 \frac{r_4^2 r_2^2}{r_3^2} \dot{q}_3^2.$$

Potential energy

Here the potential energy V is

$$V = V_e + \sum_{i=1}^{4} V_{m_i}, \quad V_e = 0,$$

$V_{m_1} = \text{const}, \ V_{m_2} = \text{const}, \ V_{m_3} = \text{const}, \ V_{m_4} = \text{const},$

which gives $V = \text{const}$.

Non-potential forces

The generalized non-potential forces are

$Q_{non\text{-}pot,1} = u - R_1 i_1 = u - R_1 \dot{q}_1$, u is the applied voltage,

$Q_{non\text{-}pot,2} = -u_b - R_2 i_2 = -u_b - R_2 \dot{q}_2$,

u_b is the counter-electromotive force in volts,

$Q_{non\text{-}pot,3} = \tau_3 - f_{fric-3} \dot{\varphi}_3 = \tau_3 - f_{fric-3} \dot{q}_3$, $\tau_3 = k i_2 = k \dot{q}_2$

is a twisting moment.

Lagrange equations

Using the obtained formulas for T, V, and $L = T - V$ we are able to derive the dynamic Lagrange equations:

$$\frac{d}{dt} \frac{\partial}{\partial \dot{q}_i} L - \frac{\partial}{\partial q_i} L = Q_{non\text{-}pot,i}, \ i = 1, 2, 3,$$

which can be represented in the standard matrix format (12.5)

$$D(q) \ddot{q} + C(q, \dot{q}) \dot{q} + g(q) = \tau,$$

where

$$D(q) = \begin{Vmatrix} L_1 & -\mu & 0 \\ -\mu & L_2 & 0 \\ 0 & 0 & \frac{1}{2}\left[m_1 r_1^2 + m_2 r_2^2 + m_3 r_3^2 + m_4 \frac{r_4^2 r_2^2}{r_3^2} \right] \end{Vmatrix},$$

$$C(q, \dot{q}) = \begin{Vmatrix} R_1 & 0 & 0 \\ 0 & R_2 & 0 \\ 0 & -k & f_{fric-3} \end{Vmatrix}, \ g(q) = \begin{Vmatrix} 0 \\ 0 \\ 0 \end{Vmatrix}, \ \tau = \begin{Vmatrix} u \\ -u_b \\ 0 \end{Vmatrix}.$$

12.9 Stanford/JPL robot manipulator

Consider now the Stanford/JPL manipulator represented in Fig. 12.9.

Generalized coordinates

The generalized coordinates for this mechanical system are as follows:

$q_1 := \varphi_1, \quad q_2 := \varphi_2, \quad q_3 := x, \quad q_4 := y.$

Collection of electromechanical models 415

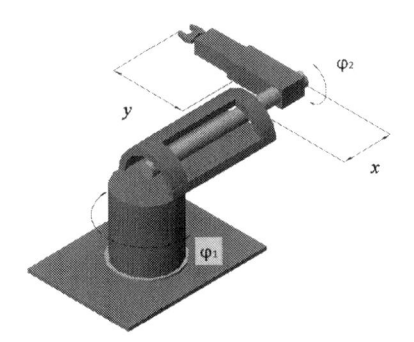

Figure 12.9 The Stanford/JPL manipulator.

Kinetic energy

The kinetic energy $T = \sum_{i=1}^{5} T_{m_i}$ of this system consists of the terms T_{m_i}, which can be calculated based on the König formula:

$$T_{m_i} = T_{m_i,0} + T_{m_i,rot-0} + 2m_i \left(\mathbf{v}_{m_i-c.i.-0}, \mathbf{v}_0\right),$$

$$T_{m_i,0} = \frac{1}{2} m_i \|\mathbf{v}_0\|^2, \quad T_{m_i,rot-0} = \frac{1}{2} \left(\boldsymbol{\omega}, I_{i,0}\boldsymbol{\omega}\right),$$

which in our case leads to the following expressions:

$$T_{m_1} = T_{m_1,0} + T_{m_1,rot-0} + m_1 \left(\mathbf{v}_{m_2-c.i.-0}, \mathbf{v}_0\right)$$

$$= T_{m_1,rot-0} = \frac{1}{2} I_{1_{zz}} \dot{\varphi}_1^2 = \frac{1}{2} I_{1_{zz}} \dot{q}_1^2,$$

$$T_{m_2} = T_{m_2,0} + T_{m_2,rot-0} + m_2 \left(\mathbf{v}_{m_2-c.i.-0}, \mathbf{v}_0\right) = T_{m_2,rot-0} =$$

$$\frac{1}{2} \begin{pmatrix} 0 \\ \dot{\varphi}_1 \\ 0 \end{pmatrix}^\top \begin{Vmatrix} m_2 d_2^2 + I_{2_{xx}} & 0 & 0 \\ 0 & m_2 d_2^2 + I_{2_{yy}} & -I_{2_{yz}} \\ 0 & -I_{2_{zy}} & I_{2_{zz}} \end{Vmatrix} \begin{pmatrix} 0 \\ \dot{\varphi}_1 \\ 0 \end{pmatrix}$$

$$= \frac{1}{2} \left(m_2 d_2^2 + I_{2_{yy}}\right) \dot{q}_1^2,$$

$$T_{m_3} = T_{m_3,0} + T_{m_3,rot-0} + m_3 \left(\mathbf{v}_{m_3-c.i.-0}, \mathbf{v}_0\right)$$

$$= T_{m_3,0} + T_{m_3,rot-0} = \frac{1}{2} m_3 \left(\dot{x}^2 + (x + d_3)^2 \dot{\varphi}_1^2\right) +$$

$$\frac{1}{2} \begin{pmatrix} 0 \\ \dot{\varphi}_1 \\ \dot{\varphi}_2 \end{pmatrix}^\top \begin{Vmatrix} I_{3_{xx}} & 0 & 0 \\ 0 & I_{3_{yy}} & 0 \\ 0 & 0 & I_{3_{zz}} \end{Vmatrix} \begin{pmatrix} 0 \\ \dot{\varphi}_1 \\ \dot{\varphi}_2 \end{pmatrix} =$$

$$\frac{1}{2} \left[m_3 \left(\dot{q}_3^2 + (q_3 + d_3)^2 \dot{q}_1^2\right) + I_{3_{yy}} \dot{q}_1^2 + I_{3_{zz}} \dot{q}_2^2 \right],$$

$$T_{m_4} = T_{m_4,0} + T_{m_4,rot-0} + m_4 \left(\mathbf{v}_{m_4-c.i.-0}, \mathbf{v}_0\right) = \frac{1}{2} m_4 \left(\dot{x}^2 + (x + l_3)^2 \dot{\varphi}_1^2\right) +$$

$$\frac{1}{2}\begin{pmatrix}\dot\varphi_1\sin\varphi_2\\ \dot\varphi_1\cos\varphi_2\\ \dot\varphi_2\end{pmatrix}^{\mathsf T}\begin{Vmatrix}I_{4_{xx}} & 0 & 0\\ 0 & m_4 d_4^2+I_{4_{yy}} & 0\\ 0 & 0 & m_4 d_4^2+I_{4_{zz}}\end{Vmatrix}\begin{pmatrix}\dot\varphi_1\sin\varphi_2\\ \dot\varphi_1\cos\varphi_2\\ \dot\varphi_2\end{pmatrix}+$$

$$m_4\begin{pmatrix}-d_4\cos\varphi_1\cos\varphi_2\dot\varphi_1+d_4\sin\varphi_1\sin\varphi_2\dot\varphi_2\\ -d_4\cos\varphi_2\dot\varphi_2\\ d_4\sin\varphi_1\cos\varphi_2\dot\varphi_1+d_4\cos\varphi_1\sin\varphi_2\dot\varphi_2\end{pmatrix}^{\mathsf T}\times$$

$$\begin{pmatrix}\dot x\cos\varphi_1-(x+l_3)\sin\varphi_1\dot\varphi_1\\ 0\\ -\dot x\sin\varphi_1-(x+l_3)\cos\varphi_1\dot\varphi_1\end{pmatrix}=\frac{1}{2}\left(m_4 d_4^2+I_{4_{zz}}\right)\dot q_2^2+m_4\dot q_3^2+$$

$$\frac{1}{2}\left[I_{4_{xx}}\sin^2 q_2+m_4(q_3+l_3)^2+\left(m_4 d_4^2+I_{4_{yy}}\right)\cos^2 q_2\right]\dot q_1^2$$

$$-m_4 d_4\left[\cos q_2\dot q_1\dot q_3-(q_3+l_3)\sin q_2\dot q_1\dot q_2\right],$$

$$T_{m_5}=T_{m_5,0}+T_{m_5,rot-0}+m_4\left(\mathbf{v}_{m_5-c.i.-0},\mathbf{v}_0\right)=T_{m_5,0}+T_{m_5,rot-0}=\frac{1}{2}m_4\times$$

$$\left\|\begin{pmatrix}-(l_4+y)\cos\varphi_1\cos\varphi_2\dot\varphi_1+(l_4+y)\sin\varphi_1\sin\varphi_2\dot\varphi_2-\dot y\sin\varphi_1\cos\varphi_2\\ -(l_4+y)\cos\varphi_2\dot\varphi_2-\dot y\sin\varphi_2\\ (l_4+y)\sin\varphi_1\cos\varphi_2\dot\varphi_1+(l_4+y)\cos\varphi_1\sin\varphi_2\dot\varphi_2-\dot y\cos\varphi_1\cos\varphi_2\end{pmatrix}\right.$$

$$+\begin{pmatrix}\dot x\cos\varphi_1-(x+l_3)\sin\varphi_1\dot\varphi_1\\ 0\\ -\dot x\sin\varphi_1-(x+l_3)\cos\varphi_1\dot\varphi_1\end{pmatrix}\Bigg\|^2+$$

$$\frac{1}{2}\begin{pmatrix}\dot\varphi_1\sin\varphi_2\\ \dot\varphi_1\cos\varphi_2\\ \dot\varphi_2\end{pmatrix}^{\mathsf T}\begin{Vmatrix}I_{4_{xx}} & 0 & 0\\ 0 & I_{4_{yy}} & 0\\ 0 & 0 & I_{4_{zz}}\end{Vmatrix}\begin{pmatrix}\dot\varphi_1\sin\varphi_2\\ \dot\varphi_1\cos\varphi_2\\ \dot\varphi_2\end{pmatrix}=$$

$$\frac{1}{2}\left[m_5(q_3+l_3)^2+I_{5_{xx}}\sin^2 q_2+\left(m_5(l_4+q_4)^2+I_{5_{yy}}\right)\cos^2 q_2\right]\dot q_1^2$$

$$+\frac{1}{2}\left[I_{5_{zz}}+m_5(l_4+q_4)^2\right]\dot q_2^2+\frac{1}{2}m_5\dot q_4^2+\frac{1}{2}m_5\dot q_5^2$$

$$-m_5(l_4+q_4)\left[\dot q_1\dot q_3\cos q_2-(q_3+l_3)\dot q_1\dot q_2\sin q_2\right]+$$

$$m_5\left[(q_3+l_3)\dot q_1\dot q_4\cos q_2\right].$$

Potential energy

The potential energy $V=\sum_{i=1}^{5}V_{m_i}$ has components

$$V_{m_1}=\text{const},\ V_{m_2}=\text{const},\ V_{m_3}=\text{const},$$
$$V_{m_4}=m_4 g d_4\sin\varphi_2=m_4 g d_4\sin q_2,$$
$$V_{m_5}=m_5 g(y+l_4)\sin\varphi_2=m_5 g(q_4+l_4)\sin q_2,$$

which gives

$$V=m_4 g d_4\sin q_2+m_5 g(q_4+l_4)\sin q_2+\text{const}.$$

Non-potential forces

The non-potential generalized forces are given by the following formulas:

$$Q_{non\text{-}pot,1} = \tau_1 - f_{fric-1}\dot{\varphi}_1 = \tau_1 - f_{fric-1}\dot{q}_1,$$

τ_1 is a torsion force,

$$Q_{non\text{-}pot,2} = \tau_2 - f_{fric-2}\dot{\varphi}_2 = \tau_2 - f_{fric-2}\dot{q}_2,$$

τ_2 is a torsion force,

$$Q_{non\text{-}pot,3} = F_3 - f_{fric-3}\dot{x} = F_3 - f_{fric-3}\dot{q}_3,$$

F_3 is a force acting on the horizontal movement,

$$Q_{non\text{-}pot,4} = F_4 - f_{fric-4}\dot{y} = F_4 - f_{fric-4}\dot{q}_4,$$

F_4 is a force acting on the transversal movement.

Lagrange equations

Based on the obtained expressions for T, V, and $L = T - V$ we are able to derive the Lagrange equations:

$$\frac{\partial}{\partial \dot{q}_4} L = m_5 (q_3 + l_3) \dot{q}_1 \cos q_2 + m_5 \dot{q}_4,$$

which in the standard matrix format (12.5) is

$$D(q)\ddot{q} + C(q,\dot{q})\dot{q} + g(q) = \tau,$$

where the matrix

$$D(q) = \begin{Vmatrix} D_{11}(q) & D_{12}(q) & D_{13}(q) & D_{14}(q) \\ D_{21}(q) & D_{22}(q) & 0 & 0 \\ D_{31}(q) & 0 & m_3 + m_4 + m_5 & 0 \\ D_{41}(q) & 0 & 0 & m_5 \end{Vmatrix}$$

contains the following block elements:

$$D_{11}(q) = I_{1_{zz}} + I_{2_{yy}} + I_{3_{yy}} + m_2 d_2^2 + m_3 (q_3 + d_3)^2 +$$
$$I_{4_{xx}} \sin^2 q_2 + m_4 (q_3 + l_3)^2 + \left(m_4 d_4^2 + I_{4_{yy}}\right) \cos^2 q_2 +$$
$$m_5 (q_3 + l_3)^2 + I_{5_{xx}} \sin^2 q_2 + \left(m_5 (l_4 + q_4)^2 + I_{5_{yy}}\right) \cos^2 q_2,$$

$$D_{21}(q) = [m_4 d_4 + m_5 (l_4 + q_4)](q_3 + l_3) \sin q_2,$$
$$D_{31}(q) = -(m_4 d_4 + m_5 (l_4 + q_4)) \cos q_2,$$
$$D_{41}(q) = m_5 (q_3 + l_3) \cos q_2,$$
$$D_{12}(q) = (m_4 d_4 + m_5 (l_4 + q_4))(q_3 + l_3) \sin q_2,$$
$$D_{22}(q) = I_{3_{zz}} + m_4 d_4^2 + I_{4_{zz}} + I_{5_{zz}} + m_5 (l_4 + q_4)^2,$$
$$D_{13}(q) = -(m_4 d_4 + m_5 (l_4 + q_4)) \cos q_2,$$

$$D_{14}(q) = m_5(q_3 + l_3)\cos q_2,$$

and the matrix

$$C(q,\dot{q}) = \begin{Vmatrix} C_{11}(q,\dot{q}) & C_{12}(q,\dot{q}) & 0 & 0 \\ C_{21}(q,\dot{q}) & C_{22}(q,\dot{q}) & 0 & C_{24}(q,\dot{q}) \\ C_{31}(q,\dot{q}) & 0 & f_{fric-3} & -2m_5\cos q_2\dot{q}_1 \\ C_{41}(q,\dot{q}) & C_{42}(q,\dot{q}) & 2m_5\cos q_2\dot{q}_1 & f_{fric-4} \end{Vmatrix}$$

has the elements

$$C_{11}(q,\dot{q}) = f_{fric-1} + 2\left[\left(-m_5(l_4+q_4)^2 - I_{5_{yy}}\right)\sin q_2 \cos q_2\right]\dot{q}_2 +$$
$$2\left[\left(I_{4_{xx}} - m_4 d_4^2 - I_{4_{yy}} + I_{5_{xx}}\right)\sin q_2 \cos q_2\right]\dot{q}_2 +$$
$$2[m_4(q_3+l_3) + m_3(q_3+d_3)]\dot{q}_3 +$$
$$2m_5(q_3+l_3)\dot{q}_3 + 2m_5(l_4+q_4)\dot{q}_4\cos^2 q_2,$$
$$C_{21}(q,\dot{q}) = m_5(q_3+l_3)\dot{q}_4 \sin q_2 -$$
$$\left[\left(I_{4_{xx}} - m_4 d_4^2 - I_{4_{yy}} + I_{5_{xx}}\right)\sin q_2 \cos q_2\right]\dot{q}_1 -$$
$$\left[\left(-m_5(l_4+q_4)^2 - I_{5_{yy}}\right)\sin q_2 \cos q_2\right]\dot{q}_1,$$
$$C_{31}(q,\dot{q}) = -2[m_3(q_3+d_3) + m_4(q_3+l_3)]\dot{q}_1 - 2m_5(q_3+l_3)\dot{q}_1,$$
$$C_{41}(q,\dot{q}) = -2m_5(q_3+l_3)\sin q_2\dot{q}_2 - m_5(l_4+q_4)\dot{q}_1\cos^2 q_2,$$
$$C_{12}(q,\dot{q}) = +2[(m_4 d_4 + m_5(l_4+q_4))\sin q_2]\dot{q}_3,$$
$$+ [m_4 d_4 + m_5(l_4+q_4)](q_3+l_3)\cos q_2\dot{q}_2,$$
$$C_{22}(q,\dot{q}) = f_{fric-2} + 2m_5(l_4+q_4)\dot{q}_4,$$
$$C_{42}(q,\dot{q}) = -m_5(l_4+q_4)\dot{q}_2,$$
$$C_{24}(q,\dot{q}) = m_5(q_3+l_3)\dot{q}_1 \sin q_2,$$

with

$$g(q) = \begin{Vmatrix} 0 \\ m_4 g d_4 \cos q_2 + m_5 g(q_4+l_4)\cos q_2 \\ 0 \\ m_5 g \sin q_2 \end{Vmatrix}, \quad \tau = \begin{Vmatrix} \tau_1 \\ \tau_2 \\ F_3 \\ F_4 \end{Vmatrix}.$$

12.10 Unimate 2000 manipulator

Let us consider here the robot manipulator Unimate 2000, represented in Fig. 12.10.

Collection of electromechanical models

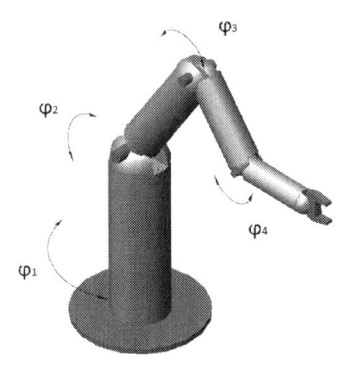

Figure 12.10 Robot manipulator Unimate 2000.

Generalized coordinates

The generalized coordinates of this robot are as follows:

$$q_1 := \varphi_1, \quad q_2 := \varphi_2, \quad q_3 := \varphi_3, \quad q_4 := \varphi_4.$$

Kinetic energy

The kinetic energy $T = \sum_{i=1}^{4} T_{m_i}$ contains the elements T_{m_i}, which are calculated as in the previous sections:

$$T_{m_i} = T_{m_i,0} + T_{m_i,rot-0} + 2m_i \left(\mathbf{v}_{m_i-c.i.-0}, \mathbf{v}_0\right),$$

$$T_{m_i,0} = \frac{1}{2} m_i \|\mathbf{v}_0\|^2, \quad T_{m_i,rot-0} = \frac{1}{2} \left(\boldsymbol{\omega}, I_{i,0}\boldsymbol{\omega}\right),$$

where

$$T_{m_1} = T_{m_1,0} + T_{m_1,rot-0} + m_1 \left(\mathbf{v}_{m_1-c.i.-0}, \mathbf{v}_0\right) =$$

$$T_{m_1,rot-0} = \frac{1}{2} \dot{\varphi}_1^2 \left(I_{1_{xx}}\right) = \frac{1}{2} \left(I_{1_{xx}}\right) \dot{q}_1^2,$$

$$T_{m_2} = T_{m_2,0} + T_{m_2,rot-0} + m_2 \left(\mathbf{v}_{m_2-c.i.-0}, \mathbf{v}_0\right) = T_{m_2,rot-0} =$$

$$\frac{1}{2} \begin{pmatrix} \dot{\varphi}_1 \sin \varphi_2 \\ \dot{\varphi}_1 \cos \varphi_2 \\ \dot{\varphi}_2 \end{pmatrix}^T \begin{Vmatrix} I_{2xx} & 0 & 0 \\ 0 & m_2 d_2^2 + I_{2yy} & 0 \\ 0 & 0 & m_2 d_2^2 + I_{2zz} \end{Vmatrix} \begin{pmatrix} \dot{\varphi}_1 \sin \varphi_2 \\ \dot{\varphi}_1 \cos \varphi_2 \\ \dot{\varphi}_2 \end{pmatrix}$$

$$= \frac{1}{2} \left[I_{2xx} \dot{q}_1^2 \sin^2 q_2 + \left(m_2 d_2^2 + I_{2yy}\right) \dot{q}_1^2 \cos^2 q_2 + \left(m_2 d_2^2 + I_{2zz}\right) \dot{q}_2^2 \right],$$

$$T_{m_3} = T_{m_3,0} + T_{m_3,rot-0} + m_2 \left(\mathbf{v}_{m_3-c.i.-0}, \mathbf{v}_0\right) =$$

$$\frac{1}{2} m_3 \left[l_2^2 \left(\cos^2 \varphi_2\right) \dot{\varphi}_1^2 + l_2^2 \dot{\varphi}_2^2 \right] + \frac{1}{2} \begin{pmatrix} \dot{\varphi}_1 \sin (\varphi_2 + \varphi_3) \\ \dot{\varphi}_1 \cos (\varphi_2 + \varphi_3) \\ \dot{\varphi}_2 + \dot{\varphi}_3 \end{pmatrix}^T \times$$

$$\left\| \begin{array}{ccc} I_{3xx} & 0 & 0 \\ 0 & m_3 d_3^2 + I_{3yy} & 0 \\ 0 & 0 & m_3 d_3^2 + I_{3zz} \end{array} \right\| \begin{pmatrix} \dot{\varphi}_1 \sin(\varphi_2 + \varphi_3) \\ \dot{\varphi}_1 \cos(\varphi_2 + \varphi_3) \\ \dot{\varphi}_2 + \dot{\varphi}_3 \end{pmatrix}$$

$$+ m_3 \begin{pmatrix} \dot{\varphi}_1 d_3 \sin \varphi_1 \cos(\varphi_2 + \varphi_3) + (\dot{\varphi}_2 + \dot{\varphi}_3) d_3 \cos \varphi_1 \sin(\varphi_2 + \varphi_3) \\ -(\dot{\varphi}_2 + \dot{\varphi}_3) d_3 \cos(\varphi_2 + \varphi_3) \\ \dot{\varphi}_1 d_3 \cos \varphi_1 \cos(\varphi_2 + \varphi_3) - (\dot{\varphi}_2 + \dot{\varphi}_3) d_3 \sin \varphi_1 \sin(\varphi_2 + \varphi_3) \end{pmatrix}^{\mathsf{T}} \times$$

$$\begin{pmatrix} -\dot{\varphi}_2 l_2 \cos \varphi_1 \sin \varphi_2 - \dot{\varphi}_1 l_2 \sin \varphi_1 \cos \varphi_2 \\ \dot{\varphi}_2 l_2 \cos \varphi_2 \\ -\dot{\varphi}_1 l_2 \cos \varphi_1 \cos \varphi_2 + \dot{\varphi}_2 l_2 \sin \varphi_1 \sin \varphi_2 \end{pmatrix} = \frac{1}{2} m_3 \left[l_2^2 \cos^2 q_2 \dot{q}_1^2 + l_2^2 \dot{q}_2^2 \right] +$$

$$\frac{1}{2} \left[I_{3xx} \sin^2(q_2 + q_3) + \left(m_3 d_3^2 + I_{3yy} \right) \cos^2(q_2 + q_3) \right] \dot{q}_1^2 +$$

$$\frac{1}{2} \left(m_3 d_3^2 + I_{3zz} \right) (\dot{q}_2 + \dot{q}_3)^2 -$$

$$m_3 \left[d_3 l_2 \cos q_2 \dot{q}_1^2 \cos(q_2 + q_3) + d_3 l_2 (\dot{q}_2 + \dot{q}_3) \dot{q}_2 \cos q_3 \right],$$

$$T_{m_4} = T_{m_4,0} + T_{m_4,rot-0} + m_4 \left(\mathbf{v}_{m_4-c.i.-0}, \mathbf{v}_0 \right) =$$

$$\frac{1}{2} m_4 \left[\dot{\varphi}_1^2 \left[l_3 \cos(\varphi_2 + \varphi_3) - l_2 \cos \varphi_2 \right]^2 + \dot{\varphi}_2^2 l_2^2 \right] +$$

$$\frac{1}{2} m_4 \left(\dot{\varphi}_2^2 + \dot{\varphi}_3^2 + 2 \dot{\varphi}_2 \dot{\varphi}_3 \right) l_3^2 \sin^2(\varphi_2 + \varphi_3) -$$

$$\frac{1}{2} m_4 \left[2 \dot{\varphi}_2 (\dot{\varphi}_2 + \dot{\varphi}_3) l_2 l_3 \sin \varphi_2 \sin(\varphi_2 + \varphi_3) \right] +$$

$$\frac{1}{2} \begin{pmatrix} \dot{\varphi}_1 \sin(\varphi_2 + \varphi_3 + \varphi_4) \\ \dot{\varphi}_1 \cos(\varphi_2 + \varphi_3 + \varphi_4) \\ \dot{\varphi}_2 + \dot{\varphi}_3 + \dot{\varphi}_4 \end{pmatrix}^{\mathsf{T}} \times$$

$$\left\| \begin{array}{ccc} I_{4xx} & 0 & 0 \\ 0 & m_4 d_4^2 + I_{4yy} & 0 \\ 0 & 0 & m_4 d_4^2 + I_{4zz} \end{array} \right\| \begin{pmatrix} \dot{\varphi}_1 \sin(\varphi_2 + \varphi_3 + \varphi_4) \\ \dot{\varphi}_1 \cos(\varphi_2 + \varphi_3 + \varphi_4) \\ \dot{\varphi}_2 + \dot{\varphi}_3 + \dot{\varphi}_4 \end{pmatrix} +$$

$$m_4 \begin{pmatrix} -\dot{\varphi}_1 d_4 \sin \varphi_1 \cos(\varphi_2 + \varphi_3 + \varphi_4) - \\ (\dot{\varphi}_2 + \dot{\varphi}_3 + \dot{\varphi}_4) d_4 \cos \varphi_1 \sin(\varphi_2 + \varphi_3 + \varphi_4) \\ (\dot{\varphi}_2 + \dot{\varphi}_3 + \dot{\varphi}_4) d_4 \cos(\varphi_2 + \varphi_3 + \varphi_4) \\ -\dot{\varphi}_1 d_4 \cos \varphi_1 \cos(\varphi_2 + \varphi_3 + \varphi_4) + \\ (\dot{\varphi}_2 + \dot{\varphi}_3 + \dot{\varphi}_4) d_4 \sin \varphi_1 \sin(\varphi_2 + \varphi_3 + \varphi_4) \end{pmatrix}^{\mathsf{T}}$$

$$\begin{pmatrix} -\dot{\varphi}_2 l_2 \cos \varphi_1 \sin \varphi_2 + (\dot{\varphi}_2 + \dot{\varphi}_3) l_3 \cos \varphi_1 \sin(\varphi_2 + \varphi_3) \\ +\dot{\varphi}_1 \sin \varphi_1 (l_3 \cos(\varphi_2 + \varphi_3) - l_2 \cos \varphi_2) \\ \dot{\varphi}_2 l_2 \cos \varphi_2 - (\dot{\varphi}_2 + \dot{\varphi}_3) l_3 \cos(\varphi_2 + \varphi_3) \\ \dot{\varphi}_2 l_2 \sin \varphi_1 \sin \varphi_2 - (\dot{\varphi}_2 + \dot{\varphi}_3) l_3 \sin \varphi_1 \sin(\varphi_2 + \varphi_3) \\ +\dot{\varphi}_1 \cos \varphi_1 (l_3 \cos(\varphi_2 + \varphi_3) - l_2 \cos \varphi_2) \end{pmatrix}$$

$$= \frac{1}{2} m_4 \left[\left[l_3 \cos(q_2 + q_3) - l_2 \cos q_2 \right]^2 \dot{q}_1^2 \right] + \frac{1}{2} m_4 l_2^2 \dot{q}_2^2$$

$$\frac{1}{2}\left[I_{4xx}\sin^2(q_2+q_3+q_4)+\left(m_4 d_4^2+I_{4yy}\right)\cos^2(q_2+q_3+q_4)\right]\dot{q}_1^2-$$
$$m_4[d_4\cos(q_2+q_3+q_4)][l_3\cos(q_2+q_3)-l_2\cos q_2]\dot{q}_1^2+$$
$$\frac{1}{2}m_4\left[l_3^2\sin^2(q_2+q_3)\right](\dot{q}_2+\dot{q}_3)^2-$$
$$m_4[l_2 l_3\sin q_2\sin(q_2+q_3)](\dot{q}_2+\dot{q}_3)\dot{q}_2+$$
$$\frac{1}{2}\left[m_4 d_4^2+I_{4zz}\right](\dot{q}_2+\dot{q}_3+\dot{q}_4)^2+$$
$$m_4[d_4 l_2\cos(q_3+q_4)](\dot{q}_2+\dot{q}_3+\dot{q}_4)\dot{q}_2-$$
$$m_4[d_4 l_3\cos(q_4)](\dot{q}_2+\dot{q}_3+\dot{q}_4)(\dot{q}_2+\dot{q}_3).$$

Potential energy

The potential energy V may be calculated as $V=\sum_{i=1}^{4}V_{m_i}$, with

$V_{m_1}=\text{const},$
$V_{m_2}=m_2 g\,(d_2\sin\varphi_2+\text{const})=m_2 g d_2\,(\sin q_2+\text{const}),$
$V_{m_3}=m_3 g\,(-d_3\sin(\varphi_2+\varphi_3)+l_2\sin\varphi_2+\text{const}),$
$\quad\quad=m_3 g\,(-d_3\sin(q_2+q_3)+\sin q_2+\text{const}),$
$V_{m_4}=m_4 g\,(-l_3\sin(\varphi_2+\varphi_3)+l_2\sin\varphi_2+d_4\sin(\varphi_2+\varphi_3+\varphi_4)+\text{const}),$
$\quad\quad=m_4 g\,(-l_3\sin(q_2+q_3)+l_2\sin q_2+d_4\sin(q_2+q_3+q_4)+\text{const}),$

which gives

$$V=m_2 g d_2\sin q_2+m_3 g\,(-d_3\sin(q_2+q_3)+l_2\sin q_2)+$$
$$m_4 g\,[-l_3\sin(q_2+q_3)+l_2\sin q_2+d_4\sin(q_2+q_3+q_4)]+\text{const}.$$

Non-potential forces

The generalized non-potential forces are as follows:

$$Q_{non\text{-}pot,i}=\tau_i-f_{fric-i}\dot{\varphi}_i=\tau_i-f_{fric-i}\dot{q}_i,$$
τ_i is a torsion force, $i=1,...,4$.

Lagrange equations

Based on the obtained T and V, we are able to derive the dynamic Lagrange equations:

$$\frac{d}{dt}\frac{\partial}{\partial\dot{q}_i}L-\frac{\partial}{\partial q_i}L=Q_{non\text{-}pot,i},\ i=1,...,4,\ L=T-V,$$

which can be represented in the standard matrix format (12.5)

$$D(q)\ddot{q}+C(q,\dot{q})\dot{q}+g(q)=\tau,$$

where the following matrices participate:

$$D(q) = \begin{Vmatrix} D_{11}(q) & 0 & 0 & 0 \\ 0 & D_{22}(q) & D_{23}(q) & D_{24}(q) \\ 0 & D_{32}(q) & D_{33}(q) & D_{34}(q) \\ 0 & D_{42}(q) & D_{43}(q) & m_4 d_4^2 + I_{4zz} \end{Vmatrix},$$

$D_{11}(q) = I_{1xx} + I_{2xx} \sin^2 q_2 + \left(m_2 d_2^2 + I_{2yy}\right) \cos^2 q_2 +$
$m_3 l_2^2 \cos^2 q_2 + I_{3xx} \sin^2 (q_2 + q_3) + \left(m_3 d_3^2 + I_{3yy}\right) \cos^2 (q_2 + q_3) -$
$2m_3 d_3 l_2 \cos q_2 \cos (q_2 + q_3) + m_4 \left[l_3 \cos (q_2 + q_3) - l_2 \cos q_2\right]^2 -$
$2m_4 d_4 l_3 \cos (q_2 + q_3) \cos (q_2 + q_3 + q_4) +$
$2m_4 \left[d_4 l_2 \cos q_2 \cos (q_2 + q_3 + q_4)\right] + I_{4xx} \sin^2 (q_2 + q_3 + q_4) +$
$\left(m_4 d_4^2 + I_{4yy}\right) \cos^2 (q_2 + q_3 + q_4),$

$D_{22}(q) = m_2 d_2^2 + I_{2zz} + [m_3 + m_4] l_2^2 + m_3 d_3^2 + I_{3zz} + m_4 d_4^2 + I_{4zz} +$
$m_4 l_3^2 \sin^2 (q_2 + q_3) - 2 l_2 [m_3 d_3 \cos q_3] - 2m_4 l_2 l_3 \sin q_2 \sin (q_2 + q_3)$
$+ 2m_4 d_4 l_2 \cos (q_3 + q_4) - 2m_4 d_4 l_3 \cos (q_4),$

$D_{32}(q) = m_3 d_3^2 + I_{3zz} + m_4 l_3^2 \sin^2 (q_2 + q_3) -$
$m_3 d_3 l_2 \cos q_3 - m_4 l_2 l_3 \sin q_2 \sin (q_2 + q_3) +$
$m_4 l_2 d_4 \cos (q_3 + q_4) + m_4 d_4^2 + I_{4zz} - 2m_4 d_4 l_3 \cos q_4,$

$D_{42}(q) = m_4 d_4^2 + I_{4zz} + m_4 d_4 l_2 \cos (q_3 + q_4) - m_4 d_4 l_3 \cos q_4,$

$D_{23}(q) = m_3 d_3^2 + I_{3zz} + m_4 l_3^2 \sin^2 (q_2 + q_3) - l_2 [m_3 d_3 \cos q_3] -$
$m_4 l_2 l_3 \sin q_2 \sin (q_2 + q_3) - 2m_4 d_4 l_3 \cos (q_4) +$
$\left[m_4 d_4^2 + I_{4zz}\right] + m_4 [d_4 l_2 \cos (q_3 + q_4)],$

$D_{33}(q) = m_3 d_3^2 + I_{3zz} + m_4 l_3^2 \sin^2 (q_2 + q_3) + \left[m_4 d_4^2 + I_{4zz}\right] -$
$2m_4 [d_4 l_3 \cos q_4],$

$D_{43}(q) = m_4 d_4^2 + I_{4zz} - m_4 d_4 l_3 \cos q_4,$
$D_{24}(q) = m_4 d_4^2 + I_{4zz} + m_4 d_4 l_2 \cos (q_3 + q_4) - m_4 d_4 l_3 \cos (q_4),$
$D_{34}(q) = m_4 d_4^2 + I_{4zz} - m_4 d_4 l_3 \cos q_4,$

and

$$C(q, \dot{q}) = \begin{Vmatrix} C_{11}(q, \dot{q}) & 0 & 0 & 0 \\ C_{21}(q, \dot{q}) & C_{22}(q, \dot{q}) & C_{23}(q, \dot{q}) & C_{24}(q, \dot{q}) \\ C_{31}(q, \dot{q}) & C_{32}(q, \dot{q}) & C_{33}(q, \dot{q}) & C_{34}(q, \dot{q}) \\ C_{41}(q, \dot{q}) & C_{42}(q, \dot{q}) & C_{43}(q, \dot{q}) & C_{44}(q, \dot{q}) \end{Vmatrix},$$

Collection of electromechanical models

$$C_{11}(q, \dot{q}) = f_{fric-1} + 2\left[\left(I_{2xx} - m_2 d_2^2 - I_{2yy} - m_3 l_2^2\right) \sin q_2 \cos q_2\right] \dot{q}_2 +$$
$$2\left[\left(I_{3xx} - m_3 d_3^2 - I_{3yy}\right) \sin(q_2 + q_3) \cos(q_2 + q_3)\right] (\dot{q}_2 + \dot{q}_3) +$$
$$2 m_3 d_3 l_2 \cos q_2 \sin(q_2 + q_3) \dot{q}_3 + [2 m_3 d_3 l_2 \sin(2q_2 + q_3)] \dot{q}_2 +$$
$$2 m_4 [l_3 \cos(q_2 + q_3) - l_2 \cos q_2] \times$$
$$[-l_3 (\dot{q}_2 + \dot{q}_3) \sin(q_2 + q_3) + l_2 \sin q_2 \dot{q}_2] + \left[2\left(I_{4xx} - m_4 d_4^2 - I_{4yy}\right)\right] \times$$
$$[\sin(q_2 + q_3 + q_4) \cos(q_2 + q_3 + q_4)] (\dot{q}_2 + \dot{q}_3 + \dot{q}_4) +$$
$$2 m_4 d_4 \sin(2q_2 + 2q_3 + q_4) [(l_3 - l_2) \dot{q}_2 + l_3 \dot{q}_3] -$$
$$2 m_4 d_4 \sin(q_2 + q_3 + q_4) [l_2 (\dot{q}_3 + \dot{q}_4) \cos q_2 + \cos(q_2 + q_3) \dot{q}_4],$$

$$C_{21}(q, \dot{q}) = \left[\left(-I_{2xx} + m_2 d_2^2 + I_{2yy}\right) \sin q_2 \cos q_2\right] \dot{q}_1 +$$
$$\left[m_3 l_2^2 \sin q_2 \cos q_2\right] \dot{q}_1 +$$
$$\left[I_{3yy} - I_{3xx} + m_3 d_3^2\right] [\sin(q_2 + q_3) \cos(q_2 + q_3)] \dot{q}_1 -$$
$$\left[m_3 d_3 l_2 \sin(2q_2 + q_3) - m_4 l_2^2 \cos q_2 \sin q_2\right] \dot{q}_1 -$$
$$m_4 [l_3 l_2 \sin(2q_2 + q_3)] \dot{q}_1 + m_4 \left[l_3^2 \cos(q_2 + q_3) \sin(q_2 + q_3)\right] \dot{q}_1 -$$
$$[\sin(q_2 + q_3 + q_4) \cos(q_2 + q_3 + q_4)] \left[\left(I_{4xx} - m_4 d_4^2 - I_{4yy}\right)\right] \dot{q}_1$$
$$- m_4 d_4 (l_3 + l_2) [\sin(2q_2 + 2q_3 + q_4)] \dot{q}_1,$$

$$C_{31}(q, \dot{q}) = -\left[I_{3xx} - m_3 d_3^2 - I_{3yy}\right] [\sin(q_2 + q_3) \cos(q_2 + q_3)] \dot{q}_1 -$$
$$m_3 d_3 l_2 [\cos q_2 \sin(q_2 + q_3)] \dot{q}_1 +$$
$$m_4 l_3 \sin(q_2 + q_3) [l_3 \cos(q_2 + q_3) - l_2 \cos q_2] \dot{q}_1 -$$
$$\left[I_{4xx} - m_4 d_4^2 - I_{4yy}\right] [\sin(q_2 + q_3 + q_4) \cos(q_2 + q_3 + q_4)] \dot{q}_1 -$$
$$m_4 d_4 \sin(q_2 + q_3 + q_4) [l_3 \cos(q_2 + q_3) - l_2 \cos q_2] \dot{q}_1 -$$
$$m_4 d_4 l_3 \cos(q_2 + q_3 + q_4) \sin(q_2 + q_3) \dot{q}_1,$$

$$C_{41}(q, \dot{q}) = -\left(I_{4xx} - m_4 d_4^2 - I_{4yy}\right) [\sin(q_2 + q_3 + q_4) \cos(q_2 + q_3 + q_4)] \dot{q}_1$$
$$- m_4 d_4 \sin(q_2 + q_3 + q_4) [l_3 \cos(q_2 + q_3) - l_2 \cos q_2] \dot{q}_1,$$

$$C_{22}(q, \dot{q}) = f_{fric-2} + m_4 l_3^2 [\sin(q_2 + q_3) \cos(q_2 + q_3)] (\dot{q}_2 + 2\dot{q}_3)$$
$$- m_4 l_2 l_3 \sin(2q_2 + q_3) (\dot{q}_2),$$

$$C_{32}(q, \dot{q}) = -[m_3 d_3 l_2 \sin q_3] (\dot{q}_2 + \dot{q}_3) + m_4 l_3^2 [\sin(q_2 + q_3) \cos(q_2 + q_3)] \dot{q}_2$$
$$- m_4 l_2 l_3 ([\sin(2q_2 + q_3)] \dot{q}_2 + [\sin q_2 \cos(q_2 + q_3)] (\dot{q}_2 + \dot{q}_3))$$
$$+ [m_4 d_4 l_2 \sin(q_3 + q_4)] \dot{q}_2,$$

$$C_{42}(q, \dot{q}) = m_4 [d_4 l_2 \sin(q_3 + q_4)] \dot{q}_2 - m_4 [d_4 l_3 \sin q_4] (\dot{q}_2 + \dot{q}_3 + \dot{q}_4),$$

$$C_{23}(q, \dot{q}) = m_3 l_2 d_3 [\sin q_3] (2\dot{q}_2 + \dot{q}_3) - m_4 l_2 l_3 [\sin q_2 \cos(q_2 + q_3)] (2\dot{q}_2 + \dot{q}_3)$$

$$-m_4\left[d_4l_2\sin(q_3+q_4)\right](2\dot{q}_2+\dot{q}_3+\dot{q}_4)+$$
$$m_4l_3^2\left[\sin(q_2+q_3)\cos(q_2+q_3)\right]\dot{q}_3,$$
$$C_{33}(q,\dot{q})=f_{fric-3}+m_4l_3^2\left[\sin(q_2+q_3)\cos(q_2+q_3)\right](\dot{q}_3+2\dot{q}_2)$$
$$+m_3d_3l_2\left[\sin q_3\right]\dot{q}_2-m_4l_2l_3\left[\sin q_2\cos(q_2+q_3)\right]\dot{q}_2,$$
$$C_{43}(q,\dot{q})=-m_4d_4l_3\left[\sin q_4\right](\dot{q}_2+\dot{q}_3+\dot{q}_4),$$
$$C_{24}(q,\dot{q})=m_4d_4l_3\sin q_4(2\dot{q}_2+2\dot{q}_3+\dot{q}_4)-$$
$$m_4\left[d_4l_2\sin(q_3+q_4)\right](2\dot{q}_2+\dot{q}_3+\dot{q}_4),$$
$$C_{34}(q,\dot{q})=m_4\left[d_4l_3\sin q_4\right](2\dot{q}_2+2\dot{q}_3+\dot{q}_4),$$
$$C_{44}(q,\dot{q})=f_{fric-4}+m_4\left[d_4l_3\sin q_4\right](\dot{q}_2+\dot{q}_3).$$

The vectors g and τ are as follows:

$$g(q)=\begin{Vmatrix}g_1(q)\\g_2(q)\\g_3(q)\\g_4(q)\end{Vmatrix},\ \tau=\begin{Vmatrix}\tau_1\\\tau_2\\\tau_3\\\tau_4\end{Vmatrix},$$

$$g_1(q)=0,$$
$$g_2(q)=-m_2gd_2\cos q_2+m_3g(d_3\cos(q_2+q_3)-l_2\cos q_2)+$$
$$m_4g\left[l_3\cos(q_2+q_3)-l_2\cos q_2\right]-m_4g\left[d_4\cos(q_2+q_3+q_4)\right],$$
$$g_3(q)=m_3gd_3\cos(q_2+q_3)+m_4g\left[l_3\cos(q_2+q_3)-d_4\cos(q_2+q_3+q_4)\right],$$
$$g_4(q)=-m_4gd_4\cos(q_2+q_3+q_4).$$

12.11 Robot manipulator with swivel base

Consider now the robot manipulator with swivel base of three degrees of freedom represented in Fig. 12.11.

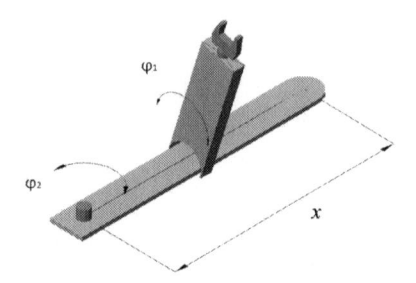

Figure 12.11 Robot manipulator with swivel base.

Generalized coordinates

The generalized coordinates of the considered mechanical system are

$$q_1 := \varphi_1, \quad q_2 := x, \quad q_3 := \varphi_2.$$

Kinetic energy

The kinetic energy $T = \sum_{i=1}^{2} T_{m_i}$ has two elements, T_{m_i} ($i = 1, 2$), which can be calculated as in the examples before:

$$T_{m_i} = T_{m_i,0} + T_{m_i,rot-0} + 2m_i \left(\mathbf{v}_{m_i-c.i.-0}, \mathbf{v}_0\right),$$

$$T_{m_i,0} = \frac{1}{2} m_i \|\mathbf{v}_0\|^2, \quad T_{m_i,rot-0} = \frac{1}{2} (\boldsymbol{\omega}, I_{i,0}\boldsymbol{\omega}).$$

In our case we have

$$T_{m_1} = T_{m_1,0} + T_{m_1,rot-0} + m_1 \left(\mathbf{v}_{m_2-c.i.-0}, \mathbf{v}_0\right) = \frac{1}{2} I_{1_{zz}} \dot{\varphi}_1^2 = \frac{1}{2} I_{1_{zz}} \dot{q}_1^2,$$

$$T_{m_2} = T_{m_2,0} + T_{m_2,rot-0} + m_2 \left(\mathbf{v}_{m_2-c.i.-0}, \mathbf{v}_0\right) = \frac{1}{2} m_2 \left(x^2 \dot{\varphi}_1^2 + \dot{x}^2\right) +$$

$$\frac{1}{2} \begin{pmatrix} \dot{\varphi}_1 \sin\varphi_2 \\ \dot{\varphi}_1 \cos\varphi_2 \\ \dot{\varphi}_2 \end{pmatrix}^T \begin{Vmatrix} I_{2_{xx}} & 0 & 0 \\ 0 & m_2 d_2^2 + I_{2_{yy}} & 0 \\ 0 & 0 & m_2 d_2^2 + I_{2_{zz}} \end{Vmatrix} \begin{pmatrix} \dot{\varphi}_1 \sin\varphi_2 \\ \dot{\varphi}_1 \cos\varphi_2 \\ \dot{\varphi}_2 \end{pmatrix} +$$

$$m_2 \begin{pmatrix} \dot{\varphi}_1 d_2 \cos\varphi_1 \cos\varphi_2 - \dot{\varphi}_2 d_2 \sin\varphi_1 \sin\varphi_2 \\ \dot{\varphi}_2 d_2 \cos\varphi_2 \\ -\dot{\varphi}_1 d_2 \sin\varphi_1 \cos\varphi_2 - \dot{\varphi}_2 d_2 \cos\varphi_1 \sin\varphi_2 \end{pmatrix}^T \begin{pmatrix} -\dot{\varphi}_1 x \sin\varphi_1 + \dot{x} \cos\varphi_1 \\ 0 \\ -\dot{\varphi}_1 x \cos\varphi_1 - \dot{x} \sin\varphi_1 \end{pmatrix}$$

$$= \frac{1}{2} \left[I_{2_{xx}} \dot{q}_1^2 \sin^2 q_3 + \left(m_2 d_2^2 + I_{2_{yy}}\right) \dot{q}_1^2 \cos^2 q_3 + \left(m_2 d_2^2 + I_{2_{zz}}\right) \dot{q}_3^2 \right] +$$

$$\frac{1}{2} m_2 \left(q_2^2 \dot{q}_1^2 + \dot{q}_2^2\right) + m_2 \left[\dot{q}_1 \dot{q}_2 d_2 \cos q_3 + \dot{q}_1 \dot{q}_3 q_2 d_2 \sin q_3\right].$$

Potential energy

The potential energy $V = \sum_{i=1}^{2} V_{m_i}$ contains

$$V_{m_1} = \text{const}, \quad V_{m_2} = mgd_2 \sin\varphi_2 = mgd_2 \sin q_3,$$

which gives

$$V = mgd_2 \sin q_3 + \text{const}.$$

Non-potential forces

The generalized forces are given by the following formulas:

$$Q_{non-pot,1} = \tau_1 - f_{fric-1} \dot{\varphi}_1 = \tau_1 - f_{fric-1} \dot{q}_1,$$

τ_1 is a torsion force,

$$Q_{non\text{-}pot,2} = F_2 - f_{fric-2}\dot{y} = F_2 - f_{fric-2}\dot{q}_2,$$

F_2 is a force of horizontal movement,

$$Q_{non\text{-}pot,3} = \tau_3 - f_{fric-3}\dot{\varphi}_2 = \tau_3 - f_{fric-3}\dot{q}_3,$$

τ_3 is a torsion force.

Lagrange equations

Using the obtained expressions for T and V, we are able to derive the dynamic Lagrange equations

$$\frac{d}{dt}\frac{\partial}{\partial \dot{q}_i}L - \frac{\partial}{\partial q_i}L = Q_{non\text{-}pot,i}, \quad i = 1, 2, 3, \quad L = T - V,$$

which in the standard matrix format (12.5) gives

$$D(q)\ddot{q} + C(q,\dot{q})\dot{q} + g(q) = \tau,$$

where

$$D(q) = \begin{Vmatrix} D_{11}(q) & [m_2 d_2 \cos q_3] & [m_2 q_2 d_2 \sin q_3] \\ [m_2 d_2 \cos q_3] & m_2 & 0 \\ [m_2 q_2 d_2 \sin q_3] & 0 & m_2 d_2^2 + I_{2zz} \end{Vmatrix},$$

with $D_{11}(q)$ equal to

$$D_{11}(q) = \left[I_{1zz} + \left(m_2 d_2^2 + I_{2yy}\right)\cos^2 q_3\right] + I_{2xx}\sin^2 q_3 + m_2 q_2^2.$$

The matrix $C(q,\dot{q})$ is as follows:

$$C(q,\dot{q}) = \begin{Vmatrix} C_{11}(q,\dot{q}) & 0 & m_2 q_2 d_2 \cos q_3 \dot{q}_3 \\ 0 & f_{fric-2} & -m_2 d_2 \sin q_3 \dot{q}_1 \\ C_{31}(q,\dot{q}) & 2m_2 d_2 \sin q_3 \dot{q}_1 & f_{fric-3} \end{Vmatrix},$$

where

$$C_{11}(q,\dot{q}) = f_{fric-1} - 2\left[m_2 d_2^2 \sin q_3 \cos q_3\right]\dot{q}_3 +$$
$$2\left[(I_{2xx} - I_{2yy})\sin q_3 \cos q_3\right]\dot{q}_3 + 2m_2 q_2 \dot{q}_2,$$
$$C_{21}(q,\dot{q}) = -m_2 d_2 \sin q_3 \dot{q}_3 - m_2 q_2 \dot{q}_1,$$
$$C_{31}(q,\dot{q}) = -\left[(I_{2xx} - I_{2yy})\sin q_3 \cos q_3\right]\dot{q}_1 + \left[m_2 d_2^2 \sin q_3 \cos q_3\right]\dot{q}_1,$$

and

$$g(q) = \begin{Vmatrix} 0 \\ 0 \\ -mg d_2 \cos q_3 \end{Vmatrix}, \quad \tau = \begin{Vmatrix} \tau_1 \\ F_2 \\ \tau_3 \end{Vmatrix}.$$

12.12 Cylindrical robot with spring

Consider the cylindrical robot with spring represented in Fig. 12.12.

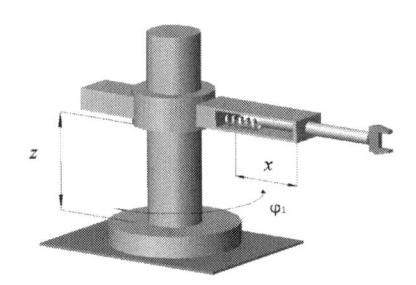

Figure 12.12 Cylindrical robot with spring.

Generalized coordinates
Select the generalized coordinates as

$$q_1 := \varphi, \quad q_2 := z, \quad q_3 := x.$$

Kinetic energy
The kinetic energy $T = \sum_{i=1}^{5} T_{m_i}$ consists of the elements T_{m_i}, which can be calculated using standard formulas:

$$T_{m_i} = T_{m_i,0} + T_{m_i,rot-0} + 2m_i \left(\mathbf{v}_{m_i-c.i.-0}, \mathbf{v}_0\right),$$

$$T_{m_i,0} = \frac{1}{2} m_i \|\mathbf{v}_0\|^2, \quad T_{m_i,rot-0} = \frac{1}{2} \left(\boldsymbol{\omega}, I_{i,0}\boldsymbol{\omega}\right).$$

For this system we have

$$T_{m_1} = T_{m_1,0} + T_{m_1,rot-0} + m_1 \left(\mathbf{v}_{m_2-c.i.-0}, \mathbf{v}_0\right)$$
$$= T_{m_1,rot-0} = \frac{1}{2} I_{1_{yy}} \dot{\varphi}^2 = \frac{1}{2} I_{1_{yy}} \dot{q}_1^2,$$

$$T_{m_2} = T_{m_2,0} + T_{m_2,rot-0} + m_2 \left(\mathbf{v}_{m_2-c.i.-0}, \mathbf{v}_0\right)$$
$$= T_{m_2,rot-0} = \frac{1}{2} I_{2_{yy}} \dot{\varphi}^2 = \frac{1}{2} I_{2_{yy}} \dot{q}_1^2,$$

$$T_{m_3} = T_{m_3,0} + T_{m_3,rot-0} + m_3 \left(\mathbf{v}_{m_3-c.i.-0}, \mathbf{v}_0\right)$$
$$= T_{m_3,0} + T_{m_3,rot-0} = \frac{1}{2} m_3 \dot{z}^2 +$$

$$\frac{1}{2} \begin{pmatrix} 0 \\ \dot{\varphi}_1 \\ 0 \end{pmatrix}^T \begin{Vmatrix} I_{3_{xx}} & 0 & 0 \\ 0 & I_{3_{yy}} & 0 \\ 0 & 0 & I_{3_{zz}} \end{Vmatrix} \begin{pmatrix} 0 \\ \dot{\varphi}_1 \\ 0 \end{pmatrix} = \frac{1}{2} \left[m_3 \dot{q}_3^2 + I_{3_{yy}} \dot{q}_1^2\right],$$

$$T_{m_4} = T_{m_4,0} + T_{m_4,rot-0} + m_4 \left(\mathbf{v}_{m_4-c.i.-0}, \mathbf{v}_0\right) = 0.$$

The mass of the spring m_4 is considered negligible. Now we have

$$T_{m_5} = T_{m_5,0} + T_{m_5,rot-0} + m_5\left(\mathbf{v}_{m_5-c.i.-0}, \mathbf{v}_0\right)$$

$$= \frac{1}{2}m_5\left(\dot{z}^2 + \dot{x}^2 + m_5x^2\dot{\varphi}^2\right) +$$

$$\frac{1}{2}\begin{pmatrix}0\\ \dot{\varphi}\\ 0\end{pmatrix}^T \begin{Vmatrix}I_{5xx} & 0 & 0\\ 0 & I_{5yy} & 0\\ 0 & 0 & I_{5zz}\end{Vmatrix}\begin{pmatrix}0\\ \dot{\varphi}\\ 0\end{pmatrix} =$$

$$\frac{1}{2}\left[m_5\left(q_2^2\dot{q}_1^2 + \dot{q}_2^2 + \dot{q}_3^2\right) + I_{5yy}\dot{q}_1^2\right].$$

Potential energy

The potential energy $V = \sum_{i=1}^{5} V_{m_i}$ has

$V_{m_1} = \text{const}, \quad V_{m_2} = \text{const}, \quad V_{m_3} = m_3 g z = m_3 g q_3,$

$V_{m_4} = \frac{1}{2}k(x - x_0)^2 = \frac{1}{2}k(q_2 - q_{2_0})^2,$

$V_{m_5} = m_5 g z = m_5 g q_3,$

which gives

$$V = m_3 g q_3 + \frac{1}{2}k(q_2 - q_{2_0})^2 + m_5 g q_3 + \text{const}.$$

Non-potential forces

The generalized non-potential forces are

$Q_{non-pot,1} = \tau_1 - f_{fric-1}\dot{\varphi} = \tau_1 - f_{fric-1}\dot{q}_1,$
τ_1 is a torsion force,
$Q_{non-pot,2} = F_2 - f_{fric-2}\dot{x} = F_2 - f_{fric-2}\dot{q}_2,$
F_2 is a force of horizontal movement,
$Q_{non-pot,3} = F_3 - f_{fric-3}\dot{y} = F_3 - f_{fric-3}\dot{q}_3,$
F_3 is a force of vertical movement.

Lagrange equations

Using the obtained expressions for T and V in the dynamic Lagrange equations

$$\frac{d}{dt}\frac{\partial}{\partial \dot{q}_i}L - \frac{\partial}{\partial q_i}L = Q_{non-pot,i}, \quad i = 1,2,3, \quad L = T - V,$$

we get the following representation in matrix format:

$$D(q)\ddot{q} + C(q,\dot{q})\dot{q} + g(q) = \tau,$$

where

$$D(q) = \begin{Vmatrix} \left[I_{1_{yy}} + I_{2_{yy}} + I_{3_{yy}} + I_{5_{yy}}\right] + m_5 q_2^2 & 0 & 0 \\ 0 & m_5 & 0 \\ 0 & 0 & m_3 + m_5 \end{Vmatrix},$$

$$C(q,\dot{q}) = \begin{Vmatrix} f_{fric-1} + 2m_5 q_2 \dot{q}_2 & 0 & 0 \\ -m_5 q_2 \dot{q}_1 & f_{fric-2} & 0 \\ 0 & 0 & f_{fric-3} \end{Vmatrix},$$

$$g(q) = \begin{Vmatrix} 0 \\ k(q_2 - q_{2_0}) \\ m_3 g + m_5 g \end{Vmatrix}, \quad \tau = \begin{Vmatrix} \tau_1 \\ F_2 \\ F_3 \end{Vmatrix}.$$

12.13 Non-ordinary manipulator with shock absorber

Consider the non-ordinary manipulator with a shock absorber which is presented in Fig. 12.13.

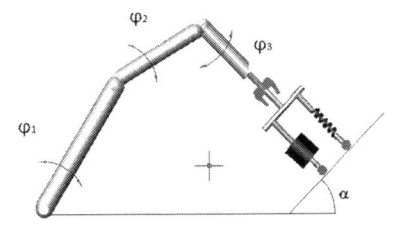

Figure 12.13 Non-ordinary manipulator with shock absorber.

Generalized coordinates

The generalized coordinates of the considered systems are

$$q_1 := x, \quad q_2 := y, \quad q_3 := \varphi_1, \quad q_4 := \varphi_2, \quad q_5 := \varphi_3.$$

Kinetic energy

The kinetic energy $T = \sum_{i=1}^{5} T_{m_i}$ has terms T_{m_i}, which can be calculated as

$$T_{m_i} = T_{m_i,0} + T_{m_i,rot-0} + 2m_i \left(\mathbf{v}_{m_i - c.i.-0}, \mathbf{v}_0\right),$$

$$T_{m_i,0} = \frac{1}{2} m_i \|\mathbf{v}_0\|^2, \quad T_{m_i,rot-0} = \frac{1}{2}\left(\boldsymbol{\omega}, I_{i,0} \boldsymbol{\omega}\right),$$

so that

$$T_{m_1} = T_{m_1,0} + T_{m_1,rot-0} + m_1 \left(\mathbf{v}_{m_1 - c.i.-0}, \mathbf{v}_0\right) =$$

$$T_{m_1,0} + T_{m_1,rot-0} = \frac{1}{2} m_1 \dot{x}^2 + \frac{1}{4} m_1 \dot{x}^2 = \frac{3}{4} m_1 \dot{q}_1^2,$$

$$T_{m_2} = T_{m_2,0} + T_{m_2,rot-0} + m_2 \left(\mathbf{v}_{m_1-c.i.-0}, \mathbf{v}_0\right) =$$
$$T_{m_2,0} + T_{m_2,rot-0} = \frac{1}{2} m_2 \dot{x}^2 + \frac{1}{2} m_2 \dot{x}^2 = \frac{3}{4} m_2 \dot{q}_1^2,$$
$$T_{m_3} = T_{m_3,0} + T_{m_3,rot-0} + m_3 \left(\mathbf{v}_{m_3-c.i.-0}, \mathbf{v}_0\right) =$$
$$T_{m_3,0} + T_{m_3,rot-0} = \frac{1}{2} m_3 \dot{x}^2 + \frac{1}{2} m_3 \dot{y}^2 + \frac{1}{2} I_{3yy} \dot{\varphi}_1^2 =$$
$$\frac{1}{2} m_3 \dot{q}_1^2 + \frac{1}{2} m_3 \dot{q}_2^2 + \frac{1}{2} I_{3yy} \dot{q}_3^2,$$
$$T_{m_4} = T_{m_4,0} + T_{m_4,rot-0} + m_4 \left(\mathbf{v}_{m_4-c.i.-0}, \mathbf{v}_0\right) = \frac{1}{2} m_4 \dot{x}^2 + \frac{1}{2} m_4 \dot{y}^2 +$$
$$\frac{1}{2} \begin{pmatrix} \dot{\varphi}_1 \cos\varphi_2 \\ \dot{\varphi}_1 \sin\varphi_2 \\ \dot{\varphi}_2 \end{pmatrix}^{\mathsf{T}} \begin{Vmatrix} I_{4xx} & 0 & 0 \\ 0 & m_4 d_4^2 + I_{4yy} & 0 \\ 0 & 0 & m_4 d_4^2 + I_{4zz} \end{Vmatrix} \begin{pmatrix} \dot{\varphi}_1 \cos\varphi_2 \\ \dot{\varphi}_1 \sin\varphi_2 \\ \dot{\varphi}_2 \end{pmatrix} +$$
$$m_4 \begin{pmatrix} \dot{\varphi}_1 d_4 \sin\varphi_1 \sin\varphi_2 + \dot{\varphi}_2 d_4 \cos\varphi_1 \cos\varphi_2 \\ -\dot{\varphi}_2 d_4 \sin\varphi_2 \\ \dot{\varphi}_1 d_4 \cos\varphi_1 \sin\varphi_2 - \dot{\varphi}_2 d_4 \sin\varphi_1 \cos\varphi_2 \end{pmatrix}^{\mathsf{T}} \begin{pmatrix} \dot{x} \\ \dot{y} \\ 0 \end{pmatrix}$$
$$= \frac{1}{2} m_4 \dot{q}_1^2 + \frac{1}{2} m_4 \dot{q}_2^2 + \frac{1}{2} \left[m_4 d_4^2 + I_{4zz}\right] \dot{q}_4^2 +$$
$$\frac{1}{2} \left[I_{4xx} \cos^2 q_4 + \left(m_4 d_4^2 + I_{4yy}\right) \sin^2 q_4\right] \dot{q}_3^2 +$$
$$m_4 \left[(d_4 \sin q_3 \sin q_4) \dot{q}_1 \dot{q}_3 + (d_4 \cos q_3 \cos q_4) \dot{q}_1 \dot{q}_4 - (d_4 \sin q_4) \dot{q}_2 \dot{q}_4\right],$$
$$T_{m_5} = T_{m_5,0} + T_{m_5,rot-0} + m_5 \left(\mathbf{v}_{m_5-c.i.-0}, \mathbf{v}_0\right) =$$
$$\frac{1}{2} m_5 \left\Vert \begin{pmatrix} \dot{\varphi}_1 l_4 \sin\varphi_1 \sin\varphi_2 + \dot{\varphi}_2 l_4 \cos\varphi_1 \cos\varphi_2 + \dot{x} \\ -\dot{\varphi}_2 l_4 \sin\varphi_2 + \dot{y} \\ \dot{\varphi}_1 l_4 \cos\varphi_1 \sin\varphi_2 - \dot{\varphi}_2 l_4 \sin\varphi_1 \cos\varphi_2 \end{pmatrix} \right\Vert^2 +$$
$$\frac{1}{2} \begin{pmatrix} \dot{\varphi}_1 \cos(\varphi_2 + \varphi_3) \\ -\dot{\varphi}_1 \sin(\varphi_2 + \varphi_3) \\ \dot{\varphi}_2 + \dot{\varphi}_3 \end{pmatrix}^{\mathsf{T}} \begin{Vmatrix} I_{5xx} & 0 & 0 \\ 0 & \begin{matrix} m_5 d_5^2 \\ +I_{5yy} \end{matrix} & 0 \\ 0 & 0 & \begin{matrix} m_5 d_5^2 \\ +I_{5zz} \end{matrix} \end{Vmatrix} \begin{pmatrix} \dot{\varphi}_1 \cos(\varphi_2 + \varphi_3) \\ -\dot{\varphi}_1 \sin(\varphi_2 + \varphi_3) \\ \dot{\varphi}_2 + \dot{\varphi}_3 \end{pmatrix} +$$
$$m_5 \begin{pmatrix} -\dot{\varphi}_1 d_5 \sin\varphi_1 \sin(\varphi_2 + \varphi_3) \\ -(\dot{\varphi}_2 + \dot{\varphi}_3) d_5 \cos\varphi_1 \cos(\varphi_2 + \varphi_3) \\ (\dot{\varphi}_2 + \dot{\varphi}_3) d_5 \sin(\varphi_2 + \varphi_3) \\ -\dot{\varphi}_1 d_5 \cos\varphi_1 \sin(\varphi_2 + \varphi_3) \\ +(\dot{\varphi}_2 + \dot{\varphi}_3) d_5 \sin\varphi_1 \cos(\varphi_2 + \varphi_3) \end{pmatrix}^{\mathsf{T}} \begin{pmatrix} \dot{\varphi}_1 l_4 \sin\varphi_1 \sin\varphi_2 \\ +\dot{\varphi}_2 l_4 \cos\varphi_1 \cos\varphi_2 + \dot{x} \\ -\dot{\varphi}_2 l_4 \sin\varphi_2 + \dot{y} \\ \dot{\varphi}_1 l_4 \cos\varphi_1 \sin\varphi_2 \\ -\dot{\varphi}_2 l_4 \sin\varphi_1 \cos\varphi_2 \end{pmatrix}$$
$$= \frac{1}{2} m_5 \left(\dot{q}_1^2 + \dot{q}_2^2\right) + \frac{1}{2} \left[m_5 l_4^2 \sin^2 q_4 + I_{5xx} \cos^2 (q_4 + q_5)\right] \dot{q}_3^2 +$$

$$\frac{1}{2}\left[\left(m_5 d_5^2 + I_{5yy}\right)\sin^2(q_4+q_5) - 2m_5 l_4 d_5 \sin q_4 \sin(q_4+q_5)\right]\dot{q}_3^2 +$$
$$\frac{1}{2}m_5 l_4^2 \dot{q}_4^2 + \frac{1}{2}\left[m_5 d_5^2 + I_{5zz}\right](\dot{q}_4+\dot{q}_5)^2 +$$
$$m_5\left[l_4 \sin q_3 \sin q_4 - d_5 \sin q_3 \sin(q_4+q_5)\right]\dot{q}_1 \dot{q}_3 + m_5 l_4 \left[\cos q_3 \cos q_4\right]\dot{q}_1 \dot{q}_4 -$$
$$m_5 l_4 \left[\sin q_4\right]\dot{q}_2 \dot{q}_4 - m_5 l_4 d_5 \left[\cos q_5\right](\dot{q}_4+\dot{q}_5)\dot{q}_4 -$$
$$m_5 d_5 \left[\cos q_3 \cos(q_4+q_5)\right](\dot{q}_4+\dot{q}_5)\dot{q}_1 + m_5 d_5 \left[\sin(q_4+q_5)\right](\dot{q}_4+\dot{q}_5)\dot{q}_2.$$

Potential energy

The potential energy $V = V_r + \sum_{i=1}^{5} V_{m_i}$ is as follows:

$$V_r = \frac{1}{2}k(y-y_0)^2 = \frac{1}{2}k(q_2-q_{20})^2,$$
$$V_{m_1} = m_1 g x \sin\alpha = m_1 g q_1 \sin\alpha, \quad V_{m_2} = m_2 g x \sin\alpha = m_2 g q_1 \sin\alpha,$$
$$V_{m_3} = m_3 g \left[x \sin\alpha + (d_3+y)\cos\alpha\right] = m_3 g \left[q_1 \sin\alpha + (d_3+q_2)\cos\alpha\right],$$
$$V_{m_4} = m_4 g \left[x \sin\alpha + (l_3+y)\cos\alpha - d_4 \cos\varphi_2 \cos\alpha - d_4 \sin\alpha \cos\varphi_1 \sin\varphi_2\right]$$
$$= m_4 g \left[q_1 \sin\alpha + (l_3+q_2)\cos\alpha - d_4 \cos q_4 \cos\alpha - d_4 \sin\alpha \cos q_3 \sin q_4\right],$$
$$V_{m_5} = m_5 g \left[x \sin\alpha + (l_3+y)\cos\alpha - l_4 \cos\varphi_2 \cos\alpha + l_4 \sin\varphi_2 \sin\alpha \cos\varphi_1\right]$$
$$+m_5 g \left[d_5 \cos(\varphi_3+\varphi_2)\cos\alpha + d_5 \sin(\varphi_3+\varphi_2)\sin\alpha \cos\varphi_1\right] =$$
$$m_5 g \left[q_1 \sin\alpha + (l_3+q_2)\cos\alpha - l_4 \cos q_4 \cos\alpha + l_4 \sin q_4 \sin\alpha \cos q_3\right]$$
$$+m_5 g \left[d_5 \cos(q_5+q_4)\cos\alpha + d_5 \sin(q_5+q_4)\sin\alpha \cos q_3\right],$$

which gives

$$V = \frac{1}{2}k(q_2-q_{20})^2 + [m_1 g + m_2 g + m_3 g + m_4 g + m_5 g]q_1 \sin\alpha +$$
$$m_3 g \left[(d_3+q_2)\cos\alpha\right] +$$
$$m_4 g \left[(l_3+q_2)\cos\alpha - d_4 \cos q_4 \cos\alpha - d_4 \sin\alpha \cos q_3 \sin q_4\right] +$$
$$m_5 g \left[(l_3+q_2)\cos\alpha - l_4 \cos q_4 \cos\alpha + l_4 \sin q_4 \sin\alpha \cos q_3\right] +$$
$$m_5 g \left[d_5 \cos(q_5+q_4)\cos\alpha + d_5 \sin(q_5+q_4)\sin\alpha \cos q_3\right].$$

Non-potential forces

The generalized non-potential forces are given by

$$Q_{non\text{-}pot,1} = F_1 - f_{fric-1}\dot{x} = F_1 - f_{fric-1}\dot{q}_1,$$

F_1 it is a force of longitudinal movement,

$$Q_{non\text{-}pot,2} = -c\dot{y} = -c\dot{q}_2,$$

c is the coefficient of viscous friction,

$$Q_{non\text{-}pot,3} = \tau_3 - f_{fric-3}\dot{\varphi}_3 = \tau_3 - f_{fric-3}\dot{q}_3,$$

τ_3 is a twisting moment,

$$Q_{non\text{-}pot,4} = \tau_4 - f_{fric\text{-}4}\dot{\varphi}_4 = \tau_4 - f_{fric\text{-}4}\dot{q}_4,$$

τ_4 is a twisting moment,

$$Q_{non\text{-}pot,5} = \tau_5 - f_{fric\text{-}5}\dot{\varphi}_5 = \tau_5 - f_{fric\text{-}5}\dot{q}_5,$$

τ_5 is a twisting moment.

Lagrange equations

Based on the obtained formulas for T and V, we are able to derive the dynamic Lagrange equations:

$$\frac{d}{dt}\frac{\partial}{\partial \dot{q}_i}L - \frac{\partial}{\partial q_i}L = Q_{non\text{-}pot,i}, \quad i = 1, ..., 5, \quad L = T - V,$$

which can be represented in the standard matrix format

$$D(q)\ddot{q} + C(q, \dot{q})\dot{q} + g(q) = \tau,$$

where

$$D(q) = \begin{Vmatrix} D_{11}(q) & 0 & D_{13}(q) & D_{14}(q) & D_{15}(q) \\ 0 & m_3 + m_4 + m_5 & 0 & D_{24}(q) & D_{25}(q) \\ D_{31}(q) & 0 & D_{33}(q) & 0 & 0 \\ D_{41}(q) & D_{42}(q) & 0 & D_{44}(q) & D_{45}(q) \\ D_{51}(q) & D_{52}(q) & 0 & D_{54}(q) & m_5 d_5^2 + I_{5zz} \end{Vmatrix},$$

$$D_{11}(q) = \left[\frac{3}{2}(m_1 + m_2) + m_3 + m_4 + m_5\right],$$

$D_{31}(q) = (m_4 d_4 + m_5 l_4)\sin q_3 \sin q_4 - m_5 d_5 \sin q_3 \sin(q_4 + q_5),$

$D_{41}(q) = (m_4 d_4 + m_5 l_4)\cos q_3 \cos q_4 - m_5 d_5 \cos q_3 \cos(q_4 + q_5),$

$D_{51}(q) = -m_5 d_5 \cos q_3 \cos(q_4 + q_5),$

$D_{42}(q) = -(m_4 d_4 + m_5 l_4)\sin q_4 + m_5 d_5 \sin(q_4 + q_5),$

$D_{52}(q) = m_5 d_5 \sin(q_4 + q_5),$

$D_{13}(q) = (m_4 d_4 + m_5 l_4)\sin q_3 \sin q_4 - m_5 d_5 \sin q_3 \sin(q_4 + q_5),$

$D_{33}(q) = \left(m_5 d_5^2 + I_{5yy}\right)\sin^2(q_4 + q_5) -$

$2m_5 l_4 d_5 \sin q_4 \sin(q_4 + q_5) + I_{3yy} + I_{4xx}\cos^2 q_4 +$

$\left(m_4 d_4^2 + I_{4yy} + m_5 l_4^2\right)\sin^2 q_4 + I_{5xx}\cos^2(q_4 + q_5),$

$D_{14}(q) = (m_4 d_4 + m_5 l_4)\cos q_3 \cos q_4 - m_5 d_5 \cos q_3 \cos(q_4 + q_5),$

$D_{24}(q) = -(m_4 d_4 + m_5 l_4)\sin q_4 + m_5 d_5 \sin(q_4 + q_5),$

$D_{44}(q) = m_4 d_4^2 + I_{4zz} + m_5 l_4^2 + m_5 d_5^2 + I_{5zz} - 2m_5 l_4 d_5 \cos q_5,$

$D_{54}(q) = m_5 d_5^2 + I_{5zz} - m_5 l_4 d_5 \cos q_5,$

$D_{15}(q) = -m_5 d_5 \cos q_3 \cos(q_4 + q_5),$

$$D_{25}(q) = m_5 d_5 \sin(q_4 + q_5),$$
$$D_{45}(q) = m_5 d_5^2 + I_{5zz} - m_5 l_4 d_5 \cos q_5,$$

and

$$C(q,\dot{q}) = \begin{Vmatrix} f_{fric-1} & 0 & C_{13}(q,\dot{q}) & C_{14}(q,\dot{q}) & C_{15}(q,\dot{q}) \\ 0 & c & 0 & C_{24}(q,\dot{q}) & C_{25}(q,\dot{q}) \\ C_{31}(q,\dot{q}) & 0 & C_{33}(q,\dot{q}) & C_{34}(q,\dot{q}) & 0 \\ 0 & C_{42}(q,\dot{q}) & C_{43}(q,\dot{q}) & f_{fric-4} & C_{45}(q,\dot{q}) \\ 0 & 0 & C_{53}(q,\dot{q}) & -[m_5 l_4 d_5 \sin q_5]\dot{q}_4 & C_{55}(q,\dot{q}) \end{Vmatrix},$$

$$C_{31}(q,\dot{q}) = -2[m_5 d_5 \sin q_3 \cos(q_4 + q_5)](\dot{q}_4 + \dot{q}_5),$$
$$C_{42}(q,\dot{q}) = -2[m_5 d_5 \cos(q_4 + q_5)](\dot{q}_4 + \dot{q}_5),$$
$$C_{13}(q,\dot{q}) = [(m_4 d_4 + m_5 l_4)\cos q_3 \sin q_4]\dot{q}_3 -$$
$$[m_5 d_5 \cos q_3 \sin(q_4 + q_5)]\dot{q}_3,$$
$$C_{33}(q,\dot{q}) = f_{fric-3} - 2[m_5 l_4 d_5 \cos q_4 \sin(q_4 + q_5)]q_4 +$$
$$2\left[\left(m_4 d_4^2 + I_{4yy} + m_5 l_4^2 - I_{4xx}\right)\sin q_4 \cos q_4\right]\dot{q}_4 +$$
$$2\left(m_5 d_5^2 + I_{5yy} - I_{5xx}\right)\sin(q_4 + q_5)\cos(q_4 + q_5)(\dot{q}_4 + \dot{q}_5) -$$
$$2[m_5 l_4 d_5 \sin q_4 \cos(q_4 + q_5)](\dot{q}_4 + \dot{q}_5),$$
$$C_{43}(q,\dot{q}) = -2[(m_4 d_4 + m_5 l_4)\sin q_3 \cos q_4]\dot{q}_1 +$$
$$2[m_5 d_5 \sin q_3 \cos(q_4 + q_5)]\dot{q}_1 -$$
$$\left[\left(m_4 d_4^2 + I_{4yy} - I_{4xx} + m_5 l_4^2\right)\sin q_4 \cos q_4\right]\dot{q}_3 -$$
$$\left(m_5 d_5^2 + I_{5yy} - I_{5xx}\right)\sin(q_4 + q_5)\cos(q_4 + q_5)\dot{q}_3 +$$
$$[m_5 l_4 d_5 \sin(2q_4 + q_5)]\dot{q}_3,$$
$$C_{53}(q,\dot{q}) = 2[m_5 d_5 \sin q_3 \cos(q_4 + q_5)]\dot{q}_1 -$$
$$\left(m_5 d_5^2 + I_{5yy} - I_{5xx}\right)\sin(q_4 + q_5)\cos(q_4 + q_5)\dot{q}_3 +$$
$$[m_5 l_4 d_5 \sin q_4 \cos(q_4 + q_5)]\dot{q}_3,$$
$$C_{14}(q,\dot{q}) = -[(m_4 d_4 + m_5 l_4)\cos q_3 \sin q_4]\dot{q}_4 +$$
$$[m_5 d_5 \cos q_3 \sin(q_4 + q_5)](\dot{q}_4 + \dot{q}_5),$$
$$C_{24}(q,\dot{q}) = -[(m_4 d_4 + m_5 l_4)\cos q_4]\dot{q}_4 + [m_5 d_5 \cos(q_4 + q_5)](\dot{q}_4 + \dot{q}_5),$$
$$C_{34}(q,\dot{q}) = 2[(m_4 d_4 + m_5 l_4)\sin q_3 \cos q_4]\dot{q}_1,$$
$$C_{15}(q,\dot{q}) = [m_5 d_5 \cos q_3 \sin(q_4 + q_5)](\dot{q}_4 + \dot{q}_5),$$
$$C_{25}(q,\dot{q}) = [m_5 d_5 \cos(q_4 + q_5)](\dot{q}_4 + \dot{q}_5),$$
$$C_{45}(q,\dot{q}) = [m_5 l_4 d_5 \sin q_5](2\dot{q}_4 + \dot{q}_5),$$
$$C_{55}(q,\dot{q}) = f_{fric-5} + [m_5 l_4 d_5 \sin q_5]\dot{q}_4,$$

$$g(q) = \begin{Vmatrix} [m_1 + m_2 + m_3 + m_4 + m_5]g\sin\alpha \\ k(q_2 - q_{20}) + [m_3 + m_4 + m_5]g\cos\alpha \\ -[(m_5l_4 - m_4d_4)\sin q_4 + m_5d_5\sin(q_5 + q_4)]g\sin\alpha\sin q_3 \\ [m_4d_4 + m_5l_4]g\sin q_4\cos\alpha + [m_5l_4 - m_4d_4]g\sin\alpha\cos q_3\cos q_4 \\ -m_5g[d_5\sin(q_5 + q_4)\cos\alpha - d_5\cos(q_5 + q_4)\sin\alpha\cos q_3] \end{Vmatrix},$$

$$\tau = \begin{Vmatrix} F_1 \\ 0 \\ \tau_3 \\ \tau_4 \\ \tau_5 \end{Vmatrix}.$$

12.14 Planar manipulator with two joints

In this section we will consider the two-joint planar manipulator represented in Fig. 12.14.

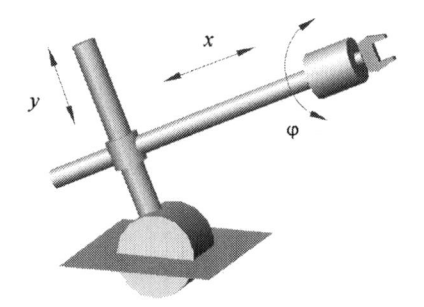

Figure 12.14 Two-joint planar manipulator.

Generalized coordinates
Select the generalized coordinates as follows:

$$q_1 := \varphi_1, \quad q_2 := y, \quad q_3 := x, \quad q_4 := \varphi_2.$$

Kinetic energy
The kinetic energy $T = \sum_{i=1}^{5} T_{m_i}$ has the terms

$$T_{m_1} = T_{m_1,0} + T_{m_1,rot-0} + m_1\left(v_{m_1-c.i.-0}, v_0\right)$$
$$= T_{m_1,rot-0} = \frac{1}{2}I_{1zz}\dot\varphi_1^2 = \frac{1}{2}I_{1zz}\dot q_1^2,$$
$$T_{m_2} = T_{m_2,0} + T_{m_2,rot-0} + m_2\left(v_{m_2-c.i.-0}, v_0\right) = T_{m_2,rot-0} =$$

$$\frac{1}{2}\begin{pmatrix}0\\0\\\dot{\varphi}_1\end{pmatrix}^\top \begin{Vmatrix} I_{2xx} & 0 & 0 \\ 0 & m_2(d_2+r_1)^2+I_{2yy} & 0 \\ 0 & 0 & m_2(d_2+r_1)^2+I_{2zz} \end{Vmatrix} \begin{pmatrix}0\\0\\\dot{\varphi}_1\end{pmatrix}$$

$$=\frac{1}{2}\left(m_2(d_2+r_1)^2+I_{2zz}\right)\dot{q}_1^2,$$

$$T_{m_3}=T_{m_3,0}+T_{m_3,rot-0}+m_3\left(\mathbf{v}_{m_3-c.i.-0},\mathbf{v}_0\right)=T_{m_3,0}+T_{m_3,rot-0}=$$

$$\frac{1}{2}m_3\left[\dot{y}^2+\dot{\varphi}_1^2(r_1+l_2+y)^2\right]+$$

$$\frac{1}{2}\begin{pmatrix}0\\0\\\dot{\varphi}_1\end{pmatrix}^\top \begin{Vmatrix} I_{3xx} & 0 & 0 \\ 0 & I_{3yy} & 0 \\ 0 & 0 & I_{3zz} \end{Vmatrix} \begin{pmatrix}0\\0\\\dot{\varphi}_1\end{pmatrix}$$

$$=\frac{1}{2}\left[m_3(r_1+l_2+q_2)^2+I_{3zz}\right]\dot{q}_1^2+\frac{1}{2}m_3\dot{q}_2^2,$$

$$T_{m_4}=T_{m_4,0}+T_{m_4,rot-0}+m_4\left(\mathbf{v}_{m_4-c.i.-0},\mathbf{v}_0\right)=$$

$$\frac{1}{2}m_4\left[\dot{y}^2+\dot{\varphi}_1^2(r_1+l_2+y)^2+\dot{x}^2\right]+$$

$$\frac{1}{2}\begin{pmatrix}0\\0\\\dot{\varphi}_1\end{pmatrix}^\top \begin{Vmatrix} I_{4xx} & 0 & 0 \\ 0 & m_4x^2+I_{4yy} & 0 \\ 0 & 0 & m_4x^2+I_{4zz} \end{Vmatrix} \begin{pmatrix}0\\0\\\dot{\varphi}_1\end{pmatrix}+$$

$$m_4\begin{pmatrix}\dot{x}\cos\varphi_1-\dot{\varphi}_1 x\sin\varphi_1\\ \dot{x}\sin\varphi_1+\dot{\varphi}_1 x\cos\varphi_1\\ 0\end{pmatrix}^\top \begin{pmatrix}-\dot{y}\sin\varphi_1-\dot{\varphi}_1(r_1+l_2+y)\cos\varphi_1\\ \dot{y}\cos\varphi_1-\dot{\varphi}_1(r_1+l_2+y)\sin\varphi_1\\ 0\end{pmatrix}$$

$$=\frac{1}{2}\left[m_4(r_1+l_2+q_2)^2+m_4 q_3^2+I_{4zz}\right]\dot{q}_1^2+$$

$$\frac{1}{2}m_4\left(\dot{q}_2^2+\dot{q}_3^2\right)+[m_4 q_3]\dot{q}_1\dot{q}_2-[m_4(r_1+l_2+q_2)]\dot{q}_1\dot{q}_3,$$

$$T_{m_5}=T_{m_5,0}+T_{m_5,rot-0}+m_5\left(\mathbf{v}_{m_5-c.i.-0},\mathbf{v}_0\right)=$$

$$\frac{1}{2}m_5\left[\dot{y}^2+\dot{\varphi}_1^2(r_1+l_2+y)^2+\dot{x}^2\right]+$$

$$\frac{1}{2}\begin{pmatrix}\dot{\varphi}_2\\0\\\dot{\varphi}_1\end{pmatrix}^\top \begin{Vmatrix} I_{5xx} & 0 & 0 \\ 0 & m_5(x+d_4+d_5)^2+I_{5yy} & 0 \\ 0 & 0 & m_5(x+d_4+d_5)^2+I_{5zz} \end{Vmatrix} \begin{pmatrix}\dot{\varphi}_2\\0\\\dot{\varphi}_1\end{pmatrix}$$

$$+m_5\begin{pmatrix}\dot{x}\cos\varphi_1-\dot{\varphi}_1(x+d_4+d_5)\sin\varphi_1\\ \dot{x}\sin\varphi_1+\dot{\varphi}_1(x+d_4+d_5)\cos\varphi_1\\ 0\end{pmatrix}^\top \times$$

$$\begin{pmatrix}-\dot{y}\sin\varphi_1-\dot{\varphi}_1(r_1+l_2+y)\cos\varphi_1\\ \dot{y}\cos\varphi_1-\dot{\varphi}_1(r_1+l_2+y)\sin\varphi_1\\ 0\end{pmatrix}$$

$$=\frac{1}{2}\left[m_5(r_1+l_2+q_2)^2+m_5(q_3+d_4+d_5)^2+I_{5zz}\right]\dot{q}_1^2+\frac{1}{2}[m_5]\dot{q}_2^2+$$

$$\frac{1}{2}m_5\dot{q}_3^2 + [m_5(q_3+d_4+d_5)]\dot{q}_1\dot{q}_2 - [m_5(r_1+l_2+q_2)]\dot{q}_1\dot{q}_3 + \frac{1}{2}[I_{5xx}]\dot{q}_4^2.$$

Potential energy

The potential energy $V = \sum_{i=1}^{5} V_{m_i}$ contains

$V_{m_1} = \text{const}$,
$V_{m_2} = m_2 g(r_1+d_2)\cos\varphi_1 = m_2 g(r_1+d_2)\cos q_1$,
$V_{m_3} = m_3 g(r_1+d_2+y)\cos\varphi_1 = m_3 g(r_1+d_2+q_2)\cos q_1$,
$V_{m_4} = m_4 g[x\sin\varphi_1 + (r_1+l_2+y)\cos\varphi_1] =$
$\quad m_4 g[q_3\sin q_1 + (r_1+l_2+q_2)\cos q_1]$,
$V_{m_5} = m_5 g[(x+d_4+d_5)\sin\varphi_1 + (r_1+l_2+y)\cos\varphi_1] =$
$\quad m_5 g[(q_3+d_4+d_5)\sin q_1 + (r_1+l_2+q_2)\cos q_1]$,

which gives

$V = m_2 g(r_1+d_2)\cos q_1 + m_3 g(r_1+l_2+q_2)\cos q_1 +$
$\quad m_4 g[q_3\sin q_1 + (r_1+l_2+q_2)\cos q_1] +$
$\quad m_5 g[(q_3+d_4+d_5)\sin q_1 + (r_1+l_2+q_2)\cos q_1] + \text{const}.$

Non-potential forces

The generalized non-potential forces are given by the following formulas:

$Q_{non\text{-}pot,1} = \tau_1 - f_{fric-1}\dot{\varphi}_1 = \tau_1 - f_{fric-1}\dot{q}_1$,
τ_1 is a twisting moment,
$Q_{non\text{-}pot,2} = F_2 - f_{fric-2}\dot{y} = F_2 - f_{fric-2}\dot{q}_2$,
F_2 is a force of longitudinal movement,
$Q_{non\text{-}pot,3} = F_3 - f_{fric-3}\dot{x}_3 = F_3 - f_{fric-3}\dot{q}_3$,
F_3 is a force of transverse movement,
$Q_{non\text{-}pot,4} = \tau_4 - f_{fric-4}\dot{\varphi}_2 = \tau_4 - f_{fric-4}\dot{q}_4$,
τ_4 is a twisting moment.

Lagrange equations

Using the obtained expression for T and V, we are able to derive the Lagrange equations

$$\frac{d}{dt}\frac{\partial}{\partial \dot{q}_i}L - \frac{\partial}{\partial q_i}L = Q_{non\text{-}pot,i}, \quad i=1,...,4, \quad L = T - V,$$

which may be represented in the standard matrix format:

$$D(q)\ddot{q} + C(q,\dot{q})\dot{q} + g(q) = \tau,$$

where

$$D(q) = \begin{Vmatrix} D_{11}(q) & \begin{matrix} m_4 q_3 \\ +m_5(q_3+d_4+d_5) \end{matrix} & \begin{matrix} -m_4(r_1+l_2+q_2) \\ -m_5(r_1+l_2+q_2) \end{matrix} & 0 \\ D_{21}(q) & m_3+m_4+m_5 & 0 & 0 \\ D_{31}(q) & 0 & m_4+m_5 & 0 \\ 0 & 0 & 0 & I_{5xx} \end{Vmatrix},$$

$D_{11}(q) = I_{1zz} + I_{2zz} + I_{3zz} + m_2(d_2+r_1)^2 + m_3(r_1+l_2+q_2)^2 +$
$\quad m_4(r_1+l_2+q_2)^2 + m_4 q_3^2 + I_{4zz} + I_{5zz} +$
$\quad m_5(r_1+l_2+q_2)^2 + m_5(q_3+d_4+d_5)^2,$

$D_{21}(q) = m_4 q_3 + m_5(q_3+d_4+d_5),$

$D_{31}(q) = -m_4(r_1+l_2+q_2) - m_5(r_1+l_2+q_2),$

and

$$C(q, \dot{q}) = \begin{Vmatrix} C_{11}(q,\dot{q}) & 0 & C_{13}(q,\dot{q}) & 0 \\ C_{21}(q,\dot{q}) & f_{fric-2} & 0 & 0 \\ C_{31}(q,\dot{q}) & 0 & f_{fric-3} & 0 \\ 0 & 0 & 0 & f_{fric-4} \end{Vmatrix},$$

$C_{11}(q,\dot{q}) = f_{fric-1} + + 2(m_3+m_4+m_5)(r_1+l_2+q_2)\dot{q}_2,$
$C_{21}(q,\dot{q}) = -[(m_3+m_4+m_5)(r_1+l_2+q_2)]\dot{q}_1 + 2[m_4+m_5]\dot{q}_3,$
$C_{31}(q,\dot{q}) = -m_4 q_3 \dot{q}_1 - m_5(q_3+d_4+d_5) - 2[m_4+m_5]\dot{q}_2,$
$C_{13}(q,\dot{q}) = 2m_4 q_3 \dot{q}_1 + 2m_5(q_3+d_4+d_5)\dot{q}_1,$

$$g(q) = \begin{Vmatrix} g_1(q) \\ [m_3+m_4+m_5]g\cos q_1 \\ [m_4+m_5]g\sin q_1 \\ 0 \end{Vmatrix}, \quad \tau = \begin{Vmatrix} \tau_1 \\ F_2 \\ F_3 \\ \tau_4 \end{Vmatrix},$$

$g_1(q) = -m_2 g(r_1+d_2)\sin q_1 - m_3 g(r_1+l_2+q_2)\sin q_1 +$
$\quad m_4 g [q_3 \cos q_1 - (r_1+l_2+q_2)\sin q_1] +$
$\quad m_5 g [(q_3+d_4+d_5)\cos q_1 - (r_1+l_2+q_2)\sin q_1].$

12.15 Double "crank-turn" swivel manipulator

Consider a double "crank-turn" swivel manipulator, represented in Fig. 12.15.

Generalized coordinates

The generalized coordinates are

$$q_1 := \varphi_1, \quad q_2 := \varphi_2, \quad q_3 := \varphi_3, \quad q_4 := x.$$

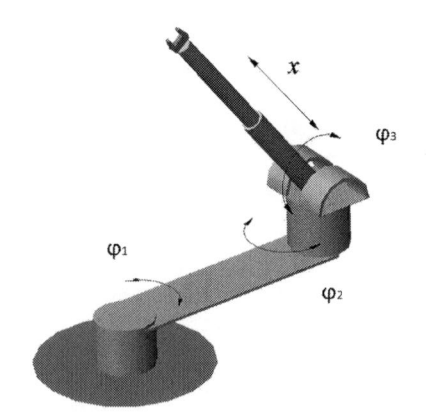

Figure 12.15 "Crank-turn" robot manipulator.

Kinetic energy

The kinetic energy $T = \sum_{i=1}^{5} T_{m_i}$ contains

$$T_{m_1} = T_{m_1,0} + T_{m_1,rot-0} + m_1 \left(\mathbf{v}_{m_1-c.i.-0}, \mathbf{v}_0\right) = \frac{1}{2} I_{1yy} \dot{\varphi}_1 = \frac{1}{2} I_{1yy} \dot{q}_1^2,$$

$$T_{m_2} = T_{m_2,0} + T_{m_2,rot-0} + m_2 \left(\mathbf{v}_{m_1-c.i.-0}, \mathbf{v}_0\right) = T_{m_2,rot-0} =$$

$$\frac{1}{2} \begin{pmatrix} 0 \\ \dot{\varphi}_1 \\ 0 \end{pmatrix}^T \begin{Vmatrix} I_{2xx} & 0 & 0 \\ 0 & m_2 d_2^2 + I_{2yy} & 0 \\ 0 & 0 & m_2 d_2^2 + I_{2zz} \end{Vmatrix} \begin{pmatrix} 0 \\ \dot{\varphi}_1 \\ 0 \end{pmatrix} =$$

$$\frac{1}{2} \left[m_2 d_2^2 + I_{2yy} \right] \dot{q}_1^2,$$

$$T_{m_3} = T_{m_3,0} + T_{m_3,rot-0} + m_3 \left(\mathbf{v}_{m_3-c.i.-0}, \mathbf{v}_0\right) =$$

$$T_{m_3,0} + T_{m_3,rot-0} = \frac{1}{2} m_3 l_2^2 \dot{\varphi}_1^2 +$$

$$\frac{1}{2} \begin{pmatrix} 0 \\ \dot{\varphi}_1 + \dot{\varphi}_2 \\ 0 \end{pmatrix}^T \begin{Vmatrix} I_{3xx} & 0 & 0 \\ 0 & I_{3yy} & 0 \\ 0 & 0 & I_{3zz} \end{Vmatrix} \begin{pmatrix} 0 \\ \dot{\varphi}_1 + \dot{\varphi}_2 \\ 0 \end{pmatrix} =$$

$$\frac{1}{2} m_3 l_2^2 \dot{q}_1^2 + \frac{1}{2} I_{3yy} \left(\dot{q}_1 + \dot{q}_2\right)^2,$$

$$T_{m_4} = T_{m_4,0} + T_{m_4,rot-0} + m_4 \left(\mathbf{v}_{m_4-c.i.-0}, \mathbf{v}_0\right) = \frac{1}{2} m_4 l_2^2 \dot{\varphi}_1^2 +$$

$$\frac{1}{2} \begin{pmatrix} -(\dot{\varphi}_1 + \dot{\varphi}_2) \cos \varphi_3 \\ (\dot{\varphi}_1 + \dot{\varphi}_2) \sin \varphi_3 \\ \dot{\varphi}_3 \end{pmatrix}^T \begin{Vmatrix} I_{4xx} & 0 & 0 \\ 0 & m_4 d_4^2 + I_{4yy} & 0 \\ 0 & 0 & m_4 d_4^2 + I_{4zz} \end{Vmatrix} \begin{pmatrix} -(\dot{\varphi}_1 + \dot{\varphi}_2) \cos \varphi_3 \\ (\dot{\varphi}_1 + \dot{\varphi}_2) \sin \varphi_3 \\ \dot{\varphi}_3 \end{pmatrix}$$

$$+m_4 \begin{pmatrix} (\dot{\varphi}_1+\dot{\varphi}_2)d_4\sin\varphi_3\sin(\varphi_1+\varphi_2) \\ +\dot{\varphi}_3 d_4\cos\varphi_3\cos(\varphi_1+\varphi_2) \\ -\dot{\varphi}_3 d_4\sin\varphi_3 \\ (\dot{\varphi}_1+\dot{\varphi}_2)d_4\sin\varphi_3\cos(\varphi_1+\varphi_2) \\ -\dot{\varphi}_3 d_4\cos\varphi_3\sin(\varphi_1+\varphi_2) \end{pmatrix}^\top \begin{pmatrix} -\dot{\varphi}_1 l_2\sin\varphi_1 \\ 0 \\ -\dot{\varphi}_1 l_2\cos\varphi_1 \end{pmatrix} =$$

$$\frac{1}{2}\left[m_4 l_2^2\right]\dot{q}_1^2 + \frac{1}{2}\left[m_4 d_4^2 + I_{4zz}\right]\dot{q}_3^2 + [m_4 l_2 d_4 \sin q_2 \cos q_3]\dot{q}_1\dot{q}_3 +$$
$$\frac{1}{2}\left[\left(m_4 d_4^2 + I_{4yy}\right)\sin^2 q_3 + I_{4xx}\cos^2 q_3\right](\dot{q}_1+\dot{q}_2)^2 -$$
$$m_4 [l_2 d_4 \cos q_2 \sin q_3](\dot{q}_1+\dot{q}_2)\dot{q}_1,$$

$$T_{m_5} = T_{m_5,0} + T_{m_5,rot-0} + m_5 \left(\mathbf{v}_{m_5-c.i.-0}, \mathbf{v}_0\right) = T_{m_5,0} + T_{m_5,rot-0} =$$

$$\frac{1}{2}m_5 \left\| \begin{pmatrix} (\dot{\varphi}_1+\dot{\varphi}_2)(l_4+x)\sin\varphi_3\sin(\varphi_1+\varphi_2) - \dot{\varphi}_1 l_2\sin\varphi_1 \\ +\dot{\varphi}_3(l_4+x)\cos\varphi_3\cos(\varphi_1+\varphi_2) - \dot{x}\sin\varphi_3\cos(\varphi_1+\varphi_2) \\ -\dot{\varphi}_3(l_4+x)\sin\varphi_3 - \dot{x}\cos\varphi_3 \\ (\dot{\varphi}_1+\dot{\varphi}_2)(l_4+x)\sin\varphi_3\cos(\varphi_1+\varphi_2) - \dot{\varphi}_1 l_2\cos\varphi_1 \\ -\dot{\varphi}_3(l_4+x)\cos\varphi_3\sin(\varphi_1+\varphi_2) + \dot{x}\sin\varphi_3\sin(\varphi_1+\varphi_2) \end{pmatrix} \right\|^2 +$$

$$\frac{1}{2}\begin{pmatrix} -(\dot{\varphi}_1+\dot{\varphi}_2)\cos\varphi_3 \\ (\dot{\varphi}_1+\dot{\varphi}_2)\sin\varphi_3 \\ \dot{\varphi}_3 \end{pmatrix}^\top \left\| \begin{matrix} I_{5xx} & 0 & 0 \\ 0 & I_{5yy} & 0 \\ 0 & 0 & I_{5zz} \end{matrix} \right\| \begin{pmatrix} -(\dot{\varphi}_1+\dot{\varphi}_2)\cos\varphi_3 \\ (\dot{\varphi}_1+\dot{\varphi}_2)\sin\varphi_3 \\ \dot{\varphi}_3 \end{pmatrix} =$$

$$\frac{1}{2}m_5 l_2^2 \dot{q}_1^2 + \frac{1}{2}\left[I_{5zz}+m_5(l_4+q_4)^2\right]\dot{q}_3^2 + \frac{1}{2}m_5 \dot{q}_4^2 +$$
$$\frac{1}{2}\left[I_{5xx}\cos^2 q_3 + \left(I_{5yy}+m_5(l_4+q_4)^2\right)\sin^2 q_3\right](\dot{q}_1+\dot{q}_2)^2 +$$
$$[(l_4+q_4)l_2\sin q_2\cos q_3]\dot{q}_1\dot{q}_3 - m_5[l_2\sin q_3\sin q_2]\dot{q}_1\dot{q}_4 -$$
$$m_5[(l_4+q_4)l_2\cos q_2\sin q_3]\dot{q}_1(\dot{q}_1+\dot{q}_2).$$

Potential energy

The potential energy $V = \sum_{i=1}^{5} V_{m_i}$ contains

$V_{m_1} = \text{const}, \quad V_{m_2} = \text{const}, \quad V_{m_3} = \text{const},$
$V_{m_4} = -m_4 g(d_4\cos\varphi_3 + \text{const}) = -m_4 g(d_4\cos q_3 + \text{const}),$
$V_{m_5} = -m_4 g((l_4+x)\cos\varphi_3 + \text{const}) = -m_4 g((l_4+q_4)\cos q_3 + \text{const}),$

which gives

$$V = -m_4 g\left[(d_4+l_4)\cos q_3 + q_4\right]\cos q_3 + \text{const}.$$

Non-potential forces

The generalized non-potential forces are as follows:

$$Q_{non\text{-}pot,1} = \tau_1 - f_{fric-1}\dot{\varphi}_1 = \tau_1 - f_{fric-1}\dot{q}_1,$$

τ_1 is a twisting moment,

$Q_{non\text{-}pot,2} = \tau_2 - f_{fric-2}\dot{\varphi}_2 = \tau_2 - f_{fric-2}\dot{q}_2,$

τ_2 is a twisting moment,

$Q_{non\text{-}pot,3} = \tau_3 - f_{fric-3}\dot{\varphi}_3 = \tau_3 - f_{fric-3}\dot{q}_3,$

τ_3 is a twisting moment,

$Q_{non\text{-}pot,4} = F_4 - f_{fric-4}\dot{x} = F_4 - f_{fric-4}\dot{q}_4,$

F_4 is a force of longitudinal movement.

Lagrange equations

The obtained formulas for T and V allow to derive the Lagrange dynamic equations

$$\frac{d}{dt}\frac{\partial}{\partial \dot{q}_i}L - \frac{\partial}{\partial q_i}L = Q_{non\text{-}pot,i}, \quad i = 1, ..., 4, \quad L = T - V,$$

which can be represented in the following standard matrix format:

$$D(q)\ddot{q} + C(q,\dot{q})\dot{q} + g(q) = \tau,$$

where

$$D(q) = \begin{Vmatrix} D_{11}(q) & D_{12}(q) & D_{13}(q) & -m_5 l_2 \sin q_2 \sin q_3 \\ D_{21}(q) & D_{22}(q) & 0 & 0 \\ D_{31}(q) & 0 & D_{33}(q) & 0 \\ -m_5 l_2 \sin q_2 \sin q_3 & 0 & 0 & m_5 \end{Vmatrix},$$

$D_{11}(q) = I_{1yy} + m_2 d_2^2 + I_{2yy} + I_{3yy} + (m_3 + m_4 + m_5)l_2^2 +$
$\left(m_4 d_4^2 + I_{4yy} + I_{5yy} + m_5(l_4 + q_4)^2\right)\sin^2 q_3 + (I_{4xx} + I_{5xx})\cos^2 q_3 -$
$2l_2(m_4 d_4 + m_5(l_4 + q_4))\cos q_2 \sin q_3,$

$D_{21}(q) = I_{3yy} - l_2(m_4 d_4 + m_5(l_4 + q_4))\cos q_2 \sin q_3 +$
$\left(m_4 d_4^2 + I_{4yy} + I_{5yy} + m_5(l_4 + q_4)^2\right)\sin^2 q_3 + (I_{4xx} + I_{5xx})\cos^2 q_3,$

$D_{31}(q) = l_2(m_4 d_4 + m_5(l_4 + q_4))\sin q_2 \cos q_3,$

$D_{12}(q) = I_{3yy} - l_2(m_4 d_4 + m_5(l_4 + q_4))\cos q_2 \sin q_3 +$
$\left(m_4 d_4^2 + I_{4yy} + I_{5yy} + m_5(l_4 + q_4)^2\right)\sin^2 q_3 + (I_{4xx} + I_{5xx})\cos^2 q_3,$

$D_{22}(q) = +I_{3yy} + (I_{4xx} + I_{5xx})\cos^2 q_3 +$
$\left(m_4 d_4^2 + I_{4yy} + I_{5yy} + m_5(l_4 + q_4)^2\right)\sin^2 q_3,$

$D_{13}(q) = l_2(m_4 d_4 + m_5(l_4 + q_4))\sin q_2 \cos q_3,$

$D_{33}(q) = +m_4 d_4^2 + I_{4zz} + I_{5zz} + m_5(l_4 + q_4)^2,$

and

$$C(q,\dot{q}) = \begin{Vmatrix} C_{11}(q,\dot{q}) & C_{12}(q,\dot{q}) & C_{13}(q,\dot{q}) & C_{14}(q,\dot{q}) \\ C_{21}(q,\dot{q}) & C_{22}(q,\dot{q}) & C_{23}(q,\dot{q}) & C_{24}(q,\dot{q}) \\ C_{31}(q,\dot{q}) & C_{32}(q,\dot{q}) & C_{33}(q,\dot{q}) & 0 \\ C_{41}(q,\dot{q}) & C_{42}(q,\dot{q}) & C_{43}(q,\dot{q}) & f_{fric-4} \end{Vmatrix},$$

$C_{11}(q,\dot{q}) = f_{fric-1} - 2l_2(m_4d_4 + m_5(l_4 + q_4))\dot{q}_3 \cos q_2 \cos q_3,$

$C_{21}(q,\dot{q}) = -2l_2(m_4d_4 + m_5(l_4 + q_4))\dot{q}_3 \cos q_2 \cos q_3 -$
$\quad l_2(m_4d_4 + m_5(l_4 + q_4)) \sin q_2 \sin q_3 (\dot{q}_1 + \dot{q}_2),$

$C_{31}(q,\dot{q}) = 2l_2(m_4d_4 + m_5(l_4 + q_4)) \cos q_2 \cos q_3 \dot{q}_2 +$
$\quad 2l_2 m_5 \sin q_2 \cos q_3 \dot{q}_4 + l_2(m_4d_4 + m_5(l_4 + q_4)l_2)\dot{q}_1 \cos q_2 \cos q_3 -$
$\quad \left(I_{5yy} + m_5(l_4 + q_4)^2 - I_{5xx} + m_4 d_4^2 + I_{4yy} - I_{4xx}\right)(\dot{q}_1 + \dot{q}_2) \sin q_3 \cos q_3,$

$C_{41}(q,\dot{q}) = [m_5 l_2 \cos q_2 \sin q_3] \dot{q}_1 - \left[m_5(l_4 + q_4) \sin^2 q_3\right](\dot{q}_1 + \dot{q}_2),$

$C_{12}(q,\dot{q}) = l_2(m_4 d_4 + m_5(l_4 + q_4))(2\dot{q}_1 + \dot{q}_2) \sin q_2 \sin q_3,$

$C_{22}(q,\dot{q}) = f_{fric-2} + l_2(m_4 d_4 + m_5(l_4 + q_4))\dot{q}_2 \sin q_2 \sin q_3,$

$C_{32}(q,\dot{q}) = -\left(I_{5yy} + m_5(l_4 + q_4)^2 - I_{5xx} + m_4 d_4^2 + I_{4yy} - I_{4xx}\right) \times$
$\quad (\dot{q}_2 + \dot{q}_1) \sin q_3 \cos q_3,$

$C_{42}(q,\dot{q}) = -\left[m_5(l_4 + q_4) \sin^2 q_3\right](\dot{q}_1 + \dot{q}_2),$

$C_{13}(q,\dot{q}) = -l_2(m_4 d_4 + m_5(l_4 + q_4)) \sin q_2 \sin q_3 \dot{q}_3 +$
$\quad 2\left(m_4 d_4^2 + I_{4yy} + I_{5yy} + m_5(l_4 + q_4)^2 - I_{4xx} - I_{5xx}\right)(\dot{q}_1 + \dot{q}_2) \sin q_3 \cos q_3,$

$C_{23}(q,\dot{q}) = 2\left(m_4 d_4^2 + I_{4yy} + I_{5yy} + m_5(l_4 + q_4)^2 - I_{4xx} - I_{5xx}\right) \times$
$\quad (\dot{q}_1 + \dot{q}_2) \sin q_3 \cos q_3,$

$C_{33}(q,\dot{q}) = f_{fric-3} + 2m_5(l_4 + q_4)\dot{q}_4,$

$C_{43}(q,\dot{q}) = -2m_5 l_2 \sin q_2 \cos q_3 \dot{q}_1 - [m_5(l_4 + q_4)]\dot{q}_3,$

$C_{14}(q,\dot{q}) = 2(m_5(l_4 + q_4)) \sin^2 q_3 (\dot{q}_1 + \dot{q}_2)\dot{q}_4 -$
$\quad 2l_2 m_5 [\cos q_2 \sin q_3](\dot{q}_1 + \dot{q}_2),$

$C_{24}(q,\dot{q}) = 2m_5(l_4 + q_4) \sin^2 q_3 (\dot{q}_1 + \dot{q}_2),$

with

$$g(q) = \begin{Vmatrix} 0 \\ 0 \\ m_4 g d_4 \sin q_3 + m_4 g(l_4 + q_4) \sin q_3 \\ -m_4 g \cos q_3 \end{Vmatrix}, \quad \tau = \begin{Vmatrix} \tau_1 \\ \tau_2 \\ \tau_3 \\ F_4 \end{Vmatrix}.$$

12.16 Robot manipulator of multicylinder type

Let us consider a robot manipulator of multicylinder type, depicted in Fig. 12.16.

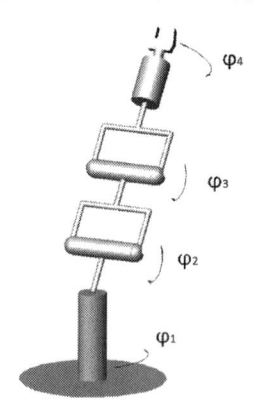

Figure 12.16 Robot manipulator of multicylinder type.

Generalized coordinates

The generalized coordinates are

$$q_1 := \varphi_1, \quad q_2 := \varphi_2, \quad q_3 := \varphi_3, \quad q_4 := \varphi_4.$$

Kinetic energy

The kinetic energy $T = \sum_{i=1}^{4} T_{m_i}$ contains the terms

$$T_{m_1} = T_{m_1,0} + T_{m_1,rot-0} + m_1 \left(\mathbf{v}_{m_1-c.i.-0}, \mathbf{v}_0\right) =$$

$$T_{m_1,rot-0} = \frac{1}{2} I_{1yy} \dot\varphi_1^2 = \frac{1}{2} I_{1yy} \dot q_1^2,$$

$$T_{m_2} = T_{m_2,0} + T_{m_2,rot-0} + m_2 \left(\mathbf{v}_{m_2-c.i.-0}, \mathbf{v}_0\right) = T_{m_2,rot-0} =$$

$$\frac{1}{2} \begin{pmatrix} \dot\varphi_2 \\ \dot\varphi_1 \\ 0 \end{pmatrix}^T \begin{Vmatrix} I_{2xx} & 0 & 0 \\ 0 & I_{2yy} & 0 \\ 0 & 0 & I_{2zz} \end{Vmatrix} \begin{pmatrix} \dot\varphi_2 \\ \dot\varphi_1 \\ 0 \end{pmatrix} = \frac{1}{2} I_{2yy} \dot q_1^2 + \frac{1}{2} I_{2xx} \dot q_2^2,$$

$$T_{m_3} = T_{m_3,0} + T_{m_3,rot-0} + m_2 \left(\mathbf{v}_{m_3-c.i.-0}, \mathbf{v}_0\right) =$$

$$\frac{1}{2} m_3 \left\Vert \begin{pmatrix} -\dot\varphi_2 a \cos\varphi_1 \sin\varphi_2 - \dot\varphi_1 a \sin\varphi_1 \cos\varphi_2 \\ \dot\varphi_2 a \cos\varphi_2 \\ -\dot\varphi_1 a \cos\varphi_1 \cos\varphi_2 + \dot\varphi_2 a \sin\varphi_1 \sin\varphi_2 \end{pmatrix} \right\Vert^2 +$$

$$\frac{1}{2} \begin{pmatrix} \dot\varphi_2 + \dot\varphi_3 \\ \dot\varphi_1 \\ 0 \end{pmatrix}^T \begin{Vmatrix} I_{3xx} & 0 & 0 \\ 0 & I_{3yy} & 0 \\ 0 & 0 & I_{3zz} \end{Vmatrix} \begin{pmatrix} \dot\varphi_2 + \dot\varphi_3 \\ \dot\varphi_1 \\ 0 \end{pmatrix} =$$

$$\frac{1}{2} m_3 \left[a^2 \cos^2 q_2 \dot q_1^2 + a^2 \dot q_2^2 \right] + \frac{1}{2} \left[I_{3yy} \dot q_1^2 + I_{3xx} (\dot q_2 + \dot q_3)^2 \right],$$

$$T_{m4} = T_{m4,0} + T_{m4,rot-0} + m_4 \left(\mathbf{v}_{m4-c.i.-0}, \mathbf{v}_0\right) =$$

$$\frac{1}{2} m_4 \left[a^2 \cos^2 \varphi_2 \dot{\varphi}_1^2 + a^2 \dot{\varphi}_2^2\right] + \frac{1}{2} \mathbf{w}_4^T \begin{Vmatrix} I_{4xx} & 0 & 0 \\ 0 & \begin{array}{c} m_4 b^2 \\ +I_{4yy} \end{array} & 0 \\ 0 & 0 & \begin{array}{c} m_4 b^2 \\ +I_{4zz} \end{array} \end{Vmatrix} \mathbf{w}_4 +$$

$$m_4 \begin{pmatrix} \dot{\varphi}_1 b \sin \varphi_1 \cos (\varphi_2 + \varphi_3) + \\ (\dot{\varphi}_2 + \dot{\varphi}_3) b \cos \varphi_1 \sin (\varphi_2 + \varphi_3) \\ -(\dot{\varphi}_2 + \dot{\varphi}_3) b \cos (\varphi_2 + \varphi_3) \\ \dot{\varphi}_1 b \cos \varphi_1 \cos (\varphi_2 + \varphi_3) - \\ (\dot{\varphi}_2 + \dot{\varphi}_3) b \sin \varphi_1 \sin (\varphi_2 + \varphi_3) \end{pmatrix}^T \begin{pmatrix} -\dot{\varphi}_2 a \cos \varphi_1 \sin \varphi_2 - \\ \dot{\varphi}_1 a \sin \varphi_1 \cos \varphi_2 \\ \dot{\varphi}_2 a \cos \varphi_2 \\ -\dot{\varphi}_1 a \cos \varphi_1 \cos \varphi_2 + \\ \dot{\varphi}_2 a \sin \varphi_1 \sin \varphi_2 \end{pmatrix} =$$

$$\frac{1}{2} \left[I_{4xx} \sin^2 (q_2 + q_3) + \left(m_4 b^2 + I_{4yy}\right) \cos^2 (q_2 + q_3)\right] \dot{q}_1^2 +$$
$$\frac{1}{2} m_4 a^2 \dot{q}_2^2 + \frac{1}{2} m_4 \left[a^2 \cos^2 q_2 - 2ab \cos q_2 \cos (q_2 + q_3)\right] \dot{q}_1^2 +$$
$$\frac{1}{2} I_{4xx} \left[2 \sin (q_2 + q_3) \dot{q}_1 \dot{q}_4 + \dot{q}_4^2\right] +$$
$$\frac{1}{2} \left[\left(m_4 b^2 + I_{4zz}\right)\right] (\dot{q}_2 + \dot{q}_3)^2 - [m_4 ab \cos q_3] (\dot{q}_2 + \dot{q}_3) \dot{q}_2,$$

where

$$\mathbf{w}_4 := \begin{pmatrix} \dot{\varphi}_1 \sin (\varphi_2 + \varphi_3) + \dot{\varphi}_4 \\ \dot{\varphi}_1 \cos (\varphi_2 + \varphi_3) \\ \dot{\varphi}_2 + \dot{\varphi}_3 \end{pmatrix}.$$

Potential energy

The potential energy $V = \sum_{i=1}^{4} V_{m_i}$ is given by

$$V_{m_1} = \text{const}, \quad V_{m_2} = \text{const},$$
$$V_{m_3} = m_3 g \left(a \sin \varphi_2 + \text{const}\right) = m_3 g \left(a \sin q_2 + \text{const}\right),$$
$$V_{m_4} = m_4 g \left(-b \sin (\varphi_2 + \varphi_3) + a \sin \varphi_2 + \text{const}\right) =$$
$$m_4 g \left(-b \sin (q_2 + q_3) + a \sin q_2 + \text{const}\right),$$

which implies

$$V = m_3 g a \sin q_2 + m_4 g \left[a \sin q_2 - b \sin (q_2 + q_3)\right] + \text{const}.$$

Non-potential forces

The generalized non-potential forces are

$$Q_{non\text{-}pot,i} = \tau_i - f_{fric-i} \dot{\varphi}_i = \tau_i - f_{fric-i} \dot{q}_i,$$
τ_i is a twisting moment, $i = 1, ..., 4$.

Lagrange equations

Based on the obtained equations for T and V, we are able to represent the dynamic model of the considered system in the Lagrangian form:

$$\frac{d}{dt}\frac{\partial}{\partial \dot{q}_i}L - \frac{\partial}{\partial q_i}L = Q_{non\text{-}pot,i}, \quad i = 1, ..., 4, \quad L = T - V,$$

which in the standard matrix format represents the *dynamic model* of the considered system:

$$D(q)\ddot{q} + C(q,\dot{q})\dot{q} + g(q) = \tau,$$

where

$$D(q) = \begin{Vmatrix} D_{11}(q) & 0 & 0 & I_{4xx}\sin(q_2+q_3) \\ 0 & D_{22}(q) & D_{23}(q) & 0 \\ 0 & D_{32}(q) & D_{33}(q) & 0 \\ I_{4xx}\sin(q_2+q_3) & 0 & 0 & I_{4xx} \end{Vmatrix},$$

$$D_{11}(q) = I_{1yy} + I_{2yy} + I_{3yy} + m_3 a^2 \cos^2 q_2 +$$
$$I_{4xx}\sin^2(q_2+q_3) + \left(m_4 b^2 + I_{4yy}\right)\cos^2(q_2+q_3) +$$
$$m_4 a^2 \cos^2 q_2 - m_4 2ab \cos q_2 \cos(q_2+q_3),$$

$$D_{22}(q) = I_{2xx} + (m_3 + m_4)a^2 +$$
$$2\left[I_{3xx} + m_4 b^2 + I_{4zz}\right] - 2m_4 ab \cos q_3,$$

$$D_{32}(q) = I_{3xx} + m_4 b^2 + I_{4zz} - m_4 ab \cos q_3,$$

$$D_{23}(q) = +I_{3xx} + m_4 b^2 + I_{4zz} - m_4 ab \cos q_3,$$

$$D_{33}(q) = I_{3xx} + m_4 b^2 + I_{4zz},$$

and

$$C(q,\dot{q}) = \begin{Vmatrix} C_{11}(q,\dot{q}) & 0 & 0 & C_{14}(q,\dot{q}) \\ C_{21}(q,\dot{q}) & f_{fric-2} & C_{23}(q,\dot{q}) & -I_{4xx}\cos(q_2+q_3)\dot{q}_1 \\ C_{31}(q,\dot{q}) & C_{32}(q,\dot{q}) & f_{fric-3} & -I_{4xx}\cos(q_2+q_3)\dot{q}_1 \\ C_{41}(q,\dot{q}) & 0 & 0 & f_{fric-4} \end{Vmatrix},$$

$$C_{11}(q,\dot{q}) = f_{fric-1} - 2\left(m_3 a^2 + m_4 a^2\right)\sin q_2 \cos q_2 \dot{q}_2 +$$
$$2m_4 ab \sin q_2 \cos(2q_2+q_3)\dot{q}_2 + 2m_4 ab \cos q_2 \sin(q_2+q_3)\dot{q}_3 +$$
$$2\left(I_{4xx} - m_4 b^2 - I_{4yy}\right)\sin(q_2+q_3)\cos(q_2+q_3)(\dot{q}_2+\dot{q}_3),$$

$$C_{21}(q,\dot{q}) = \left(m_3 a^2 + a^2 m_4\right)\sin q_2 \cos q_2 \dot{q}_1 - m_4 ab \sin(2q_2+q_3)\dot{q}_1$$
$$-\left(I_{4xx} - m_4 b^2 - I_{4yy}\right)\sin(q_2+q_3)\cos(q_2+q_3)\dot{q}_1,$$

$$C_{31}(q,\dot{q}) = -\left(I_{4xx} - m_4 b^2 - I_{4yy}\right)\dot{q}_1 \sin(q_2+q_3)\cos(q_2+q_3)$$

$$-m_4ab\cos q_2 \sin(q_2+q_3)\dot{q}_1,$$
$$C_{41}(q,\dot{q}) = I_{4xx}\cos(q_2+q_3)(\dot{q}_2+\dot{q}_3),$$
$$C_{32}(q,\dot{q}) = -m_4ab\sin q_3 \dot{q}_2,$$
$$C_{23}(q,\dot{q}) = m_4ab\sin q_3 (2\dot{q}_2+\dot{q}_3),$$
$$C_{14}(q,\dot{q}) = I_{4xx}\cos(q_2+q_3)(\dot{q}_2+\dot{q}_3),$$

$$g(q) = \left\| \begin{array}{c} 0 \\ m_3ga\cos q_2 + m_4g(a\cos q_2 - b\cos(q_2+q_3)) \\ -m_4gb\cos(q_2+q_3) \\ 0 \end{array} \right\|, \quad \tau = \left\| \begin{array}{c} \tau_1 \\ \tau_2 \\ \tau_3 \\ \tau_4 \end{array} \right\|.$$

12.17 Arm manipulator with springs

Consider an arm manipulator with springs, as depicted in Fig. 12.17.

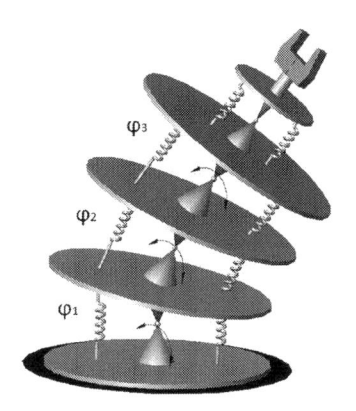

Figure 12.17 Arm manipulator with springs.

Generalized coordinates

The generalized coordinates are

$$q_1 := \varphi_1, \quad q_2 := \varphi_2, \quad q_3 := \varphi_3, \quad q_4 := \varphi_4, \quad q_5 := \varphi_5$$

(in Fig. 12.17, $\theta_i = \varphi_i$ ($i = 1, ..., 4$)).

Kinetic energy

The kinetic energy $T = \sum_{i=0}^{4} T_{m_i}$ is calculated as

$$T_{m_0} = T_{m_0,0} + T_{m_0,rot-0} + m_0\left(\mathbf{v}_{m_0-c.i.-0}, \mathbf{v}_0\right)$$
$$= T_{m_0,rot-0} = \frac{1}{2}\frac{m_0 r_0}{2}\dot{\varphi}_1^2 = \frac{1}{4}m_0 r_0 \dot{q}_1^2,$$

$$T_{m_1} = T_{m_1,0} + T_{m_1,rot-0} + m_1 \left(\mathbf{v}_{m_1-c.i.-0}, \mathbf{v}_0\right) = T_{m_1,rot-0} =$$

$$\frac{1}{2} \begin{pmatrix} \dot{\varphi}_1 \sin \varphi_2 \\ \dot{\varphi}_1 \cos \varphi_2 \\ \dot{\varphi}_2 \end{pmatrix}^\top \begin{Vmatrix} I_{1xx} & 0 & 0 \\ 0 & m_1 \left(\frac{l}{2}\right)^2 + I_{1yy} & 0 \\ 0 & 0 & m_1 \left(\frac{l}{2}\right)^2 + I_{1zz} \end{Vmatrix} \begin{pmatrix} \dot{\varphi}_1 \sin \varphi_2 \\ \dot{\varphi}_1 \cos \varphi_2 \\ \dot{\varphi}_2 \end{pmatrix}$$

$$= \frac{1}{2} \left[I_{1xx} \sin^2 q_2 + \left(m_1 \frac{l^2}{4} + I_{1yy}\right) \cos^2 q_2 \right] \dot{q}_1^2 + \frac{1}{2} \left[m_1 \frac{l^2}{4} + I_{1zz}\right] \dot{q}_2^2,$$

$$T_{m_2} = T_{m_2,0} + T_{m_2,rot-0} + m_2 \left(\mathbf{v}_{m_1-c.i.-0}, \mathbf{v}_0\right) =$$

$$\frac{1}{2} m_2 \left\| \begin{pmatrix} -\dot{\varphi}_1 l \sin \varphi_1 \cos \varphi_2 - \dot{\varphi}_2 l \cos \varphi_1 \sin \varphi_2 \\ \dot{\varphi}_2 l \cos \varphi_2 \\ -\dot{\varphi}_1 l \cos \varphi_1 \cos \varphi_2 + \dot{\varphi}_2 l \sin \varphi_1 \sin \varphi_2 \end{pmatrix} \right\|^2 +$$

$$\frac{1}{2} \mathbf{w}_2^\top \begin{Vmatrix} I_{2xx} & 0 & 0 \\ 0 & m_2 \frac{l^2}{4} + I_{2yy} & 0 \\ 0 & 0 & m_2 \frac{l^2}{4} + I_{2zz} \end{Vmatrix} \mathbf{w}_2 +$$

$$m_2 \begin{pmatrix} \dot{\varphi}_1 \frac{l}{2} \sin \varphi_1 \cos(\varphi_2 + \varphi_3) + (\dot{\varphi}_2 + \dot{\varphi}_3) \frac{l}{2} \cos \varphi_1 \sin(\varphi_2 + \varphi_3) \\ -(\dot{\varphi}_2 + \dot{\varphi}_3) \frac{l}{2} \cos(\varphi_2 + \varphi_3) \\ \dot{\varphi}_1 \frac{l}{2} \cos \varphi_1 \cos(\varphi_2 + \varphi_3) - (\dot{\varphi}_2 + \dot{\varphi}_3) \frac{l}{2} \sin \varphi_1 \sin(\varphi_2 + \varphi_3) \end{pmatrix}^\top \times$$

$$\begin{pmatrix} -\dot{\varphi}_1 l \sin \varphi_1 \cos \varphi_2 - \dot{\varphi}_2 l \cos \varphi_1 \sin \varphi_2 \\ \dot{\varphi}_2 l \cos \varphi_2 \\ -\dot{\varphi}_1 l \cos \varphi_1 \cos \varphi_2 + \dot{\varphi}_2 l \sin \varphi_1 \sin \varphi_2 \end{pmatrix} =$$

$$\frac{1}{2} \left[m_2 l^2 \cos^2 q_2 + I_{2xx} \sin^2 (q_2 + q_3) + \left(m_2 \frac{l^2}{4} + I_{2yy}\right) \cos^2 (q_2 + q_3) \right] \dot{q}_1^2 -$$

$$\frac{1}{2} \left[m_2 l^2 \cos q_2 \cos(q_2 + q_3) \right] \dot{q}_1^2 + \frac{1}{2} \left[m_2 l^2 \right] \dot{q}_2^2 +$$

$$\frac{1}{2} \left[m_2 \frac{l^2}{4} + I_{2zz} \right] (\dot{q}_2 + \dot{q}_3)^2 -$$

$$\frac{1}{2} \left[m_2 l^2 \cos q_3 \right] (\dot{q}_2 + \dot{q}_3) \dot{q}_2,$$

where

$$\mathbf{w}_2 := \begin{pmatrix} \dot{\varphi}_1 \sin(\varphi_2 + \varphi_3) \\ \dot{\varphi}_1 \cos(\varphi_2 + \varphi_3) \\ \dot{\varphi}_2 + \dot{\varphi}_3 \end{pmatrix},$$

$$T_{m_3} = T_{m_3,0} + T_{m_3,rot-0} + m_3 \left(\mathbf{v}_{m_3-c.i.-0}, \mathbf{v}_0\right) =$$

$$\frac{1}{2}m_3 \left\| \begin{pmatrix} -\dot{\varphi}_1 l \sin\varphi_1 \cos\varphi_2 - \dot{\varphi}_2 l \cos\varphi_1 \sin\varphi_2 \\ +\dot{\varphi}_1 l \sin\varphi_1 \cos(\varphi_2+\varphi_3) + (\dot{\varphi}_2+\dot{\varphi}_3) l \cos\varphi_1 \sin(\varphi_2+\varphi_3) \\ \dot{\varphi}_2 l \cos\varphi_2 - (\dot{\varphi}_2+\dot{\varphi}_3) l \cos(\varphi_2+\varphi_3) \\ -\dot{\varphi}_1 l \cos\varphi_1 \cos\varphi_2 + \dot{\varphi}_2 l \sin\varphi_1 \sin\varphi_2 \\ +\dot{\varphi}_1 l \cos\varphi_1 \cos(\varphi_2+\varphi_3) - (\dot{\varphi}_2+\dot{\varphi}_3) l \sin\varphi_1 \sin(\varphi_2+\varphi_3) \end{pmatrix} \right\|^2 +$$

$$\frac{1}{2}\mathbf{w}_3^\top \left\| \begin{matrix} I_{3xx} & 0 & 0 \\ 0 & m_3\left(\frac{l}{2}\right)^2 + I_{3yy} & 0 \\ 0 & 0 & m_3\left(\frac{l}{2}\right)^2 + I_{3zz} \end{matrix} \right\| \mathbf{w}_3 +$$

$$m_3 \begin{pmatrix} -\dot{\varphi}_1 \frac{l}{2} \sin\varphi_1 \cos(\varphi_2+\varphi_3+\varphi_4) \\ -(\dot{\varphi}_2+\dot{\varphi}_3+\dot{\varphi}_4)\frac{l}{2} \cos\varphi_1 \sin(\varphi_2+\varphi_3+\varphi_4) \\ (\dot{\varphi}_2+\dot{\varphi}_3+\dot{\varphi}_4)\frac{l}{2} \cos(\varphi_2+\varphi_3+\varphi_4) \\ -\dot{\varphi}_1 \frac{l}{2} \cos\varphi_1 \cos(\varphi_2+\varphi_3+\varphi_4) \\ +(\dot{\varphi}_2+\dot{\varphi}_3+\dot{\varphi}_4)\frac{l}{2} \sin\varphi_1 \sin(\varphi_2+\varphi_3+\varphi_4) \end{pmatrix}^\top \times$$

$$\begin{pmatrix} -\dot{\varphi}_1 l \sin\varphi_1 \cos\varphi_2 - \dot{\varphi}_2 l \cos\varphi_1 \sin\varphi_2 \\ +\dot{\varphi}_1 l \sin\varphi_1 \cos(\varphi_2+\varphi_3) \\ +(\dot{\varphi}_2+\dot{\varphi}_3) l \cos\varphi_1 \sin(\varphi_2+\varphi_3) \\ \dot{\varphi}_2 l \cos\varphi_2 - (\dot{\varphi}_2+\dot{\varphi}_3) l \cos(\varphi_2+\varphi_3) \\ -\dot{\varphi}_1 l \cos\varphi_1 \cos\varphi_2 + \dot{\varphi}_2 l \sin\varphi_1 \sin\varphi_2 \\ +\dot{\varphi}_1 l \cos\varphi_1 \cos(\varphi_2+\varphi_3) \\ -(\dot{\varphi}_2+\dot{\varphi}_3) l \sin\varphi_1 \sin(\varphi_2+\varphi_3) \end{pmatrix} =$$

$$\frac{1}{2}m_3 l^2 \left[(\cos q_2 - \cos(q_2+q_3))\cos(q_2+q_3+q_4) - 2\cos q_2 \cos(q_2+q_3)\right]\dot{q}_1^2 +$$

$$\frac{1}{2}\left[m_3 l^2 \left(\cos^2 q_2 + \cos^2(q_2+q_3)\right)\right]\dot{q}_1^2 +$$

$$\frac{1}{2}\left[I_{3xx}\sin^2(q_2+q_3+q_4) + \left(m_3\frac{l^2}{4}+I_{3yy}\right)\cos^2(q_2+q_3+q_4)\right]\dot{q}_1^2 +$$

$$\frac{1}{2}\left[m_3 l^2\right]\dot{q}_2^2 + \frac{1}{2}\left[m_3 l^2\right](\dot{q}_2+\dot{q}_3)^2 + \frac{1}{2}\left[m_3\frac{l^2}{4}+I_{3zz}\right](\dot{q}_2+\dot{q}_3+\dot{q}_4)^2 -$$

$$\left[m_3 l^2 \cos q_3\right](\dot{q}_2+\dot{q}_3)\dot{q}_2 + \frac{1}{2}m_3\left[l^2\cos(q_3+q_4)\right](\dot{q}_2+\dot{q}_3+\dot{q}_4)\dot{q}_2 -$$

$$\frac{1}{2}m_3\left[l^2\cos q_4\right](\dot{q}_2+\dot{q}_3+\dot{q}_4)(\dot{q}_2+\dot{q}_3),$$

with

$$\mathbf{w}_3 := \begin{pmatrix} \dot{\varphi}_1 \sin(\varphi_2 + \varphi_3 + \varphi_4) \\ \dot{\varphi}_1 \cos(\varphi_2 + \varphi_3 + \varphi_4) \\ \dot{\varphi}_2 + \dot{\varphi}_3 + \dot{\varphi}_4 \end{pmatrix},$$

$$T_{m_4} = T_{m_4,0} + T_{m_4,rot-0} + m_4\left(\mathbf{v}_{m_4-c.i.-0}, \mathbf{v}_0\right) = \frac{1}{2} m_4 \left\| \begin{pmatrix} a_1 \\ a_2 \\ a_3 \end{pmatrix} \right\|^2 +$$

$$\frac{1}{2} \mathbf{w}_4^\top \left\| \begin{matrix} I_{4xx} & 0 & 0 \\ 0 & m_4\left(\frac{l}{2}\right)^2 + I_{4yy} & 0 \\ 0 & 0 & m_4\left(\frac{l}{2}\right)^2 + I_{4zz} \end{matrix} \right\| \mathbf{w}_4 +$$

$$m_4 \begin{pmatrix} s_1 \\ s_2 \\ s_3 \end{pmatrix}^\top \begin{pmatrix} s_4 \\ s_5 \\ s_6 \end{pmatrix} =$$

$$-m_4 l^2 \left[\begin{matrix} \cos q_2 \cos(q_2 + q_3 + q_4 + q_5) - \\ \cos(q_2 + q_3) \cos(q_2 + q_3 + q_4 + q_5) \end{matrix} \right] \frac{\dot{q}_1^2}{2} -$$

$$m_4 l^2 \left[\cos(q_2 + q_3 + q_4) \cos(q_2 + q_3 + q_4 + q_5) \right] \frac{\dot{q}_1^2}{2} +$$

$$m_4 l^2 \left[\cos^2 q_2 + \cos^2(q_2 + q_3) \right] \frac{\dot{q}_1^2}{2} +$$

$$\left[\begin{matrix} I_{4xx} \sin^2(q_2 + q_3 + q_4 + q_5) \\ + \left(m_4 \frac{l^2}{4} + I_{4yy} \right) \cos^2(q_2 + q_3 + q_4 + q_5) \end{matrix} \right] \frac{\dot{q}_1^2}{2} +$$

$$m_4 l^2 \left[\cos q_2 \left(\cos(q_2 + q_3 + q_4) - \cos(q_2 + q_3) \right) \right] \dot{q}_1^2 -$$

$$m_4 l^2 \left[\cos(q_2 + q_3) \cos(q_2 + q_3 + q_4) \right] \dot{q}_1^2 +$$

$$\frac{1}{2} \left[m_4 l^2 \cos^2(q_2 + q_3 + q_4) \right] \dot{q}_1^2 + \frac{1}{2} \left[m_4 l^2 \right] \dot{q}_2^2 + \frac{1}{2} \left[m_4 l^2 \right] (\dot{q}_2 + \dot{q}_3)^2 +$$

$$\frac{1}{2} \left[m_4 l^2 \right] (\dot{q}_2 + \dot{q}_3 + \dot{q}_4)^2 + \frac{1}{2} \left[m_4 \frac{l^2}{4} + I_{4zz} \right] (\dot{q}_2 + \dot{q}_3 + \dot{q}_4 + \dot{q}_5)^2 -$$

$$\left[m_4 l^2 \cos q_3 \right] (\dot{q}_2 + \dot{q}_3) \dot{q}_2 + m_4 \left[l^2 \cos(q_3 + q_4) \right] (\dot{q}_2 + \dot{q}_3 + \dot{q}_4) \dot{q}_2 -$$

$$m_4 \frac{1}{2} \left[l^2 \cos(q_3 + q_4 + q_5) \right] (\dot{q}_2 + \dot{q}_3 + \dot{q}_4 + \dot{q}_5) \dot{q}_2 +$$

$$m_4 \frac{1}{2} \left[l^2 \cos(q_4 + q_5) \right] (\dot{q}_2 + \dot{q}_3 + \dot{q}_4 + \dot{q}_5) (\dot{q}_2 + \dot{q}_3) -$$

$$m_4 \frac{1}{2} \left[l^2 \cos q_5 \right] (\dot{q}_2 + \dot{q}_3 + \dot{q}_4 + \dot{q}_5) (\dot{q}_2 + \dot{q}_3 + \dot{q}_4) -$$

$$m_4 \left[l^2 \cos q_4 \right] (\dot{q}_2 + \dot{q}_3 + \dot{q}_4) (\dot{q}_2 + \dot{q}_3),$$

where

$$a_1 = -\dot\varphi_1 l \sin\varphi_1 \cos\varphi_2 - \dot\varphi_2 l \cos\varphi_1 \sin\varphi_2 +$$
$$\dot\varphi_1 l \sin\varphi_1 \cos(\varphi_2+\varphi_3) + (\dot\varphi_2+\dot\varphi_3) l \cos\varphi_1 \sin(\varphi_2+\varphi_3) -$$
$$\dot\varphi_1 l \sin\varphi_1 \cos(\varphi_2+\varphi_3+\varphi_4) -$$
$$(\dot\varphi_2+\dot\varphi_3+\dot\varphi_4) l \cos\varphi_1 \sin(\varphi_2+\varphi_3+\varphi_4),$$
$$a_2 = \dot\varphi_2 l \cos\varphi_2 - (\dot\varphi_2+\dot\varphi_3) l \cos(\varphi_2+\varphi_3) +$$
$$(\dot\varphi_2+\dot\varphi_3+\dot\varphi_4) l \cos(\varphi_2+\varphi_3+\varphi_4),$$
$$a_3 = -\dot\varphi_1 l \cos\varphi_1 \cos\varphi_2 + \dot\varphi_2 l \sin\varphi_1 \sin\varphi_2 +$$
$$\dot\varphi_1 l \cos\varphi_1 \cos(\varphi_2+\varphi_3) - (\dot\varphi_2+\dot\varphi_3) l \sin\varphi_1 \sin(\varphi_2+\varphi_3) -$$
$$\dot\varphi_1 l \cos\varphi_1 \cos(\varphi_2+\varphi_3+\varphi_4) +$$
$$(\dot\varphi_2+\dot\varphi_3+\dot\varphi_4) l \sin\varphi_1 \sin(\varphi_2+\varphi_3+\varphi_4),$$

$$\mathbf{w}_4 := \begin{pmatrix} \dot\varphi_1 \sin(\varphi_2+\varphi_3+\varphi_4+\varphi_5) \\ \dot\varphi_1 \cos(\varphi_2+\varphi_3+\varphi_4+\varphi_5) \\ \dot\varphi_2+\dot\varphi_3+\dot\varphi_4+\dot\varphi_5 \end{pmatrix},$$

$$s_1 = \dot\varphi_1 \frac{l}{2} \sin\varphi_1 \cos(\varphi_2+\varphi_3+\varphi_4+\varphi_5) +$$
$$(\dot\varphi_2+\dot\varphi_3+\dot\varphi_4+\dot\varphi_5) \frac{l}{2} \cos\varphi_1 \sin(\varphi_2+\varphi_3+\varphi_4+\varphi_5),$$
$$s_2 = -(\dot\varphi_2+\dot\varphi_3+\dot\varphi_4+\dot\varphi_5) \frac{l}{2} \cos(\varphi_2+\varphi_3+\varphi_4+\varphi_5),$$
$$s_3 = \dot\varphi_1 \frac{l}{2} \cos\varphi_1 \cos(\varphi_2+\varphi_3+\varphi_4+\varphi_5)$$
$$-(\dot\varphi_2+\dot\varphi_3+\dot\varphi_4+\dot\varphi_5) \frac{l}{2} \sin\varphi_1 \sin(\varphi_2+\varphi_3+\varphi_4+\varphi_5),$$
$$s_4 = -\dot\varphi_1 l \sin\varphi_1 \cos\varphi_2 - \dot\varphi_2 l \cos\varphi_1 \sin\varphi_2 + \dot\varphi_1 l \sin\varphi_1 \cos(\varphi_2+\varphi_3) +$$
$$(\dot\varphi_2+\dot\varphi_3) l \cos\varphi_1 \sin(\varphi_2+\varphi_3) - \dot\varphi_1 l \sin\varphi_1 \cos(\varphi_2+\varphi_3+\varphi_4) -$$
$$(\dot\varphi_2+\dot\varphi_3+\dot\varphi_4) l \cos\varphi_1 \sin(\varphi_2+\varphi_3+\varphi_4),$$
$$s_5 = \dot\varphi_2 l \cos\varphi_2 - (\dot\varphi_2+\dot\varphi_3) l \cos(\varphi_2+\varphi_3) +$$
$$(\dot\varphi_2+\dot\varphi_3+\dot\varphi_4) l \cos(\varphi_2+\varphi_3+\varphi_4),$$
$$s_6 = -\dot\varphi_1 l \cos\varphi_1 \cos\varphi_2 + \dot\varphi_2 l \sin\varphi_1 \sin\varphi_2 +$$
$$\dot\varphi_1 l \cos\varphi_1 \cos(\varphi_2+\varphi_3) - (\dot\varphi_2+\dot\varphi_3) l \sin\varphi_1 \sin(\varphi_2+\varphi_3) -$$
$$\dot\varphi_1 l \cos\varphi_1 \cos(\varphi_2+\varphi_3+\varphi_4) +$$
$$(\dot\varphi_2+\dot\varphi_3+\dot\varphi_4) l \sin\varphi_1 \sin(\varphi_2+\varphi_3+\varphi_4).$$

Potential energy

The potential energy V is

$$V = \sum_{i=1}^{8} V_{r_i} + \sum_{i=0}^{4} V_{m_i},$$

$$V_{r_1} = \frac{1}{2}c_1\left[\left(2r_1^2(1-\sin\varphi_2) + \frac{l^2}{2}(1+\sin\varphi_2) + 2r_1 l\cos\varphi_2\right)^{1/2} - l\right]^2$$

$$= \frac{1}{2}c_1\left[\left(2r_1^2(1-\sin q_2) + \frac{l^2}{2}(1+\sin q_2) + 2r_1 l\cos q_2\right)^{1/2} - l\right]^2,$$

$$V_{r_2} = \frac{1}{2}c_1\left[\left(2r_1^2(1-\sin\varphi_2) + \frac{l^2}{2}(1+\sin\varphi_2) - 2r_1 l\cos\varphi_2\right)^{1/2} - l\right]^2$$

$$= \frac{1}{2}c_1\left[\left(2r_1^2(1-\sin q_2) + \frac{l^2}{2}(1+\sin q_2) - 2r_1 l\cos q_2\right)^{1/2} - l\right]^2,$$

$$V_{r_3} = \frac{1}{2}c_2\left[\left(2r_2^2(1-\cos\varphi_3) + \frac{l^2}{2}(1+\cos\varphi_3) + 2r_2 l\sin\varphi_3\right)^{1/2} - l\right]^2$$

$$= \frac{1}{2}c_2\left[\left(2r_2^2(1-\cos q_3) + \frac{l^2}{2}(1+\cos q_3) + 2r_2 l\sin q_3\right)^{1/2} - l\right]^2,$$

$$V_{r_4} = \frac{1}{2}c_2\left[\left(2r_2^2(1-\cos\varphi_3) + \frac{l^2}{2}(1+\cos\varphi_3) - 2r_2 l\sin\varphi_3\right)^{1/2} - l\right]^2$$

$$= \frac{1}{2}c_2\left[\left(2r_2^2(1-\cos q_3) + \frac{l^2}{2}(1+\cos q_3) - 2r_2 l\sin q_3\right)^{1/2} - l\right]^2,$$

$$V_{r_5} = \frac{1}{2}c_3\left[\left(2r_3^2(1-\cos\varphi_4) + \frac{l^2}{2}(1+\cos\varphi_4) + 2r_3 l\sin\varphi_4\right)^{1/2} - l\right]^2$$

$$= \frac{1}{2}c_3\left[\left(2r_3^2(1-\cos q_4) + \frac{l^2}{2}(1+\cos q_4) + 2r_3 l\sin q_4\right)^{1/2} - l\right]^2,$$

$$V_{r_6} = \frac{1}{2}c_3\left[\left(2r_3^2(1-\cos\varphi_4) + \frac{l^2}{2}(1+\cos\varphi_4) - 2r_3 l\sin\varphi_4\right)^{1/2} - l\right]^2$$

$$= \frac{1}{2}c_3\left[\left(2r_3^2(1-\cos q_4) + \frac{l^2}{2}(1+\cos q_4) - 2r_3 l\sin q_4\right)^{1/2} - l\right]^2,$$

$$V_{r_7} = \frac{1}{2}c_4\left[\left(2r_4^2(1-\cos\varphi_5) + \frac{l^2}{2}(1+\cos\varphi_5) + 2r_4 l\sin\varphi_5\right)^{1/2} - l\right]^2$$

$$= \frac{1}{2}c_4\left[\left(2r_4^2(1-\cos q_5) + \frac{l^2}{2}(1+\cos q_5) + 2r_4 l\sin q_5\right)^{1/2} - l\right]^2,$$

Collection of electromechanical models

$$V_{r_8} = \frac{1}{2}c_4\left[\left(2r_4^2(1-\cos\varphi_5) + \frac{l^2}{2}(1+\cos\varphi_5) - 2r_4 l\sin\varphi_5\right)^{1/2} - l\right]^2$$

$$= \frac{1}{2}c_4\left[\left(2r_4^2(1-\cos q_5) + \frac{l^2}{2}(1+\cos q_5) - 2r_4 l\sin q_5\right)^{1/2} - l\right]^2,$$

$$V_{m_0} = 0,$$

$$V_{m_1} = m_1 g\left[\frac{l}{2}\sin\varphi_2 + \frac{l}{2}\right] = \frac{1}{2}m_1 gl\,[\sin q_2 + 1],$$

$$V_{m_2} = m_2 g\left[-\frac{l}{2}\sin(\varphi_2+\varphi_3) + l\sin\varphi_2 + \frac{l}{2}\right] =$$

$$\frac{1}{2}m_2 gl\,[2\sin q_2 - \sin(q_2+q_3) + 1],$$

$$V_{m_3} = m_3 g\left[\frac{l}{2}\sin(\varphi_2+\varphi_3+\varphi_4) - l\sin(\varphi_2+\varphi_3) + l\sin\varphi_2 + \frac{l}{2}\right]$$

$$= m_3 g\frac{1}{2}l\,[\sin(q_2+q_3+q_4) - 2\sin(q_2+q_3) + 2\sin q_2 + 1],$$

$$V_{m_4} = \frac{1}{2}m_4 g\,[2[\sin(q_2+q_3+q_4) - \sin(q_2+q_3) + \sin q_2]$$
$$-\sin(q_2+q_3+q_4+q_5) + 1],$$

which gives

$$V = \frac{c_1}{2}\left[\left(2r_1^2(1-\sin q_2) + \frac{l^2}{2}(1+\sin q_2) + 2r_1 l\cos q_2\right)^{1/2} - l\right]^2 +$$

$$\frac{c_1}{2}\left[\left(2r_1^2(1-\sin q_2) + \frac{l^2}{2}(1+\sin q_2) - 2r_1 l\cos q_2\right)^{1/2} - l\right]^2 +$$

$$\frac{c_2}{2}\left[\left(2r_2^2(1-\cos q_3) + \frac{l^2}{2}(1+\cos q_3) + 2r_2 l\sin q_3\right)^{1/2} - l\right]^2 +$$

$$\frac{c_2}{2}\left[\left(2r_2^2(1-\cos q_3) + \frac{l^2}{2}(1+\cos q_3) - 2r_2 l\sin q_3\right)^{1/2} - l\right]^2 +$$

$$\frac{c_3}{2}\left[\left(2r_3^2(1-\cos q_4) + \frac{l^2}{2}(1+\cos q_4) + 2r_3 l\sin q_4\right)^{1/2} - l\right]^2 +$$

$$\frac{c_3}{2}\left[\left(2r_3^2(1-\cos q_4) + \frac{l^2}{2}(1+\cos q_4) - 2r_3 l\sin q_4\right)^{1/2} - l\right]^2 +$$

$$\frac{c_4}{2}\left[\left(2r_4^2(1-\cos q_5) + \frac{l^2}{2}(1+\cos q_5) + 2r_4 l\sin q_5\right)^{1/2} - l\right]^2 +$$

$$\frac{1}{2}c_4\left[\left(2r_4^2\left(1-\cos q_5\right)+\frac{l^2}{2}\left(1+\cos q_5\right)-2r_4 l\sin q_5\right)^{1/2}-l\right]^2+$$

$$\frac{m_1 gl}{2}[\sin q_2+1]+\frac{1}{2}m_2 gl[2\sin q_2-\sin(q_2+q_3)+1]+$$

$$\frac{m_3 gl}{2}[\sin(q_2+q_3+q_4)-2\sin(q_2+q_3)+2\sin q_2+1]+$$

$$\frac{m_4 g}{2}[2[\sin(q_2+q_3+q_4)-\sin(q_2+q_3)+\sin q_2]-$$

$$\sin(q_2+q_3+q_4+q_5)+1].$$

Non-potential forces
The generalized non-potential forces are

$$Q_{non\text{-}pot,i}=\tau_i-f_{fric-i}\dot{\varphi}_i=\tau_i-f_{fric-i}\dot{q}_i,$$
τ_i is a twisting moment, $i=1,...,5$.

Lagrange equations
Based on the obtained formulas for T and V, we are able to derive the dynamic equation of the considered system in the Lagrange form:

$$\frac{d}{dt}\frac{\partial}{\partial \dot{q}_i}L-\frac{\partial}{\partial q_i}L=Q_{non\text{-}pot,i},\ i=1,...,5,\ L=T-V,$$

which allows to represent the dynamic model of the considered system in the standard matrix format:

$$D(q)\ddot{q}+C(q,\dot{q})\dot{q}+g(q)=\tau,$$

where

$$D(q)=\begin{Vmatrix} d_{11} & d_{12} & d_{13} & d_{14} & d_{15} \\ d_{21} & d_{22} & d_{23} & d_{24} & d_{25} \\ d_{31} & d_{32} & d_{33} & d_{34} & d_{35} \\ d_{41} & d_{42} & d_{43} & d_{44} & d_{45} \\ d_{51} & d_{52} & d_{53} & d_{54} & d_{55} \end{Vmatrix},$$

$$d_{11}=\left[\frac{1}{2}m_0 r_0+I_{1xx}\sin^2 q_2+\left(m_1\frac{l^2}{4}+I_{1yy}\right)\cos^2 q_2\right]+$$

$$\left[m_2 l^2\cos^2 q_2+I_{2xx}\sin^2(q_2+q_3)+\left(m_2\frac{l^2}{4}+I_{2yy}\right)\cos^2(q_2+q_3)\right]+$$

$$\left[I_{3xx}\sin^2(q_2+q_3+q_4)+\left(m_3\frac{l^2}{4}+I_{3yy}\right)\cos^2(q_2+q_3+q_4)\right]+$$

$$\left[I_{4xx}\sin^2(q_2+q_3+q_4+q_5)+\left(m_4\frac{l^2}{4}+I_{4yy}\right)\cos^2(q_2+q_3+q_4+q_5)\right]+$$

$$l^2\left[m_3\left(\cos^2 q_2 + \cos^2(q_2+q_3)\right) - m_2\cos q_2\cos(q_2+q_3)\right] +$$
$$m_3 l^2\left[(\cos q_2 - \cos(q_2+q_3))\cos(q_2+q_3+q_4) - 2\cos q_2\cos(q_2+q_3)\right] -$$
$$m_4 l^2\left[\cos q_2\cos(q_2+q_3+q_4+q_5) - \cos(q_2+q_3)\cos(q_2+q_3+q_4+q_5)\right] +$$
$$m_4 l^2\left[\cos^2(q_2+q_3+q_4) - \cos(q_2+q_3+q_4)\cos(q_2+q_3+q_4+q_5)\right] +$$
$$m_4 l^2\left[\cos^2 q_2 + \cos^2(q_2+q_3) - 2\cos(q_2+q_3)\cos(q_2+q_3+q_4)\right] +$$
$$2m_4 l^2\left[\cos q_2(\cos(q_2+q_3+q_4) - \cos(q_2+q_3))\right],$$

$d_{12} = d_{13} = d_{14} = d_{15} = 0, \ d_{21} = 0,$

$$d_{22} = \left[\left(\frac{1}{4}m_1 + m_2 + m_3 + m_4\right)l^2 + I_{1zz}\right] +$$
$$\left[\left(\frac{1}{4}m_2 + m_3 + m_4\right)l^2 + I_{2zz}\right] - \left[(m_2 + 2m_3 + m_4)l^2\cos q_3\right] +$$
$$\left[\left(\frac{1}{4}m_3 + m_4\right)l^2 + I_{3zz}\right] + \left[m_4\frac{l^2}{4} + I_{4zz}\right] +$$
$$\left[(m_3 + 2m_4)l^2\cos(q_3+q_4) - (m_3 + m_4)l^2\cos q_4\right] +$$
$$m_4 l^2\left[\cos(q_4+q_5) - \cos(q_3+q_4+q_5) - \cos q_5\right],$$

$$d_{23} = \frac{l^2}{2}\left[m_4\cos(q_4+q_5) - (m_3+m_4)\cos q_4\right] +$$
$$\left[\left(\frac{1}{4}m_2 + m_3 + m_4\right)l^2 + I_{2zz}\right] - \frac{(m_2+2m_3+m_4)l^2}{2}\left[\cos q_3\right] +$$
$$\left[\left(\frac{1}{4}m_3 + m_4\right)l^2 + I_{3zz}\right] + \left[m_4\frac{l^2}{4} + I_{4zz}\right] +$$
$$\frac{1}{2}\left[(m_3+2m_4)l^2\cos(q_3+q_4) - (m_3+m_4)l^2\cos q_4\right] +$$
$$\frac{m_4 l^2}{2}\left[\cos(q_4+q_5) - \cos(q_3+q_4+q_5) - \cos q_5\right] - \frac{m_4}{2}\left[l^2\cos q_5\right],$$

$$d_{24} = \left[\left(\frac{1}{4}m_3 + m_4\right)l^2 + I_{3zz}\right] + \left[m_4\frac{l^2}{4} + I_{4zz}\right] +$$
$$\frac{1}{2}\left[(m_3+2m_4)l^2\cos(q_3+q_4) - (m_3+m_4)l^2\cos q_4\right] +$$
$$\frac{m_4 l^2}{2}\left[\cos(q_4+q_5) - \cos(q_3+q_4+q_5) - \cos q_5\right] - \frac{m_4}{2}\left[l^2\cos q_5\right],$$

$$d_{25} = \left[m_4\frac{l^2}{4} + I_{4zz}\right] + \frac{m_4 l^2}{2}\left[\cos(q_4+q_5) - \cos(q_3+q_4+q_5) - \cos q_5\right],$$

$d_{31} = 0,$

$$d_{32} = -\frac{1}{2}\left[(m_2 + 2m_3 + 2m_4)l^2\cos q_3 + (m_3 + 2m_4)\cos q_4\right] +$$

$$\frac{l^2}{2}[(m_3+2m_4)\cos(q_3+q_4)+m_4\cos(q_4+q_5)-m_4\cos(q_3+q_4+q_5)]+$$

$$\left[\left(\frac{1}{4}m_2+m_3+m_4\right)l^2+I_{2zz}\right]+\left[\left(\frac{1}{4}m_3+m_4\right)l^2+I_{3zz}\right]-$$

$$\frac{l^2(m_3+2m_4)}{2}\cos q_4+\left[m_4\frac{l^2}{4}+I_{4zz}\right]-\frac{m_4 l^2}{2}\cos q_5+$$

$$\frac{m_4 l^2}{2}[\cos(q_4+q_5)-\cos q_5],$$

$$d_{33}=\left[\left(\frac{1}{4}m_2+m_3+m_4\right)l^2+I_{2zz}\right]+\left[\left(\frac{1}{4}m_3+m_4\right)l^2+I_{3zz}\right]-$$

$$l^2(m_3+2m_4)\cos q_4+\left[m_4\frac{l^2}{4}+I_{4zz}\right]+m_4 l^2[\cos(q_4+q_5)-\cos q_5],$$

$$d_{34}=\left[\left(\frac{1}{4}m_3+m_4\right)l^2+I_{3zz}\right]-\frac{1}{2}l^2[(m_3+2m_4)\cos q_4]+$$

$$\left[m_4\frac{l^2}{4}+I_{4zz}\right]-\frac{m_4}{2}\left[l^2\cos q_5\right]+\frac{1}{2}m_4 l^2[\cos(q_4+q_5)-\cos q_5],$$

$$d_{35}=\left[m_4\frac{l^2}{4}+I_{4zz}\right]+\frac{1}{2}m_4 l^2[\cos(q_4+q_5)-\cos q_5],$$

$$d_{41}=0,$$

$$d_{42}=\frac{l^2}{2}[(m_3+2m_4)\cos(q_3+q_4)-m_4\cos(q_3+q_4+q_5)]-$$

$$\frac{l^2}{2}[(m_3+2m_4)\cos q_4+m_4\cos q_5-m_4\cos(q_4+q_5)]+$$

$$\left[\left(\frac{1}{4}m_3+m_4\right)l^2+I_{3zz}\right]+\left[m_4\frac{l^2}{4}+I_{4zz}\right]-m_4\frac{1}{2}\left[l^2\cos q_5\right],$$

$$d_{43}=-\frac{l^2}{2}[(m_3+2m_4)\cos q_4+m_4\cos q_5-m_4\cos(q_4+q_5)]+$$

$$\left[\left(\frac{1}{4}m_3+m_4\right)l^2+I_{3zz}\right]+\left[m_4\frac{l^2}{4}+I_{4zz}\right]-\frac{m_4}{2}\left[l^2\cos q_5\right],$$

$$d_{44}=\left[\left(\frac{1}{4}m_3+m_4\right)l^2+I_{3zz}\right]+\left[m_4\frac{l^2}{4}+I_{4zz}\right]-m_4\left[l^2\cos q_5\right],$$

$$d_{45}=\left[m_4\frac{l^2}{4}+I_{4zz}\right]-\frac{1}{2}m_4\left[l^2\cos q_5\right],$$

$$d_{51}=0,$$

$$d_{52}=-\frac{m_4 l^2}{2}\cos(q_3+q_4+q_5)+\frac{1 m_4 l^2}{2}\cos(q_4+q_5)-$$

$$\frac{m_4 l^2}{2}\cos q_5+\left[m_4\frac{l^2}{4}+I_{4zz}\right],$$

Collection of electromechanical models

$$d_{53} = \frac{m_4 l^2}{2} \cos(q_4 + q_5) - \frac{m_4 l^2}{2} \cos q_5 + m_4 \frac{l^2}{4} + I_{4zz},$$

$$d_{54} = -\frac{1}{2} m_4 l^2 \cos q_5 + m_4 \frac{l^2}{4} + I_{4zz},$$

$$d_{55} = m_4 \frac{l^2}{4} + I_{4zz},$$

and

$$C(q, \dot{q}) = \begin{Vmatrix} c_{11} & c_{12} & c_{13} & c_{14} & c_{15} \\ c_{21} & c_{22} & c_{23} & c_{24} & c_{25} \\ c_{31} & c_{32} & c_{33} & c_{34} & c_{35} \\ c_{41} & c_{42} & c_{43} & c_{44} & c_{45} \\ c_{51} & c_{52} & c_{53} & c_{54} & c_{55} \end{Vmatrix},$$

$c_{11} = f_{fric-1} +$

$2\left[\left(I_{1xx} - m_1 \frac{l^2}{4} - I_{1yy} + l^2(m_4 - m_3 - m_2)\right) \sin q_2 \cos q_2\right] \dot{q}_2 +$

$l^2 [(m_2 + 2m_3 + 2m_4) \sin q_2 \cos(q_2 + q_3)] \dot{q}_2,$

$l^2 [m_4 \sin q_2 \cos(q_2 + q_3 + q_4 + q_5) - (m_3 + m_4) \sin q_2 \cos(q_2 + q_3 + q_4)] \dot{q}_2 +$

$2\left[\left(I_{4xx} - m_4 \frac{l^2}{4} - I_{4yy}\right) \sin(q_2 + q_3 + q_4 + q_5) \cos(q_2 + q_3 + q_4 + q_5)\right] \times$

$(\dot{q}_2 + \dot{q}_3 + \dot{q}_4 + \dot{q}_5) +$

$2\left[\left(I_{2xx} - m_2 \frac{l^2}{4} - I_{2yy} + l^2(m_4 - m_3)\right) \sin(q_2 + q_3) \cos(q_2 + q_3)\right] (\dot{q}_2 + \dot{q}_3) +$

$l^2 [(m_2 + 2m_3 + 2m_4) \cos q_2 \sin(q_2 + q_3)] (\dot{q}_2 + \dot{q}_3) +$

$l^2 [(m_3 + 2m_4) \sin(q_2 + q_3) \cos(q_2 + q_3 + q_4)] (\dot{q}_2 + \dot{q}_3) +$

$2\left(I_{3xx} - m_3 \frac{l^2}{4} - I_{3yy}\right) (\dot{q}_2 + \dot{q}_3 + \dot{q}_4) \sin(q_2 + q_3 + q_4) \cos(q_2 + q_3 + q_4) -$

$l^2 (m_3 + 2m_4) (\dot{q}_2 + \dot{q}_3 + \dot{q}_4) \cos q_2 \sin(q_2 + q_3 + q_4) +$

$l^2 (m_3 + 2m_4) (\dot{q}_2 + \dot{q}_3 + \dot{q}_4) \cos(q_2 + q_3) \sin(q_2 + q_3 + q_4) +$

$2m_4 l^2 (\dot{q}_2 + \dot{q}_3 + \dot{q}_4) \sin(q_2 + q_3 + q_4) \cos(q_2 + q_3 + q_4) +$

$m_4 l^2 [\sin(q_2 + q_3 + q_4) \cos(q_2 + q_3 + q_4 + q_5)] (\dot{q}_2 + \dot{q}_3 + \dot{q}_4)$

$m_4 l^2 [\cos(q_2 + q_3) \sin(q_2 + q_3 + q_4 + q_5)] (\dot{q}_2 + \dot{q}_3 + \dot{q}_4 + \dot{q}_5) +$

$m_4 l^2 [\cos(q_2 + q_3 + q_4) \sin(q_2 + q_3 + q_4 + q_5)] (\dot{q}_2 + \dot{q}_3 + \dot{q}_4 + \dot{q}_5),$

$c_{12} = c_{13} = c_{14} = c_{15} = 0,$

$c_{21} = -\left[\left(I_{1xx} - m_1 \frac{l^2}{4} - I_{1yy}\right) \sin q_2 \cos q_2 + \frac{1}{2} m_2 l^2 \sin(2q_2 + q_3)\right] \dot{q}_1 -$

$$\left[m_2 l^2 \sin q_2 \cos q_2 + \left(I_{2xx} - m_2 \frac{l^2}{4} - I_{2yy} \right) \sin (q_2 + q_3) \cos (q_2 + q_3) \right] \dot{q}_1 -$$

$$\left[\left(I_{3xx} - m_3 \frac{l^2}{4} - I_{3yy} \right) \sin (q_2 + q_3 + q_4) \cos (q_2 + q_3 + q_4) \right] \dot{q}_1 -$$

$$\frac{m_3 l^2}{2} [2 \sin (2q_2 + q_3) + \sin (q_3 + q_4) + \sin (2q_2 + 2q_3 + q_4)] \dot{q}_1 -$$

$$m_3 l^2 [\sin q_2 \cos q_2 + \sin (q_2 + q_3) \cos (q_2 + q_3)] \dot{q}_1 -$$

$$\frac{m_4 l^2}{2} [\sin (2q_2 + q_3 + q_4 + q_5) - \sin (2q_2 + 2q_3 + q_4 + q_5)] \dot{q}_1 -$$

$$\frac{m_4 l^2}{2} [\sin (2q_2 + 2q_3 + 2q_4 + q_5)] \dot{q}_1 +$$

$$m_4 l^2 [\sin q_2 \cos q_2 + \sin (q_2 + q_3) \cos (q_2 + q_3)] \dot{q}_1 -$$

$$\left[\left(I_{4xx} - m_4 \frac{l^2}{4} - I_{4yy} \right) \sin (q_2 + q_3 + q_4 + q_5) \cos (q_2 + q_3 + q_4 + q_5) \right] \dot{q}_1 -$$

$$m_4 l^2 [\sin (2q_2 + q_3) - \sin (2q_2 + q_3 + q_4) + \sin (2q_2 + 2q_3 + q_4)] \dot{q}_1 +$$

$$m_4 l^2 [\sin (q_2 + q_3 + q_4) \cos (q_2 + q_3 + q_4)] \dot{q}_1,$$

$$c_{22} = f_{fric-2},$$

$$c_{23} = \frac{1}{2} l^2 [(m_3 + m_4) \sin q_4] \dot{q}_4 -$$

$$\frac{1}{2} l^2 [(m_3 + 2m_4) \sin (q_3 + q_4)] (2\dot{q}_2 + \dot{q}_3 + \dot{q}_4) +$$

$$\frac{(m_2 + 2m_3 + m_4) l^2}{2} [\sin q_3] (2\dot{q}_2 + \dot{q}_3) - \frac{l^2 m_4}{2} [\sin (q_4 + q_5)] (\dot{q}_4 + \dot{q}_5) +$$

$$\frac{1 m_4 l^2}{2} [\sin (q_3 + q_4 + q_5)] (2\dot{q}_2 + \dot{q}_3 + \dot{q}_4 + \dot{q}_5),$$

$$c_{24} = \frac{l^2 (m_3 + m_4)}{2} [\sin q_4 - (m_3 + 2m_4) \sin (q_3 + q_4)] (2\dot{q}_2 + \dot{q}_3 + \dot{q}_4) +$$

$$\frac{m_4 l^2}{2} [\sin (q_3 + q_4 + q_5) - \sin (q_4 + q_5)] (2\dot{q}_2 + \dot{q}_3 + \dot{q}_4 + \dot{q}_5),$$

$$c_{25} = \frac{m_4 l^2}{2} [\sin q_5 + \sin (q_3 + q_4 + q_5) - \sin (q_4 + q_5)] (2\dot{q}_2 + \dot{q}_3 + \dot{q}_4 + \dot{q}_5) +$$

$$\frac{m_4 l^2}{2} [\sin q_5] (\dot{q}_3 + \dot{q}_4),$$

$$c_{31} = - \left[\left(I_{2xx} - m_2 \frac{l^2}{4} - I_{2yy} \right) \sin (q_2 + q_3) \cos (q_2 + q_3) \right] \dot{q}_1 -$$

$$\left[\left(I_{3xx} - m_3 \frac{l^2}{4} - I_{3yy} \right) \sin (q_2 + q_3 + q_4) \cos (q_2 + q_3 + q_4) \right] \dot{q}_1 -$$

$$\left[\left(I_{4xx} - m_4 \frac{l^2}{4} - I_{4yy} \right) \sin (q_2 + q_3 + q_4 + q_5) \cos (q_2 + q_3 + q_4 + q_5) \right] \dot{q}_1 -$$

$$\frac{l^2(m_2+2m_3+2m_4)}{2}[\cos q_2 \sin(q_2+q_3)]\dot{q}_1-$$

$$\frac{l^2(m_3+2m_4)}{2}[\sin(2q_2+2q_3+q_4)]\dot{q}_1+$$

$$l^2[(m_3+m_4)\sin(q_2+q_3)\cos(q_2+q_3)]\dot{q}_1-$$

$$\frac{l^2(2m_4-m_3)}{2}[\cos q_2 \sin(q_2+q_3+q_4)]\dot{q}_1-$$

$$\frac{m_4 l^2}{2}[\cos q_2 \sin(q_2+q_3+q_4+q_5)]\dot{q}_1+m_4 l^2[\sin q_2 \cos q_2]\dot{q}_1-$$

$$\frac{m_4 l^2}{2}[\sin(2q_2+2q_3+2q_4+q_5)-\sin(2q_2+2q_3+q_4+q_5)]\dot{q}_1+$$

$$m_4 l^2[\sin(q_2+q_3+q_4)\cos(q_2+q_3+q_4)]\dot{q}_1,$$

$$c_{32}=\frac{(m_3+2m_4)}{2}\left[\sin(q_3+q_4)-(m_2+2m_3+2m_4)l^2\sin q_3\right]\dot{q}_2-$$

$$\frac{m_4 l^2}{2}[\sin(q_3+q_4+q_5)]\dot{q}_2+\frac{(m_3+2m_4)}{2}[\sin q_4]\dot{q}_4-$$

$$\frac{m_4 l^2}{2}[\sin(q_4+q_5)](\dot{q}_4+\dot{q}_5),$$

$$c_{33}=f_{fric-3},$$

$$c_{34}=\frac{(m_3+2m_4)l^2}{2}[\sin q_4](\dot{q}_2+2\dot{q}_3+\dot{q}_4)-$$

$$\frac{m_4 l^2}{2}[\sin(q_4+q_5)](\dot{q}_2+2\dot{q}_3+\dot{q}_4+\dot{q}_5),$$

$$c_{35}=\frac{m_4 l^2}{2}[\sin q_5](\dot{q}_2+\dot{q}_4)+$$

$$\frac{m_4 l^2}{2}[\sin q_5-\sin(q_4+q_5)](\dot{q}_2+2\dot{q}_3+\dot{q}_4+\dot{q}_5),$$

$$c_{41}=-\left[\left(I_{3xx}-m_3\frac{l^2}{4}-I_{3yy}\right)\sin(q_2+q_3+q_4)\cos(q_2+q_3+q_4)\right]\dot{q}_1-$$

$$\left[\left(I_{4xx}-m_4\frac{l^2}{4}-I_{4yy}\right)\sin(q_2+q_3+q_4+q_5)\cos(q_2+q_3+q_4+q_5)\right]\dot{q}_1+$$

$$\frac{(m_3+2m_4)l^2}{2}[(\cos q_2-\cos(q_2+q_3))\sin(q_2+q_3+q_4)]\dot{q}_1-$$

$$\frac{m_4 l^2}{2}[(\cos q_2-\cos(q_2+q_3))\sin(q_2+q_3+q_4+q_5)]\dot{q}_1-$$

$$\frac{1}{2}m_4 l^2[\sin(2q_2+2q_3+2q_4+q_5)]\dot{q}_1+$$

$$\frac{m_4 l^2}{2}[\sin(q_2+q_3+q_4)\cos(q_2+q_3+q_4)]\dot{q}_1,$$

$$c_{42} = \frac{(m_3 + 2m_4)\,l^2}{2}\,[\sin(q_3+q_4)]\,\dot{q}_2 - \frac{(m_3+2m_4)\,l^2}{2}\,[\sin q_4]\,(\dot{q}_2+\dot{q}_3) +$$

$$\frac{m_4 l^2}{2}\,[\sin(q_4+q_5)]\,(\dot{q}_4+\dot{q}_5) - \frac{m_4 l^2}{2}\,[\sin(q_3+q_4+q_5)]\,\dot{q}_5,$$

$$c_{43} = -\frac{(m_3+2m_4)\,l^2}{2}\,[\sin q_4]\,(\dot{q}_2+\dot{q}_3) +$$

$$\frac{m_4 l^2}{2}\,[\sin(q_4+q_5)]\,(\dot{q}_2+\dot{q}_3+\dot{q}_4),$$

$$c_{44} = f_{fric-4} - \frac{m_4 l^2}{2}\,[\sin(q_4+q_5)]\,(\dot{q}_2+\dot{q}_3),$$

$$c_{45} = -\frac{m_4 l^2}{2}\,[\sin(q_4+q_5)]\,\dot{q}_2 + \frac{m_4 l^2}{2}\,[\sin q_5]\,(2\dot{q}_2+2\dot{q}_3+2\dot{q}_4+\dot{q}_5),$$

$$c_{51} = -\left[\left(I_{4xx} - \frac{m_4 l^2}{4} - I_{4yy}\right)\sin(q_2+q_3+q_4+q_5)\cos(q_2+q_3+q_4+q_5)\right]\dot{q}_1 -$$

$$\frac{m_4 l^2}{2}\,[(\cos q_2 - \cos(q_2+q_3) + \cos(q_2+q_3+q_4)) \times$$

$$\sin(q_2+q_3+q_4+q_5)]\,\dot{q}_1,$$

$$c_{52} = -\frac{m_4 l^2}{2}\,[\sin(q_3+q_4+q_5)]\,\dot{q}_2 + \frac{m_4 l^2}{2}\,[\sin(q_4+q_5)]\,(\dot{q}_2+\dot{q}_3)$$

$$-\frac{1}{2}m_4 l^2\,[\sin q_5]\,(\dot{q}_2+\dot{q}_3+\dot{q}_4),$$

$$c_{53} = \frac{1}{2}m_4\left[l^2 \sin(q_4+q_5)\right](\dot{q}_2+\dot{q}_3) - \frac{1}{2}m_4\left[l^2 \sin q_5\right](\dot{q}_2+\dot{q}_3+\dot{q}_4),$$

$$c_{54} = -\frac{1}{2}m_4\left[l^2 \sin q_5\right](\dot{q}_2+\dot{q}_3+\dot{q}_4),$$

$$c_{55} = f_{fric-5},$$

$$g(q) = \begin{Vmatrix} g_1 \\ g_2 \\ g_3 \\ g_4 \\ g_5 \end{Vmatrix},\ \tau = \begin{Vmatrix} \tau_1 \\ \tau_2 \\ \tau_3 \\ \tau_4 \\ \tau_5 \end{Vmatrix},$$

$$g_1 = 0,$$

$$g_2 = c_1\left[\left(2r_1^2(1-\sin q_2) + \frac{l^2}{2}(1+\sin q_2) + 2r_1 l \cos q_2\right)^{1/2} - l\right] \times$$

$$\frac{1}{2}\left[\left(2r_1^2(1-\sin q_2) + \frac{l^2}{2}(1+\sin q_2) + 2r_1 l \cos q_2\right)^{-1/2}\right] \times$$

$$\left(\left(\frac{l^2}{2} - 2r_1^2\right)\cos q_2 - 2r_1 l \sin q_2\right) +$$

$$c_1 \left[\left(2r_1^2 (1 - \sin q_2) + \frac{l^2}{2}(1 + \sin q_2) - 2r_1 l \cos q_2 \right)^{1/2} - l \right] \times$$

$$\frac{1}{2} \left[\left(2r_1^2 (1 - \sin q_2) + \frac{l^2}{2}(1 + \sin q_2) - 2r_1 l \cos q_2 \right)^{-1/2} \right] \times$$

$$\left(\left(\frac{l^2}{2} - 2r_1^2 \right) \cos q_2 + 2r_1 l \sin q_2 \right)$$

$$+ \frac{(m_1 + 2m_2 + 2m_3 + 2m_4) gl}{2} [\cos q_2] - \frac{m_4 gl}{2} \cos(q_2 + q_3 + q_4 + q_5) -$$

$$\frac{(m_2 + 2m_3 + 2 + 2m_4) gl}{2} \cos(q_2 + q_3) + \frac{(m_3 + 2m_4) gl}{2} \cos(q_2 + q_3 + q_4),$$

$$g_3 = c_2 \left[\left(2r_2^2 (1 - \sin q_3) + \frac{l^2}{2}(1 + \sin q_3) + 2r_2 l \cos q_3 \right)^{1/2} - l \right] \times$$

$$\frac{1}{2} \left[\left(2r_2^2 (1 - \sin q_3) + \frac{l^2}{2}(1 + \sin q_3) + 2r_2 l \cos q_3 \right)^{-1/2} \right] \times$$

$$\left(\left(\frac{l^2}{2} - 2r_2^2 \right) \cos q_3 - 2r_2 l \sin q_3 \right) +$$

$$c_2 \left[\left(2r_2^2 (1 - \sin q_3) + \frac{l^2}{2}(1 + \sin q_3) - 2r_2 l \cos q_3 \right)^{1/2} - l \right] \times$$

$$\frac{1}{2} \left[\left(2r_2^2 (1 - \sin q_3) + \frac{l^2}{2}(1 + \sin q_3) - 2r_2 l \cos q_3 \right)^{-1/2} \right] \times$$

$$\left(\left(\frac{l^2}{2} - 2r_2^2 \right) \cos q_3 + 2r_2 l \sin q_3 \right) -$$

$$\frac{(m_2 + 2m_3 + 2m_4) gl}{2} \cos(q_2 + q_3) + \frac{(m_3 + 2m_4) gl}{2} \cos(q_2 + q_3 + q_4) -$$

$$\frac{m_4 gl}{2} \cos(q_2 + q_3 + q_4 + q_5),$$

$$g_4 = c_3 \left[\left(2r_3^2 (1 - \sin q_4) + \frac{l^2}{2}(1 + \sin q_4) + 2r_3 l \cos q_4 \right)^{1/2} - l \right] \times$$

$$\frac{1}{2} \left[\left(2r_3^2 (1 - \sin q_4) + \frac{l^2}{2}(1 + \sin q_4) + 2r_3 l \cos q_4 \right)^{-1/2} \right] \times$$

$$\left(\left(\frac{l^2}{2} - 2r_3^2 \right) \cos q_4 - 2r_3 l \sin q_4 \right) +$$

$$c_3 \left[\left(2r_3^2 (1 - \sin q_4) + \frac{l^2}{2}(1 + \sin q_4) - 2r_3 l \cos q_4 \right)^{1/2} - l \right] \times$$

$$\frac{1}{2}\left[\left(2r_3^2(1-\sin q_4)+\frac{l^2}{2}(1+\sin q_4)-2r_3l\cos q_4\right)^{-1/2}\right]\times$$

$$\left(\left(\frac{l^2}{2}-2r_3^2\right)\cos q_4+2r_3l\sin q_4\right)+$$

$$\frac{(m_3+2m_4)gl}{2}[\cos(q_2+q_3+q_4)]-\frac{m_4gl}{2}\cos(q_2+q_3+q_4+q_5),$$

$$g_5=c_4\left[\left(2r_4^2(1-\sin q_5)+\frac{l^2}{2}(1+\sin q_5)+2r_4l\cos q_5\right)^{1/2}-l\right]\times$$

$$\frac{1}{2}\left[\left(2r_4^2(1-\sin q_5)+\frac{l^2}{2}(1+\sin q_5)+2r_4l\cos q_5\right)^{-1/2}\right]\times$$

$$\left(\left(\frac{l^2}{2}-2r_4^2\right)\cos q_5-2r_4l\sin q_5\right)+$$

$$c_4\left[\left(2r_4^2(1-\sin q_5)+\frac{l^2}{2}(1+\sin q_5)-2r_4l\cos q_5\right)^{1/2}-l\right]\times$$

$$\frac{1}{2}\left[\left(2r_4^2(1-\sin q_5)+\frac{l^2}{2}(1+\sin q_5)-2r_4l\cos q_5\right)^{-1/2}\right]\times$$

$$\left(\left(\frac{l^2}{2}-2r_4^2\right)\cos q_5+2r_4l\sin q_5\right)-\frac{m_4gl}{2}\cos(q_2+q_3+q_4+q_5).$$

12.18 Articulated robot manipulator 2

Consider the robot depicted in Fig. 12.18.

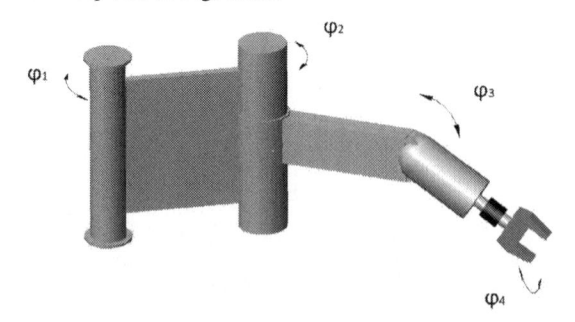

Figure 12.18 Articulated robot.

Generalized coordinates

The generalized coordinates for this mechanical system are as follows:

$q_1:=\varphi_1,\quad q_2:=\varphi_2,\quad q_3:=\varphi_3,\quad q_4:=\varphi_4.$

Kinetic energy

The kinetic energy $T = \sum_{i=1}^{5} T_{m_i}$ of this system is given by the following expression:

$$T_{m_1} = T_{m_1,0} + T_{m_1,rot-0} + m_1 \left(\mathbf{v}_{m_1-c.i.-0}, \mathbf{v}_0\right) =$$

$$T_{m_1,rot-0} = \frac{1}{2} I_{1yy} \dot{\varphi}_1^2 = \frac{1}{2} I_{1yy} \dot{q}_1^2,$$

$$T_{m_2} = T_{m_2,0} + T_{m_2,rot-0} + m_2 \left(\mathbf{v}_{m_2-c.i.-0}, \mathbf{v}_0\right) = T_{m_2,0} + T_{m_2,rot-0} =$$

$$\frac{m_2}{2} (l_1 + r_2)^2 \dot{\varphi}_1^2 + \frac{1}{2} \begin{pmatrix} 0 \\ \dot{\varphi}_1 \\ 0 \end{pmatrix}^\mathsf{T} \begin{Vmatrix} I_{2xx} & 0 & 0 \\ 0 & I_{2yy} & 0 \\ 0 & 0 & I_{2zz} \end{Vmatrix} \begin{pmatrix} 0 \\ \dot{\varphi}_1 \\ 0 \end{pmatrix} =$$

$$\frac{1}{2} \left[m_2 (l_1 + r_2)^2 + I_{2yy} \right] \dot{q}_1^2,$$

$$T_{m_3} = T_{m_3,0} + T_{m_3,rot-0} + m_2 \left(\mathbf{v}_{m_3-c.i.-0}, \mathbf{v}_0\right) = \frac{1}{2} m_3 (l_1 + r_2)^2 \dot{\varphi}_1^2 +$$

$$\frac{1}{2} \begin{pmatrix} 0 \\ \dot{\varphi}_1 + \dot{\varphi}_2 \\ 0 \end{pmatrix}^\mathsf{T} \begin{Vmatrix} I_{3xx} & 0 & 0 \\ 0 & m_3(r_2+d_3)^2 + I_{3yy} & 0 \\ 0 & 0 & I_{3zz} \end{Vmatrix} \begin{pmatrix} 0 \\ \dot{\varphi}_1 + \dot{\varphi}_2 \\ 0 \end{pmatrix} +$$

$$m_3 \begin{pmatrix} -(\dot{\varphi}_1 + \dot{\varphi}_2)(r_2 + d_3) \cos(\varphi_1 + \varphi_2) \\ 0 \\ -(\dot{\varphi}_1 + \dot{\varphi}_2)(r_2 + d_3) \sin(\varphi_1 + \varphi_2) \end{pmatrix}^\mathsf{T} \begin{pmatrix} -\dot{\varphi}_1 (l_1 + r_2) \sin \varphi_1 \\ 0 \\ -\dot{\varphi}_1 (l_1 + r_2) \cos \varphi_1 \end{pmatrix} =$$

$$\frac{m_3}{2} \left[(l_1 + r_2)^2 \right] \dot{q}_1^2 + \frac{1}{2} \left[m_3 (r_2 + d_3)^2 + I_{3yy} \right] (\dot{q}_1 + \dot{q}_2)^2 +$$

$$m_3 \left[(r_2 + d_3)(l_1 + r_2) \sin(2q_1 + q_2) \right] (\dot{q}_1 + \dot{q}_2) \dot{q}_1,$$

$$T_{m_4} = T_{m_4,0} + T_{m_4,rot-0} + m_4 \left(\mathbf{v}_{m_4-c.i.-0}, \mathbf{v}_0\right) =$$

$$\frac{m_4}{2} \left[\dot{\varphi}_1^2 (l_1 + r_2)^2 + (\dot{\varphi}_1 + \dot{\varphi}_2)^2 (r_2 + d_3)^2 \right] +$$

$$m_4 \left[\dot{\varphi}_1 (\dot{\varphi}_1 + \dot{\varphi}_2)(l_1 + r_2)(r_2 + d_3) \sin(2\varphi_1 + \varphi_2) \right] +$$

$$\frac{1}{2} \mathbf{w}^\mathsf{T} \begin{Vmatrix} I_{4xx} & 0 & 0 \\ 0 & m_4 d_4^2 + I_{4yy} & 0 \\ 0 & 0 & m_4 d_4^2 + I_{4zz} \end{Vmatrix} \mathbf{w} + m_4 \mathbf{a}_4^\mathsf{T} \mathbf{b}_4 =$$

$$\frac{1}{2} \left[m_4 (r_2 + d_3)^2 + I_{4xx} \sin^2 q_3 + \left(m_4 d_4^2 + I_{4yy} \right) \cos^2 q_3 \right] (\dot{q}_1 + \dot{q}_2)^2 +$$

$$[m_4 (l_1 + r_2)(r_2 + d_3) \sin(2q_1 + q_2)] (\dot{q}_1 + \dot{q}_2) \dot{q}_1 -$$

$$[m_4 d_4 (l_1 + r_2) \cos(2q_1 + q_2) \cos q_3] (\dot{q}_1 + \dot{q}_2) \dot{q}_1 + \frac{m_4}{2} \left[(l_1 + r_2)^2 \right] \dot{q}_1^2 +$$

$$\frac{1}{2} \left[m_4 d_4^2 + I_{4zz} \right] \dot{q}_3^2 - m_4 [d_4 (l_1 + r_2) \sin(2q_1 + q_2) \sin q_3] \dot{q}_1 \dot{q}_3 -$$

$$m_4 [d_4 (r_2 + d_3) \sin q_3] (\dot{q}_1 + \dot{q}_2) \dot{q}_3,$$

where

$$\mathbf{w}_4 = \begin{pmatrix} -(\dot{\varphi}_1 + \dot{\varphi}_2)\sin\varphi_3 \\ -(\dot{\varphi}_1 + \dot{\varphi}_2)\cos\varphi_3 \\ \dot{\varphi}_3 \end{pmatrix},$$

$$\mathbf{a}_4 = \begin{pmatrix} -(\ddot{\varphi}_1 + \ddot{\varphi}_2)d_4\sin(\varphi_1 + \varphi_2)\cos\varphi_3 + \dot{\varphi}_3 d_4\cos(\varphi_1 + \varphi_2)\sin\varphi_3 \\ \dot{\varphi}_3 d_4 \cos\varphi_3 \\ (\ddot{\varphi}_1 + \ddot{\varphi}_2)d_4\cos(\varphi_1 + \varphi_2)\cos\varphi_3 + \dot{\varphi}_3 d_4\sin(\varphi_1 + \varphi_2)\sin\varphi_3 \end{pmatrix},$$

$$\mathbf{b}_4 = \begin{pmatrix} -\dot{\varphi}_1(l_1 + r_2)\sin\varphi_1 - (\dot{\varphi}_1 + \dot{\varphi}_2)(r_2 + d_3)\cos(\varphi_1 + \varphi_2) \\ 0 \\ -\dot{\varphi}_1(l_1 + r_2)\cos\varphi_1 - (\dot{\varphi}_1 + \dot{\varphi}_2)(r_2 + d_3)\sin(\varphi_1 + \varphi_2) \end{pmatrix},$$

$$T_{m_5} = T_{m_5,0} + T_{m_5,rot-0} + m_5\left(\mathbf{v}_{m_5-c.i.-0}, \mathbf{v}_0\right) =$$

$$\frac{m_5}{2}\left[\dot{\varphi}_1^2(l_1 + r_2)^2 + (\dot{\varphi}_1 + \dot{\varphi}_2)^2(r_2 + d_3)^2\right] +$$

$$m_5\left[\dot{\varphi}_1(\dot{\varphi}_1 + \dot{\varphi}_2)(l_1 + r_2)(r_2 + d_3)\sin(2\varphi_1 + \varphi_2)\right] +$$

$$\frac{1}{2}\mathbf{w}_5^T \begin{Vmatrix} I_{5xx} & 0 & 0 \\ 0 & m_5(l_4 + d_5)^2 + I_{5yy} & 0 \\ 0 & 0 & m_5(l_4 + d_5)^2 + I_{5zz} \end{Vmatrix} \mathbf{w}_5 + m_4\mathbf{a}_5^T \mathbf{b}_5 =$$

$$\frac{1}{2}\left[m_5(l_1 + r_2)^2\right]\dot{q}_1^2 + \frac{1}{2}\left[\left(m_5(l_4 + d_5)^2 + I_{5zz}\right)\right]\dot{q}_3^2 + \frac{1}{2}[I_{5xx}]\dot{q}_4^2 +$$

$$\frac{1}{2}\left[m_5(r_2 + d_3)^2 + I_{5xx}\sin^2 q_3 + \left(m_5(l_4 + d_5)^2 + I_{5yy}\right)\cos^2 q_3\right](\dot{q}_1 + \dot{q}_2)^2 +$$

$$m_5\left[(l_1 + r_2)(r_2 + d_3)\sin(2q_1 + q_2)\right](\dot{q}_1 + \dot{q}_2)\dot{q}_1 -$$

$$[m_5(l_4 + d_5)(l_1 + r_2)\cos(2q_1 + q_2)\cos q_3](\dot{q}_1 + \dot{q}_2)\dot{q}_1 -$$

$$I_{5xx}[\sin q_3](\dot{q}_1 + \dot{q}_2)\dot{q}_4 - m_5(l_4 + d_5)(l_1 + r_2)[\sin(2q_1 + q_2)\sin q_3]\dot{q}_1\dot{q}_3 -$$

$$m_5(l_4 + d_5)(r_2 + d_3)[\sin q_3](\dot{q}_1 + \dot{q}_2)\dot{q}_3,$$

with

$$\mathbf{w}_5 = \begin{pmatrix} -(\dot{\varphi}_1 + \dot{\varphi}_2)\sin\varphi_3 \\ +\dot{\varphi}_4 \\ -(\dot{\varphi}_1 + \dot{\varphi}_2)\cos\varphi_3 \\ \dot{\varphi}_3 \end{pmatrix},$$

$$\mathbf{a}_5 = \begin{pmatrix} -(\ddot{\varphi}_1 + \ddot{\varphi}_2)(l_4 + d_5)\sin(\varphi_1 + \varphi_2)\cos\varphi_3 + \\ \dot{\varphi}_3(l_4 + d_5)\cos(\varphi_1 + \varphi_2)\sin\varphi_3 \\ \dot{\varphi}_3(l_4 + d_5)\cos\varphi_3 \\ (\ddot{\varphi}_1 + \ddot{\varphi}_2)(l_4 + d_5)\cos(\varphi_1 + \varphi_2)\cos\varphi_3 + \\ \dot{\varphi}_3(l_4 + d_5)\sin(\varphi_1 + \varphi_2)\sin\varphi_3 \end{pmatrix},$$

$$\mathbf{b}_5 = \begin{pmatrix} -\dot{\varphi}_1(l_1 + r_2)\sin\varphi_1 - (\dot{\varphi}_1 + \dot{\varphi}_2)(r_2 + d_3)\cos(\varphi_1 + \varphi_2) \\ 0 \\ -\dot{\varphi}_1(l_1 + r_2)\cos\varphi_1 - (\dot{\varphi}_1 + \dot{\varphi}_2)(r_2 + d_3)\sin(\varphi_1 + \varphi_2) \end{pmatrix}.$$

Potential energy

The potential energy $V = \sum_{i=1}^{4} V_{m_i}$ contains

$V_{m_1} = \text{const}, \ V_{m_2} = \text{const}, \ V_{m_3} = \text{const},$
$V_{m_4} = m_4 g d_4 \sin \varphi_3 = m_4 g d_4 \sin q_3,$
$V_{m_5} = m_5 g (l_4 + d_5) \sin \varphi_3 = m_5 g (l_4 + d_5) \sin q_3,$

so that

$$V = [m_4 g d_4 + m_5 g (l_4 + d_5)] \sin q_3 + \text{const}.$$

Non-potential forces

The generalized forces are given by the following formulas:

$$Q_{non\text{-}pot,i} = \tau_i - f_{fric-1} \dot{\varphi}_i = \tau_i - f_{fric-1} \dot{q}_i,$$
τ_i is a twisting moment, $i = 1, ..., 4$.

Lagrange equations

Based on the expressions for T and V, we can derive the Lagrange equations for the system:

$$\frac{d}{dt} \frac{\partial}{\partial \dot{q}_i} L - \frac{\partial}{\partial q_i} L = Q_{non\text{-}pot,i}, \ i = 1, ..., 4, \ L = T - V,$$

which can be represented in the standard matrix format

$$D(q) \ddot{q} + C(q, \dot{q}) \dot{q} + g(q) = \tau,$$

where

$$D(q) = \begin{Vmatrix} D_{11}(q) & D_{12}(q) & D_{13}(q) & -I_{5xx} \sin q_3 \\ D_{21}(q) & D_{22}(q) & D_{23}(q) & -I_{5xx} \sin q_3 \\ D_{31}(q) & D_{32}(q) & D_{33}(q) & 0 \\ -I_{5xx} \sin q_3 & -I_{5xx} \sin q_3 & 0 & I_{5xx} \end{Vmatrix},$$

$D_{11}(q) = I_{1yy} + (m_2 + m_3 + m_4 + m_5)(l_1 + r_2)^2 + I_{2yy} +$
$\quad (m_3 + m_4 + m_5)(r_2 + d_3)^2 + I_{3yy} + (I_{4xx} + I_{5xx}) \sin^2 q_3 +$
$\quad \left(m_4 d_4^2 + I_{4yy} + m_5 (l_4 + d_5)^2 + I_{5yy} \right) \cos^2 q_3 +$
$\quad 2 (m_3 + m_4 + m_5)(l_1 + r_2)(r_2 + d_3) \sin(2q_1 + q_2) -$
$\quad 2 (l_1 + r_2)(m_4 d_4 + m_5 (l_4 + d_5)) \cos(2q_1 + q_2) \cos q_3,$

$D_{21}(q) = (m_3 + m_4 + m_5)(l_1 + r_2)(r_2 + d_3) \sin(2q_1 + q_2) -$
$\quad (l_1 + r_2)(m_4 d_4 + m_5 (l_4 + d_5)) \cos(2q_1 + q_2) \cos q_3 +$
$\quad m_3 (r_2 + d_3) + I_{3yy} + m_4 (r_2 + d_3)^2 + m_5 (r_2 + d_3)^2 +$

$$(I_{4xx} + I_{5xx}) \sin^2 q_3 + \left(m_4 d_4^2 + I_{4yy} + m_5 (l_4 + d_5)^2 + I_{5yy}\right) \cos^2 q_3,$$
$$D_{31}(q) = -(l_1 + r_2)(m_4 d_4 + m_5 (l_4 + d_5)) \sin(2q_1 + q_2) \sin q_3 -$$
$$(m_4 d_4 + m_5 (l_4 + d_5))(r_2 + d_3) \sin q_3,$$
$$D_{12}(q) = (m_3 + m_4 + m_5)(r_2 + d_3)^2 + I_{3yy} + (I_{4xx} + I_{5xx}) \sin^2 q_3 +$$
$$\left(m_4 d_4^2 + I_{4yy} + m_5 (l_4 + d_5)^2 + I_{5yy}\right) \cos^2 q_3 +$$
$$(m_3 + m_4 + m_5)(l_1 + r_2)(r_2 + d_3) \sin(2q_1 + q_2) -$$
$$(l_1 + r_2)(m_4 d_4 + m_5 (l_4 + d_5)) \cos(2q_1 + q_2) \cos q_3,$$
$$D_{22}(q) = m_3 (r_2 + d_3) + I_{3yy} + m_4 (r_2 + d_3)^2 + m_5 (r_2 + d_3)^2 +$$
$$(I_{4xx} + I_{5xx}) \sin^2 q_3 + \left(m_4 d_4^2 + I_{4yy} + m_5 (l_4 + d_5)^2 + I_{5yy}\right) \cos^2 q_3,$$
$$D_{32}(q) = -[m_4 d_4 + m_5 (l_4 + d_5)](r_2 + d_3) \sin q_3,$$
$$D_{13}(q) = -(l_1 + r_2)[m_4 d_4 + m_5 (l_4 + d_5)] \sin(2q_1 + q_2) \sin q_3 -$$
$$[m_4 d_4 + m_5 (l_4 + d_5)](r_2 + d_3) \sin q_3,$$
$$D_{23}(q) = -[m_4 d_4 + m_5 (l_4 + d_5)](r_2 + d_3) \sin q_3,$$
$$D_{33}(q) = m_4 d_4^2 + I_{4zz} + m_5 (l_4 + d_5)^2 + I_{5zz},$$

and

$$C(q, \dot{q}) = \begin{Vmatrix} C_{11}(q, \dot{q}) & C_{12}(q, \dot{q}) & C_{13}(q, \dot{q}) & -I_{5xx} \cos q_3 \dot{q}_3 \\ C_{21}(q, \dot{q}) & f_{fric-2} & C_{23}(q, \dot{q}) & -I_{5xx} \cos q_3 \dot{q}_3 \\ C_{31}(q, \dot{q}) & C_{32}(q, \dot{q}) & f_{fric-3} & I_{5xx} \cos q_3 (\dot{q}_1 + \dot{q}_2) \\ -I_{5xx} \cos q_3 \dot{q}_3 & -I_{5xx} \cos q_3 \dot{q}_3 & 0 & f_{fric-4} \end{Vmatrix},$$

with

$$C_{11}(q, \dot{q}) = f_{fric-1} +$$
$$2(m_3 + m_4 + m_5)(l_1 + r_2)(r_2 + d_3) \cos(2q_1 + q_2)(\dot{q}_1 + \dot{q}_2) +$$
$$2(l_1 + r_2)[m_4 d_4 + m_5 (l_4 + d_5)] \sin(2q_1 + q_2) \cos q_3 (\dot{q}_1 + \dot{q}_2) +$$
$$2(l_1 + r_2)[m_4 d_4 + m_5 (l_4 + d_5)] \cos(2q_1 + q_2) \sin q_3 \dot{q}_3,$$
$$C_{21}(q, \dot{q}) = (m_3 + m_4 + m_5)(l_1 + r_2)(r_2 + d_3) \dot{q}_1 \cos(2q_1 + q_2) +$$
$$(l_1 + r_2)(m_4 d_4 + m_5 (l_4 + d_5)) \dot{q}_1 \sin(2q_1 + q_2) \cos q_3,$$
$$C_{31}(q, \dot{q}) = -(l_1 + r_2)(m_4 d_4 + m_5 (l_4 + d_5)) \cos(2q_1 + q_2) \sin q_3 (3\dot{q}_1 + 2\dot{q}_2) -$$
$$\left(I_{4xx} - m_4 d_4^2 - I_{4yy} + I_{5xx} - m_5 (l_4 + d_5)^2 - I_{5yy}\right) \sin q_3 \cos q_3 (\dot{q}_1 + 2\dot{q}_2),$$
$$C_{12}(q, \dot{q}) = (m_3 + m_4 + m_5)(l_1 + r_2)(r_2 + d_3) \cos(2q_1 + q_2) \dot{q}_2 +$$
$$(l_1 + r_2)[m_4 d_4 + m_5 (l_4 + d_5)] \dot{q}_2 \sin(2q_1 + q_2) \cos q_3,$$
$$C_{32}(q, \dot{q}) = -\left(I_{4xx} - m_4 d_4^2 - I_{4yy} + I_{5xx} - m_5 (l_4 + d_5)^2 - I_{5yy}\right) \times$$
$$\sin q_3 \cos q_3 (\dot{q}_2 + 2\dot{q}_1),$$

Collection of electromechanical models 465

$$C_{13}(q,\dot{q}) = 2\left(I_{4xx} + I_{5xx} - m_4 d_4^2\right)[\sin q_3 \cos q_3](\dot{q}_1 + \dot{q}_2) -$$
$$2\left(I_{4yy} + m_5 d_4^2 + I_{5yy}\right)[\sin q_3 \cos q_3](\dot{q}_1 + \dot{q}_2) -$$
$$(l_1 + r_2)[m_4 d_4 + m_5 (l_4 + d_5)]\sin(2q_1 + q_2)\cos q_3 \dot{q}_3 -$$
$$(r_2 + d_3)[m_4 d_4 + m_5 (l_4 + d_5)]\dot{q}_3 \cos q_3,$$
$$C_{23}(q,\dot{q}) = 2\left(I_{4xx} + I_{5xx} - m_4 d_4^2 - I_{4yy} - m_5 d_4^2 - I_{5yy}\right) \times$$
$$\sin q_3 \cos q_3 (\dot{q}_1 + \dot{q}_2) - (r_2 + d_3)(m_4 d_4 + m_5 (l_4 + d_5))\dot{q}_3 \cos q_3,$$
$$g(q) = \begin{Vmatrix} 0 \\ 0 \\ -m_4 g d_4 \cos q_3 - m_5 g (l_4 + d_5) \cos q_3 \\ 0 \end{Vmatrix}, \quad \tau = \begin{Vmatrix} \tau_1 \\ \tau_2 \\ \tau_3 \\ \tau_4 \end{Vmatrix}.$$

12.19 Maker 110

Consider the robot *Maker 110* represented in Fig. 12.19.

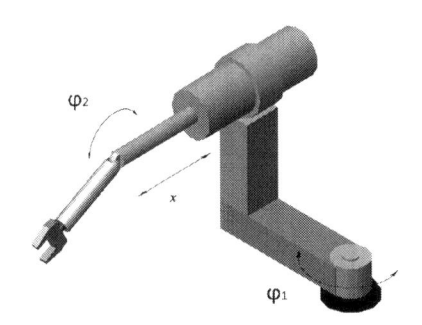

Figure 12.19 "Robot-Maker 110."

Generalized coordinates

The generalized coordinates are

$$q_1 := \varphi_1, \quad q_2 := x, \quad q_3 := \varphi_2.$$

Kinetic energy

The kinetic energy $T = \sum_{i=1}^{6} T_{m_i}$ consists of

$$T_{m_1} = T_{m_1,0} + T_{m_1,rot-0} + m_1 \left(\mathbf{v}_{m_1-c.i.-0}, \mathbf{v}_0\right) = \frac{1}{2} I_{1yy} \dot{\varphi}_1^2 = \frac{1}{2} I_{1yy} \dot{q}_1^2,$$
$$T_{m_2} = T_{m_2,0} + T_{m_2,rot-0} + m_2 \left(\mathbf{v}_{m_1-c.i.-0}, \mathbf{v}_0\right) = T_{m_2,rot-0} =$$

$$\frac{1}{2}\begin{pmatrix}0\\\dot\varphi_1\\0\end{pmatrix}^\top\begin{Vmatrix}I_{2xx} & 0 & 0\\0 & m_2d_2^2+I_{2yy} & 0\\0 & 0 & m_2d_2^2+I_{2zz}\end{Vmatrix}\begin{pmatrix}0\\\dot\varphi_1\\0\end{pmatrix}=$$

$$\frac{1}{2}\left(m_2d_2^2+I_{2yy}\right)\dot q_1^2,$$

$$T_{m3}=T_{m3,0}+T_{m3,rot-0}+m_3\left(\mathbf{v}_{m3-c.i.-0},\mathbf{v}_0\right)=T_{m3,rot-0}=$$

$$\frac{1}{2}\begin{pmatrix}0\\\dot\varphi_1\\0\end{pmatrix}^\top\begin{Vmatrix}I_{3xx} & 0 & 0\\0 & m_3l_2^2+I_{3yy} & 0\\0 & 0 & m_3l_2^2+I_{3zz}\end{Vmatrix}\begin{pmatrix}0\\\dot\varphi_1\\0\end{pmatrix}=$$

$$\frac{1}{2}\left[m_3l_2^2+I_{3yy}\right]\dot q_1^2,$$

$$T_{m4}=T_{m4,0}+T_{m4,rot-0}+m_4\left(\mathbf{v}_{m4-c.i.-0},\mathbf{v}_0\right)=T_{m4,rot-0}=$$

$$\frac{1}{2}\begin{pmatrix}0\\\dot\varphi_1\\0\end{pmatrix}^\top\begin{Vmatrix}m_4l_2^2+I_{4xx} & 0 & 0\\0 & m_4l_2^2+I_{4yy} & 0\\0 & 0 & I_{4zz}\end{Vmatrix}\begin{pmatrix}0\\\dot\varphi_1\\0\end{pmatrix}=$$

$$\frac{1}{2}\left[m_4l_2^2+I_{4yy}\right]\dot q_1^2,$$

$$T_{m5}=T_{m5,0}+T_{m5,rot-0}+m_5\left(\mathbf{v}_{m5-c.i.-0},\mathbf{v}_0\right)=T_{m5,0}+T_{m5,rot-0}=$$

$$\frac{1}{2}m_5\left\|\begin{pmatrix}-\dot x\sin\varphi_1-\dot\varphi_1 x\cos\varphi_1-\dot\varphi_1 l_2\sin\varphi_1\\0\\-\dot x\cos\varphi_1+\dot\varphi_1 x\sin\varphi_1-\dot\varphi_1 l_2\cos\varphi_1\end{pmatrix}\right\|^2+$$

$$\frac{1}{2}\begin{pmatrix}0\\\dot\varphi_1\\0\end{pmatrix}^\top\begin{Vmatrix}I_{5xx} & 0 & 0\\0 & I_{5yy} & 0\\0 & 0 & I_{5zz}\end{Vmatrix}+\begin{pmatrix}0\\\dot\varphi_1\\0\end{pmatrix}=$$

$$\frac{1}{2}\left[m_5\left(l_2^2+q_2^2\right)+I_{5yy}\right]\dot q_1^2+\frac{1}{2}m_5\dot q_2^2+m_5l_2\dot q_1\dot q_2,$$

$$T_{m6}=T_{m6,0}+T_{m6,rot-0}+m_6\left(\mathbf{v}_{m6-c.i.-0},\mathbf{v}_0\right)=$$

$$\frac{1}{2}m_6\left\|\begin{pmatrix}-\dot x\sin\varphi_1-\dot\varphi_1(x+d_5)\cos\varphi_1-\dot\varphi_1 l_2\sin\varphi_1\\0\\-\dot x\cos\varphi_1+\dot\varphi_1(x+d_5)\sin\varphi_1-\dot\varphi_1 l_2\cos\varphi_1\end{pmatrix}\right\|^2+$$

$$\frac{1}{2}\begin{pmatrix}\dot\varphi_1\sin\varphi_2\\-\dot\varphi_1\cos\varphi_2\\\dot\varphi_2\end{pmatrix}^\top\begin{Vmatrix}I_{6xx} & 0 & 0\\0 & m_6d_6^2+I_{6yy} & 0\\0 & 0 & m_6d_6^2+I_{6zz}\end{Vmatrix}\begin{pmatrix}\dot\varphi_1\sin\varphi_2\\-\dot\varphi_1\cos\varphi_2\\\dot\varphi_2\end{pmatrix}+$$

$$m_6\mathbf{a}_6^\top\mathbf{b}_6=\frac{m_6}{2}\left[l_2^2+(q_2+d_5)^2\right]\dot q_1^2+\frac{1}{2}m_6\dot q_2^2+\frac{1}{2}\left[m_6d_6^2+I_{6yy}\right]\dot q_3^2+$$

$$\frac{1}{2}\left[I_{6xx}\sin^2 q_3+\left(m_6d_6^2+I_{6yy}\right)\cos^2 q_3+m_6(q_2+d_5)d_6\cos q_3\right]\dot q_1^2+$$

$$m_6l_2\dot q_1\dot q_2-\left[m_6d_6\sin q_3\right]\dot q_2\dot q_3-\left[m_6d_6l_2\sin q_3\right]\dot q_1\dot q_3,$$

where

$$\mathbf{a}_6 = \begin{pmatrix} -\dot\varphi_1 d_6 \cos\varphi_1 \cos\varphi_2 + \dot\varphi_2 d_6 \sin\varphi_1 \sin\varphi_2 \\ -\dot\varphi_2 d_6 \cos\varphi_2 \\ \dot\varphi_1 d_6 \sin\varphi_1 \cos\varphi_2 + \dot\varphi_2 d_6 \cos\varphi_1 \sin\varphi_2 \end{pmatrix},$$

$$\mathbf{b}_6 = \begin{pmatrix} -\dot x \sin\varphi_1 - \dot\varphi_1 (x+d_5)\cos\varphi_1 - \dot\varphi_1 l_2 \sin\varphi_1 \\ 0 \\ -\dot x \cos\varphi_1 + \dot\varphi_1 (x+d_5)\sin\varphi_1 - \dot\varphi_1 l_2 \cos\varphi_1 \end{pmatrix}.$$

Potential energy

The potential energy $V = \sum_{i=1}^{5} V_{m_i}$ contains

$V_{m_1} = \text{const}$, $V_{m_2} = \text{const}$, $V_{m_3} = \text{const}$, $V_{m_4} = \text{const}$, $V_{m_5} = \text{const}$,
$V_{m_6} = m_6 g d_6 \sin\varphi_2 = m_6 g d_6 \sin q_3$,

which gives

$$V = m_6 g d_6 \sin q_3 + \text{const}.$$

Non-potential forces

The non-potential forces are

$Q_{non\text{-}pot,1} = \tau_1 - f_{fric-1}\dot\varphi_1 = \tau_1 - f_{fric-1}\dot q_1$,

τ_1 is a twisting moment,

$Q_{non\text{-}pot,2} = F_2 - f_{fric-2}\dot x = F_2 - f_{fric-2}\dot q_2$,

F_2 is a force of horizontal motion,

$Q_{non\text{-}pot,3} = \tau_3 - f_{fric-3}\dot\varphi_3 = \tau_3 - f_{fric-3}\dot q_3$,

τ_3 is a twisting moment.

Lagrange equations

Based on the obtained formulas for T and V, we are able to get the Lagrange equations

$$\frac{d}{dt}\frac{\partial}{\partial \dot q_i}L - \frac{\partial}{\partial q_i}L = Q_{non\text{-}pot,i}, \quad i = 1, 2, 3, \quad L = T - V,$$

which can be rewritten in the following standard matrix format, representing the dynamic model of the considered system:

$$D(q)\ddot q + C(q,\dot q)\dot q + g(q) = \tau,$$

where

$$D(q) = \begin{Vmatrix} D_{11}(q) & [m_5+m_6]l_2 & -m_6 d_6 l_2 \sin q_3 \\ [m_5+m_6]l_2 & [m_5+m_6] & -m_6 d_6 \sin q_3 \\ -m_6 d_6 l_2 \sin q_3 & -m_6 d_6 \sin q_3 & m_6 d_6^2 + I_{6yy} \end{Vmatrix},$$

$$D_{11}(q) = I_{1yy} + m_2 d_2^2 + I_{2yy} + m_3 l_2^2 + I_{3yy} +$$
$$m_4 l_2^2 + I_{4yy} + m_5 \left(l_2^2 + q_2^2\right) + I_{5yy} + m_6 \left(l_2^2 + (q_2 + d_5)^2\right) +$$
$$m_6 (q_2 + d_5) d_6 \cos q_3 + I_{6xx} \sin^2 q_3 + \left(m_6 d_6^2 + I_{6yy}\right) \cos^2 q_3,$$

$$C(q, \dot{q}) = \begin{Vmatrix} C_{11}(q, \dot{q}) & m_6 d_6 \cos q_3 \dot{q}_1 & 0 \\ C_{21}(q, \dot{q}) & f_{fric-2} & -m_6 d_6 \cos q_3 \dot{q}_3 \\ C_{31}(q, \dot{q}) & 0 & f_{fric-3} \end{Vmatrix},$$

$$C_{11}(q, \dot{q}) = f_{fric-1} + 2 (m_6 (q_2 + d_5) + m_5 q_2) \dot{q}_2 +$$
$$2 \left(I_{6xx} - m_6 d_6^2 - I_{6yy}\right) \dot{q}_3 \sin q_3 \cos q_3 - m_6 (q_2 + d_5) d_6 \dot{q}_3 \sin q_3,$$

$$C_{21}(q, \dot{q}) = -[m_5 q_2 + m_6 (q_2 + d_5)] \dot{q}_1 - \left[\frac{1}{2} m_6 d_6 \cos q_3\right] \dot{q}_1,$$

$$C_{31}(q, \dot{q}) = -\left[\left(I_{6xx} - m_6 d_6^2 - I_{6yy}\right) \sin q_3 \cos q_3\right] \dot{q}_1 +$$
$$[m_6 (q_2 + d_5) d_6 \sin q_3] \dot{q}_1,$$

$$g(q) = \begin{Vmatrix} 0 \\ 0 \\ m_6 g d_6 \cos q_3 \end{Vmatrix}, \quad \tau = \begin{Vmatrix} \tau_1 \\ F_2 \\ \tau_3 \end{Vmatrix}.$$

12.20 Manipulator on a horizontal platform

Consider the following manipulator located on a horizontal platform, which is represented in Fig. 12.20.

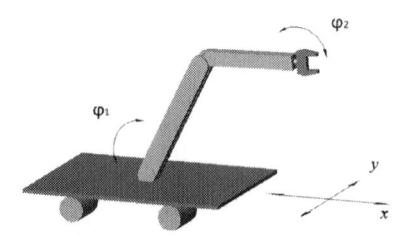

Figure 12.20 Manipulator on a horizontal platform.

Generalized coordinates

The generalized coordinates for this mechanical system are as follows:

$$q_1 := x, \quad q_2 := y, \quad q_3 := \varphi_1, \quad q_4 := \varphi_2.$$

Kinetic energy

The kinetic energy $T = \sum_{i=1}^{3} T_{m_i}$ of this system is given by the following expressions:

$$T_{m_1} = T_{m_1,0} + T_{m_1,rot-0} + m_1\left(\mathbf{v}_{m_1-c.i.-0}, \mathbf{v}_0\right) =$$
$$T_{m_1,0} = \frac{1}{2}m_1\left(\dot{x}^2 + \dot{y}^2\right) = \frac{1}{2}m_1\left(\dot{q}_1^2 + \dot{q}_2^2\right),$$
$$T_{m_2} = T_{m_2,0} + T_{m_2,rot-0} + m_2\left(\mathbf{v}_{m_2-c.i.-0}, \mathbf{v}_0\right) = \frac{1}{2}m_2\left(\dot{x}^2 + \dot{y}^2\right) +$$
$$\frac{1}{2}\begin{pmatrix}0\\ \dot{\varphi}_1\\ 0\end{pmatrix}^T \begin{Vmatrix} I_{2xx} & 0 & 0 \\ 0 & m_2 d_2^2 + I_{2yy} & 0 \\ 0 & 0 & m_2 d_2^2 + I_{2zz} \end{Vmatrix} \begin{pmatrix}0\\ \dot{\varphi}_1\\ 0\end{pmatrix} +$$
$$\begin{pmatrix}-\dot{\varphi}_1 d_2 \sin\varphi_1\\ 0\\ -\dot{\varphi}_1 d_2 \cos\varphi_1\end{pmatrix}^T \begin{pmatrix}\dot{x}\\ \dot{y}\\ 0\end{pmatrix} =$$
$$\frac{1}{2}\left[m_2\left(\dot{q}_1^2 + \dot{q}_2^2\right) + \left(m_2 d_2^2 + I_{2yy}\right)\dot{q}_3^2 - 2m_2 d_2 \sin q_3 \dot{q}_1 \dot{q}_3\right],$$
$$T_{m_3} = T_{m_3,0} + T_{m_3,rot-0} + m_2\left(\mathbf{v}_{m_3-c.i.-0}, \mathbf{v}_0\right) =$$
$$\frac{1}{2}m_3\left[\dot{x}^2 + \dot{y}^2 + \dot{\varphi}_1^2 l_2^2 - 2\dot{x}\dot{\varphi}_1 l_2 \sin\varphi_1\right] +$$
$$\frac{1}{2}\begin{pmatrix}0\\ \dot{\varphi}_1 + \dot{\varphi}_2\\ 0\end{pmatrix}^T \begin{Vmatrix} I_{3xx} & 0 & 0 \\ 0 & m_3 d_3^2 + I_{3yy} & 0 \\ 0 & 0 & m_3 d_3^2 + I_{3zz} \end{Vmatrix} \begin{pmatrix}0\\ \dot{\varphi}_2 + \dot{\varphi}_3\\ 0\end{pmatrix} +$$
$$m_3 \begin{pmatrix}-(\dot{\varphi}_1 + \dot{\varphi}_2)d_3 \sin(\varphi_1 + \varphi_2)\\ 0\\ -(\dot{\varphi}_1 + \dot{\varphi}_2)d_3 \cos(\varphi_1 + \varphi_2)\end{pmatrix}^T \begin{pmatrix}\dot{x} - \dot{\varphi}_1 l_2 \sin\varphi_1\\ \dot{y}\\ -\dot{\varphi}_1 l_2 \cos\varphi_1\end{pmatrix} =$$
$$\frac{1}{2}\left[m_3\left(\dot{q}_1^2 + \dot{q}_2^2 + l_2^2 \dot{q}_3^2 - 2l_2 \sin q_3 \dot{q}_1 \dot{q}_3\right) + \left(m_3 d_3^2 + I_{3yy}\right)(\dot{q}_3 + \dot{q}_4)^2\right] +$$
$$m_3\left[d_3 \sin(q_3 + q_4)(\dot{q}_3 + \dot{q}_4)\dot{q}_1 + d_3 l_2 \cos q_4 (\dot{q}_3 + \dot{q}_4)\dot{q}_3\right].$$

Potential energy
The potential energy $V = \sum_{i=1}^{3} V_i$ contains

$$V_{m_1} = \text{const},$$
$$V_{m_2} = m_2 g\,(d_2 \sin\varphi_1 + \text{const}) = m_2 g d_2 (\sin q_3 + \text{const}),$$
$$V_{m_3} = m_3 g\,(-d_3 \sin(\varphi_1 + \varphi_2) + l_2 \sin\varphi_1 + \text{const}) =$$
$$m_3 g\,(-d_3 \sin(q_3 + q_4) + l_2 \sin q_3 + \text{const}),$$

which gives

$$V = m_2 g d_2 \sin q_3 - m_3 g\,(d_3 \sin(q_3 + q_4) - l_2 \sin q_3) + \text{const}.$$

Non-potential forces
The non-potential forces are given by

$$Q_{non\text{-}pot,1} = F_1 - f_{fric-1}\dot{x} = F_1 - f_{fric-1}\dot{q}_1,$$

F_1 is a force of horizontal motion,
$$Q_{non\text{-}pot,2} = F_2 - f_{fric-2}\dot{y} = F_2 - f_{fric-2}\dot{q}_2,$$
F_2 is a force of transversal motion,
$$Q_{non\text{-}pot,3} = \tau_3 - f_{fric-3}\dot{\varphi}_1 = \tau_3 - f_{fric-1}\dot{q}_3,$$
τ_3 is a twisting moment,
$$Q_{non\text{-}pot,4} = \tau_4 - f_{fric-4}\dot{\varphi}_2 = \tau_4 - f_{fric-4}\dot{q}_4,$$
τ_4 is a twisting moment.

Lagrange equations

Using the obtained expressions for T and V, we are able to derive the Lagrange equations

$$\frac{d}{dt}\frac{\partial}{\partial \dot{q}_i}L - \frac{\partial}{\partial q_i}L = Q_{non\text{-}pot,i}, \ i = 1, ..., 4, \ L = T - V,$$

which leads to the following dynamic model:

$$D(q)\ddot{q} + C(q,\dot{q})\dot{q} + g(q) = \tau,$$

where

$$D(q) = \begin{Vmatrix} D_{11}(q) & 0 & D_{13}(q) & m_3 d_3 \sin(q_3 + q_4) \\ 0 & [m_1 + m_2 + m_3] & 0 & 0 \\ D_{31}(q) & 0 & D_{33}(q) & D_{34}(q) \\ m_3 d_3 \sin(q_3 + q_4) & 0 & D_{43}(q) & m_3 d_3^2 + I_{3yy} \end{Vmatrix},$$

$D_{11}(q) = [m_1 + m_2 + m_3]$,
$D_{31}(q) = (m_2 d_2 + m_3 l_2) \sin q_3 + m_3 d_3 \sin(q_3 + q_4)$,
$D_{13}(q) = -(m_2 d_2 + m_3 l_2) \sin q_3 + m_3 d_3 \sin(q_3 + q_4)$,
$D_{33}(q) = m_2 d_2^2 + I_{2yy} + m_3 l_2^2 + m_3 d_3^2 + I_{3yy} + 2 m_3 d_3 l_2 \cos q_4$,
$D_{43}(q) = +m_3 d_3^2 + I_{3yy} + m_3 d_3 l_2 \cos q_4$,
$D_{34}(q) = +m_3 d_3^2 + I_{3yy} + m_3 d_3 l_2 \cos q_4$,

and

$$C(q,\dot{q}) = \begin{Vmatrix} f_{fric-1} & 0 & C_{13}(q,\dot{q}) & C_{14}(q,\dot{q}) \\ 0 & f_{fric-2} & 0 & 0 \\ 0 & 0 & f_{fric-3} & C_{24}(q,\dot{q}) \\ 0 & 0 & m_3 d_3 l_2 \sin q_4 \dot{q}_3 & f_{fric-4} \end{Vmatrix},$$

$C_{13}(q,\dot{q}) = -(m_2 d_2 + m_3 l_2) \cos q_3 \dot{q}_3 + m_3 d_3 \cos(q_3 + q_4)(\dot{q}_3 + 2\dot{q}_4)$,
$C_{14}(q,\dot{q}) = m_3 d_3 \cos(q_3 + q_4)(\dot{q}_4 + 2\dot{q}_3)$,
$C_{24}(q,\dot{q}) = -m_3 d_3 l_2 \sin q_4 (2\dot{q}_3 + \dot{q}_4)$,

$$g(q) = \begin{Vmatrix} 0 \\ 0 \\ m_2 g d_2 \cos q_3 - m_3 g \left(d_3 \cos (q_3 + q_4) - l_2 \cos q_3 \right) \\ -m_3 g d_3 \cos (q_3 + q_4) \end{Vmatrix}, \quad \tau = \begin{Vmatrix} F_1 \\ F_2 \\ \tau_3 \\ \tau_4 \end{Vmatrix}.$$

12.21 Two-arm planar manipulator

Consider the two-arm planar manipulator represented in Fig. 12.21.

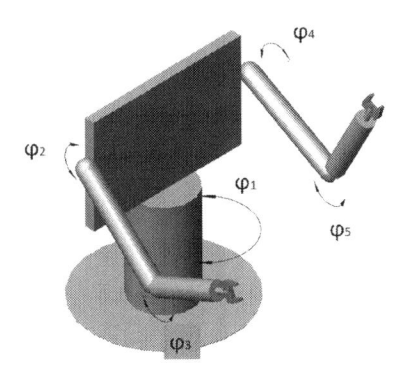

Figure 12.21 Two-arms planar manipulator.

Generalized coordinates

The generalized coordinates are

$$q_1 := \varphi_1, \quad q_2 := \varphi_2, \quad q_3 := \varphi_3, \quad q_4 := \varphi_4, \quad q_5 := \varphi_5.$$

Kinetic energy

The kinetic energy $T = \sum_{i=1}^{6} T_{m_i}$ consists of

$$T_{m_1} = T_{m_1,0} + T_{m_1,rot-0} + m_1 \left(\mathbf{v}_{m_1-c.i.-0}, \mathbf{v}_0 \right) =$$
$$T_{m_1,rot-0} = \frac{1}{2} I_{1yy} \dot{\varphi}_1^2 = \frac{1}{2} I_{1yy} \dot{q}_1^2,$$
$$T_{m_2} = T_{m_2,0} + T_{m_2,rot-0} + m_2 \left(\mathbf{v}_{m_1-c.i.-0}, \mathbf{v}_0 \right) =$$
$$T_{m_2,rot-0} = \frac{1}{2} I_{2yy} \dot{\varphi}_1^2 = \frac{1}{2} I_{2yy} \dot{q}_1^2,$$
$$T_{m_3} = T_{m_3,0} + T_{m_3,rot-0} + m_3 \left(\mathbf{v}_{m_3-c.i.-0}, \mathbf{v}_0 \right) = \frac{1}{2} m_3 \left(\frac{l_2}{2} \right)^2 \dot{\varphi}_1^2 +$$
$$\frac{1}{2} \begin{pmatrix} 0 \\ \dot{\varphi}_1 + \dot{\varphi}_2 \\ 0 \end{pmatrix}^T \begin{Vmatrix} I_{3xx} & 0 & 0 \\ 0 & m_3 d_3^2 + I_{3yy} & 0 \\ 0 & 0 & m_3 d_3^2 + I_{3zz} \end{Vmatrix} \begin{pmatrix} 0 \\ \dot{\varphi}_1 + \dot{\varphi}_2 \\ 0 \end{pmatrix} +$$

$$m_3 \begin{pmatrix} -(\dot{\varphi}_1 + \dot{\varphi}_2) d_3 \sin(\varphi_1 + \varphi_2) \\ 0 \\ -(\dot{\varphi}_1 + \dot{\varphi}_2) d_3 \cos(\varphi_1 + \varphi_2) \end{pmatrix}^\top \begin{pmatrix} \dot{\varphi}_1 \frac{l_2}{2} \sin \varphi_1 \\ 0 \\ \dot{\varphi}_1 \frac{l_2}{2} \cos \varphi_1 \end{pmatrix} =$$

$$\frac{m_3 l_2^2}{8} \dot{q}_1^2 + \frac{1}{2} \left[m_3 d_3^2 + I_{3yy} \right] (\dot{q}_1 + \dot{q}_2)^2 - \frac{m_3}{2} [l_2 d_3 \cos q_2] (\dot{q}_1 + \dot{q}_2) \dot{q}_1,$$

$$T_{m_4} = T_{m_4,0} + T_{m_4,rot-0} + m_4 \left(\mathbf{v}_{m_4-c.i.-0}, \mathbf{v}_0 \right) =$$

$$\frac{1}{2} m_4 \left\| \begin{pmatrix} -(\dot{\varphi}_1 + \dot{\varphi}_2) l_3 \sin(\varphi_1 + \varphi_2) + \dot{\varphi}_1 \frac{l_2}{2} \sin \varphi_1 \\ 0 \\ -(\dot{\varphi}_1 + \dot{\varphi}_2) l_3 \cos(\varphi_1 + \varphi_2) + \dot{\varphi}_1 \frac{l_2}{2} \cos \varphi_1 \end{pmatrix} \right\|^2 +$$

$$\frac{1}{2} \mathbf{w}_4^\top \begin{Vmatrix} I_{4xx} & 0 & 0 \\ 0 & m_4 d_4^2 + I_{4yy} & 0 \\ 0 & 0 & m_4 d_4^2 + I_{4zz} \end{Vmatrix} \mathbf{w}_4 + m_4 \mathbf{a}_4^\top \mathbf{b}_4 =$$

$$\frac{m_4}{8} l_2^2 \dot{q}_1^2 + \frac{1}{2} m_4 l_3^2 (\dot{q}_1 + \dot{q}_2)^2 + \frac{1}{2} \left[m_4 d_4^2 + I_{4yy} \right] (\dot{q}_1 + \dot{q}_2 + \dot{q}_3)^2 -$$

$$\frac{m_4}{2} [l_2 l_3 \cos q_2] (\dot{q}_1 + \dot{q}_2) \dot{q}_1 - m_4 [l_3 d_4 \cos q_3] (\dot{q}_1 + \dot{q}_2 + \dot{q}_3) (\dot{q}_1 + \dot{q}_2) +$$

$$\frac{m_4}{2} [l_2 d_4 \cos(q_2 + q_3)] (\dot{q}_1 + \dot{q}_2 + \dot{q}_3) \dot{q}_1,$$

where

$$\mathbf{w}_4 = \begin{pmatrix} 0 \\ (\dot{\varphi}_1 + \dot{\varphi}_2 + \dot{\varphi}_3) \\ 0 \end{pmatrix}, \quad \mathbf{a}_4 = \begin{pmatrix} (\dot{\varphi}_1 + \dot{\varphi}_2 + \dot{\varphi}_3) d_4 \sin(\varphi_1 + \varphi_2 + \varphi_3) \\ 0 \\ (\dot{\varphi}_1 + \dot{\varphi}_2 + \dot{\varphi}_3) d_4 \cos(\varphi_1 + \varphi_2 + \varphi_3) \end{pmatrix},$$

$$\mathbf{b}_4 = \begin{pmatrix} -(\dot{\varphi}_1 + \dot{\varphi}_2) l_3 \sin(\varphi_1 + \varphi_2) + \dot{\varphi}_1 \frac{l_2}{2} \sin \varphi_1 \\ 0 \\ -(\dot{\varphi}_1 + \dot{\varphi}_2) l_3 \cos(\varphi_1 + \varphi_2) + \dot{\varphi}_1 \frac{l_2}{2} \cos \varphi_1 \end{pmatrix},$$

$$T_{m_5} = T_{m_5,0} + T_{m_5,rot-0} + m_5 \left(\mathbf{v}_{m_5-c.i.-0}, \mathbf{v}_0 \right) = \frac{1}{2} m_5 \left(\frac{l_2}{2} \right)^2 \dot{\varphi}_1^2 +$$

$$\frac{1}{2} \begin{pmatrix} 0 \\ \dot{\varphi}_1 + \dot{\varphi}_4 \\ 0 \end{pmatrix}^\top \begin{Vmatrix} I_{5xx} & 0 & 0 \\ 0 & m_5 d_5^2 + I_{5yy} & 0 \\ 0 & 0 & m_5 d_5^2 + I_{5zz} \end{Vmatrix} \begin{pmatrix} 0 \\ \dot{\varphi}_1 + \dot{\varphi}_4 \\ 0 \end{pmatrix} +$$

$$m_5 \begin{pmatrix} -(\dot{\varphi}_1 + \dot{\varphi}_4) d_5 \sin(\varphi_4 - \varphi_1) \\ 0 \\ (\dot{\varphi}_1 + \dot{\varphi}_4) d_5 \cos(\varphi_4 - \varphi_1) \end{pmatrix}^\top \begin{pmatrix} -\dot{\varphi}_1 \frac{l_2}{2} \sin \varphi_1 \\ 0 \\ -\dot{\varphi}_1 \frac{l_2}{2} \cos \varphi_1 \end{pmatrix} =$$

$$\frac{m_5 l_2^2}{8}\dot{q}_1^2 + \frac{[m_5 d_5^2 + I_{5yy}]}{2}(\dot{q}_1 + \dot{q}_4)^2 - \frac{m_5}{2}[l_2 d_5 \cos q_4](\dot{q}_1 + \dot{q}_4)\dot{q}_1,$$

$$T_{m_6} = T_{m_6,0} + T_{m_6,rot-0} + m_4 \left(\mathbf{v}_{m_6-c.i.-0}, \mathbf{v}_0\right) =$$

$$\frac{m_6}{2}\left\|\begin{pmatrix} -(\dot{\varphi}_1 + \dot{\varphi}_4) l_5 \sin(\varphi_4 - \varphi_1) - \dot{\varphi}_1 \frac{l_2}{2} \sin\varphi_1 \\ 0 \\ (\dot{\varphi}_1 + \dot{\varphi}_4) l_5 \cos(\varphi_4 - \varphi_1) - \dot{\varphi}_1 \frac{l_2}{2} \cos\varphi_1 \end{pmatrix}\right\|^2 +$$

$$\frac{1}{2}\mathbf{w}_6^\top \begin{Vmatrix} I_{4xx} & 0 & 0 \\ 0 & m_6 d_6^2 + I_{6yy} & 0 \\ 0 & 0 & m_6 d_6^2 + I_{6zz} \end{Vmatrix} \mathbf{w}_6 + m_6 \mathbf{a}_6^\top \mathbf{b}_6 =$$

$$\frac{1}{8}m_6 l_2^2 \dot{q}_1^2 + \frac{1}{2}m_6 l_5^2 (\dot{q}_1 + \dot{q}_4)^2 + \frac{1}{2}\left[m_6 d_6^2 + I_{6yy}\right](\dot{q}_1 + \dot{q}_4 + \dot{q}_5)^2 -$$

$$\frac{m_6}{2}[l_2 l_5 \cos q_4](\dot{q}_1 + \dot{q}_4)\dot{q}_1 - m_6 [l_5 d_6 \cos q_5](\dot{q}_1 + \dot{q}_4 + \dot{q}_5)(\dot{q}_1 + \dot{q}_4) +$$

$$\frac{1}{2}m_6[l_2 d_6 \cos(q_4 + q_5)](\dot{q}_1 + \dot{q}_4 + \dot{q}_5)\dot{q}_1,$$

with

$$\mathbf{w}_6 = \begin{pmatrix} 0 \\ (\dot{\varphi}_1 + \dot{\varphi}_4 + \dot{\varphi}_5) \\ 0 \end{pmatrix}, \quad \mathbf{a}_6 = \begin{pmatrix} (\dot{\varphi}_1 + \dot{\varphi}_4 + \dot{\varphi}_5) d_6 \sin(\varphi_4 + \varphi_5 - \varphi_1) \\ 0 \\ -(\dot{\varphi}_1 + \dot{\varphi}_4 + \dot{\varphi}_5) d_6 \cos(\varphi_4 + \varphi_5 - \varphi_1) \end{pmatrix},$$

$$\mathbf{b}_6 = \begin{pmatrix} -(\dot{\varphi}_1 + \dot{\varphi}_4) l_5 \sin(\varphi_4 - \varphi_1) - \dot{\varphi}_1 \frac{l_2}{2}\sin\varphi_1 \\ 0 \\ (\dot{\varphi}_1 + \dot{\varphi}_4) l_5 \cos(\varphi_4 - \varphi_1) - \dot{\varphi}_1 \frac{l_2}{2}\cos\varphi_1 \end{pmatrix}.$$

Potential energy

The potential energy V is calculated as

$$V = \sum_{i=1}^{6} V_{m_i},$$

$V_{m_1} = \text{const}, \ V_{m_2} = \text{const}, \ V_{m_3} = \text{const},$
$V_{m_4} = \text{const}, \ V_{m_5} = \text{const}, \ V_{m_6} = \text{const},$

which gives

$$V = \text{const}.$$

Non-potential forces

The non-potential forces are

$$Q_{non\text{-}pot,i} = \tau_i - f_{fric-i}\dot{\varphi}_i = \tau_i - f_{fric-i}\dot{q}_i,$$

τ_i is a twisting moment, $i = 1, ..., 5$.

Lagrange equations

Based on the obtained expressions for T and V, we are able to derive the Lagrange equations

$$\frac{d}{dt}\frac{\partial}{\partial \dot{q}_i}L - \frac{\partial}{\partial q_i}L = Q_{non\text{-}pot,i}, \quad i = 1, ..., 5, \quad L = T - V,$$

which lead to the following dynamic model:

$$D(q)\ddot{q} + C(q,\dot{q})\dot{q} + g(q) = \tau,$$

where

$$D(q) = \begin{Vmatrix} D_{11}(q) & D_{12}(q) & D_{13}(q) & D_{14}(q) & D_{15}(q) \\ D_{21}(q) & D_{22}(q) & D_{23}(q) & 0 & 0 \\ D_{31}(q) & D_{32}(q) & m_4 d_4^2 + I_{4yy} & 0 & 0 \\ D_{41}(q) & 0 & 0 & D_{44}(q) & D_{45}(q) \\ D_{51}(q) & 0 & 0 & D_{54}(q) & m_6 d_6^2 + I_{6yy} \end{Vmatrix},$$

$D_{11}(q) = I_{1yy} + I_{2yy} + \frac{1}{4}(m_3 + m_4 + m_5 + m_6)l_2^2 +$
$\quad m_3 d_3^2 + I_{3yy} + m_4 l_3^2 - 2m_4 l_3 d_4 \cos q_3 - (m_3 d_3 + m_4 l_3) l_2 \cos q_2 +$
$\quad m_4 l_2 d_4 \cos(q_2 + q_3) + m_4 d_4^2 + I_{4yy} + m_5 d_5^2 + I_{5yy} + m_6 l_5^2 -$
$\quad 2m_6 l_5 d_6 \cos q_5 - (m_5 d_5 + m_6 l_5) l_2 \cos q_4 + m_6 l_2 d_6 \cos(q_4 + q_5) +$
$\quad m_6 d_6^2 + I_{6yy},$

$D_{21}(q) = -\frac{1}{2}(m_3 d_3 + m_4 l_3) l_2 \cos q_2 + \frac{1}{2} m_4 l_2 d_4 \cos(q_2 + q_3) +$
$\quad m_3 d_3^2 + I_{3yy} + m_4 l_3^2 - 2m_4 l_3 d_4 \cos q_3 + m_4 d_4^2 + I_{4yy},$

$D_{31}(q) = \frac{1}{2} m_4 l_2 d_4 \cos(q_2 + q_3) - m_4 l_3 d_4 \cos q_3 + m_4 d_4^2 + I_{4yy},$

$D_{41}(q) = -\frac{1}{2}(m_5 d_5 + m_6 l_5) l_2 \cos q_4 + \frac{1}{2} m_6 l_2 d_6 \cos(q_4 + q_5) +$
$\quad m_5 d_5^2 + I_{5yy} + m_6 l_5^2 - 2m_6 l_5 d_6 \cos q_5 + m_6 d_6^2 + I_{6yy},$

$D_{51}(q) = \frac{1}{2} m_6 l_2 d_6 \cos(q_4 + q_5) + m_6 d_6^2 + I_{6yy} - m_6 l_5 d_6 \cos q_5,$

$D_{12}(q) = m_3 d_3^2 + I_{3yy} + m_4 l_3^2 - 2m_4 l_3 d_4 \cos q_3 -$
$\quad \frac{(m_3 d_3 + m_4 l_3) l_2}{2} \cos q_2 + \frac{m_4 l_2 d_4}{2} \cos(q_2 + q_3) + m_4 d_4^2 + I_{4yy},$

$D_{22}(q) = m_3 d_3^2 + I_{3yy} + m_4 l_3^2 - 2m_4 l_3 d_4 \cos q_3 + m_4 d_4^2 + I_{4yy},$

$D_{32}(q) = -m_4 l_3 d_4 \cos q_3 + m_4 d_4^2 + I_{4yy},$

$D_{13}(q) = \frac{1}{2} m_4 l_2 d_4 \cos(q_2 + q_3) + m_4 d_4^2 + I_{4yy} - m_4 l_3 d_4 \cos q_3,$

$D_{23}(q) = m_4 d_4^2 + I_{4yy} - m_4 l_3 d_4 \cos q_3,$

$$D_{14}(q) = m_5 d_5^2 + I_{5yy} + m_6 l_5^2 - 2m_6 l_5 d_6 \cos q_5 -$$
$$\frac{(m_5 d_5 + m_6 l_5) l_2}{2} \cos q_4 + \frac{m_6 l_2 d_6}{2} \cos(q_4 + q_5) + m_6 d_6^2 + I_{6yy},$$
$$D_{44}(q) = m_5 d_5^2 + I_{5yy} + m_6 l_5^2 - 2m_6 l_5 d_6 \cos q_5 + m_6 d_6^2 + I_{6yy},$$
$$D_{54}(q) = m_6 d_6^2 + I_{6yy} - m_6 l_5 d_6 \cos q_5,$$
$$D_{15}(q) = \frac{1}{2} m_6 l_2 d_6 \cos(q_4 + q_5) + m_6 d_6^2 + I_{6yy} - m_6 l_5 d_6 \cos q_5,$$
$$D_{45}(q) = m_6 d_6^2 + I_{6yy} - m_6 l_5 d_6 \cos q_5,$$

and

$$C(q,\dot{q}) = \begin{Vmatrix} f_{fric-1} & C_{12}(q,\dot{q}) & C_{13}(q,\dot{q}) & C_{14}(q,\dot{q}) & C_{15}(q,\dot{q}) \\ C_{21}(q,\dot{q}) & f_{fric-2} & C_{23}(q,\dot{q}) & 0 & 0 \\ C_{31}(q,\dot{q}) & C_{32}(q,\dot{q}) & f_{fric-3} & 0 & 0 \\ C_{41}(q,\dot{q}) & 0 & 0 & f_{fric-4} & C_{45}(q,\dot{q}) \\ C_{51}(q,\dot{q}) & 0 & 0 & C_{54}(q,\dot{q}) & f_{fric-5} \end{Vmatrix},$$

$$C_{21}(q,\dot{q}) = -\frac{(m_3 d_3 + m_4 l_3) l_2}{2} [\sin q_2] \dot{q}_1 + \frac{m_4 l_2 d_4}{2} [\sin(q_2 + q_3)] \dot{q}_1,$$

$$C_{31}(q,\dot{q}) = -m_4 l_3 d_4 \sin q_3 (\dot{q}_1 + \dot{q}_2) + \frac{m_4 l_2 d_4}{2} [\sin(q_2 + q_3)] \dot{q}_1,$$

$$C_{41}(q,\dot{q}) = -\frac{(m_5 d_5 + m_6 l_5) l_2}{2} [\sin q_4] \dot{q}_1 + \frac{m_6 l_2 d_6}{2} [\sin(q_4 + q_5)] \dot{q}_1,$$

$$C_{51}(q,\dot{q}) = -m_6 l_5 d_6 \sin q_5 (\dot{q}_1 + \dot{q}_4) + \frac{m_6 l_2 d_6}{2} [\sin(q_4 + q_5)] \dot{q}_1,$$

$$C_{12}(q,\dot{q}) = \frac{(m_3 d_3 + m_4 l_3) l_2}{2} \sin q_2 (2\dot{q}_1 + \dot{q}_2) -$$
$$\frac{m_4 l_2 d_4}{2} \sin(q_2 + q_3)(2\dot{q}_1 + \dot{q}_2 + \dot{q}_3),$$

$$C_{32}(q,\dot{q}) = -m_4 l_3 d_4 \sin q_3 (\dot{q}_1 + \dot{q}_2),$$

$$C_{13}(q,\dot{q}) = m_4 l_3 d_4 \sin q_3 (2\dot{q}_1 + 2\dot{q}_2 + \dot{q}_3) -$$
$$\frac{m_4 l_2 d_4}{2} [\sin(q_2 + q_3)](2\dot{q}_1 + \dot{q}_2 + \dot{q}_3),$$

$$C_{23}(q,\dot{q}) = m_4 l_3 d_4 \sin q_3 (2\dot{q}_1 + 2\dot{q}_2 + \dot{q}_3),$$

$$C_{14}(q,\dot{q}) = \frac{(m_5 d_5 + m_6 l_5) l_2}{2} [\sin q_4](2\dot{q}_1 + \dot{q}_4) -$$
$$\frac{m_6 l_2 d_6}{2} [\sin(q_4 + q_5)](2\dot{q}_1 + \dot{q}_4 + \dot{q}_5),$$

$$C_{54}(q,\dot{q}) = -m_6 l_5 d_6 \sin q_5 (\dot{q}_1 + \dot{q}_4),$$

$$C_{15}(q,\dot{q}) = [m_6 l_5 d_6 \sin q_5](2\dot{q}_1 + 2\dot{q}_4 + \dot{q}_5) -$$
$$\frac{m_6 l_2 d_6}{2} [\sin(q_4 + q_5)](2\dot{q}_1 + \dot{q}_4 + \dot{q}_5),$$

$$C_{45}(q, \dot{q}) = m_6 l_5 d_6 \sin q_5 \, (2\dot{q}_1 + 2\dot{q}_4 + \dot{q}_5),$$

$$g(q) = \begin{Vmatrix} 0 \\ 0 \\ 0 \\ 0 \\ 0 \end{Vmatrix}, \quad \tau = \begin{Vmatrix} \tau_1 \\ \tau_2 \\ \tau_3 \\ \tau_4 \\ \tau_5 \end{Vmatrix}.$$

12.22 Manipulator with three degrees of freedom

Consider the manipulator with three degrees of freedom represented in Fig. 12.22.

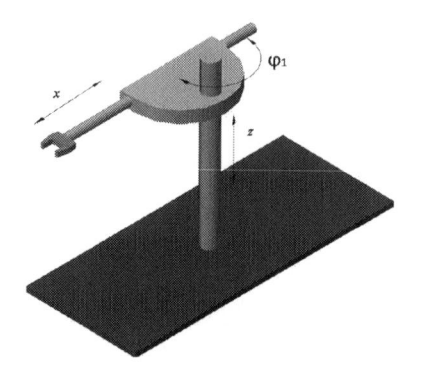

Figure 12.22 Manipulator with three degrees of freedom.

Generalized coordinates
The generalized coordinates are selected as

$$q_1 := z, \quad q_2 := \varphi, \quad q_3 := x.$$

Kinetic energy
The kinetic energy $T = \sum_{i=1}^{2} T_{m_i}$ for this system is as follows:

$$T_{m_1} = T_{m_1,0} + T_{m_1,rot-0} + m_1 \left(\mathbf{v}_{m_2-c.i.-0}, \mathbf{v}_0 \right) = T_{m_1,0} + T_{m_1,rot-0} = \frac{1}{2} m_1 \dot{z}^2$$

$$+ \frac{1}{2} \begin{pmatrix} 0 \\ \dot{\varphi} \\ 0 \end{pmatrix}^T \begin{Vmatrix} I_{1_{xx}} & 0 & 0 \\ 0 & I_{1_{yy}} & 0 \\ 0 & 0 & I_{1_{zz}} \end{Vmatrix} \begin{pmatrix} 0 \\ \dot{\varphi} \\ 0 \end{pmatrix} = \frac{1}{2} \left[m_1 \dot{q}_1^2 + I_{1_{yy}} \dot{q}_2^2 \right],$$

$$T_{m_2} = T_{m_2,rot-0} + m_2 \left(\mathbf{v}_{m_2-c.i.-0}, \mathbf{v}_0 \right) = T_{m_2,0} + T_{m_2,rot-0} =$$
$$\frac{1}{2} m_2 \left[\dot{x}^2 + \dot{z}^2 + \left(a^2 + (b+x)^2 \right) \dot{\varphi}^2 \right] +$$

$$\frac{1}{2}\begin{pmatrix}0\\\dot{\varphi}\\0\end{pmatrix}^T \begin{Vmatrix} I_{2_{xx}} & 0 & 0 \\ 0 & I_{2_{yy}} & 0 \\ 0 & 0 & I_{2_{zz}} \end{Vmatrix} \begin{pmatrix}0\\\dot{\varphi}\\0\end{pmatrix} =$$
$$\frac{m_2}{2}\left[\dot{q}_3^2 + \dot{q}_1^2 + \left(a^2 + (b+q_3)^2\right)\dot{q}_2^2\right].$$

Potential energy

The potential energy $V = \sum_{i=1}^{2} V_{m_i}$ contains

$$V_{m_1} = m_1 g z = m_1 g q_1, \quad V_{m_2} = m_2 g z = m_2 g q_1,$$

which gives

$$V = (m_1 + m_2) g q_1.$$

Non-potential forces

The non-potential forces are

$$Q_{non\text{-}pot,1} = F_1 - f_{fric-1}\dot{y} = F_1 - f_{fric-1}\dot{q}_1,$$
F_1 is a force of vertical motion,
$$Q_{non\text{-}pot,2} = \tau_2 - f_{fric-2}\dot{\varphi} = \tau_2 - f_{fric-2}\dot{q}_2,$$
τ_2 is a twisting moment,
$$Q_{non\text{-}pot,3} = F_3 - f_{fric-3}\dot{x} = F_3 - f_{fric-3}\dot{q}_3,$$
F_3 is a force of horizontal motion.

Lagrange equations

The obtained expressions for T and V allow us to derive the following Lagrange equations:

$$\frac{d}{dt}\frac{\partial}{\partial \dot{q}_i}L - \frac{\partial}{\partial q_i}L = Q_{non\text{-}pot,i}, \quad i = 1, 2, 3, \quad L = T - V,$$

which leads to the following dynamic model:

$$D(q)\ddot{q} + C(q,\dot{q})\dot{q} + g(q) = \tau,$$

where

$$D(q) = \begin{Vmatrix} [m_1 + m_2] & 0 & 0 \\ 0 & [I_{1_{yy}} + m_2(a^2 + (b+q_3)^2)] & 0 \\ 0 & 0 & m_2 \end{Vmatrix},$$

$$C(q,\dot{q}) = \begin{Vmatrix} f_{fric-1} & 0 & 0 \\ 0 & f_{fric-2} + 2[m_2(b+q_3)]\dot{q}_3 & 0 \\ 0 & -m_2[b+q_3]\dot{q}_2 & f_{fric-3} \end{Vmatrix},$$

$$g(q) = \begin{Vmatrix} [m_1 + m_2]g \\ 0 \\ 0 \end{Vmatrix}, \quad \tau = \begin{Vmatrix} F_1 \\ \tau_2 \\ F_3 \end{Vmatrix}.$$

12.23 CD motor with load

Consider now a CD motor with load, as depicted in Fig. 12.23.

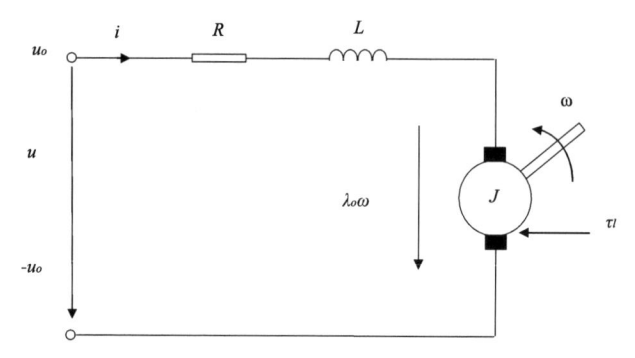

Figure 12.23 CD motor with load.

Here:
- i is the armor current,
- u is the terminal voltage,
- L is the armature inductance,
- λ_0 is the counter-electric force constant,
- R is the armor resistance,
- u is the switch position control,
- ω is the angular velocity,
- τ_l is the load torque,
- J is the motor and load inertia, and
- k is the torque constant.

Generalized coordinates

The generalized coordinates are

$$q_1 = q, \quad q_2 = \varphi,$$

so that

$$i := \dot{q}_1 = \dot{q}, \quad \omega := \dot{q}_2 = \dot{\varphi}. \tag{12.8}$$

Kinetic and potential energies and non-potential forces
Using Table 6.1 we obtain

$$T_1 = \frac{1}{2}L_{ind}\dot{q}_1^2, \quad V_1 = 0, \quad Q_{1,non\text{-}pot} = u - R\dot{q}_1 - \lambda\dot{q}_2,$$
$$T_2 = \frac{1}{2}J\dot{q}_2^2, \quad V_2 = 0, \quad Q_{2,non\text{-}pot} = k\dot{q}_1 - \tau_l.$$

Lagrange equations
In this case the Lagrange equations

$$\frac{d}{dt}\frac{\partial}{\partial \dot{q}_i}L - \frac{\partial}{\partial q_i}L = Q_{i,non\text{-}pot}, \quad i = 1, 2,$$
$$L = T - V, \quad T = \sum_{i=1}^{2} T_i, \quad V = \sum_{i=1}^{2} V_i$$

are as follows:

$$\frac{\partial}{\partial \dot{q}_1}L = \frac{\partial}{\partial \dot{q}_1}T_1 = L_{ind}\dot{q}_1, \quad \frac{\partial}{\partial q_1}L = 0,$$
$$\frac{\partial}{\partial \dot{q}_2}L = \frac{\partial}{\partial \dot{q}_2}T_2 = J\dot{q}_2, \quad \frac{\partial}{\partial q_2}L = 0,$$

and, as a result,

$$L_{ind}\ddot{q}_1 = u - R\dot{q}_1 - \lambda\dot{q}_2,$$
$$J\ddot{q}_2 = k\dot{q}_1 - \tau_l,$$

or using (12.8), we finally obtain

$$L_{ind}\frac{d}{dt}i = u - Ri - \lambda\omega,$$
$$J\frac{d}{dt}\omega = ki - \tau_l.$$

12.24 Models of power converters with switching-mode power supply

Four non-isolated switching-mode (SM) DC-to-DC converter topologies are known: buck, boost, buck-boost, and Ćuk. Here we will consider only the first two. The input is on the left side, and the output with load is on the right side. The switch is typically a MOSFET, IGBT, or BJT transistor.

The buck (step-down, Fig. 12.24) or boost (step-up, Fig. 12.25) converter is a type of DC-to-DC converter that has an output voltage magnitude that is either greater than or less than the input voltage magnitude. It is equivalent to a "flyback converter" using a single inductor instead of a transformer.

Two different topologies are called *buck* or *boost* converter. Both of them can produce a range of output voltages, ranging from much larger (in absolute magnitude) than the input voltage, down to almost zero.

12.24.1 Buck type DC-DC converter

The buck type DC-DC converter has a counter-circuit, as shown in Fig. 12.24, where:

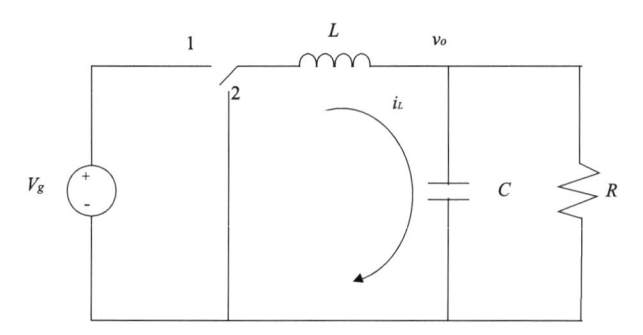

Figure 12.24 *Buck* type DC-DC converter.

- i_L is the coil current,
- v_0 is the capacitor voltage,
- L is the converter inductance,
- V_g is the voltage source,
- R is the load, and
- u is the switch position control.

Generalized coordinates

The following coordinates are defined:

$$\dot{q}_1 = i_L, \quad q_2 = v_0.$$

Kinetic and potential energies and non-potential forces

Using Table 6.1, the kinetic and potential energies are defined as

$$T_1 = \frac{L\dot{q}_1^2}{2}, \quad V_1 = \frac{q_1^2}{2C},$$
$$T_2 = \frac{C\dot{q}_2^2}{2}, \quad V_2 = 0,$$
$$T = T_1 + T_2, \quad V = V_1 + V_2,$$

Collection of electromechanical models

and

$$Q_{1,non\text{-}pot} = V_g u,$$
$$Q_{2,non\text{-}pot} = \frac{d}{dt}\dot{q}_1 - \frac{\dot{q}_2}{R},$$

where u takes the values

$$u = \begin{cases} 1, & \text{switch in position 1,} \\ 0, & \text{switch in position 2.} \end{cases}$$

Lagrange equations

Lagrange's equations are given by

$$\frac{d}{dt}\frac{\partial(T-V)}{\partial\dot{q}_i} - \frac{\partial(T-V)}{\partial q_i} =$$
$$\frac{d}{dt}\frac{\partial T}{\partial\dot{q}_i} + \frac{\partial V}{\partial q_i} = Q_{i,non\text{-}pot}, \quad i = 1, 2. \qquad (12.9)$$

The following dynamic equations are derived from (12.9):

$$L\frac{d}{dt}\dot{q}_1 + q_2 = V_g u,$$
$$C\ddot{q}_2 = \frac{d}{dt}\dot{q}_1 - \frac{\dot{q}_2}{R},$$

or equivalently,

$$L\frac{d}{dt}i_L = -v_0 + V_g u,$$
$$C\ddot{v}_0 = \frac{d}{dt}i_L - \frac{\dot{v}_0}{R}.$$

Integrating the second equation we obtain

$$L\frac{d}{dt}i_L = -v_0 + V_g u,$$
$$C\dot{v}_0 = i_L - \frac{v_0}{R},$$

if the relationship

$$C\dot{v}_0(0) = i_L(0) - \frac{v_0(0)}{R}$$

is taken into account.

12.24.2 Boost type DC-DC converter

A *boost* type DC-DC converter is shown in Fig. 12.25, where again:

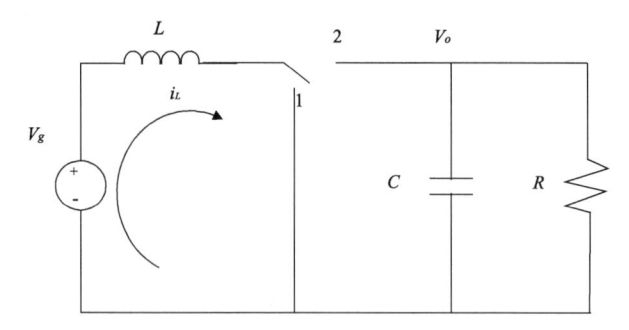

Figure 12.25 *Boost* type DC-DC converter.

- i_L is the coil current,
- v_0 is the capacitor voltage,
- L is the converter inductance,
- V_g is the voltage source,
- R is the load, and
- u is the switch position control.

Generalized coordinates

Analogously, define the generalized coordinates as

$$\dot{q}_1 = i_L, \quad q_2 = v_0.$$

Kinetic and potential energies and non-potential forces

Using Table 6.1, the kinetic and potential energies are defined as

$$T_1 = \frac{L\dot{q}_1^2}{2}, \quad V_1 = \frac{q_1^2}{2C},$$
$$T_2 = \frac{C\dot{q}_2^2}{2}, \quad V_2 = \frac{q_2^2}{2L}u,$$
$$T = T_1 + T_2, \quad V = V_1 + V_2,$$

and

$$Q_{1,non\text{-}pot} = -q_2 u + V_g,$$
$$Q_{2,non\text{-}pot} = \frac{d}{dt}\dot{q}_1 - \frac{\dot{q}_2}{R},$$

Collection of electromechanical models

where u takes the values

$$u = \begin{cases} 1, & \text{switch in position 1,} \\ 0, & \text{switch in position 2.} \end{cases}$$

Lagrange equations

Lagrange's equations are given by

$$\frac{d}{dt}\frac{\partial (T-V)}{\partial \dot{q}_i} - \frac{\partial (T-V)}{\partial q_i} =$$
$$\frac{d}{dt}\frac{\partial T}{\partial \dot{q}_i} + \frac{\partial V}{\partial q_i} = Q_{i,non\text{-}pot}, \quad i = 1, 2.$$

The following dynamic equations are derived from the relations above:

$$L\frac{d}{dt}\dot{q}_1 = -(1-u)q_2 + V_g,$$
$$C\frac{d}{dt}\dot{q}_2 = \frac{d}{dt}\dot{q}_1 - \frac{q_2 u}{L} - \frac{\dot{q}_2}{R}.$$

Integration of the second equation gives

$$L\frac{d}{dt}\dot{q}_1 = -(1-u)q_2 + V_g,$$
$$C\frac{d}{dt}q_2 = \dot{q}_1 - \dot{q}_1 u - \frac{q_2}{R},$$

or equivalently,

$$L\frac{d}{dt}i_L = -(1-u)v_0 + V_g,$$
$$C\frac{d}{dt}v_0 = (1-u)i_L - \frac{q_2}{R}.$$

12.25 Induction motor

A model of an induction motor can be represented by the circuit shown in Fig. 12.26, where:

- i_s is the current in the stator,
- i_r is the rotor current,
- R_s is the stator resistance,
- R_r is the rotor resistance,
- L_{1s} is the inductance in the stator,
- L_{1r} is the rotor inductance,
- L_m is the mutual inductance,
- $p\lambda_s$ is the magnetic flux in the stator,

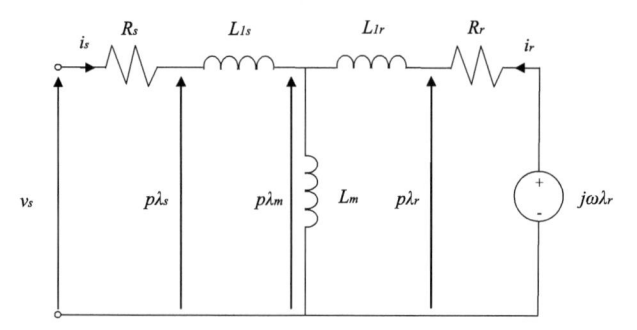

Figure 12.26 Model of an induction motor.

- $p\lambda_r$ is the magnetic flux in the rotor,
- v_s is the voltage applied to the stator, and
- $j\omega\lambda_r$ is the induced voltage in the rotor.

Generalized coordinates

Expressing the coordinates in terms of the currents that appear in the figure we have

$$q_1 = i_s, \quad q_2 = i_r.$$

Kinetic energy

Using the relation (see Chapter 6)

$$\Phi = Li,$$

we are able to derive the kinetic energies in stator, rotor, and mutual induction, which are

$$T_s = \frac{L_{1s} i_s^2}{2},$$

$$T_r = \frac{L_{1r} i_r^2}{2},$$

$$T_m = \frac{L_m (i_m + i_r)^2}{2}.$$

Potential energy

Here (in the no capacity case)

$$V = 0.$$

Non-potential forces

Moreover

$$Q_{s,non\text{-}pot} = -v_s + i_s (R_s + j\omega L_{1s}) + (i_s + i_r) j\omega L_m,$$

$$Q_{r,non\text{-}pot} = -v_r + i_r (R_r + j\omega L_{1r}) + (i_s + i_r) j\omega L_m.$$

Lagrange equations

Applying the Lagrange formula

$$\frac{d}{dt}\frac{\partial T}{\partial \dot{q}_i} - \frac{\partial T}{\partial q_i} = Q_{i,non\text{-}pot}, \quad i=1,2,$$

to the two branches, as shown in the circuit of Fig. 12.26, the following equations are obtained:

$$i_s(R_s + j\omega L_{1s}) + (i_s + i_r)j\omega L_m = v_s - L_{1s}\frac{di_s}{dt} - L_m\frac{d(i_r + i_s)}{dt},$$

$$i_r(R_r + j\omega L_{1r}) + (i_s + i_r)j\omega L_m = v_r - L_{1r}\frac{di_r}{dt} - L_m\frac{d(i_r + i_s)}{dt}.$$

Defining the vectors

$$\mathbf{i} = \begin{pmatrix} i_s \\ i_r \end{pmatrix}, \quad \mathbf{v} = \begin{pmatrix} v_s \\ v_r \end{pmatrix},$$

we can rewrite the previous relations in the vector format

$$\frac{d}{dt}\begin{pmatrix} i_s \\ i_r \end{pmatrix} = \frac{1}{L_{\sigma^2}}\mathbf{Ai} + \frac{1}{L_{\sigma^2}}\mathbf{Bv},$$

where

$$\mathbf{A} = \begin{pmatrix} -L_r(R_s + j\omega L_m + j\omega L_s) + j\omega L_m^2 & -j\omega L_m L_r + L_m(R_r + j\omega L_m + j\omega L_r) \\ -j\omega L_s L_m + L_R(R_s + j\omega L_m + j\omega L_s) & j\omega L_R L_m - L_S(R_r + j\omega L_m + j\omega L_r) \end{pmatrix},$$

$$\mathbf{B} = \begin{pmatrix} -L_{1r} & L_m \\ L_R & -L_S \end{pmatrix}.$$

Here

$$L_R = L_m + L_{1r},$$
$$L_S = L_m + L_{1s},$$
$$L_{\sigma^2} = L_{1r}L_s - L_m^2.$$

Flow equations are represented by

$$\frac{d}{dt}p\lambda_s = v_s - R_s i_s,$$
$$\frac{d}{dt}p\lambda_r = v_r - R_r i_r.$$

Bibliography

Abraham, R., Marsden, J.E., 1978. Foundations of Mechanics, 2nd edn. Benjamin/Cummings, Reading, MA.
Aizerman, M.A., 1980. Classical Mechanics. Nauka, Moscow (in Russian).
Arnold, V.I., 1989. Mathematical Methods of Classical Mechanics, 2nd edn. Graduate Texts in Mathematics, vol. 60. Springer-Verlag, New York. Translated by K. Vogtmann and A. Weinstein.
Arya, A.P., 1998. Introduction to Classical Mechanics, 2nd edn. Prentice Hall, Upper Saddle River, NJ.
Barger, V., Olsson, M.G., 1973. Classical Mechanics: A Modern Perspective. McGraw-Hill Series in Fundamentals of Physics. McGraw-Hill, New York.
Barger, V., Olsson, M.G., et al., 1995. Classical Mechanics: A Modern Perspective, Vol. 2. McGraw-Hill, New York.
Bartlett, J.H., 1975. Classical and Modern Mechanics. The University of Alabama Press, Alabama.
Becker, R.A., 1954. Introduction to Theoretical Mechanics, Vol. 420. McGraw-Hill, New York.
Bhatia, V.B., 1997. Classical Mechanics: With Introduction to Nonlinear Oscillations and Chaos. Alpha Science Int'l Ltd.
Boltyanski, V., Poznyak, A., 2012. The Robust Maximum Principle. Birkhäuser, Springer Sciences.
Burghes, D.N., Downs, A.M., 1975. Modern Introduction to Classical Mechanics and Control. Mathematics and Its Applications. E. Horwood/Halsted, Chichester, Eng./New York.
Chetayev, N., 1965. Stability of Motion, 3rd edn. Nauka, Moscow (in Russian).
Chow, T.L., 1995. Classical Mechanics. Wiley, New York.
Corben, H., Stehle, P., 1960. Classical Mechanics, 2nd edn. Wiley, New York. General Note: Papers from a conference held at Tufts University in August 1979, sponsored by the National Science Foundation and the Conference Board of the Mathematical Sciences.
Deriglazov, A., 2016. Classical Mechanics. Springer.
Desloge, E.A., 1982. Classical Mechanics, Vol. 1. John Wiley & Sons.
Devaney, R.L., Nitecki, Z. (Eds.), 1981. Classical Mechanics and Dynamical Systems. Lecture Notes in Pure and Applied Mathematics, vol. 70. Marcel Dekker, New York.
Fowles, G.R., 1986. Analytical Mechanics, 4th edn. Saunders Golden Sunburst Series. Saunders College Pub., Philadelphia.
Fowles, G.R., Cassiday, G.L., et al., 2005. Analytical Mechanics. Thomson Brooks/Cole, Belmont, CA.
Gantmakher, F.R., 1970. Lectures in Analytical Mechanics F. Gantmacher. Mir Publishers, Moscow. Translated from the Russian by George Yankovsky. Translation of Lektsii po analiticheskoi mekhanike (romanized from).
Goldstein, H., 1980. Classical Mechanics. Addison-Wesley Series in Physics, 2nd edn. Addison-Wesley.
Hestenes, D., 1986. New Foundations for Classical Mechanics. Fundamental Theories of Physics. D. Reidel, Dordrecht.

Kharitonov, V., 1978. Asymptotic stability of a family of linear differential equations. Differential Equations 14 (11), 2086–2088 (in Russian).
Kibble, T.W.B., 1973. Classical Mechanics, 2nd edn. European Physics Series. McGraw-Hill, London.
Kibble, T.W., Berkshire, F.H., 2004. Classical Mechanics. World Scientific Publishing Company.
Kittel, C., Knight, W.D., Ruderman, M.A., 1968. Mecanica. Berkeley Physics Course, vol. 1. Reverte, Alabama. Traducción, Aguilar Peris, J.y Juan de la Rubia.
Kotkin, G.L., Serbo, V.G., 1980. Problemas de mecanica clasica. Mir, Moscu.
Kwatny, H.G., Blankenship, G.L., 2000. Nonlinear Control and Analytical Mechanics: A Computational Approach. Control Engineering. Birkhäuser, Boston.
Landau, L., Lifshitz, E., 1969. Mechanics. A Course Theoretical Physics, vol. 1. Pergamon Press.
Lawden, D.F., 1974. Mecanica analitica. Seleccion de problemas resueltos. Serie Limusa, vol. 4. Limusa, Mexico. Tr. Ricardo Vinos.
Lee, H.C., 1947. The universal integral invariants of Hamiltonian systems and application to the theory of canonical transformations. Proceedings of Edinburgh. Section A. Mathematics 62 (3), 237–246.
Levi, M., 2014. Classical Mechanics with Calculus of Variations and Optimal Control: An Intuitive Introduction, Vol. 69. American Mathematical Soc.
Malkin, I., 1952. Theory of Stability of Motion. Gostekhizdat, Moscow (in Russian).
Marsden, J.E., 1992. Lectures on Mechanics. London Mathematical Society Lecture Note Series, vol. 174. Cambridge University Press, Cambridge (England), New York.
Matzner, R.A., Shepley, L.C., 1991. Classical Mechanics. Prentice-Hall.
Poznyak, A., 2008. Advanced Mathematical Tools for Automatic Control Engineers: Volume 1: Deterministic Systems. Elsevier Science.
Pyatnickii, E., Truhan, N., Khanukaev, Y., Yakovenko, G., 1996. Collection of Problems on Analytical Mechanics, 2nd edn. Nauka-Fizmatlit, Moscow (in Russian).
Rutherford, D.E., 1951. Classical Mechanics. University Mathematical Texts. Oliver and Boyd, Edinburgh.
Seely, F.B., Newton, E., Jones, P.G., 1958. Analytical Mechanics for Engineers, 5th edn. John Wiley, New York.
Symon, K.R., 1968. Mecanica. Aguilar, Madrid. Tr. por Albino Yusta Almanza.
Takwale, R., Puranik, P., 1979. Introduction to Classical Mechanics. Tata McGraw-Hill Education.
Titherington, D., Rimmer, J.G., 1973. Mecanica. McGraw-Hill, Mexico. Tr. Antonio Abaunza de la Escosura ... [et al.].
Torres del Castillo, G.F., 2018. An Introduction to Hamiltonian Mechanics. Birkhäuser Advanced Texts. Basler Lehrbücher. Springer Nature Switzerland AG Part of Springer Nature.

Index

A
Acceleration, 11
  absolute, 40, 46
  angular, 40, 47
  Coriolis, 47
  normal, 18
  relative, 47
  tangential, 18
  tending to the axis, 40
  translation, 47
Affine plant, 372
Algebraic complement, 64
Amplitude, 247
Anticommutativity, 4
Asymptotic stability criterion, 287
Average value, 100
Axis of rotation, 34

B
Basic mechanic theorem, 206
Bernoulli's equation, 149
Boost type DC-DC converter, 482
Brackets of Lagrange, 354
Buck type DC-DC converter, 480

C
Canonical transformation, 338
Capacitance, 208
CD motor, gear train and load, 412
CD motor with load, 478
Center of mass, 93
  dynamic properties, 103
Center of velocities, 42
Central moments of inertia, 162
Centrifugal moments, 155
Characteristic
  amplitude, 279
  amplitude-phase, 278
  phase, 279
Characteristic equation, 248
Characteristic polynomial, 248

Coefficients of Lamé, 10
Commutativity, 2
Complementary slackness condition, 227
Complete integral, 362
Component, 2
Coordinate curve, 10
Coordinates
  Cartesian, 9
Cost functional, 370
Criterion of parallelism, 2, 4
Criterion of polynomial robust stability, 305
Cyclic coordinate, 323

D
Degrees of freedom, 192
Differential kinematic equations, 84
Differential kinematic equations in Euler angles, 85
Dirichlet's conditions, 274
Dissipative systems, 286
Distributivity, 2, 4
Dynamic reactions, 182
Dynamics of systems with variable mass, 142

E
Earnshaw theorem, 243
Eigenvalues, 161
Eigenvectors, 161
Energy
  kinetic, 91, 100
  potential, 100
Equilibrium, 221, 222
  asymptotically stable, 286
Euler's dynamic equations, 153, 175
Expenditure, 143

F
Faraday's law, 208
Feynman–Kac formula, 376
First integral, 321

Force
  potential, 96
  reactive, 143
  rolling, 124
  total, 92
Force work
  total, 96
Forces
  external, 92
  internal, 92
Fourier transformation, 274
Fourier's transformation
  inverse, 274
Free transformations, 356
Frequency, 247
Frequency characteristic matrix, 276
Friction
  coefficient, 124
  dynamic, 123
  force, 123
  static, 123
Function
  homogeneous, 101

G
Generalized coordinate system, 9
Generalized coordinates, 8, 192
Generalized forces, 193
Generalized impulses, 312
Geometric and mass symmetry, 107
Gyroscope, 179

H
Hamilton–Jacobi
  theorem, 361
Hamilton–Jacobi–Bellman (HJB)
  equation, 371
Hamiltonian, 312
  canonical form, 317
  system, 317
Hamiltonian systems
  stationary, 329
Hamiltonian variables, 312
Hamilton's
  theorem, 316
Hessian, 234
Holonomic system, 192
Hurwitz
  matrix, 289
  polynomial, 289
Huygens formula, 202

I
Impulse, 91
Inductance, 208
Induction motor, 483
Inertial center, *see* Center of mass
Inertial Coriolis force, 133, 135
Inertial translation force, 133, 135
Integral Poincaré–Cartan invariant, 344
Integral universal Poincaré invariant, 344
Internal (scalar) product, 2

J
Jacobi
  identity, 6
Jacobi–Poisson
  theorem, 327
Jacobian matrix, 9

K
Kelly formula, 151
Kelly problem, 149
Kharitonov's theorem, 305
Kinetic energy, 13
Kirchhoff's laws, 208
Kronecker symbol, 3
Kronecker's symbol, 12

L
Lagrange multipliers, 160
Lagrange multipliers vector, 227
Lagrange variables, 312
Lagrange–Dirichlet theorem, 232
Lagrange's equation, 198
Lagrange's error, 254
Lagrange's function, 198
Lagrange's lemma, 195
Lagrangian function, 227
Law of mesh, 208
Law of node, 209
Lee Hwa Chung
  theorem of, 345
Lee–Poisson bracket, 322
Liénard–Chipart
  criterion, 291
Linear oscillator, 230
Local bases, 10
Local equilibrium, 232

Locally stable equilibrium, 232
Locally stable state, 229

M
Magnitude (length) of a vector, 3
Mass, 92
Matrix
  positive semi-definite, 158
Mechanical constraints, 189
Meshchersky equation, 143
Mikhailov
  criterion, 296
Moment of forces with respect to pole, 93
Moment of inertia, 105
Moment of the impulse, 91
Moments of inertia
  principal, 155
Moments of the inertial Coriolis forces
  inertial Coriolis, 135
Moments of the inertial translation forces
  inertial translation, 135
Multiplicative inverse, 75

N
Newton's laws, 91
Normal coordinates, 253
Normal form, 204
Nyquist hodograph, 278

O
ODE, 372
Ohm's law, 208
Optimality
  sufficient conditions, 371
Orthogonality criterion, 2, 4
Oscillations, 245

P
Parameters
  of Rodríguez–Hamilton, 67
Partial basic solution, 248
Phase shifting, 276
Pivot, 32
Poincaré
  theorem, 347
Poincaré theorem, 343
Poisson's equation, 86
Pole, 90
Possible position, 191

Possible transfer, 193
Power, 98
Principle of mechanical energy
  conservation, 100
Principle of "zero-excluding", 305

Q
Quaternion, 71
  conjugated, 73
  norm, 74
  normalized, 77
  proper, 82

R
Regular precession, 179
Relative kinetic energy, 137
Relative momentum of the impulse, 134
Resistance, 208
Restriction
  stationary, 191
Restrictions
  ideal, 194
Rigid body, 32
Ring with division, 75
Rizal's formula, 94
Robot
  arm manipulator with springs, 445
  articulated, 460
  articulated robot manipulator, 396
  Cincinnati Milacron T^3 manipulator, 404
  cylindrical manipulator with two
    prismatic joints (PJ) and one
    rotating (R), 384
  cylindrical with springs, 427
  double type crank swivel crane, 437
  Maker 110, 465
  manipulator on a horizontal platform, 468
  manipulator scaffolding type, 389
  manipulator Unimate 2000, 418
  manipulator with swivel base, 424
  manipulator with three degrees of
    freedom, 476
  multicylindric manipulator, 442
  non-ordinary manipulator with shock
    absorber, 429
  rectangular (Cartesian) manipulator
    robot, 387
  spherical (polar) manipulator with three
    rotating joints, 392
  Stanford/JPL manipulator, 414

two-arm planar manipulator, 471
two-joint planar manipulator, 434
universal programmable manipulator (PUMA), 399
Robust stability, 304
Rotation, 59
  description, 60
  finite, 60
  matrix, 61
  plus specular reflection, 62
  pure, 62
Rotation angle
  nutation, 60
  precision, 60
  proper, 60
Rotation angles
  natural, 61
  of Euler, 60
Rotation matrix
  proper, 70
Rotations
  elementary, 60
Routh–Hurwitz
  criterion, 290

S
Silvester's lemma, 158
Simultaneous transformation of pair of quadratic forms, 251
Slater's condition, 227
Stationary systems, 207
Steiner theorem, 114
Stodola
  necessity condition of stability, 288
System
  absolute, 39
  conservative, 99
  inertial, 131
  material points, 90
  non-inertial, 131
  relative, 39
System states, 229

T
Tensor
  of inertia, 156
Tensor of inertia, 153
Theorem
  the Chetayev theorem on instability, 237
  the first Lyapunov theorem on instability, 237
  the second Lyapunov theorem on instability, 237
Transformation
  one-to-one, 9
Triple scalar product, 5, 6
Triple vector product, 5, 6
Tsiolkovsky's rocket formula, 145

U
Unit vectors, 5
Unitary vectors, 3
Unstable equilibrium position, 230

V
Value function, 370
Vector
  unitary, 10
Vector product, 4
Vectors, 2
Velocity, 11
  absolute, 40
  angular, 34
  translation, 46
Verification rule, 371
Virial of a system, 100
Virtual displacements, 222
Virtual possible transfer, 193

W
Work
  elementary, 96

Z
Zhukovski's theorem, 180

Made in the USA
Coppell, TX
27 April 2023